CHICAGO PUBLIC LIBRARY
R00582 09404

D0205326

INTRODUCTION TO OPTICAL ELECTRONICS

INTRODUCTION TO OPTICAL ELECTRONICS

Kenneth A. Jones
U.S. Army

1817
HARPER & ROW, PUBLISHERS, New York
Cambridge, Philadelphia, San Francisco, Washington,
London, Mexico City, São Paulo, Singapore, Sydney

Sponsoring Editor: Peter Richardson
Project Editor: Ellen MacElree
Cover Design: Miriam Recio
Cover Photo: Simon, Phototake
Text Art: RDL Artset, Ltd.
Production: Paula Roppolo
Compositor: Tapsco, Incorporated
Printer and Binder: The Maple Press Company

INTRODUCTION TO OPTICAL ELECTRONICS

Library of Congress Cataloging in Publication Data

Jones, Kenneth A.
Introduction to optical electronics.

Includes index.
1. Optoelectronics. I. Title.
TA1750.J66 1987 621.38′0414 86-22855
ISBN 0-06-043444-9

87 88 89 90 9 8 7 6 5 4 3 2 1

To my kind,
understanding,
and supportive parents
Elizabeth McB.
and
N. Quentin Jones

CONTENTS

3 OPTICAL SPECTRA OF ATOMS, MOLECULES, AND SOLIDS 90

4 DETECTORS 143

5 OTHER PHOTODETECTORS AND NOISE 215

6 LIGHT EMITTING DIODES 244

7 LASER DIODES 282

APPENDICES 409

PREFACE

Optical electronics draws from many different disciplines in the electrical engineering curriculum. It is, therefore, a good senior level course because it can tie together many loose ends and also introduce the student to a fascinating new applications area. The student who wrestled with the laws of reflection, transmission, and absorption in his or her E & M course now will be able to learn these laws better by applying them, for example, to transmission through a lens, down an optical fiber, or through a Fabry–Perot cavity. The concept of resonance, which is encountered in circuits and systems, is further stretched while studying oscillating dipoles and the energy stored in and lost by an optical cavity. Convolution rears its ugly head (ugly only to some) first in a signals course and then again when the Fabry–Perot etalon is used as a spectrum analyzer. Some students have told me that they did not understand convolution until they studied the Fabry–Perot cavity. Also, studying semiconductor light emitters and detectors can consolidate what is learned about the *pn* junction in a traditional semiconductor course.

Knowledge obtained in a modern physics course can also be applied to optical electronics. The difficult transition from classical electromagnetic waves to photons and vice versa is made less difficult by numerous practical examples such as computing the photon flux associated with a given Poynting vector and the number of photoelectrons generated per unit time by a known incident optical power. Light-emission and absorption processes also provide relevant examples for quantum mechanics even though the quantum mechanics in this textbook is strictly of the "dirt-ball" variety.

An effort has been made to write each chapter as a separate capsule so that it is not necessary to start the text on page one and continue consecutively through the book. For example, when optical gain is discussed in Chapter 7, it is not absolutely essential to know the derivation of the reflectivity done in Chapter 1. However, the instructor and the interested student would be able to refer back to the derivation if they so desired. Similar comments can be made about tuning the wavelength of a quaternary semiconductor optical emitter by knowing how the size of the atom affects the energy gap, or the use of a Brewster window on a helium-neon laser. I leave it to the instructor to deal with this breadth versus depth problem.

The text is divided into three topic areas: transmission, detection, and emission. The text discussion begins with light transmission because it builds on electromagnetic theory, which is material required of most junior level electrical engineers. The first two chapters, "Wave Propagation" and "Optical Fibers," deal with light transmission. The laws of reflection are reviewed in Chapter 1 and then are applied to geometric optics. In Chapter 2, scattering and dispersion and how they affect data transmission in step and graded index fibers are discussed, as well as coupling light into the fiber. Chapter 1 is primarily a condensed review of the electromagnetic theory that is needed for the understanding of optical electronics. When I teach the optical electronics course, I use this chapter as a primary reference for the interested student who wants to review topics such as reflection at a boundary and why metals are so highly reflecting.

Chapter 3, "Optical Properties of Atoms, Molecules and Solids," covers the modern physics concepts that are used in the chapters that follow. The discrete spectra of atoms, the electronic, vibrational, and rotational spectra of molecules, and the band structure of solids are discussed. Much of the material on atomic and molecular structure can be ignored if only semiconductor light emitters are going to be covered in the course. Systems covered later include the electron energy levels of helium and neon, the vibrational–rotational spectra of CO_2, and the electronic states of doped and undoped semiconductors and insulators.

Photodetectors are discussed before semiconductor emitters because they are simpler, and insight into the emission process is gained from them. In Chapter 4 photoconductors are described, and this is done by introducing the important ideas of intrinsic and extrinsic semiconductivity, the Fermi energy, and generation and recombination. The *pn* junction, photodiodes, *pin* diodes, and avalanche photodiodes are also described in some detail. Other detectors not usually used in optical communication systems, as well as a thumbnail sketch of noise, are covered in the following chapter. The detectors include photovoltaic detectors, Schottky barrier diodes, phototransistors, photodarlingtons, and bolometers.

Semiconductor light emission is the subject of the next two chapters. Chapter 6 deals with LEDs and Chapter 7 deals with laser diodes. The problems associated with producing light efficiently, modulating the signal, and coupling the light into a fiber are described in some detail. Also, the threshold current and carrier confinement by heterostructures—concepts unique to semiconductor lasers—are studied. An argument can be made for covering optical cavities in Chapter 8 before discussing semiconductor lasers in Chapter 7. I chose to discuss semiconductor lasers first, since it has been my experience that engineers learn the subject matter better if they know what it is going to be used for. The energy buildup inside the cavity, longitudinal mode structure, and the like, can have more meaning to an engineer who has some idea of how a laser works. Much the same argument can be made for teaching a student electronics before exposing him or her to semiconductor physics and the discussion of the elusive hole!

The optical cavity introduced in Chapter 7 is described in more detail in Chapter 8. The student will now be more aware of its importance and should therefore be able to better digest the basic concepts. These concepts as well as blackbody radiation and the Einstein coefficients are illustrated in Chapter 9 by briefly describing solid state and gas lasers.

We look to the future in Chapter 10 where single mode and single frequency semiconductor lasers, quantum-well lasers, optical bistable devices, and integrated optics make their cameo appearance.

My thanks to the many people who contributed to making this a better book. The following people read part or all of the manuscript and made numerous valuable suggestions: Drs. Chi H. Lee, A. V. Nurmikko, Cardinal Warde, James V. Masi, R. E. Nahory, Gregory E. Stillman, Douglas A. Davids, Thomas Wong, Jorge J. Rocca, and Sigrid McAffee. I am grateful to Ms. Erica Swanson, Pamela Williams, Betty Hutcheson, and Anna Penner for the competent typing of the manuscript. I also thank the many students who have read the manuscript and made a number of good suggestions.

Kenneth A. Jones

INTRODUCTION TO OPTICAL ELECTRONICS

Chapter 1

Propagation of Plane Waves

1-1 INTRODUCTION

Maxwell's equations, the basis of all electromagnetic theory, are the relationships between the charge density, ρ, electric field, $\boldsymbol{\xi}$, electric displacement, **D**, magnetic field, **H**, and magnetic induction, **B**. The continuity equation is an expression of the conservation of charge.

Maxwell's equations are important to this text because their solution for a certain set of conditions leads to the mathematical description of a plane wave—the sinusoidal variation in time and position of $\boldsymbol{\xi}$ and **H** with the amplitude of each being constant in the plane normal to the direction of propagation.

In a vacuum $\boldsymbol{\xi} \perp \mathbf{H} \perp \mathbf{k}$, $\boldsymbol{\xi}$, and **H** are in phase, and $\xi/H = \sqrt{\mu_0/\epsilon_0}$. The wave vector, **k**, is parallel to the direction of propagation; μ_0 is the magnetic permeability of free space; and ϵ_0 is the electric permittivity of free space. These relationships are important because they are used to determine the fraction of the incident light intensity that is reflected at an interface. The magnitude of the intensity is given by the magnitude of the Poynting vector $S = |\boldsymbol{\xi} \times \mathbf{H}|$.

Under normal operating conditions when stimulated emission can be ignored, an electromagnetic wave gives up energy to a material when they interact. Much of the energy stored in the material as kinetic and potential energy is transferred back to the wave, but the energy dissipated as heat, radiated in directions other than the direction of propagation, and the like, of the wave, is not. As a result, $\boldsymbol{\xi}$ and **H** are attenuated as the wave moves into the material. This attenuation is always accompanied by a phase lag of **H** and **P** behind $\boldsymbol{\xi}$. **P** is the polarization, and it is used to describe the modification on $\boldsymbol{\xi}$ produced by the material. The magnetic counterpart is the magnetization, **M**, which is used to describe the modifications on **B** produced by the material. The stronger the electrical interaction with the material is, the larger is **P**, and this translates into a larger electric permittivity, ϵ. The larger electric permittivity, in turn, translates into a larger index of refraction.

The larger the difference in the index of refraction of two media, the larger is the fraction of the light that is reflected at the interface. For normal incidence and no loss the indices of refraction completely determine the fraction reflected but, when there is loss, the fraction reflected increases with the amount of loss. For off normal incidence the fraction reflected also depends on the angle of incidence as well as the direction of polarization.

Refraction, the bending of a light ray at an interface, also occurs at an interface where the index of refraction changes. Thus, not only can the direction of polarization of the plane wave be changed by reflection at an interface, the direction of propagation can be changed as well. It is this ability to guide a ray that will be emphasized in this book—especially in the next chapter on optical fibers.

The waveguides can have flat interfaces as they do in flat mirrors and prisms, and their primary purpose is to redirect a beam. A prism can also be used to separate waves with different wavelengths by bending blue rays more than it does red. This is called dispersion, and it illustrates that the index of refraction is a function of the wavelength. The surfaces can also be curved as they are in spherical mirrors and lenses. Their primary function is to collimate a beam as they do for light emitted from a light-emitting diode (LED), or to concentrate it as they do for some detectors. They also have the ability to magnify the size of an object.

1-2 FUNDAMENTAL ELECTROMAGNETIC EQUATIONS

In this section we briefly review Maxwell's equations and the continuity equation. Only those aspects that will be used later will be emphasized.

In differential form, Maxwell's equations are

$$\nabla \cdot \mathbf{D} = \rho \tag{1-1a}$$

$$\nabla \times \xi = -\frac{\partial \mathbf{B}}{\partial t} \tag{1-2a}$$

$$\nabla \cdot \mathbf{B} = 0 \tag{1-3a}$$

$$\nabla \times \mathbf{H} = \mathbf{J} + \frac{\partial \mathbf{D}}{\partial t} \tag{1-4a}$$

where **J** is the current density. In integral form they are

$$\int_S \mathbf{D} \cdot d\mathbf{S} = \int_V \rho \, dv \tag{1-1b}$$

$$\oint \xi \cdot d\mathbf{l} = \int_A -\frac{\partial \mathbf{B}}{\partial t} \cdot d\mathbf{A} \tag{1-2b}$$

$$\int_S \mathbf{B} \cdot d\mathbf{S} = 0 \tag{1-3b}$$

$$\oint \mathbf{H} \cdot d\mathbf{l} = \int_A \left(\mathbf{J} + \frac{\partial \mathbf{D}}{\partial t} \right) \cdot d\mathbf{A} \tag{1-4b}$$

The integral form can be illustrated using lines of force and this is done in Fig. 1-1. In Fig. 1-1*a* one can see that lines of **D** originate on a positive electrical charge and terminate on a negative one. Thus, the integral of the product of the component of **D** parallel to the unit normal to the surface is equal to the total net electrical charge in the volume enclosed by the surface.

The illustration in Fig. 1-1*b* shows that the integral around the loop of the component of ξ tangent to the line is equal to the negative of the change with time of the component of **B** normal to the area enclosed by the loop. Thus, if the normal component of **B** directed out from the plane of the paper is decreasing with time, the integral is positive.

Because there are no isolated magnetic monopoles, the lines of **B** have no point from which to originate or terminate. Thus, the number of lines of **B** going into a volume element must equal the number coming out. This is illustrated in Fig. 1-1*c*.

There are two sources of **H.** They are the current density, **J,** and the displacement current density $\partial \mathbf{D}/\partial t$. The product of the normal components of these two current densities and the area is equal to the line integral of **H** around the loop enclosing the area. This is shown in Fig. 1-1*d*.

The continuity equation is an expression of the conservation of charge and

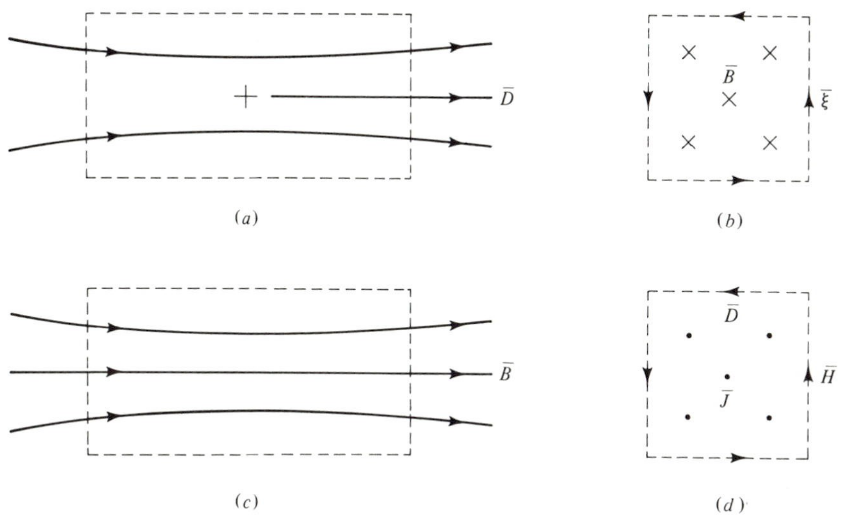

Figure 1-1 Gaussian representation of Maxwell's equations. (a) $\int_S \bar{D} \cdot d\bar{S} = \int_V \rho \, dv$. (b) $\oint \bar{\xi} \cdot d\bar{l} = \int_A - \partial\bar{B}/\partial t \cdot d\bar{A}$. (c) $\int_S \bar{B} \cdot d\bar{S} = 0$. (d) $\oint \bar{H} \cdot d\bar{l} = \int_A [(\partial\bar{D}/\partial t) + \bar{J}] \cdot d\bar{A}$.

can be better understood with the aid of Fig. 1-2. For flow in the z direction the flow of charge per unit time into the volume element, $\Delta x\ \Delta y\ \Delta z$, is

$$\text{charge flow in per unit time} = J(z)\Delta x\ \Delta y \tag{1-5a}$$

The rate of flow of charge out of the other end of the volume element per unit time is

$$\text{charge flow out per unit time} = J(z + \Delta z)\Delta x\ \Delta y \tag{1-5b}$$

The net rate of change of flow into the volume element is thus

$$\frac{\partial \rho}{\partial t}\Delta x\ \Delta y\ \Delta z = [J(z) - J(z + \Delta z)]\Delta x\ \Delta y = -\frac{\partial J(z)}{\partial z}\Delta x\ \Delta y\ \Delta z \tag{1-5c}$$

since
$$J(z + \Delta z) = J(z) + \frac{\partial J(z)}{\partial z}\Delta z \tag{1-5d}$$

In three dimensions Eq. 1-5c is written

$$\frac{\partial \rho}{\partial t} = -\nabla \cdot \mathbf{J} \tag{1-5e}$$

1-3 PLANE WAVES

1-3.1 No Loss

The electric field is a function of time and position. We now determine the functional dependence by first finding a differential equation for ξ and then solving it for some special conditions.

Taking the curl of both sides of Eq. 1-2a and using an identity from vector calculus

$$\nabla \times \nabla \times \xi = \nabla\nabla \cdot \xi - \nabla^2\xi = \nabla \times \left(-\frac{\partial \mathbf{B}}{\partial t}\right) \tag{1-6a}$$

The relationship between $\mathbf{D}$ and ξ, which will be discussed in more detail later, is

$$\mathbf{D} = \epsilon\xi \tag{1-7}$$

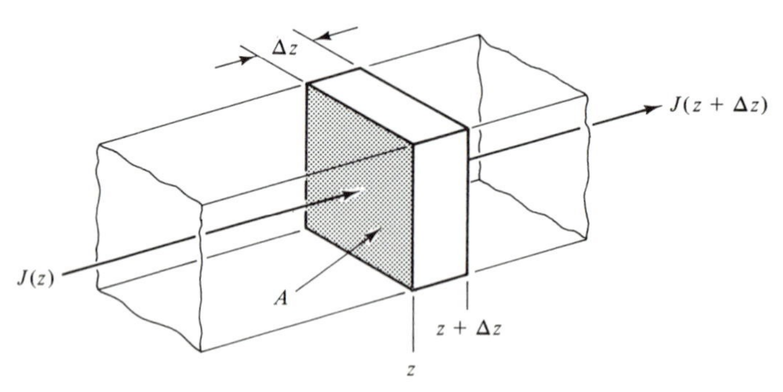

Figure 1-2 Current flow into and out of the volume element, $A\ \Delta z$.

In this chapter we only consider the cases where the electric permittivity can be treated as if it were a scalar quantity independent of position. Thus, from Eqs. 1-1a and 1-7, when the charge density, $\rho = 0$, $\nabla \cdot \xi = 0$. Equation 1-6a now reduces to

$$\nabla^2 \xi = \nabla \times \frac{\partial \mathbf{B}}{\partial t} \tag{1-6b}$$

The relationship between **B** and **H**, which also will be discussed in more detail later, is

$$\mathbf{B} = \mu \mathbf{H} \tag{1-8}$$

As was true with ϵ, μ in this chapter will be treated as a scalar quantity independent of position. By combining Eqs. 1-8 and 1-4a, Eq. 1-6b becomes

$$\nabla^2 \xi = \frac{\partial}{\partial t}\left(\mu \mathbf{J} + \mu \frac{\partial \mathbf{D}}{\partial t}\right) \tag{1-6c}$$

By substituting $\epsilon\xi$ for **D**, using the relationship

$$\mathbf{J} = \sigma \xi \tag{1-9}$$

where σ is the electrical conductivity, and only considering the cases of μ, σ, and ϵ being independent of the time, Eq. 1-6c finally becomes

$$\nabla^2 \xi = \mu\sigma \frac{\partial \xi}{\partial t} + \mu\epsilon \frac{\partial^2 \xi}{\partial t^2} \tag{1-6d}$$

For the plane wave solution ξ varies sinusoidally with t, and it is a function of only one position variable, which in our case we will initially assume is z. Thus

$$\xi(x, y, z, t) = \xi(z)e^{j\omega t} \tag{1-10}$$

As is often true, $\xi(z)e^{j\omega t}$ is the shorthand form of $\text{Re}\{\xi(z)e^{j\omega t}\}$ where Re means "the real part of." When Eq. 1-10 is substituted into Eq. 1-6d, it becomes

$$\frac{\partial^2 \xi(z)}{\partial z^2} = j\omega\mu\sigma\xi(z) - \omega^2\mu\epsilon\xi(z) \tag{1-11a}$$

and when there is no loss ($\sigma = 0$ and ϵ is real) Eq. 1-11a reduces to

$$\frac{\partial^2 \xi(z)}{\partial z^2} = -\omega^2\mu\epsilon\xi(z) \tag{1-11b}$$

ω has units of time^{-1} so that $1/\sqrt{\mu\epsilon}$ must have units of velocity. It is, in fact, the phase velocity, c, of the light wave. Thus

$$c = 1/\sqrt{\mu\epsilon} \tag{1-12}$$

Since $\omega^2\mu\epsilon$ has a positive value, the solution to Eq. 1-11b is

$$\xi(z) = \xi_0 e^{\pm jkz} \tag{1-13}$$

Substituting Eqs. 1-12 and 1-13 into 1-11b to find an expression for k yields

$$k = \frac{\omega}{c} = \frac{2\pi}{\lambda} \tag{1-14}$$

k is the wave vector and is parallel to the direction of propagation, and λ is the wavelength of the light wave.

The complete expression for $\xi(z, t)$ when there is no loss is

$$\xi(z, t) = \xi_0 e^{j(\omega t \pm kz)} \tag{1-15}$$

This is the equation for a wave traveling in the $+z(-k)$ or $-z(+k)$ direction with a phase velocity of ω/k. By using similar arguments, it can be shown that

$$H(z, t) = H_0 e^{j(\omega t \pm kz)} \tag{1-16}$$

when there is no loss. For this condition, then, $\boldsymbol{\xi}$ and $\mathbf{H}$ are in phase.

We now wish to determine the directions of $\boldsymbol{\xi}$ and $\mathbf{H}$, and the ratio of their magnitudes. This ratio is called the intrinsic wave impedance. From Eq. 1-2a

$$\hat{j}\,\frac{\partial \xi_x}{\partial z} - \hat{i}\,\frac{\partial \xi_y}{\partial z} = -\hat{i}\,\frac{\partial B_x}{\partial t} - \hat{j}\,\frac{\partial B_y}{\partial t} - \hat{k}\,\frac{\partial B_z}{\partial t} \tag{1-17a}$$

because $\boldsymbol{\xi}$ is a function of z only. The terms $\hat{i}$, $\hat{j}$, and $\hat{k}$ are unit vectors, and the subscripts designate the vector components. If we arbitrarily choose the x direction to be parallel to $\boldsymbol{\xi}$, $B_x = B_z = 0$ and

$$\frac{\partial \xi_x}{\partial z} = -\frac{\partial B_y}{\partial t} \tag{1-17b}$$

Substituting in Eqs. 1-8, 1-15, and 1-16 yields

$$\xi_{0x} = \pm\frac{\omega}{k}\,\mu H_{0y} = \pm\sqrt{\frac{\mu}{\epsilon}}\,H_{0y} \tag{1-17c}$$

The plus sign is for a wave moving in the $+z$ direction and the minus sign is for a wave moving in the $-z$ direction. One can see from this that $\boldsymbol{\xi} \perp \mathbf{H} \perp \mathbf{k}$ (see Fig. 1-3), and the vector $\boldsymbol{\xi} \times \mathbf{H}$ is parallel to the direction of propagation.

1-3.2 Poynting Vector

The vector $\boldsymbol{\xi} \times \mathbf{H} = \mathbf{S}$, where $\mathbf{S}$ is the Poynting vector. For a plane wave it is parallel to the direction of propagation and has a magnitude equal to the light intensity, that is, the energy crossing unit area per unit time.

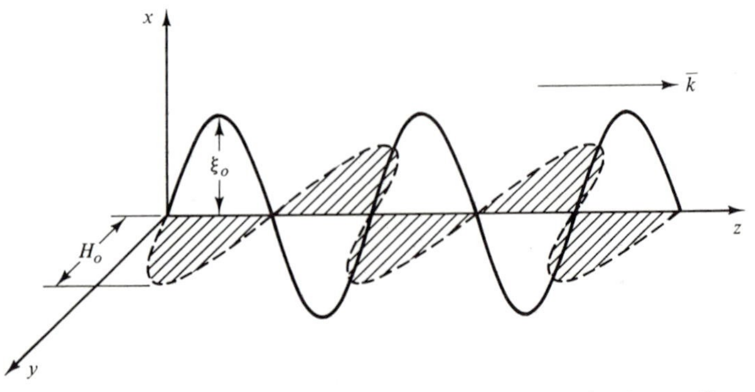

Figure 1-3 A plane electromagnetic wave propagating in the $+z$ direction.

To show this we must first find the energy of the electric field between the plates of the parallel plate capacitor with a charge, q_0, on each plate shown in Fig. 1-4*a*. The energy can be found by calculating the work necessary to charge the capacitor by moving a charge, dq, against the potential difference, V_c, of the capacitor. Thus

$$E_{\xi_q} = \int_0^{q_0} V_c dq = \int_0^{q_0} \frac{q}{C} dq = \frac{q_0^2}{2C} \qquad (1\text{-}18\text{a})$$

where q is the charge on a plate and C is the capacitance. The lines of **D** are normal to the plates so from Eq. 1-1a

$$q_0 = DA = \epsilon \xi A \qquad (1\text{-}19)$$

where A is the plate area. The voltage across the capacitor plates is $V_c = \xi d$ where d is the plate spacing. Thus

$$C = \frac{\epsilon A}{d} \qquad (1\text{-}20)$$

Substituting Eqs. 1-19 and 1-20 into Eq. 1-18a yields

$$E_{\xi_q} = \frac{1}{2} \epsilon \xi^2 A d \qquad (1\text{-}18\text{b})$$

Since Ad is the volume between the plates,

$$\frac{E_{\xi_q}}{\text{vol}} = \frac{1}{2} \epsilon \xi^2 \qquad (1\text{-}18\text{c})$$

One performs a similar operation to find the magnetic energy in the solenoid in Fig. 1-4*b*. This is done by calculating the work done on the charge, dq, when it is moved against a potential difference, V_L, created in an inductor while the current is building up to its steady state value, i_0. Thus

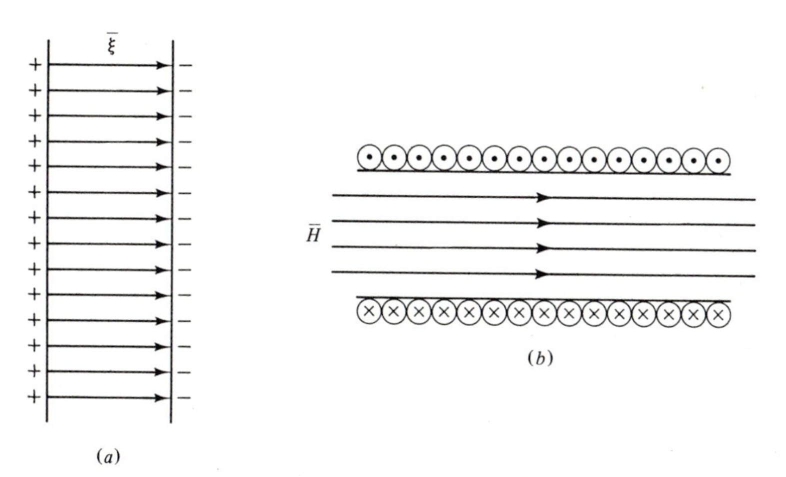

Figure 1-4 (*a*) The electric field in an ideal parallel plate capacitor. (*b*) The magnetic field in an ideal solenoid.

$$E_H = \int_0^{i_0} L \frac{di}{dt} dq = \frac{1}{2} L i_0^2 \tag{1-21a}$$

where i is the current through the solenoid. From Eq. 1-4b

$$i_0 = \frac{Hl}{N} \tag{1-22}$$

for a steady current where l is the solenoid lengths and N is the total number of turns of the wire. The inductance is defined as the total magnetic flux per unit current so that

$$L = \frac{N\mu HA}{i} \tag{1-23}$$

Substituting Eqs. 1-22 and 1-23 into Eq. 1-21a yields

$$E_H = \frac{1}{2} \mu H^2 A l \tag{1-21b}$$

or

$$\frac{E_H}{\text{vol}} = \frac{1}{2} \mu H^2 \tag{1-21c}$$

Consider the electromagnetic energy in the volume element $A\ \Delta z$ shown in Fig. 1-5*a*. The energy in the element is

$$dE = \left(\frac{1}{2} \epsilon \xi_x^2 + \frac{1}{2} \mu H_y^2\right) A\ \Delta z \tag{1-24a}$$

Thus, the energy reduction per unit time is

$$-\frac{\partial E}{\partial t} = \left(-\epsilon \xi_x \frac{\partial \xi_x}{\partial t} - \mu H_y \frac{\partial H_y}{\partial t}\right) A\ \Delta z \tag{1-24b}$$

Using Eqs. 1-7, 1-4a, 1-8, and 1-2a, Eq. 1-24b becomes

$$-\frac{\partial E}{\partial t} = \left(\xi_x \frac{\partial H_y}{\partial z} + H_y \frac{\partial \xi_x}{\partial z}\right) A\ \Delta z = \frac{\partial}{\partial z} (\xi_x H_y) A\ \Delta z \tag{1-24c}$$

This equation is similar to the conservation of charge equation, Eq. 1-5, only in this case energy is conserved. The term $\frac{\partial}{\partial z} (\xi_x H_y) \Delta z$ represents the net energy flux per unit area, $S(z + \Delta z) - S(z)$, out of the volume element. Thus, as we stated previously, the magnitude of S is equal to the light intensity.

That S is the light intensity can be derived in another way. The electromagnetic energy in the volume $c\ \Delta t A$ shown in Fig. 1-5*b* crosses the area, A, in time, Δt, since the waves move at the speed of light! Thus, the light intensity, I, is

$$I = \frac{\frac{1}{2}(\epsilon \xi_x^2 + \mu H_y^2) c A\ \Delta t}{A\ \Delta t} = \frac{c}{2} (\epsilon \xi_x^2 + \mu H_y^2) \tag{1-25a}$$

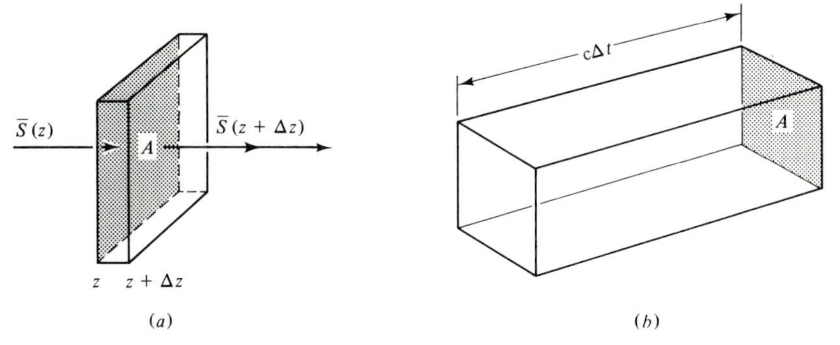

Figure 1-5 (*a*) The electromagnetic energy flowing into and out of the volume element $A\,\Delta z$. (*b*) The volume containing the electromagnetic energy that will flow across area A in time Δt.

Using Eq. 1-17c, this reduces to

$$I = c\epsilon\xi_x^2 \tag{1-25b}$$

Since

$$S = \xi_x H_y = \sqrt{\left(\frac{\epsilon}{\mu}\right)}\xi_x^2 = c\epsilon\xi_x^2 \tag{1-26}$$

$S = I$.

Usually, we are more interested in the average value of S rather than its instantaneous value. Since ξ and H vary sinusoidally,

$$\langle S \rangle = c\epsilon\langle \xi_x^2 \rangle = \frac{1}{2}c\epsilon\xi_{0x}^2 \tag{1-27}$$

One should have a feeling for the magnitudes of these quantities. They are computed in Example 1-1.

EXAMPLE 1-1

(a) The average rate at which the earth receives radiant energy from the sun at noon is 2.2 cal/min · cm². Find the amplitudes of S, ξ, D, H, and B for a wave with this intensity. **(b)** The wavelength of the most intense radiation from the sun is λ = 5000 Å (green light). Find ω and k.

(a) From the fact that 1 cal = 4.18 J

$$\langle S \rangle = \frac{(2.2)(4.18)(10^4)}{60} = 1.53 \times 10^3 \text{ W/m}^2$$

From Eq. 1-27

$$\xi_0 = \left(\frac{2\langle S \rangle}{c_0\epsilon_0}\right)^{1/2} = \left[\frac{(2)(1.53 \times 10^3)}{(3.00 \times 10^8)(8.86 \times 10^{-12})}\right]^{1/2}$$

$$= 1.07 \times 10^3 \text{ V/m}$$

$$D_0 = \epsilon_0\xi = (8.86 \times 10^{-12})(1.07 \times 10^3) = 9.48 \times 10^{-9} \text{ C/m}^2$$

$$H_0 = \left(\frac{\epsilon_0}{\mu_0}\right)^{1/2} \xi_0 = \frac{1.07 \times 10^3}{3.77 \times 10^2} = 2.84 \text{ A/m}$$

$$B_0 = \mu_0 H = (4\pi \times 10^{-7})(2.84) = 3.57 \times 10^{-6} \text{ W/m}^2$$

(b) $$k = \frac{2\pi}{\lambda} = 6.28/5 \times 10^{-7} = 1.256 \times 10^7 \text{ m}^{-1}$$

$$\omega = kc = (1.256 \times 10^7)(3.00 \times 10^8) = 3.768 \times 10^{15} \text{ rad/sec}$$

1-3.3 Polarization and Magnetization

The polarization is used to describe how the electric field is modified by a material, and the magnetization is used to describe how the magnetic field is modified by a material.

The flux source of D is the free charge on the capacitor plate so that it is not modified in the material placed between the plates of a parallel plate capacitor in Fig. 1-6. From Eq. 1-19

$$D = \frac{q}{A} = \sigma_f \tag{1-28}$$

where σ_f is the free charge per unit area. The flux source of $\epsilon_0\xi$ is the total charge, which includes the free charge on the capacitor plate and the induced charge on the material. The sign of the induced charge is opposite to that of the free charge, hence, ξ is reduced inside of the material. Mathematically

$$\int_S \epsilon_0 \xi \cdot dS = \int_V (\rho_f - \rho_p) dV \tag{1-29a}$$

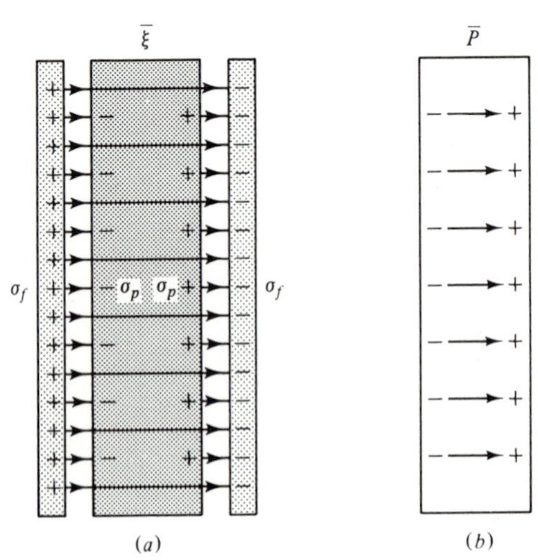

Figure 1-6 (*a*) The electric field in a parallel plate capacitor with a dielectric placed between the plates. (*b*) The polarization field inside of the dielectric.

which for a parallel plate capacitor is

$$\epsilon_0\xi = \sigma_f - \sigma_p \quad (1\text{-}29b)$$

where σ_p is the induced polarization charge per unit area on the material. By rearranging terms and using Eq. 1-28, Eq. 1-29b can be written

$$\sigma_f = D = \epsilon_0\xi + \sigma_p = \epsilon_0\xi + P \quad (1\text{-}30a)$$

where the magnitude of the polarization, P, is σ_p.

Equation 1-30a can also be written in vector form:

$$\mathbf{D} = \epsilon_0\boldsymbol{\xi} + \mathbf{P} \quad (1\text{-}30b)$$

The direction of **P** is from a negative polarization charge toward a positive polarization charge, and **P** is sometimes called the dipole moment per unit volume. A dipole, **p**, has the magnitude of the charge of a positive-negative pair times the distance of separation between the pair. For a dielectric of thickness, d,

$$p = (\sigma_p A)d = PV \quad (1\text{-}31)$$

Earlier we said

$$\mathbf{D} = \epsilon\xi \quad (1\text{-}7)$$

Comparing this with Eq. 1-30a, we see that

$$\epsilon = \epsilon_0 + \frac{P}{\xi} \quad (1\text{-}32a)$$

Because P is usually created by an electric field inducing a polarization charge, and for most materials P is proportional to ξ,

$$P = \chi\epsilon_0\xi \quad (1\text{-}33)$$

where χ is the electric susceptibility. Substituting Eq. 1-33 into Eq. 1-32a yields

$$\epsilon = \epsilon_0(1 + \chi) \quad (1\text{-}32b)$$

The relative permittivity, ϵ_r, also sometimes called the dielectric constant, is

$$\epsilon_r = \frac{\epsilon}{\epsilon_0} = 1 + \chi \quad (1\text{-}34)$$

and it is a measure of how strongly the electric field interacts with the material.

Now let us turn briefly to the magnetization. The circulation source of H in the solenoid of Fig. 1.4b is the free current in the wires. Thus, Eq. 1-22 can be rewritten as

$$H = \frac{Ni}{l} = j_f \quad (1\text{-}35)$$

where j_f is the free current per unit length.

The source of the magnetization, M, is the individual magnetic dipoles of the material. The electron moving in its orbit around the nucleus in Fig. 1-7a generates a magnetic moment just as a current in a loop does. (The electron moves in a direction opposite to that of the current.) The individual current

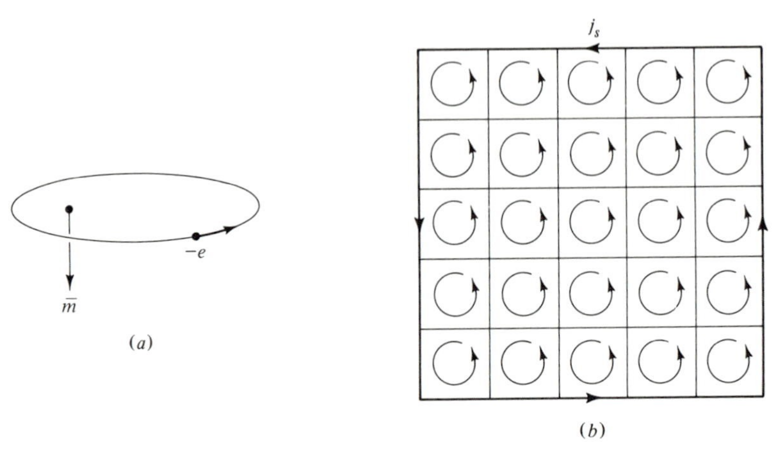

Figure 1-7 (*a*) The magnetic dipole, $\bar{m}$, created by an electron in an orbit around an atom. (*b*) The currents in the loops around the individual atoms, and their sum, the surface current, j_s.

loops in Fig. 1-7*b* add together to form a surface current, and this surface current per unit length, j_s, is the source of M. The circulation sources of the magnetic induction are both the free and surface currents. Thus

$$B = \mu_0(j_f + j_s) = \mu_0(H + M) \tag{1-36}$$

From Eq. 1-8

$$\mu = \mu_0 \left(\frac{1 + M}{H}\right) \tag{1-37a}$$

The magnetic dipoles of the material are aligned by the magnetic field so that often M is proportional to H. Thus

$$M = \chi_m H \tag{1-38a}$$

so that

$$\mu = \mu_0(1 + \chi_m) \tag{1-37b}$$

χ_m is the magnetic susceptibility. It can be negative as it is for diamagnetic materials (-10^{-5}), small and positive as it is for paramagnetic materials ($\sim 10^{-3}$), or large and positive as it is for ferro- and ferrimagnetic materials, such as iron and ferrites. We will not be concerned with ferro- or ferrimagnetic materials in this book, so that for our applications $\mu \sim \mu_0$.

To conclude this section we now rewrite Eq. 1-24b as

$$-\frac{\partial E}{\partial t} = -\left[\frac{\partial}{\partial t}\left(\frac{1}{2}\,\epsilon_0 \xi^2\right) + \xi\,\frac{\partial P}{\partial t} + \frac{\partial}{\partial t}\left(\frac{1}{2}\,\mu_0 H^2\right) + \mu_0 H\,\frac{\partial M}{\partial t}\right] A\,\Delta z \tag{1-24d}$$

The first term on the right-hand side is the self-electric energy of the wave; the second term is the electrical interaction energy between the wave and the material; the third term is the self-magnetic energy of the wave; and the fourth term is the magnetic interaction energy between the wave and the material.

The electrical energy given up to the material can be given back to the wave much as a capacitor and inductor feed energy back and forth in a simple LC circuit. When the capacitor is charging, the inductor is discharging and vice versa. However, when there is a resistor in the circuit, energy is dissipated in the form of heat, and the amplitude of the charge on the capacitor plate decays.

We learn in the next section that when there is loss due to absorption in the material, the amplitude of the wave will decay.

1-3.4 With Loss

First, we consider the situation where the conductivity is not zero and the electric permittivity is real. Then we discuss the situation when the dielectric constant is complex.

To solve the differential Eq. 1-6d, k must be replaced by the propagation constant, β, which is complex. Thus

$$\xi(z, t) = \xi_0 e^{j(wt \pm \beta z)} \tag{1-38b}$$

Substituting Eq. 1-38b into Eq. 1-6d yields

$$\beta^2 = \omega^2 \mu \epsilon - j\omega\mu\sigma \tag{1-39}$$

β can be written as

$$\beta = k - \frac{j\alpha}{2} \tag{1-40}$$

where k again is the wave vector, and α is the absorption coefficient. The factor of $\frac{1}{2}$ is used because we are ultimately interested in the decay of the intensity. Squaring (1-40) and equating the real and imaginary parts of Eqs. 1-39 and 1-40 gives us the following results:

$$k^2 = \frac{\mu\epsilon\omega^2}{2}\left\{\left[1 + \left(\frac{\sigma}{\omega\epsilon}\right)^2\right]^{1/2} + 1\right\} \tag{1-41}$$

and

$$\left(\frac{\alpha}{2}\right)^2 = \frac{\mu\epsilon\omega^2}{2}\left\{\left[1 + \left(\frac{\sigma}{\omega\epsilon}\right)^2\right]^{1/2} - 1\right\} \tag{1-42}$$

When $\sigma = 0$, k reduces to its no loss value, ω/c. The effect of σ is to increase k and, therefore, to decrease the wavelength and the phase velocity. Increasing σ also increases α and, therefore, the attenuation of the wave as it moves into the material. This can be seen by substituting Eq. 1-40 into Eq. 1-38, which gives for the wave moving in the $+z$ direction

$$\xi(z, t) = \xi_0 e^{-\alpha z/2 + j(\omega t - kz)} \tag{1-43}$$

Thus, $2/\alpha$ is the distance into the material for which the amplitude is reduced to $1/e$ of its original value. This distance is called the penetration depth.

We now quantify the amount of energy lost per unit volume per unit time by reexamining the Poynting vector when $\sigma \neq 0$.

$$\nabla \cdot \xi \times \mathbf{H} = \mathbf{H} \cdot \nabla \times \xi - \xi \cdot \nabla \times \mathbf{H} = -\mathbf{H} \cdot \frac{dB}{dt} - \xi \cdot \frac{dD}{dt} - \xi \cdot J \tag{1-44}$$

This equation can be derived by using Eqs. 1-2a and 1-4a. Recalling from Eq. 1-9 that $J = \sigma\xi$, the last term in Eq. 1-44 is $\sigma\xi^2$. This simply is the joule heating (i^2R heating) per unit volume.

To obtain a perspective on the magnitude of the different parameters discussed, let us turn to Example 1-2. There we will also see that the properties of a metal are vastly different from those of a dielectric.

EXAMPLE 1-2

A good conductor such as copper has a conductivity of 5.8×10^7 s/m, and a dielectric such as glass has a conductivity of 10^{-6} s/m. Find values for the wave vector, k, and the attenuation coefficient, $\alpha/2$, for **(a)** copper when $\omega = 10^9$ rad/sec, **(b)** glass when $\omega = 10^9$ rad/sec, **(c)** copper when $\omega = 10^{15}$ rad/sec, and **(d)** glass when $\omega = 10^{15}$ rad/sec. Assume $\epsilon = 4\epsilon_0$ and $\mu = \mu_0$ for both materials.

$$k^2 = \frac{1}{2}\mu\epsilon\omega^2\left\{\left[1 + \left(\frac{\sigma}{\omega\epsilon}\right)^2\right]^{1/2} + 1\right\}$$

$$\simeq \frac{1}{2}\mu\,\omega\sigma \quad \text{for} \quad \sigma \gg \omega\epsilon$$

$$\simeq \frac{\omega^2}{c^2}\left[1 + \frac{1}{4}\left(\frac{\sigma}{\omega\epsilon}\right)^2\right] \quad \simeq \left(\frac{\omega}{c}\right) \quad \text{for} \quad \sigma \ll \omega\epsilon$$

$$\left(\frac{\alpha}{2}\right)^2 = \frac{1}{2}\mu\epsilon\omega^2\left\{\left[1 + \left(\frac{\sigma}{\omega\epsilon}\right)^2\right]^{1/2} - 1\right\}$$

$$\simeq \frac{1}{2}\mu\omega\sigma \quad \text{for} \quad \sigma \gg \omega\epsilon$$

$$\simeq \mu\sigma^2/4\epsilon \quad \text{for} \quad \sigma \ll \omega\epsilon$$

(a) $$\frac{\sigma}{\omega\epsilon} = \frac{5.8 \times 10^7}{(10^9)(4)(8.86 \times 10^{-12})} = 1.64 \times 10^9 \gg 1$$

$$k \simeq \left(\frac{\mu\omega\sigma}{2}\right)^{1/2} = \left[\frac{(4\pi \times 10^{-7})(10^9)(5.8 \times 10^7)}{2}\right]^{1/2} = 1.91 \times 10^5\ \text{m}^{-1}$$

$$\frac{\alpha}{2} \cong 1.91 \times 10^5\ \text{m}^{-1}$$

Note that the penetration depth of the wave is only $\sim 5 \times 10^{-3}$ mm.

(b) $$\frac{\sigma}{\omega\epsilon} = \frac{10^{-6}}{(10^9)(4)(8.86 \times 10^{-12})} = 2.82 \times 10^{-5} \ll 1$$

$$k = \frac{\omega}{c} = \frac{10^9}{1.5 \times 10^8} = 6.67\ \text{m}^{-1}$$

Notice how much shorter the wavelength is in the metal

$$\frac{\alpha}{2} \cong \left(\frac{\mu_0}{4\epsilon}\right)^{1/2}\sigma = \frac{(3.77 \times 10^2)(10^{-6})}{4} = 9.43 \times 10^{-5}\ \text{m}^{-1}$$

(c) $$\frac{\sigma}{\omega\epsilon} = 1.64 \times 10^3 \gg 1$$

$$k \cong (10^3)(1.91 \times 10^5) = 1.91 \times 10^8 \text{ m}^{-1}$$

$$\frac{\alpha}{2} \cong k = 1.91 \times 10^8 \text{ m}^{-1}$$

(d) $$\frac{\sigma}{\omega\epsilon} = 2.82 \times 10^{-11} \ll 1$$

$$k = \frac{\omega}{c} = 6.67 \times 10^{-6} \text{ m}^{-1}$$

$$\frac{\alpha}{2} \text{ (independent of the frequency)} \cong 9.43 \times 10^{-5} \text{ m}^{-1}$$

The primary loss in a dielectric is through direct absorption rather than joule heating. The absorption process will be described in considerable detail later. In this chapter we represent it by a complex electric permittivity

$$\epsilon = \epsilon' - j\epsilon'' \tag{1-45}$$

where ϵ' is the real part of ϵ and ϵ'' is the imaginary part.

The wave equation is now written

$$\frac{\partial^2 \xi}{\partial z^2} = \mu\epsilon' \frac{\partial^2 \xi}{\partial t^2} - j\mu\epsilon'' \frac{\partial^2 \xi}{\partial t^2} + \mu\sigma \frac{\partial \xi}{\partial t} \tag{1-6e}$$

Now, if we substitute Eq. 1-38 into Eq. 1-6e, we find that

$$\beta^2 = \omega^2\mu\epsilon' - j\omega\mu(\sigma + \omega\epsilon'')$$

By substituting $\sigma' = \sigma + \omega\epsilon''$ for σ and ϵ' for ϵ in Eqs. 1-41 and 1-42, one can find the complete expression for k and $\alpha/2$.

If ϵ is complex, χ must also be complex. From Eq. 1-32b

$$\epsilon = \epsilon' - j\epsilon'' = \epsilon_0(1 + \chi' - j\chi'') \tag{1-32b}$$

Correlating the real and imaginary parts, one finds that

$$\epsilon'' = \epsilon_0\chi'' \tag{1-46}$$

Because χ is complex, P lags behind ξ where the lag angle is given by

$$\tan\phi = \frac{\chi''}{\chi'} \tag{1-47a}$$

Whenever there is a lag, there is a loss associated with it. We show this by considering the electrical interaction term—the second term in Eq. 1-24d.

The power lost per unit volume is

$$\frac{P}{\text{vol}} = \frac{1}{T}\int_0^T \xi \cdot \frac{\partial P}{\partial t}\, dt = \frac{1}{T}\int_0^T \xi(z) \sin\omega t P(z) \cos(\omega t - \phi)\omega\, dt \tag{1-48a}$$

where $T = 2\pi/\omega$ is the time for one cycle. The term inside the integral is the energy lost per cycle, and $1/T$ is the number of cycles per unit time. Equation 1-48a integrates to

$$\frac{P}{\text{vol}} = \frac{1}{2}\,\omega\epsilon_0\chi\xi(z)\sin\phi \tag{1-48b}$$

after Eq. 1-33 has been substituted into it. Using the complex plane, one sees that

$$\sin\phi = \frac{\chi''}{\chi} \tag{1-47b}$$

so that

$$\frac{P}{\text{vol}} = \frac{1}{2}\,\omega\epsilon_0\chi''\xi^2(z) = \frac{1}{2}\,\omega\epsilon''\xi^2(z) \tag{1-48c}$$

Because P lags ξ, the P versus ξ curve forms a hysteresis loop as is shown in Fig. 1-8. The area under this curve is the loss per cycle.

When there is loss, **H** also lags behind $\boldsymbol{\xi}$. Again using Eq. 1-2a and recalling that $\boldsymbol{\xi}$ is parallel to the x direction and that **H** is parallel to the y direction,

$$-j\beta\xi_{0x}e^{j(\omega t-\beta z)} = -j\omega\mu H_{0y}e^{j(\omega t-\beta z-\phi)} \tag{1-49}$$

Using the expression for β in Eq. 1-39b, we find that the intrinsic wave impedance, η, is

$$\eta = \xi_{0x}/H_{0y} = \frac{\sqrt{\mu/\epsilon'}}{\left[1 - j\left(\dfrac{\sigma'}{\omega\epsilon'}\right)\right]^{1/2}} \tag{1-50a}$$

Thus, the magnitude of η is

$$|\eta| = \frac{\sqrt{\mu/\epsilon'}}{\left[1 + \left(\dfrac{\sigma'}{\omega\epsilon'}\right)^2\right]^{1/4}} \tag{1-50b}$$

It can also be shown that the lag angle is

$$\tan 2\phi = \frac{\sigma'}{\omega\epsilon'} \tag{1-51}$$

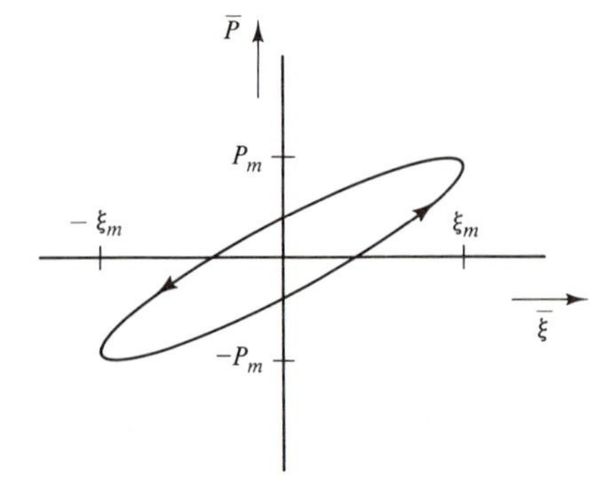

Figure 1-8 A $\bar{P}$ versus $\bar{\xi}$ hysteresis curve illustrating there is loss when $\bar{P}$ lags $\bar{\xi}$.

To gain insight into the values of the losses and the lag angles in a metal and a dielectric, we turn to Example 1-3.

EXAMPLE 1-3

(a) For a metal $\sigma \gg \omega\epsilon'$ at all frequencies. Calculate $|\eta|$ and ϕ for copper for $\omega = 10^9$ and 10^{15} rad/sec. The properties of copper are listed in Example 1-2; **(b)** compute $\alpha/2$ for a glass dielectric for $\omega = 10^9$ and 10^{15} rad/sec if $\epsilon''(10^9) = 5 \times 10^{-4}\epsilon'$ and $\epsilon''(10^{15}) = 5 \times 10^{-7}\epsilon'$, $\epsilon' = 4\epsilon_0$ for both the metal and the glass.

(a) For $\sigma' \gg \omega\epsilon'$

$$\eta = \sqrt{\frac{\mu}{\epsilon'}}\left(\frac{\omega\epsilon'}{\sigma'}\right)^{1/2} = \sqrt{\frac{\mu\omega}{\sigma'}}$$

$$\eta(10^9) = \left[\frac{(4\pi \times 10^{-7})(10^9)}{5.8 \times 10^7}\right]^{1/2} = 4.65 \times 10^{-3}\ \Omega$$

This is much less than it would be with no loss (138.5 Ω)

$$\eta(10^{15}) = 4.65 \times 10^{-3}(10^6)^{1/2} = 4.65\ \Omega$$

$$\tan 2\phi(10^9) = \frac{\sigma'}{\omega\epsilon'} = 1.64 \times 10^9 \qquad \text{(from Example 1-2)}$$

$$\therefore \phi = 45°$$

$$\tan 2\phi(10^{15}) = 1.64 \times 10^3$$

$$\therefore \phi = 45°$$

(b) For $\sigma' \ll \omega\epsilon'$

$$\frac{\alpha}{2} \sim \sqrt{\mu/\epsilon'}\,\frac{\sigma'}{2} \qquad \text{(from Example 1-2)}$$

For $\sigma \ll \omega\epsilon''$

$$\sigma' = (\sigma + \omega\epsilon'') \simeq \omega\epsilon''$$

$$\therefore \frac{\alpha}{2} \simeq \frac{1}{2}\,\omega\sqrt{\mu\epsilon'}\,\frac{\epsilon''}{\epsilon'}$$

At $\omega = 10^9$

$$\omega\epsilon'' = 5 \times 10^5 \gg \sigma$$

$$\frac{\alpha}{2} = \frac{(0.5)(10^9)(5 \times 10^{-4})}{1.5 \times 10^8} = 1.67 \times 10^{-3}\ \text{m}^{-1}$$

$$\frac{\alpha}{2}(10^{15}) = \frac{(0.5)(10^{15})(5 \times 10^{-7})}{1.5 \times 10^8} = 1.67\ \text{m}^{-1}$$

$\omega = 10^{15}$ is near the visible range. Thus, you can see through a glass block 1 m thick.

Note that the primary loss in metals is the result of joule heating via the conductivity, and most of the loss in dielectrics is due to the imaginary part of the electric permittivity (absorption). Note also that the loss in dielectrics does not noticeably change the optical constants from what they are for a lossless dielectric.

1-4 PROPAGATION OF LIGHT IN UNBOUNDED MEDIA

1-4.1 Boundary Conditions at a Planar Interface

A planar interface is an abrupt boundary at which the properties change discontinuously such as at an air-material boundary. First, we consider the electric displacement for the Gaussian pill box in Fig. 1-9*a*. If there is no free surface charge at the boundary, then as the pill box is made shorter and shorter

$$\int_S \mathbf{D} \cdot d\mathbf{S} = -D_{n1}A + D_{n2}A = 0 \tag{1-52a}$$

or

$$D_{n1} = D_{n2} \tag{1-52b}$$

Thus, the normal component of **D** is continuous across a boundary containing no free charge.

For the electric field in Fig. 1-9*b*

$$\oint \xi \cdot d\mathbf{l} = \xi_{T1}\, d - \xi_{T2}\, d = -\int \frac{dB}{dt} \cdot dA = 0 \tag{1-53a}$$

as the distance normal to the interface and, therefore, the area of the loop gets smaller and smaller. Thus

$$\xi_{T1} = \xi_{T2} \tag{1-53b}$$

across an interface where ξ_T is the tangential component of ξ.

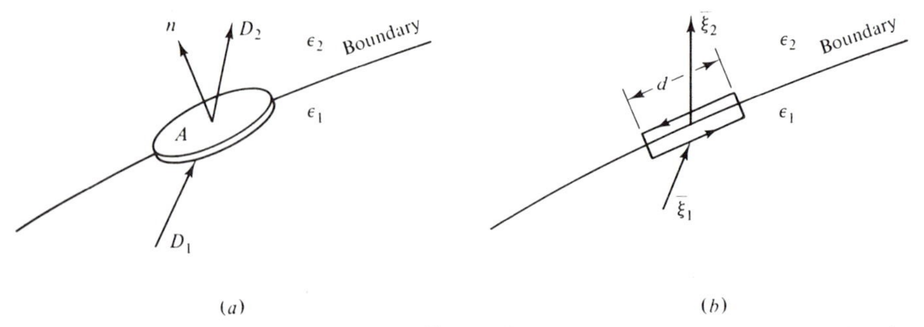

Figure 1-9 (*a*) A Gaussian surface for $\bar{D}$ at an interface that is infinitesimally thin. (*b*) *A* loop for $\bar{\xi}$ at an interface that is infinitesimally thin.

Using similar arguments, one can show that

$$\int B \cdot dA = -B_{n1}A + B_{n2}A = 0 \tag{1-54a}$$

or

$$B_{n1} = B_{n2} \tag{1-54b}$$

where B_n is the normal component of the magnetic induction vector. Also

$$\oint H \cdot d\mathbf{l} = H_{T1}\, d - H_{T2}\, d = \int_A \left(\frac{\partial D}{\partial t} + J\right) \cdot dA = 0 \tag{1-55a}$$

or

$$H_{T1} = H_{T2} \tag{1-55b}$$

where H_T is the tangential component of the magnetic field.

1-4.2 Reflection at a Single Boundary with No Loss

First, we consider reflection at normal incidence, which is illustrated in Fig. 1-10*a*, and we find that the amount of light reflected increases with the difference in the index of refraction of the media on either side of the boundary. The index of refraction is the ratio of the speed of light in a vacuum, c_0, to the speed of light in the material, c. Thus

$$n = \frac{c_0}{c} = \sqrt{\mu\epsilon/\mu_0\epsilon_0} = \sqrt{\epsilon_r} \tag{1-56}$$

for materials that are not ferro- or ferrimagnetic.

Because $\boldsymbol{\xi} \times \mathbf{H}$ is parallel to the direction of propagation, either $\boldsymbol{\xi}$ or $\mathbf{H}$ of the reflected beam must be rotated by 180° relative to the incident wave. We assume here that $\boldsymbol{\xi}$ rotates and later will check to see if our assumption is correct.

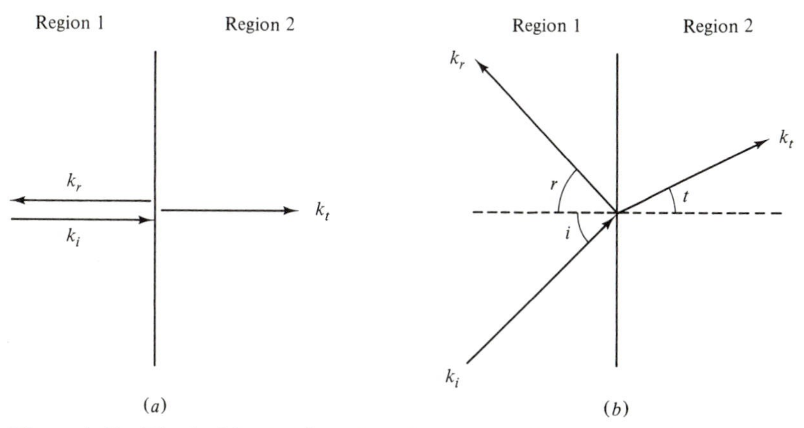

Figure 1-10 The incident, reflected, and transmitted beams for (*a*) normal incidence, and (*b*) off normal incidence.

Using the boundary condition that the tangential components of $\boldsymbol{\xi}$ and **H** are continuous across a boundary, one can write

$$\xi_i - \xi_r = \xi_t \tag{1-57}$$

and

$$H_i + H_r = H_t \tag{1-58}$$

where the subscripts i, r, and t refer respectively to the incident, reflected, and transmitted components. The negative sign in Eq. 1-57 results from the 180° rotation of the reflected beam, and the tangential component is the entire component because $\boldsymbol{\xi}$ and **H** are parallel to the interface for normal incidence. Using the intrinsic wave impedance relationships, Eq. 1-17c, and assuming that μ on both sides of the boundary is the same, one obtains

$$r = \frac{\xi_r}{\xi_i} = \frac{\sqrt{\epsilon_2} - \sqrt{\epsilon_1}}{\sqrt{\epsilon_2} + \sqrt{\epsilon_1}} = \frac{n_2 - n_1}{n_2 + n_1} \tag{1-59}$$

by eliminating the transmitted values from Eqs. 1-57 and 1-58. The term r is the reflection coefficient. We note that r is positive, as it must be, if $n_2 > n_1$. Thus, our assumption that ξ_r is rotated 180° from ξ_i is correct for this condition. If $n_1 > n_2$, then H_r, not ξ_r, is rotated by 180°. We can find the transmission coefficient, t, by eliminating the reflected values from Eqs. 1-57 and 1-58. We find that

$$t = \frac{2n_1}{n_2 + n_1} \tag{1-60}$$

EXAMPLE 1-4

Compute the fraction of the light intensity that is reflected and transmitted at normal incidence.

From Eq. 1-27 the reflected intensity is

$$I_r = \frac{1}{2} c_1 \epsilon_1 \xi_{0r}^2 = \frac{1}{2} c_1 \epsilon_1 \left(\frac{n_2 - n_1}{n_2 + n_1}\right)^2 \xi_{0i}^2$$

Thus, the reflectance, R, the fraction of the light intensity reflected, is

$$R = \left(\frac{n_2 - n_1}{n_2 + n_1}\right)^2$$

The transmitted light intensity is

$$I_t = \frac{1}{2} c_2 \epsilon_2 \xi_{0t}^2 = \frac{1}{2} c_2 \epsilon_2 \left(\frac{2n_1}{n_2 + n_1}\right)^2 \xi_{0i}^2$$

Thus, the transmittance, T, the fraction of the light intensity transmitted, is

$$T = \frac{c_2 \epsilon_2}{c_1 \epsilon_1} \left(\frac{2n_1}{n_2 + n_1}\right)^2 = \frac{4n_2 n_1}{(n_2 + n_1)^2}$$

Note that $R + T = 1$ as it must if energy is to be conserved.

The reflection and transmission of a beam that is not normal to the surface is more complicated. The reflection and transmission coefficients depend on the direction of polarization as well as the angle of incidence, and the ray is bent as it crosses the boundary. This is called refraction, and it is illustrated in Fig. 1-10*b*.

First, we find the relationships between the angle of incidence and the angle of reflection and between the angle of incidence and the angle of transmission. The equation for a plane wave is written

$$\xi = \xi_0 \exp\left\{j\omega\left[t - \frac{k}{\omega}(l_1x + l_2y + l_3z)\right]\right\} \tag{1-61a}$$

when **k** points in an arbitrary direction. The l_i are the cosines of the angle made by the **k** vector with the x, y, and z axes. With no loss of generality the axes are chosen so that the xy plane is parallel to the boundary and the y axis is normal to the plane of the incident and reflected beam (the plane of the paper). Thus, for all three beams $l_2 = 0$, and $l_1 = \sin i$ for the incident beam, $l_1 = \sin r$ for the reflected beam, and $l_1 = \sin t$ for the transmitted beam. At the interface where $z = 0$

$$\xi_i = \xi_{0i}e^{j\omega_1(t - x\sin i/c_1)} \tag{1-61b}$$

$$\xi_r = \xi_{0r}e^{j\omega_1(t - x\sin r/c_1)} \tag{1-61c}$$

$$\xi_t = \xi_{0r}e^{j\omega_2(t - x\sin t/c_2)} \tag{1-61d}$$

For the tangential component to be continuous across the interface, the phases of 1-61b, c, and d must be the same for all values of t and x.

The first result is

$$\omega_1 = \omega_2 = \omega \tag{1-62}$$

Thus, ω is continuous across a boundary; only the speed of light and the wavelength change.

From

$$x \sin\frac{i}{c_1} = x \sin\frac{r}{c_1} \tag{1-63a}$$

or

$$\sin i = \sin r \tag{1-63b}$$

Thus, the angle of reflection equals the angle of incidence. From

$$x \sin\frac{i}{c_1} = x \sin\frac{t}{c_2} \tag{1-64a}$$

or

$$n_1 \sin i = n_2 \sin t \tag{1-64b}$$

From this equation, known as Snell's law, we see that if $n_2 > n_1$, the beam is bent toward the normal, and if $n_2 < n_1$, it is bent away from the normal. For the latter condition there must, therefore, be an angle of incidence above which there is no transmitted beam. This is called the critical angle, i_c, and it is the angle for which $\sin t = 1$. From Eq. 1-64b

$$\sin i_c = \frac{n_2}{n_1} \tag{1-65}$$

As we will learn in the next chapter, this is an important relationship for step index optical fibers.

To find the fraction of light reflected, we must first break up ξ into its component parallel to the plane of incidence (the plane of the paper), and its component perpendicular to this plane (see Fig. 1-11*a*). For the perpendicular component

$$\xi_i - \xi_r = \xi_t \tag{1-66}$$

and

$$H_i \cos i + H_r \cos i = H_t \cos t \tag{1-67a}$$

If we use the intrinsic wave impedance relationship, Eq. 1-17c, assuming that μ is the same on either side of the boundary, and use Snell's law, Eq. 1-67a becomes

$$\xi_i + \xi_r = \sqrt{\frac{\epsilon_2}{\epsilon_1}\frac{\cos t}{\cos i}}\,\xi_t = \frac{\tan i}{\tan t}\,\xi_t \tag{1-67b}$$

Eliminating ξ_t from Eqs. 1-66 and 1-67b, we obtain

$$r_\perp = \frac{\xi_{r\perp}}{\xi_{i\perp}} = \frac{\sin(i-t)}{\sin(i+t)} \tag{1-68}$$

For $n_2 > n_1$, $i > t$ so that $r_\perp$ is always positive as it must be. Thus, for $n_2 > n_1$ our assumption that ξ_r is rotated 180° from ξ_i is correct.

Using a similar procedure, one can show that

$$t_\perp = \frac{\sin(i+t) - \sin(i-t)}{\sin(i+t)} \tag{1-69}$$

and

$$r_\parallel = \frac{\tan(i-t)}{\tan(i+t)} \tag{1-70}$$

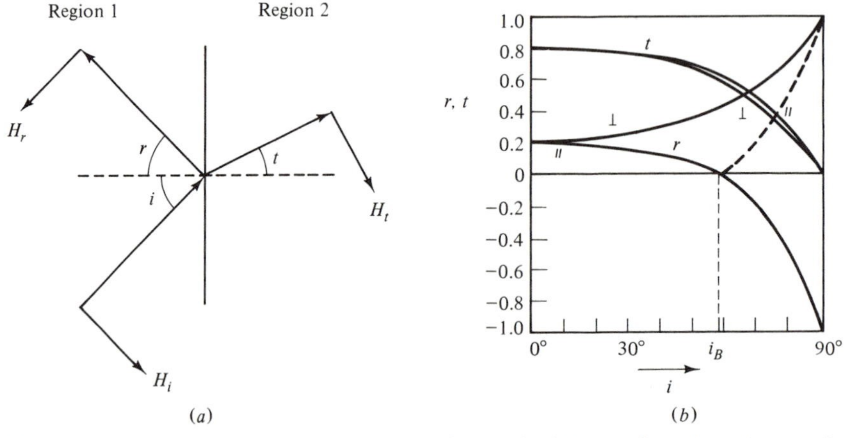

Figure 1-11 (*a*) The magnetic field vectors for the incident, reflected, and transmitted beams when the electric field is perpendicular to the plane of incidence ($\xi_\perp$). (*b*) The reflection (r) and transmission (t) coefficients plotted as a function of the incident angle when $n_2 = 1.5$ and $n_1 = 1.0$.

where $r_{\|}$ is the reflection coefficient for the component of ξ parallel to the plane of incidence.

Both $r_\perp$ and $r_{\|}$ are plotted in Fig. 1-11*b* for $n_2 > n_1$ where it is seen that $r_\perp$ increases from its value for normal incidence to 1 as i goes from 0° to 90°. However, $r_{\|}$ decreases with increasing i, and at i_B, the Brewster angle, $r_{\|} = 0$. Thus, at this angle, the reflected beam is polarized as only the perpendicular component is reflected. For $i > i_B$, $r_{\|}$ is negative. In actuality it cannot be; therefore, H_r must be rotated 180° instead of ξ_r.

The Brewster angle is found from

$$\tan i_B = \frac{\sin i_B}{\cos(\pi/2 - t)} = \frac{n_2}{n_1} \tag{1-71}$$

because from Eq. 1-70

$$i_B + t = \frac{\pi}{2} \tag{1-72}$$

which makes the ratio the same as the one for Snell's law.

EXAMPLE 1-5

Consider an incident beam passing from air into glass with an index of refraction, $n_2 = 1.5$, at an angle of 45°. **(a)** Find the transmission angle. **(b)** If the beam were moving in the other direction, what would be the critical angle? **(c)** What is the direction of polarization of the reflected beam if the electric field vector of the incident beam makes an angle of 45° with the plane of incidence? **(d)** At what angle of incidence is the direction of polarization of the reflected beam normal to the plane of incidence?

(a) From Snell's law

$$\sin t = \frac{\sin i}{n_2} = \frac{1}{\sqrt{2}\cdot 1.5} = 0.47140$$

$$t = 28.13°$$

(b) Using Eq. 1-65

$$\sin i_c = \frac{1}{n_2} = \frac{1}{1.5} = 0.66667$$

$$i_c = 41.81°$$

(c) The magnitude of the component normal to the plane of incidence is

$$\xi_{r\perp} = \xi_i \frac{\sin(i - t)}{\sin(i + t)} = 0.303\xi_i$$

The magnitude of the parallel component is

$$\xi_{r\|} = \xi_i \frac{\tan(i - t)}{\tan(i + t)} = 0.092$$

The direction of polarization makes an angle ϕ with the plane of incidence where

$$\tan \phi = \frac{\xi_{r\perp}}{\xi_{r\parallel}} = 3.297$$

$$\therefore \phi = 73.16^\circ$$

(d) The direction of polarization of the reflected beam is normal to the plane of incidence at the Brewster angle because there is no parallel component.

$$\tan i_B = \frac{n_2}{n_1} = 1.5$$

$$i_B = 56^\circ$$

1-4.3 REFLECTION FROM A SINGLE BOUNDARY WITH LOSS

For normal incidence the procedure is much the same as it is for no loss. The only difference is that the intrinsic wave impedance is complex. Thus, the analog to Eq. 1-59 is

$$r = \frac{1/\eta_2 - 1/\eta_1}{1/\eta_2 + 1/\eta_1} \tag{1-73a}$$

which reduces to

$$r = \frac{\sqrt{\epsilon_2'/\epsilon_0}\sqrt{1 - j(\sigma'/\epsilon_2')} - n_1}{\sqrt{\epsilon_2'/\epsilon_0}\sqrt{1 - j(\sigma'/\epsilon_2')} + n_1} \tag{1-73b}$$

when there is no loss on the left-hand side.

A more usual form of Eq. 1-73b is

$$r = \frac{n_2' - n_1 - jn_2''}{n_2' + n_1 - jn_2''} \tag{1-73c}$$

where n_2' is the real part of the index of refraction and n_2'' is the imaginary part. Correlating coefficients from Eqs. 1-73b and 1-73c and squaring both sides of the equation yields

$$n_2'^2 - n_2''^2 - 2jn_2'n_2'' = \frac{\epsilon_2'}{\epsilon_0}\left(1 - j\frac{\sigma'}{\omega\epsilon_2'}\right) \tag{1-74}$$

Correlating the real and imaginary parts we obtain

$$n_2'^2 = \frac{1}{2}\frac{\epsilon_2'}{\epsilon_0}\left\{\left[1 + \left(\frac{\sigma'}{\omega\epsilon_2'}\right)^2\right]^{1/2} + 1\right\} \tag{1-75a}$$

and

$$n_2''^2 = \frac{1}{2}\frac{\epsilon_2'}{\epsilon_0}\left\{\left[1 + \left(\frac{\sigma'}{\omega\epsilon_2'}\right)^2\right]^{1/2} - 1\right\} \tag{1-75b}$$

Snell's law also changes and now the transmitted beam lags the incident beam. The analog to Eq. 1-64b is

$$\sin i = \left(\frac{\beta_2}{\beta_1}\right)\sin t \tag{1-76a}$$

which reduces to

$$\sin i = \sqrt{\frac{\epsilon_2'}{\epsilon_1} - j\frac{\sigma'}{\omega\epsilon_1}}\sin t \tag{1-77}$$

if there is no loss on the left-hand side, and $\mu_1 = \mu_2$.

EXAMPLE 1-6

Compute r for normal incidence, n_2', and n_2'' for copper and glass when $\omega = 10^9$ and 10^{15} rad/sec. Their properties are listed in Examples 1-2 and 1-3.

For the copper metal

$$\eta_1 = \sqrt{\mu_0/\epsilon_0} = 377\ \Omega \gg \eta_2 = 4.65 \times 10^{-3}\ \Omega(10^9) \quad \text{and} \quad 4.65\ \Omega(10^{15})$$

$$r \sim 1$$

This is why metals are so shiny: they have a very high reflectivity.

$$n_2'^2 \simeq \frac{1}{2}\frac{\epsilon_2'}{\epsilon_0}\frac{\sigma'}{\omega\epsilon_2'}$$

For $\omega = 10^9$

$$n_2' = \left[\left(\frac{1}{2}\right)(4)1.64 \times 10^9\right]^{1/2} = 5.73 \times 10^4$$

For $\omega = 10^{15}$

$$n_2' = \frac{5.73 \times 10^4}{10^3} = 57.3$$

For $\sigma' \gg \omega\epsilon_2'$, $n_2'' \simeq n_2'$

For the glass dielectric

$$\eta_2 \simeq \sqrt{\mu/\epsilon_2'} \quad \text{when} \quad \sigma' \ll \omega\epsilon'$$

$$r = \frac{\sqrt{\epsilon_2'/\mu_0} - \sqrt{\epsilon_0/\mu_0}}{\sqrt{\epsilon_2'/\mu_0} + \sqrt{\epsilon_2'/\mu_0}} = \frac{\sqrt{\epsilon_2'/\epsilon_0} - 1}{\sqrt{\epsilon_2'/\epsilon_0} + 1} = \frac{2-1}{2+1} = 0.333$$

$$n_2'^2 = \frac{1}{2}\frac{\epsilon_2'}{\epsilon_0}\left\{\left[1 + \left(\frac{\sigma'}{\omega\epsilon_2'}\right)^2\right]^{1/2} + 1\right\} \simeq \frac{\epsilon_2'}{\epsilon_0}$$

$$n_2'^2 = \frac{\epsilon_2'}{\epsilon_0} = 2 \text{ for both frequencies}$$

$$n_2''^2 = \frac{1}{2}\frac{\epsilon_2'}{\epsilon_0}\left\{\left[1 + \left(\frac{\sigma'}{\omega\epsilon_2'}\right)^2\right]^{1/2} - 1\right\} \simeq \frac{1}{4}\frac{\epsilon_2'}{\epsilon_0}\left(\frac{\sigma'}{\omega\epsilon_2'}\right)^2 \simeq \frac{1}{4}\frac{\epsilon_2'}{\epsilon_0}\left(\frac{\epsilon_2''}{\epsilon_2'}\right)^2$$

$$n_2''(10^9) = \left[\left(\frac{1}{4}\right)(4)\right]^{1/2} 5 \times 10^{-4} = 5 \times 10^{-4}$$

$$n_2''(10^{15}) = \left[\left(\frac{1}{4}\right)4\right]^{1/2} 5 \times 10^{-7} = 5 \times 10^{-7}$$

In concluding this section we should point out that ϵ' and ϵ'' and n' and n'' are not constants. Rather, they are functions of the wavelength. One general curve can be used to describe a dielectric, and this is done in Fig. 1-12 where n is plotted as a function of the wavelength. One general curve can be used because, as we will discuss in the next chapter, a dielectric can be represented on an atomic scale as simple oscillators with different resonant frequencies.

X rays are waves with wavelengths <10 nm. At these wavelengths, $n = 1$ except at peaks designated *K, L, M* peaks in Fig. 1-12. The peaks correspond to absorption peaks for core electrons in an atom; more will be said about this in Chapter 3. For $10 < \lambda < 400$ nm the rays are ultraviolet rays. There is a broad maximum in the middle of this region that corresponds to electronic resonant frequencies. In the visible region, $400 < \lambda < 700$ nm, n decreases with increasing λ and has an average value of ~ 1.4. In the near infrared there is another maximum for $\lambda \sim 2$–10 μm, and then n decreases to a value of ~ 1.9. The peak corresponds to the resonant frequency of the ions vibrating in the lattice. The third and final broad maximum occurs at the transition between infrared radiation and radio waves: $\lambda \sim 0.1$ mm. Beyond the peak n decreases to its static value of ~ 2. This peak is due to dipole relaxation and is large only if the molecules have a permanent dipole, for example, water, H_2O.

Because of their electronic structure, the optical properties of metals are more complicated. We will only consider the reflectance, $R = rr^*$, at normal incidence for the few selected samples shown in Fig. 1-13 and listed in Table 1-1.

Silver and aluminum have the highest reflectance in the visible region—the abrupt drop off in the near ultraviolet is again associated with electronic resonance. From the table one can see that for silver there is a considerable contribution from the imaginary component, which is associated with absorption. Aluminum is often preferred to silver for mirror coatings because it is easy to evaporate, forms a protective oxide coating, and is cheaper.

An interesting application of evaporated thin metal films was the face mask of the Apollo space suits that had a gold coating which reflected 70 percent of the visible light to reduce the cooling requirements of the space suit. Objects, however, appeared to have a green tinge. How can this be explained using the reflectance curve for gold in Fig. 1-13?

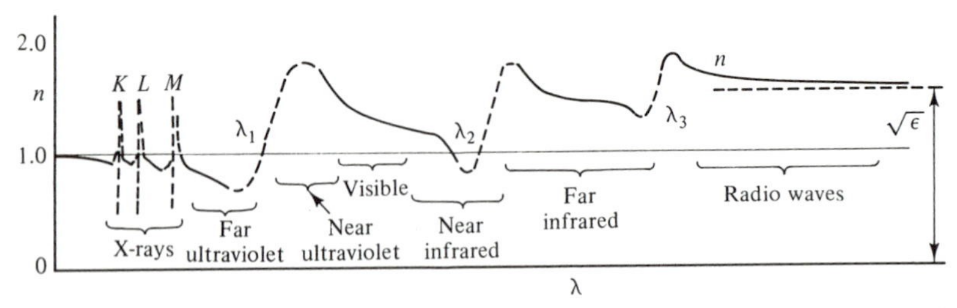

Figure 1-12 The index of refraction for a typical dielectric plotted as a function of the wavelength.

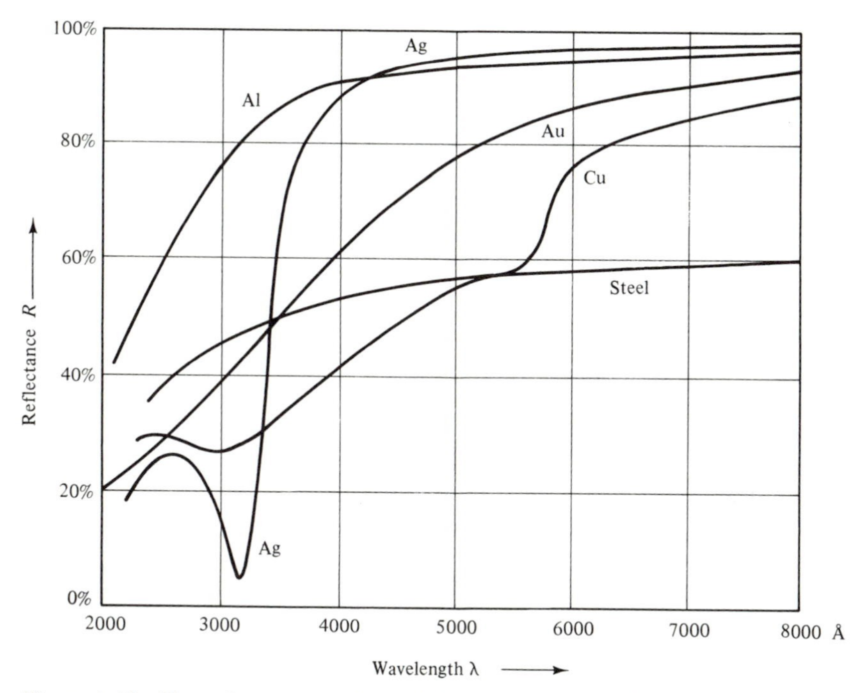

Figure 1-13 The reflectance in the visible region for a number of metals plotted as a function of the wavelength.

1-5 MIRRORS AND LENSES

1-5.1 Mirrors

For a plane mirror the reflected rays originating from the same point on the object all appear to emanate from a point that is the same distance behind the mirror as the object is in front of it. This is shown in Fig. 1-14. The image produced by these beams is a virtual image because the rays originating from a point on the object do not pass through the same point in the image plane—they only appear to. Also, it is evident that the image and object are the same size; hence, the magnification is $m = 1$.

TABLE 1-1 THE REAL AND IMAGINARY PARTS OF THE INDEX OF REFRACTION FOR SOME METALS FOR SODIUM LIGHT, $\lambda = 5893$ Å

Metal	n'	n''
Cobalt	2.120	4.040
Copper	0.617	2.630
Gold	0.37	2.82
Silver	0.177	3.638
Steel	2.485	3.433

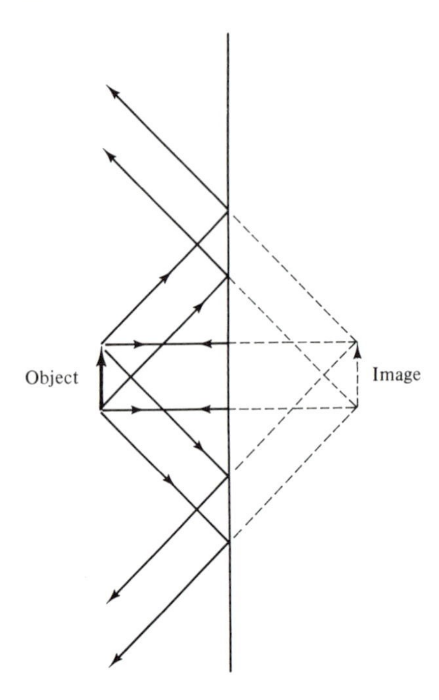

Figure 1-14 The formation of the image (virtual) of a plane mirror.

A spherical mirror with a radius of curvature, R, is shown in Fig. 1-15. First, we wish to find the point, Q, where a ray originally parallel to the axis crosses the axis when it is reflected. From the law of sines and Fig. 1-15

$$\frac{\sin(\pi - 2i)}{R} = \frac{\sin 2i}{R} = \frac{\sin i}{R - x} \tag{1-78a}$$

where x is the distance from the vertex of the mirror to the point where the ray, originally parallel to the axis, crosses the axis. Solving Eq. 1-78a for x yields

$$x = R\left(1 - \frac{1}{2\cos i}\right) \tag{1-78b}$$

For parallel rays close to the axis, often called paraxial rays, $i \sim 0°$ so that $x \simeq \frac{R}{2}$. All of the paraxial rays pass through the point, $x = \frac{R}{2}$, and this point is called the focal point of the mirror. The parallel rays further from the axis are

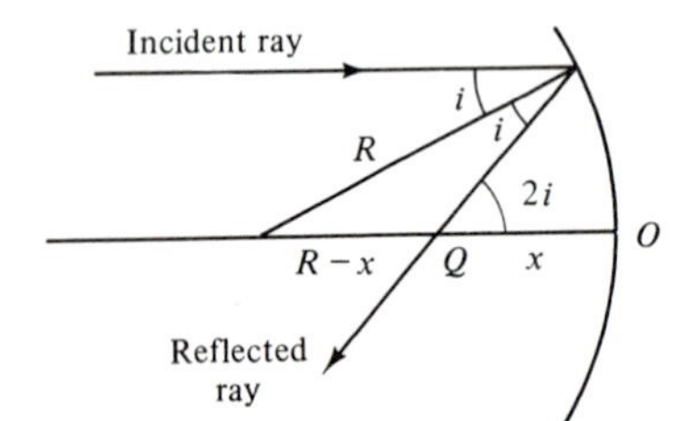

Figure 1-15 The point, Q, where an originally parallel incident ray crosses the axis after it has been reflected from a spherical mirror.

not reflected through the focal point, since now the approximation $i \sim 0°$ is no longer valid. This is the source of spherical aberration.

The image plane is the plane where rays emanating from a single point of the object are focused to a single point. Consider a ray emanating from the point of the object on the axis in Fig. 1-16*a* that is a distance, s, away from the vertex. Observe that the reflected beam is a distance, s', from the vertex. s is the object distance, and s' is the image distance. From the law of sines

$$\frac{\sin i}{s - R} = \frac{\sin \theta}{R} \tag{1-79}$$

and

$$\frac{\sin i}{R - s'} = \frac{\sin (\pi - 2i - \theta)}{R} \tag{1-80}$$

Assuming that the angles are small so that $\sin i \sim i$, one can show that

$$\frac{1}{s} + \frac{1}{s'} = \frac{2}{R} \tag{1-81}$$

Note that we did not choose a fixed value of i so that rays from the object point on the axis will cross the axis at the same point for different values of i as long as these values are small.

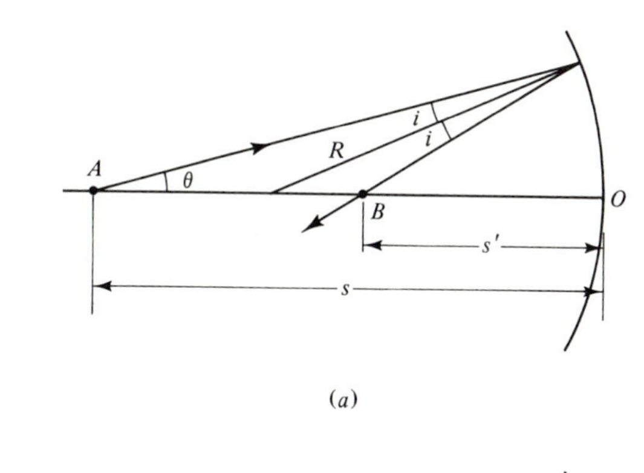

(*a*)

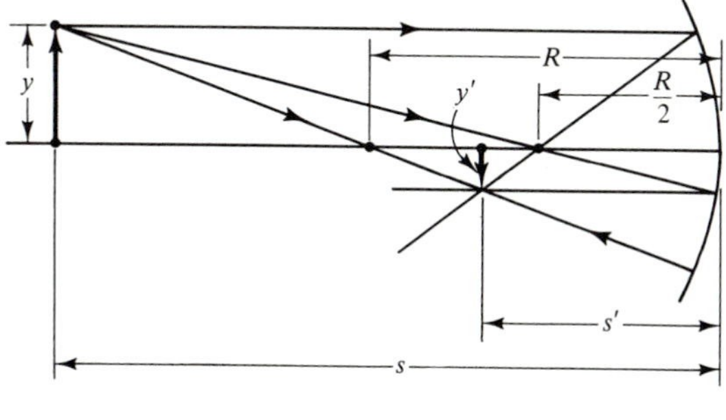

(*b*)

Figure 1-16 (*a*) Diagram used to locate the image plane for a spherical mirror at *B*. (*b*) Diagram used to determine the magnification of a spherical mirror.

The paraxial ray from the top of the object in Fig. 1-16b crosses the axis at the focal point and intersects the image plane at the top of the image. From similar triangles

$$\frac{y - y'}{s'} = \frac{-y'}{s' - R/2} \tag{1-82a}$$

which, using Eq. 1-81, reduces to

$$m = \frac{y'}{y} = -\frac{s'}{s} \tag{1-82b}$$

where y is the height of the object, y' is the height of the image, and the negative sign means the image is inverted.

EXAMPLE 1-7

Discuss the situations for a spherical mirror when **(a)** $s = \infty$; **(b)** $s > R$; **(c)** $s = R$; **(d)** $R/2 < s < R$; **(e)** $s = R/2$; and **(f)** $s < R/2$.

(a)
$$\frac{1}{\infty} + \frac{1}{s'} = \frac{2}{R}$$

$$s' = \frac{R}{2} \qquad \text{The image is at the focal point.}$$

$$m = \frac{s'}{s} = \frac{R/2}{\infty} = 0$$

(b)
$$\frac{1}{s'} = \frac{2}{R} - \frac{1}{s} > \frac{1}{R}$$

As s gets smaller s' gets larger, but $s' < s$. The magnification increases from zero at ∞ and approaches 1 as $s \longrightarrow R$

(c)
$$\frac{1}{s'} = \frac{2}{R} - \frac{1}{R} = \frac{1}{R}$$

$s' = s$. The image and object are located at the same point. This should be expected since a ray originating at R will strike the mirror at 90° and be reflected back onto itself. Also, $m = -1$.

(d)
$$\frac{1}{s'} = \frac{2}{R} - \frac{1}{s}$$

$$R < s' < \infty$$

As s gets smaller s' gets larger. The magnification increases from 1 to ∞.

(e)
$$\frac{1}{s'} = \frac{2}{R} - \frac{2}{R} = 0$$

$$s' = \infty$$

This should be expected because a ray originating at the focal point will become parallel to the axis after it has been reflected.

(f)

$$\frac{1}{s'} = \frac{2}{R} - \frac{1}{s} < 0$$

s' is negative; hence, the image, which is a virtual image, is to the right of the mirror. Also, m is positive so that the image is not inverted.

1-5.2 Lenses

Lenses, like mirrors, are used to direct the beam, but they direct the transmitted and not the reflected beam. They are also used to gather light and condense it into a smaller spot. This is useful when trying to insert light into a fiber or detect diffuse light.

Before considering lenses, we must first examine a single dielectric surface with a radius of curvature, R. If the center of curvature is to the right of the vertex as it is in Fig. 1-17, it is defined to be positive; if it is to the left, it is negative. Also, the image distance is positive to the right and negative to the left.

Again, we are interested in finding the relationship between s, s' and R, and this will be done by using Fig. 1.17a. From the law of sines

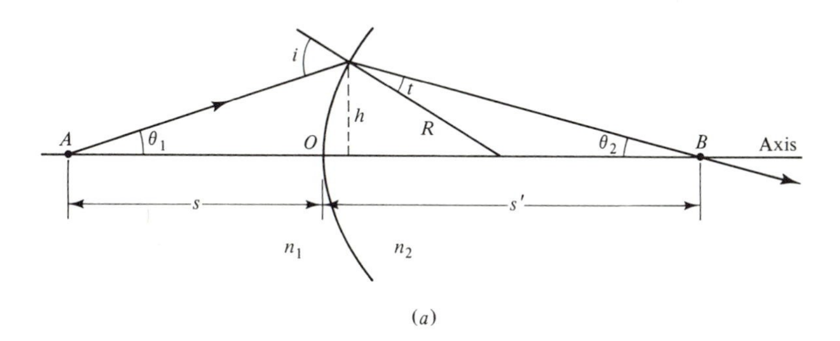

(a)

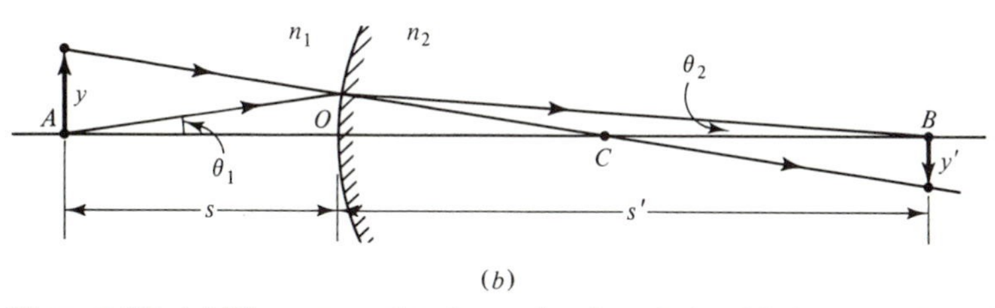

(b)

Figure 1-17 (a) Diagram used to determine the relationship between the object distance, s, the image distance, s', and the radius, R, for a spherical dielectric surface. (b) Diagram used to determine the magnification, m.

$$\frac{\sin \theta_1}{R} = \frac{\sin (\pi - i)}{s + R} = \frac{\sin i}{s + R} \tag{1-83}$$

and

$$\frac{\sin \theta_2}{R} = \frac{\sin t}{s' - R} \tag{1-84}$$

Combining these two equations with Snell's law yields

$$\frac{s' - R}{s + R} = \frac{n_1}{n_2}\frac{\sin \theta_1}{\sin \theta_2} \simeq \frac{n_1}{n_2}\frac{h/s}{h/s'} = \frac{n_1 s'}{n_2 s} \tag{1-85a}$$

Rearranging Eq. 1-85a, we find that

$$\frac{n_1}{s} + \frac{n_2}{s'} = \frac{n_2 - n_1}{R} \tag{1-85b}$$

The magnification can be obtained by using Fig. 1-17*b*. Recognizing that a ray normal to the surface is undeviated and using similar triangles

$$\frac{y'}{y} = -\frac{s' - R}{s + R} \tag{1-86a}$$

which from Eq. 1-85a is

$$\frac{y'}{y} = -\frac{n_1 s'}{n_2 s} \tag{1-86b}$$

A lens differs from a single dielectric surface in that it has two surfaces. The region to the left is still considered to be the 1 region and the region to the right is still considered to be the 2 region. Also, the image of the first surface is the object of the second surface.

Let us consider the double convex lens shown in Fig. 1-18. The object is placed at A, which is a distance $R_1/2 < s < R_1$ from the vertex of the first surface. Thus, the first image distance, $-A'O$, is negative, and it is a virtual image. The object of the second surface located at a distance $A'O'$ from the second vertex forms an image located at a distance s' from the second surface.

The relationship between the object and image distances of the lens can now be derived by using Eq. 1-85b. For the first surface

$$\frac{1}{s} + \frac{n}{-A'O} = \frac{n - 1}{R_1} \tag{1-87}$$

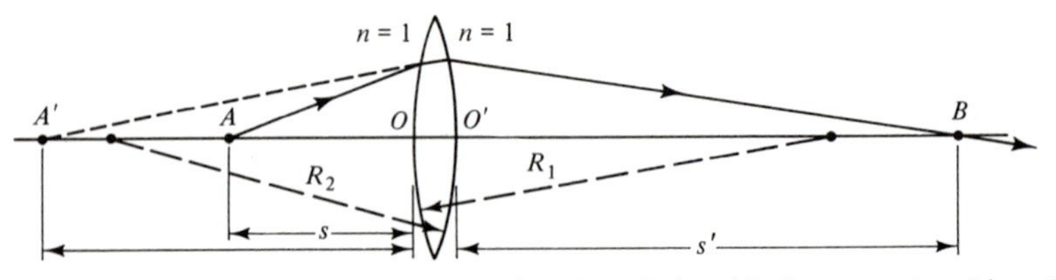

Figure 1-18 Diagram used to determine the relationship between the object distance, s, the image distance, s', and the radii of curvature, R_1 and R_2 for a double-convex lens.

because $n_1 = 1$ for air. Recalling that the region to the left is region 1, the equation for the second surface becomes

$$\frac{n}{A'O'} + \frac{1}{s'} = \frac{1-n}{R_2} \tag{1-88}$$

The radius of curvature for the second surface is negative because it is to the left of the vertex. Assuming that $A'O = A'O'$ (thin lens approximation) and eliminating it from Eqs. 1-87 and 1-88, one obtains

$$\frac{1}{s} + \frac{1}{s'} = (n-1)\left(\frac{1}{R_1} - \frac{1}{R_2}\right) = \frac{1}{f} \tag{1-89}$$

where f is the focal length of the lens.

As is shown in Fig. 1.19*a,* the focal length is again the distance from the point where the refracted paraxial rays cross the axis to the vertex. This can be seen from Eq. 1-89 by setting $s = \infty$. There is a focal point on the left side of

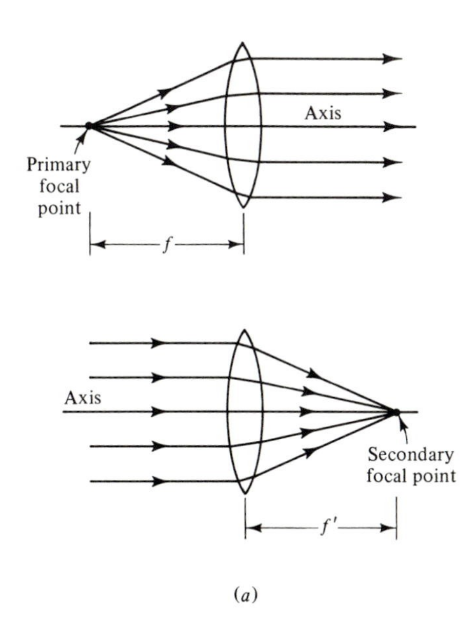

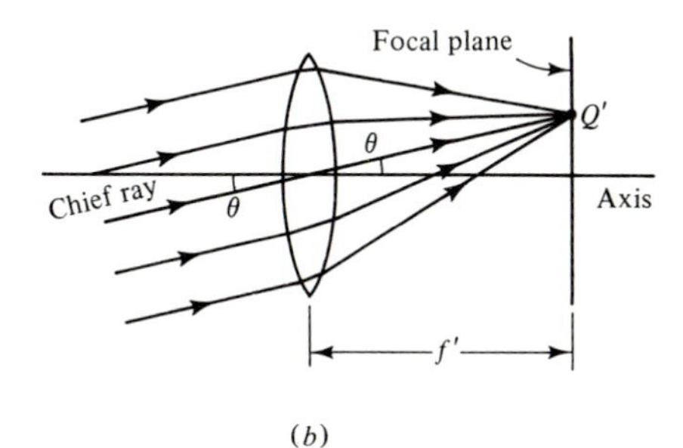

Figure 1-19 (*a*) The focal points, and (*b*) The focal plane of a double-convex lens.

the lens as well. If the object is placed there, the image is at infinity (Fig. 1-19*a*). Again, this can be confirmed by using Eq. 1-89.

The focal plane is the plane normal to the axis passing through a focal point. In Fig. 1-19*b* the rays that were originally parallel are focused to a point at Q' in the focal plane.

The magnification produced by a lens is easily found. The undeviated ray going through the center of the (thin) lens in Fig. 1-20 forms two similar triangles from which it is seen that

$$\frac{y'}{y} = -\frac{s'}{s} \tag{1-90}$$

s' can easily be found graphically by finding the point where the undeviated ray intersects the paraxial ray refracted through the focal point.

EXAMPLE 1-8

The radii of curvature of a thin double-convex lens are 20 and 30 cm. The lens forms an image 40 cm to the right of the lens when the object is placed 48 cm to the left of the lens. **(a)** What is the focal length of the lens? **(b)** What is the index of refraction of the lens material? **(c)** An object is placed 16 cm from the lens. Find the position of the image.

(a) $$\frac{1}{s} + \frac{1}{s'} = \frac{1}{f} \quad \text{or} \quad f = \frac{s's}{s'+s} = \frac{(40)(48)}{40+48} = 21.82 \text{ cm}$$

(b) $$\frac{1}{f} = (n-1)\left(\frac{1}{R_1} - \frac{1}{R_2}\right) \quad \text{or} \quad n = 1 + \frac{R_2 R_1}{f(R_2 - R_1)}$$

$$n = 1 + \frac{(-30)(20)}{21.82(-30-20)} = 1.55$$

(c) $$\frac{1}{s'} = \frac{1}{f} - \frac{1}{s} \quad \text{or} \quad s' = \frac{sf}{s-f}$$

$$s' = \frac{(16)(21.82)}{16 - 21.82} = -60.0 \text{ cm}$$

The image is a virtual image.

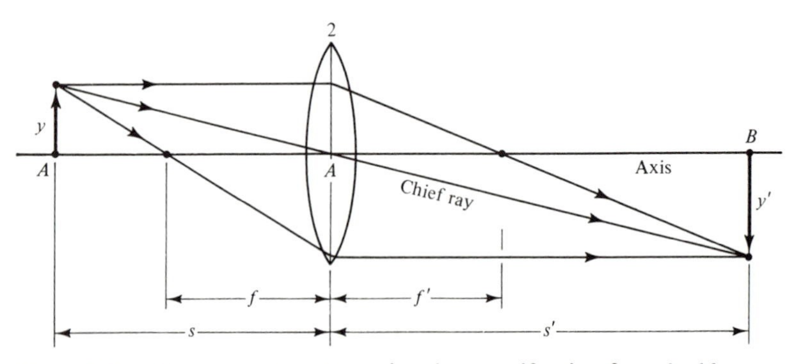

Figure 1-20 Diagram used to determine the magnification for a double-convex lens.

READING LIST

1. N. H. Frank, *Introduction to Electricity and Optics,* 2d Ed. New York: McGraw-Hill, 1950, Chapters 10, 11, 13, 15, and 16.
2. E. V. Bohn, *Introduction to Electromagnetic Fields and Waves.* Reading, MA: Addison-Wesley, 1968, Chapters 9 and 10.
3. G. Tyras, *Radiation and Propagation of Electromagnetic Waves.* New York: Academic Press, 1969, Chapters 1 and 2.
4. D. W. Dearholt and W. R. McSpadden, *Electromagnetic Wave Propagation.* New York: McGraw-Hill, 1973, Chapters 1, 3, and 7.
5. K. F. Sander and G. A. L. Reed, *Transmission and Propagation of Electromagnetic Waves.* London: Cambridge University Press, 1978, Chapters 1 to 3.
6. F. A. Jenkins and H. E. White, *Fundamentals of Optics,* 4th Ed. New York: McGraw-Hill, 1976, Chapters 1–10, 20, 24, and 25.
7. C. T. A. Johnk, *Engineering Electromagnetic Fields and Waves.* New York: John Wiley, 1975, Chapters 1–8.
8. R. Plonsey and R. E. Collin, *Principles and Applications of Electromagnetic Fields.* New York: McGraw-Hill, 1961.
9. P. Lorrain and D. R. Corson, *Electromagnetic Fields and Waves.* San Francisco, W. H. Freeman, 1970.
10. L. D. Landau and E. M. Lifshitz, *The Classical Theory of Fields,* 4th Ed. Oxford: Pergamon Press, 1975.
11. M. Born and E. Wolf, *Principles of Optics,* 5th Ed. Oxford: Pergamon Press, 1975.

PROBLEMS

1-1. A plane radio wave, traveling in the z direction with a frequency of 600 kHz, is linearly polarized with the electric vector in the x direction. It transmits an average power per unit area of 29.8 watts/m^2.

(a) What is the wavelength of this wave?

(b) What are the amplitudes of the electric and magnetic fields?

1-2. Find the equation for the electric vector ξ of a plane electromagnetic wave that satisfies the following:

(a) Its frequency is 9×10^8 Hz.

(b) It travels in a medium of refractive index 4/3.

(c) Its direction of propagation makes an angle of 30° with the x axis and lies in the plane bisecting the angle between the z and x axes, all components of $\mathbf{k}$ being positive.

(d) It is linearly polarized in the plane bisecting the angle between the $+z$ and $+x$ axes.

(e) The average value of its Poynting vector is $\frac{1}{2}\pi$ watts/m^2.

1-3. The electric field intensity of a linearly polarized plane wave traveling in a non-magnetic dielectric of dielectric constant $\epsilon_r = 2$ is $\xi = \xi_0 \exp j\omega(t - \bar{k}\cdot\mathbf{r}/\omega)$. The magnitude of ξ_0 is 50 V/m, the frequency is 10^9 Hz, and the vector k is in the $x - z$ plane at an angle of 30° with the positive z axis. The vector ξ_0 is also in the $x - z$ plane.

(a) What is the direction of propagation of the wave? What is its wavelength and what is the direction of ξ?
(b) What are the magnitudes and directions of **H, B,** and **D?**
(c) What is the direction and average value of the Poynting vector?
(d) What is the average energy density at any point of space? How is this related to your answer to part c?

1-4. A 100-W lamp radiates all the energy supplied to it uniformly in all directions.
(a) Compute the rms values of the electric and magnetic field strengths at a point 1 m from the lamp.
(b) What is the electromagnetic energy density at this point?

1-5. A GaAs LED emits infrared radiation with a wavelength of 900 nm with a uniform power of 1.6 mW over a 1 mm^2 area. Calculate:
(a) The radiation frequency, ν.
(b) The Poynting vector, S.
(c) The amplitudes of the electric and magnetic fields, ξ_0 and H_0.

1-6. A plane electromagnetic wave with a maximum value of the electric field intensity of 100 V/m is normally incident on a surface that is perfectly absorbing. The mass of the surface is 10^{-2} kg/m^2, and its specific heat is 0.2 cal/g °C. Find the rate of increase of temperature of the surface.

1-7. Prove that the superposition of two linearly polarized traveling plane waves of the same frequency gives a circularly polarized traveling wave if the two original waves are polarized at right angles to each other, have equal amplitudes of the electric field strengths, and are 90° out of phase with each other.

1-8. Two plane waves with perpendicular electric field vectors of equal magnitude, the same propagation velocity, and different phases combine to form a wave that has an amplitude and a polarization direction which can be time dependent. For a fixed position, z, construct a diagram that is the locus of points of the electric field vector at different times for each of the following values for the difference in the phase angles: $\pi/4$, $\pi/2$, $3(\pi/4)$, π, $5(\pi/4)$, $3(\pi/2)$, $7(\pi/4)$, and 2π. Example: when $\phi_x = \phi_y$, the diagram is the following figure.

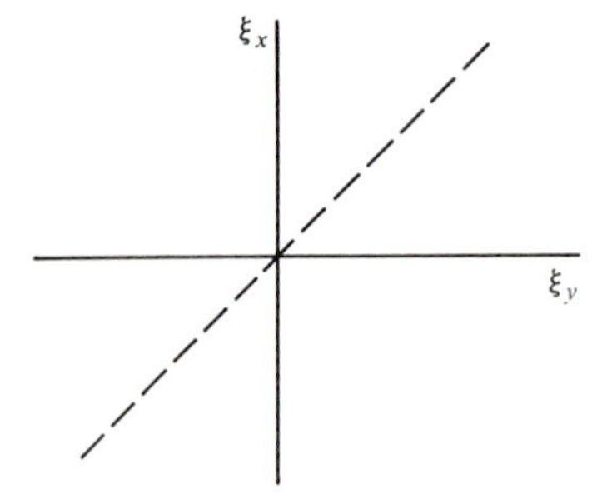

1-9. Suppose an electric field applied in a region has the value $\xi = 1$ V/m. Find the polarization field P, and the relative permittivity if the material filling the region has the dielectric susceptibility:
(a) Zero.
(b) 0.01.
(c) 10.
(d) 1000.

1-10. A parallel plate condenser with plates of area 200 cm^2 and separation 2.0 mm is immersed in oil of dielectric constant 3.0 and permanently connected to a 300-V battery.

(a) Compute the charge on the condenser plates.

(b) Find the induced dipole moment per unit volume and the electric field intensity in the oil between the plates.

1-11. A parallel plate condenser with circular plates of radius 3.0 cm, separation 3.0 mm, is connected to a 600-V battery. The plates are separated by two dielectric slabs; one of thickness 2.0 mm and dielectric constant $\epsilon_{r1} = 6$, and the other of thickness 1.0 mm and dielectric constant $\epsilon_{r2} = 3.0$. Neglecting edge effects:

(a) Compute the capacity of the condenser thus formed in microfarads.

(b) Find the magnitude of the polarization vector in each dielectric.

(c) Find the induced surface polarization charge on the interface between the two dielectrics.

1-12. A glass slab of dielectric constant $\xi_r = 10$ and a paraffin slab of dielectric constant $\xi_r = 2$, each 1.0 cm thick, are inserted into a parallel plate condenser of plate separation 3.0 cm, the slab faces being parallel to the surfaces of the condenser plates. The dielectric slabs have surfaces equal in area to those of the condenser plates, and the latter are large enough so that end effects may be neglected.

(a) If the surface charge density on the condenser plates is 1.0×10^{-6} C/m^2, what are the electric field intensities in the empty space between the condenser plates before and after the dielectrics are inserted?

(b) If the condenser plates are connected to a 432-V battery, what are the changes of the charge density on the plates and of the field ξ in the empty space occurring when the dielectrics are inserted?

1-13. A condenser is formed of two concentric, spherical metal shells of radii 2.0 and 6.0 cm. The inner sphere is covered by a wax coating 3.0 cm thick, and the remainder of the space between the spheres is filled with a liquid of dielectric constant 4.2. The dielectric constant of the wax is 2.0. Find the capacity of the condenser formed in microfarads.

1-14. If the plates of the condenser in Problem 1-13 are maintained at a potential difference of 3000 V, compute the total energy stored in the condenser.

1-15. What is the surface density of polarization charge on the wax-liquid interface of the condenser described in Problem 1-13? What is the volume density of polarization charge at any point?

1-16. A vertical cylindrical condenser of altitude 1.0 m and of capacity 5.0×10^{-11} F in air is connected through a galvanometer to a 1000-V battery. Water of dielectric constant 81 rises between the condenser plates at a uniform rate of 10 cm/sec.

(a) Find the charge on the condenser plates (not including polarization charge) when the condenser is half full of water.

(b) What is the galvanometer reading in microamperes while the water is rising?

1-17. A very long coaxial conductor pair contains a concentric, homogeneous dielectric sleeve with a permittivity ϵ as shown. Air fills the remaining regions between the conductors. Assuming positive and negative surface charges $\pm Q$ C are on every axial length l of the inner and outer conductors, respectively.

(a) Find D, ξ, and P in each region between the conductors.

(b) Find σ_f on the conductor surfaces, and the surface polarization charge density at $r = b$ and $r = c$.

(c) If $a = 1$ cm, $b = 2$ cm, $c = 3$ cm, $d = 4$ cm, $e = 4.2$ cm, $Q/1 = 10^{-2}$ μC/m, and $\epsilon = 2.1\epsilon_0$ (Teflon®), find the values of D, ξ, and P at the inner surface $r = b$ just inside the dielectric.

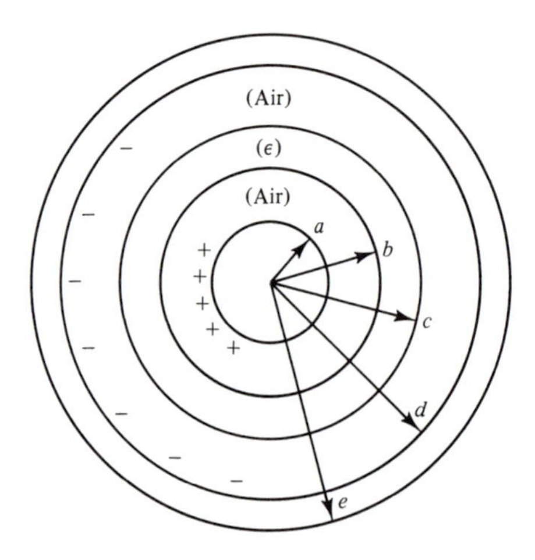

1-18. A long, straight, cylindrical conductor of radius, a, carries a steady current i uniformly distributed over its cross section. The conductivity of the conductor is σ.

(a) Find the electric field, $\boldsymbol{\xi}$, and the magnetic field, **H**, at any point within the conductor (magnitudes and directions).

(b) Derive an expression for the Poynting vector at this point, stating its direction.

(c) From your answer to part b, show that there is a flow of energy into the conductor across its surface that is equal to the rate of heating in the conductor.

1-19. Suppose a uniform plane wave with the amplitude 1000 V/m propagates in the $+z$ direction at $f = 10^8$ Hz in a conductive region having the constants $\mu = \mu_0$, $\epsilon = 4\epsilon_0$, $\sigma/\omega\epsilon = 1$.

(a) Find k, $\alpha/2$, and η for the wave.

(b) Find the associated **H** field, and sketch the wave along the z axis at $t = 0$.

(c) Find the depth of penetration, the wavelength, and the phase velocity. Compare λ and c with their values in a lossless ($\sigma = 0$) region having the same μ and ϵ values.

1-20. A uniform plane wave has an electric field amplitude of 100 V/m and propagates at $f = 10^6$ Hz in a conductive region having the parameters $\mu = \mu_0$, $\epsilon = 9\epsilon_0$, and $(\sigma/\omega\epsilon) = 0.5$.

(a) Find the values of $\alpha/2$, k, and η for this wave.

(b) Express the electric and magnetic fields in both their complex and real-time forms with the numerical values of part a inserted.

(c) What is the depth of penetration of this wave into the dissipative region? Determine the wavelength and the velocity of the wave, comparing the latter with values obtained if $\sigma = 0$.

1-21. A plane wave with a frequency of 10^8 Hz and an amplitude of $\xi_0 = 1000$ V/m at $z = 0$ propagates in a material in which $\epsilon = 4\epsilon_0$, $\mu = \mu_0$, and $\sigma = 5.56 \times 10^{-3}$ S/m.

(a) Determine the magnitudes of k, $\alpha/2$, and η $(=\xi_0/H_0)$.
(b) Find the magnitude of H and the phase angle, ϕ.
(c) Find the depth of penetration and the velocity, c.
(d) Compare the values of λ and c when $\sigma = 0$.

1-22. A plane wave in air is normally incident onto a material with the following properties: $\epsilon = 4\epsilon_0$, $\mu = \mu_0$, $\sigma = 4 \times 10^7\epsilon_0$. The angular frequency of the wave is $\omega = 10^7$ rad/sec.

(a) What is the speed of light in each medium?
(b) What are the wavelengths?
(c) What is the phase angle between ξ and H?
(d) What is the penetration depth into the material?

1-23. **(a)** Find an expression for tan ϕ where ϕ is the phase angle between H_y and ξ_x when there is joule heating.
(b) Show that $\tan 2\phi = \sigma/\epsilon\omega$.

1-24. A uniform plane wave at 10^{10} Hz propagates 12.7 m into polyethylene before its amplitude is reduced to $1/e$ of its original value. Find $\alpha/2$, k, ϵ', ϵ'', n', and n'' if $\epsilon_r = 2.26$ and $\sigma = 0$.

1-25. The light intensity in very high purity glass used for optical fibers has an attenuation of 1.0 dB/km at $\lambda = 1$ μm.

(a) What is the value of the absorption coefficient?
(b) Find n_2'' if $n_2' = 1.5$.

1-26. Make a plot of the hysteresis loop $P/P_0 = P/\chi\xi_0$ versus ξ/ξ_0 for the case $\chi'' = \chi'$.

1-27. A plane wave of light is incident on one side of a glass plate that has a thickness, d.

(a) Show that the plane wave emerging from the other side of the plate has the same direction of propagation as the incident wave.
(b) Consider a given normal of the incident wave. Prove that as the wave passes through the glass, this normal undergoes a lateral displacement given by

$$\frac{d \sin (i - t)}{\cos t}$$

where i and t are the angles of incidence and transmission at the first glass surface.

1-28. **(a)** Liquid of refractive index 1.63 stands at a height of 2.0 cm in a flat-bottomed glass vessel. The index of refraction of the glass is 1.50. Show whether or not a plane wave of light incident on the top surface of the liquid can be totally reflected at the bottom surface.
(b) A plane light wave is normally incident on the liquid surface. If the electric field vector of the incident wave has a peak value of 10 V/m, compute the intensity of the wave transmitted through the glass vessel. (Consider only one reflection at each interface.)

1-29. A plane wave of light in water of refractive index $n_1 = 4/3$ is incident at an angle i on a rectangular glass block of refractive index n_2, as shown in the figure. As the angle of incidence i is varied, it is found that the beam refracted at A is totally reflected at B for all angles i smaller than 45°. What is the index of refraction of the glass?

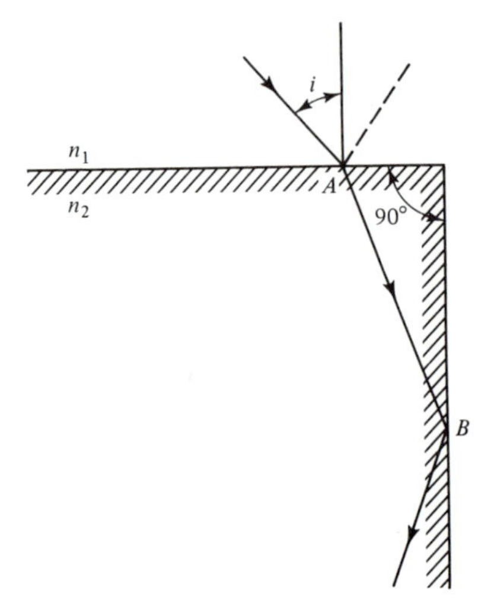

1-30. A plane wave of light is incident on a glass prism of refractive index 1.50 in air, as shown in the figure. The prism angles are all 60°, and the direction of propagation of the incident wave makes an angle of 60° with the face on which it is incident. What is the deviation angle, D, between the directions of the incident and the emerging waves?

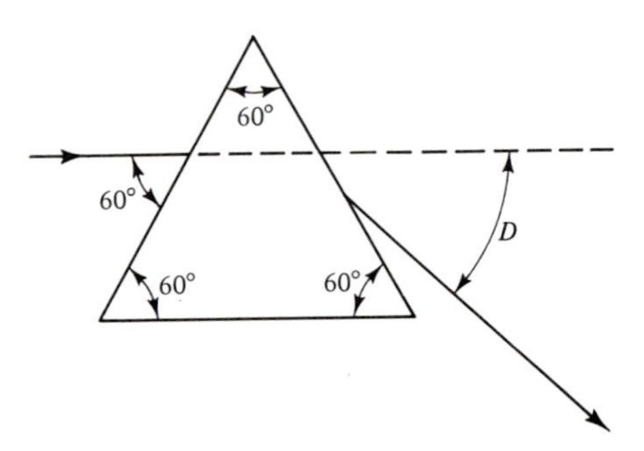

1-31. Compute the angular deviation (the angle between the incident and the emergent ray) of the ray incident as shown on the prism in the figure. The refractive index of the prism material is 1.532.

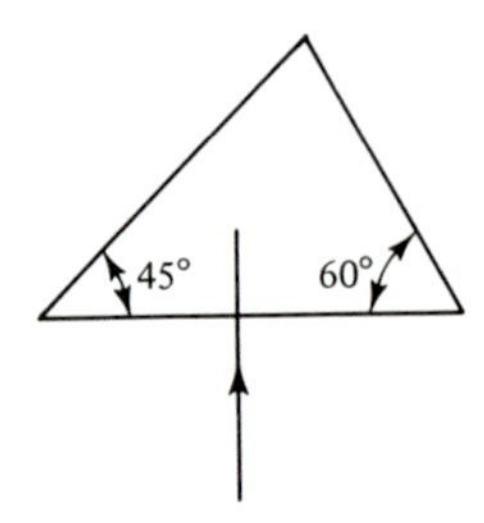

1-32. Find the largest value of the angle ϕ of the glass prism in the figure such that a light wave incident as shown will pass through the prism:

(a) When the prism is in air.

(b) When the prism is immersed in water of refractive index 1.33.

The index of refraction of the glass is 1.55, and the angle, α, is small enough so that $\cos \alpha$ is practically equal to unity.

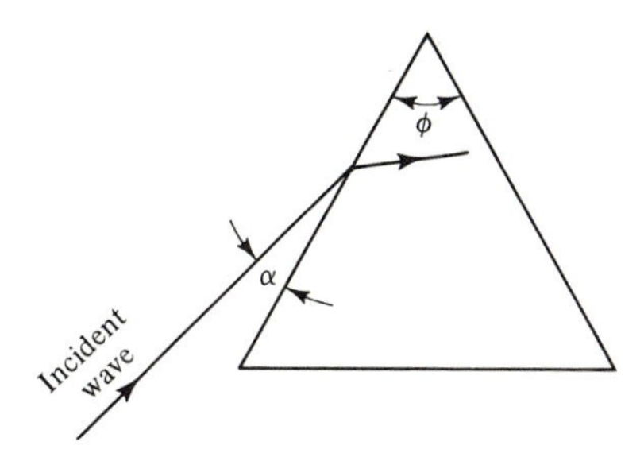

1-33. A plane light wave is incident normally on the face AB of a glass prism, as shown in the figure. The index of refraction of the glass is 1.50. Find the smallest or largest value of the angle, α, such that the wave will be totally reflected at the surface AC:

(a) If the prism is surrounded by air.

(b) If the prism is surrounded by a liquid of refractive index 1.40.

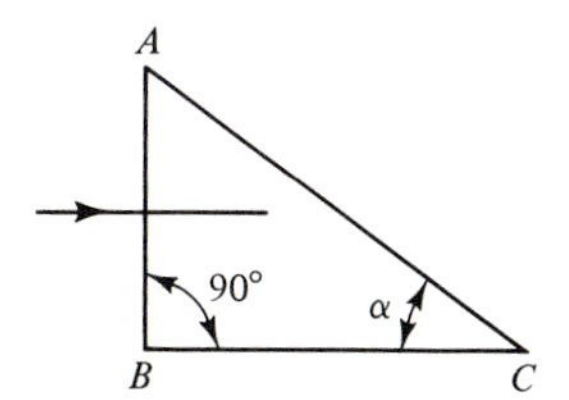

1-34. You are to find the orientation of the prism for which the angle of deviation, δ, is a minimum. One can express this condition as a transcendental equation containing the angles A and δ and the index of refraction, n.

(a) Using the figure, show that $\delta = i_1 + t_2 - A$.

(b) When δ is a minimum, show that $di_1 = -dt_2$.

(c) Using Snell's law and the fact that $i_2 + t_1 = A$, show that δ is a minimum when $(\cos i_1/\cos t_2)(\cos i_2/\cos t_1) = 1$.

(d) Using the fact that the condition described in part c is satisfied when $i_1 = t_2$ and $t_1 = i_2$, show that $\sin [(\delta + A)/2] = n \sin A/2$ when δ is a minimum.

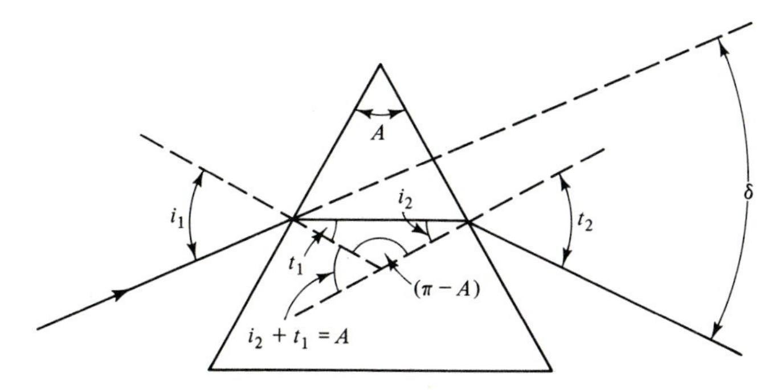

1-35. Light from a light-emitting diode (LED) is generated within a semiconductor and must be transmitted across the semiconductor interface.

(a) If $n = 3.4$ for the semiconductor, what fraction of the light intensity is transmitted at normal incidence?

(b) What is the critical angle?

(c) If it is assumed that the reflectance is constant between $i = 0$ and $i = i_c$, what fraction of the light generated is transmitted across the interface if the emission has a Lambertian distribution, that is, $I = I_0 \cos i$?

1-36. The LED in the previous problem has its surface coated with a dielectric with $n = 1.5$. Find the fraction of light transmitted across the semiconductor-dielectric interface using the assumption stated in the previous problem. Ignore the effects of light reflected from the dielectric-air interface.

1-37. Show that $r_{\|}$, $r_{\perp}$, and $t_{\perp}$ reduce to the equations given for normal incidence as $i \to 0$.

1-38. Show that $r_{\|}$, $r_{\perp}$, and $t_{\perp}$ can also be written

$$r_{\|} = \frac{n_1 \cos t - n_2 \cos i}{n_1 \cos t + n_2 \cos i}$$

$$r_{\perp} = \frac{n_1 \cos i - n_2 \cos t}{n_1 \cos i + n_2 \cos t}$$

$$t_{\perp} = \frac{2n_1 \cos i}{n_1 \cos i + n_2 \cos t}$$

1-39. Find an expression for $t_{\|}$ in terms of n_1, n_2, $\cos i$, and $\cos t$.

1-40. The angle of incidence is 45°, $n_1 = 1$, $n_2 = 1.5$; the ξ vector makes an angle of 45° with the plane of incidence; and there are no losses. Find the percentage of the energy that is reflected.

1-41. A plane wave moving in the x-z plane strikes a dielectric with an index of refraction $n = \sqrt{3}$, and the air/dielectric interface lies in the x-y plane. The wavelength in air is 1 μm, the intensity is 1 W/cm^2, and the angle of incidence is 60°.

(a) Find ω, k, k_x, and k_z on both sides of the interface.

(b) If ξ makes an angle of 45° with the x-z plane, find S_0, ξ_0, H_0, ξ_{0x}, ξ_{0y}, and ξ_{0z} of the incident beam.

(c) Find the incident angle at which all the light is reflected. Also, find the angle for which there is no reflection for the $\xi_{\|}$ component.

(d) Find ξ_r/ξ_i for the wave.

1-42. **(a)** Derive the expression for $r_{\|}$.

(b) Show that the energy flowing into the volume parallel to the z axis (that is, normal to the boundary) is equal to the energy flowing out parallel to the z axis, that is, the energy is conserved. Assume that ξ is perpendicular to the plane of incidence.

(c) Also calculate the energy flowing in parallel to the boundary and the energy flowing out. You will find that energy does not appear to be conserved. This is because surface waves are generated and their energy has not been accounted for.

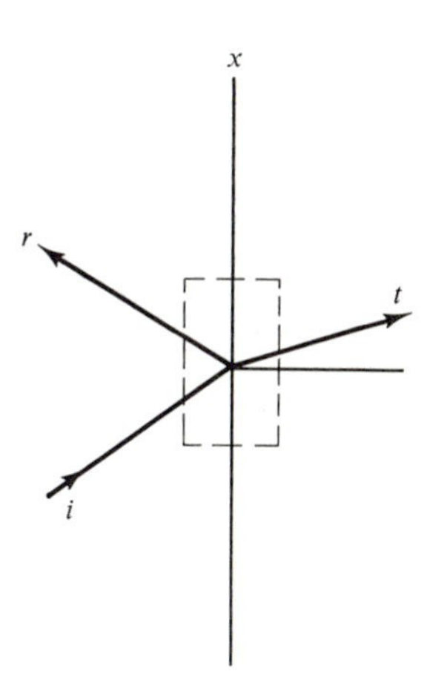

1-43. Unpolarized light strikes a smooth glass surface at an angle of 35°. Assume the glass index to be 1.750. Calculate:
- **(a)** The amplitudes.
- **(b)** The intensities of the reflected $\parallel$ and $\perp$ components.
- **(c)** Find the degree of polarization of the transmitted light, P, where $P = (I_{\parallel} - I_{\perp})/(I_{\parallel} + I_{\perp})$ if $t_{\parallel} = 2n_1 \cos i/(n_1 \cos t + n_2 \cos i)$.

1-44. Plane waves with an amplitude of 100 V/m are normally incident on seawater.
- **(a)** Find the intensity of the transmitted beam at the surface for $f = 10$ kHz and 1000 Hz if $\epsilon_r = 81$ and $\sigma = 4\ \Omega/\text{m}$.
- **(b)** How far will the beam penetrate before it has been attenuated to 5 percent of the transmitted value?
- **(c)** Comment on the effectiveness of undersea radio communication.

1-45. Compute the reflectance at normal incidence for the following materials:
- **(a)** Diamond, $n = 2.426$.
- **(b)** Quartz, $n = 1.547$.
- **(c)** Rutile, $n = 2.946$.
- **(d)** Crown glass, $n = 1.526$.
- **(e)** Metallic silver, $n' = 0.177$, $n'' = 3.638$.
- **(f)** Steel, $n' = 2.485$, $n'' = 3.433$.

1-46.
- **(a)** Derive an expression for $r_{\perp}$ when $\sigma \neq 0$. Express your answer in terms of η_1, η_2, the angle i, and the angle t.
- **(b)** Find the fraction of the energy that is reflected if $f = 10^8$ Hz, $\epsilon' = 4\epsilon_0$, $\mu = \mu_0$, and $\sigma = 5.56 \times 10^{-3}$ S/m in region 2, region 1 is free space, and $i = 45°$.
- **(c)** Compare this answer with the case $\sigma = 0$.

1-47. A point source of light is 15 cm above the silvered bottom of a flat glass dish in which water stands to a depth of 10 cm. Find the position of the image of the light source formed by the complete optical system if the refractive index of the water is 1.33.

1-48. A narrow pencil of parallel light rays is normally incident on a hollow glass sphere of inner radius 2.0 in. and outer radius 2.5 in. If the refractive index of the glass is 1.60, where will these rays be brought to a focus? Make a careful sketch of the pencil of rays.

1-49. A cube of glass of refractive index 1.50 is 1 in. on a side and has a small bubble at its center. What is the minimum radius of circular opaque disks that, when pasted on the six cube faces, will prevent the bubble from being seen?

1-50. A small air bubble inside a glass sphere (index of refraction = 1.50) of radius 6.0 cm appears to be 4.0 cm from the surface and to be of 2.0-mm diameter when viewed along a diameter passing through the center of the bubble. What is the true diameter of the bubble and where is it located relative to the center of the sphere?

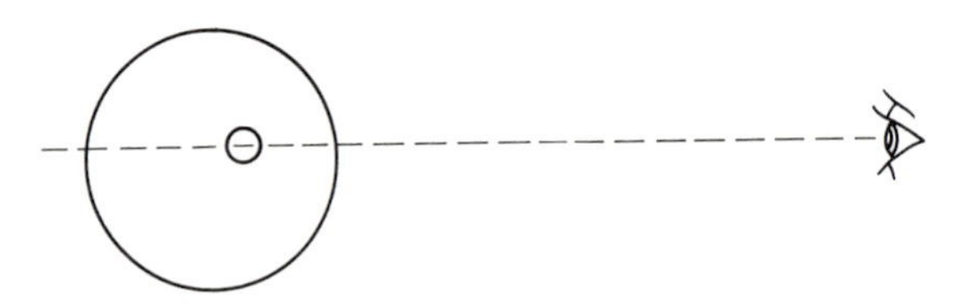

1-51. A glass "sphere" with a flat polished bottom rests on the drawing of an arrow 1.0 cm long as shown in Fig. 1-30. The radius of the spherical surface is $R = 3.0$ cm, and the refractive index of the glass is 1.50. Where does the arrow appear to be located for an observer looking vertically down into the sphere? What is the magnification? Is the image erect or inverted?

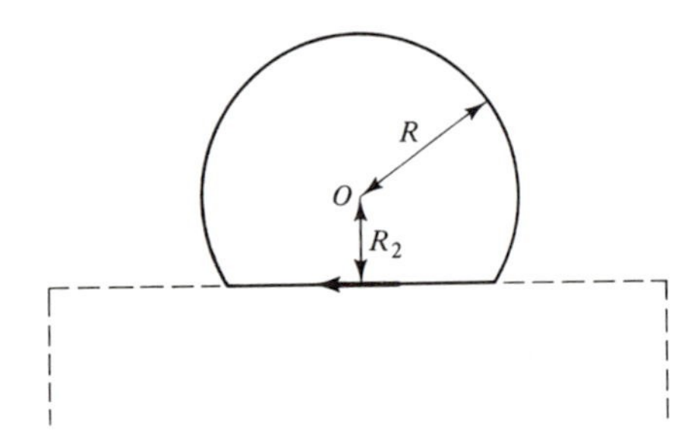

1-52. A glass hemisphere of radius 4.0 cm and refractive index 1.50 is silvered on its plane surface. Find the position of the image of an object located at a distance of 12 cm from the vertex of the curved surface on the axis. Is the image real or virtual?

1-53. A narrow pencil of parallel light rays is normally incident on a solid glass sphere of radius, R, and refractive index, n. How far from the center of the sphere are the rays brought to a focus?

1-54. A concave spherical mirror has a radius of curvature of 50 cm. Find two positions of an object such that the image will be four times as large as the object. What is the position of the image in each case? Is it real or virtual?

1-55. A convex mirror has a focal length of 10 in. Compute the position of the image of an object 6 in. in front of the mirror. What is the magnification for this case?

1-56. A glass hemisphere 20 cm in diameter and of refractive index 1.50 is silvered on its spherical surface. A small object is located at a distance of 20 cm from the plane surface on the perpendicular to this plane through the center of the sphere. Find the position of the final image of the object.

1-57. A confocal mirror configuration is one for which two spherical mirrors with the same axis and radius of curvature, R, are separated by a distance R. Trace the path of a ray that is initially parallel to the mirror axis.

1-58. The converging lens shown in the figure at the right is used to reduce the diameter of a parallel beam of light of diameter, D, so that all of the light strikes the detector of diameter, d. Find in terms of D, d, and the focal length, f, what the minimum and maximum distances are for the lens-detector length such that all of the light strikes the detector.

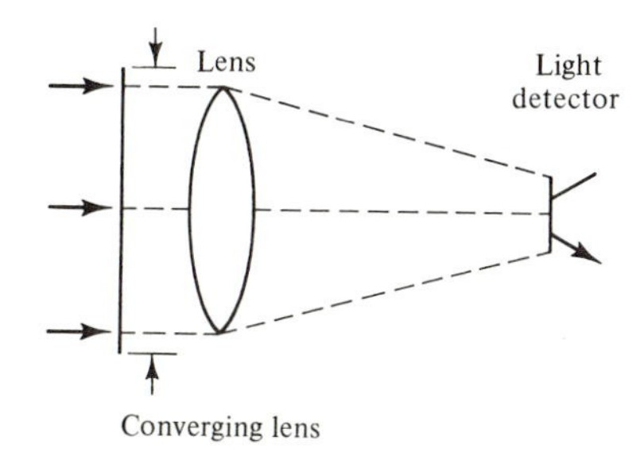

1-59. A thin convex lens of focal length 15 cm and refractive index 1.53 is immersed in water of index 1.33. Where is the image of an object placed 100 cm from the lens, the whole system being in water?

1-60. A thin glass lens of focal length 12 in. and refractive index 1.52 is immersed in water of refractive index 1.33. What is the focal length in water?

1-61. Prove that if two thin lenses are placed in contact, they are equivalent to a single lens of focal length

$$f = \frac{f_1 f_2}{f_1 + f_2}$$

1-62. Prove that the focal length f of a compound lens consisting of two coaxial thin lenses of separation d is given by

$$\frac{1}{f} = \frac{1}{f_1} + \frac{1}{f_2} - \frac{d}{f_1 f_2}$$

1-63. A periodic lens wave guide is one in which a lens or a set of lenses are mounted in a repetitive manner (see Appendix 1). If the light beam is confined to the wave guide, its path will be periodic.

(a) Consider the periodic wave guide composed of identical lenses separated from their neighbors along an axis by a distance $d = 2f$. If the incident beam striking the first lens is parallel to the axis and strikes the lens at $r = r_0$, determine the position at which it strikes the other lenses. What is the period for the periodic motion of the wave?

(b) Repeat part a for the case $d = 3f$.

(c) What happens when $d = 4f$? Explain this in mathematical terms.

Chapter 2

Optical Fibers

2-1 INTRODUCTION

Light that is totally internally reflected will propagate for long distances in optical fibers because the absorption coefficient is exceedingly small for the wavelengths used.

Although a ray is totally internally reflected, it does penetrate into the outer layer, the cladding. This effect is important because it can lead to losses in the cladding. It also reduces the time required for a ray that is not parallel to the axis to reach the end of the fiber because the ray travels faster in the cladding.

In a planar dielectric only those rays whose components normal to the core-cladding interface that form standing waves are allowed to propagate. Thus, there are a finite number of transverse modes that propagate, and for a very thin dielectric (1 to 2 μm width) only a single transverse mode can propagate. This dielectric is called a single mode dielectric; the other dielectrics are called multimode dielectrics.

Because the allowed rays have different path lengths, they will arrive at the end at different times if they travel in a core with a uniform index of refraction. This causes a pulse to broaden and, therefore, limits the data rate that can be carried by the fiber. This effect, called modal dispersion, can be greatly reduced by using a graded index fiber. The index decreases toward the edge so that the longer path length rays, which spend more time near the edge of the core, will have a greater velocity. When the index profile is optimized, it can reduce the modal dispersion by more than two orders of magnitude.

The other type of dispersion is wavelength dispersion, and it is due to the variation of the index of refraction with the wavelength. Atoms in the fiber act as oscillating dipoles that interact with the light, and the interaction is stronger for blue light than it is for red. As a result, n for blue light is larger than n for red light. The wavelength dispersion is reduced by operating at longer wavelengths $\sim 1.3\ \mu m$, and by using a laser source instead of a light-emitting diode (LED) source. The bandwidth of the laser emission is much less for a laser.

The ray theory, which is used in this chapter to describe dispersion, is only an approximate theory. In more advanced texts one would solve Maxwell's equations and find the acceptable modes for the given boundary conditions. (See Appendix B.) The errors introduced by the approximate theory, however, are in most instances small.

The rays that do propagate in the fiber are often broken down into three sets: the central ray, meridional rays, and skew rays. The central ray is parallel to the fiber axis whereas the meridional rays pass across the center of the fiber and, therefore, are reflected back and forth in a plane. The skew rays do not pass through the center and, hence, follow a three-dimensional path.

Only those rays that have a large angle of incidence ($>i_c$) at the core-cladding interface will be totally internally reflected. As a result, only a small amount of the light from the light source will be transmitted down the fiber. More light can be put into the fiber by a laser than by an LED, and more light can be put in by an edge-emitting LED than by a face-emitting LED because an edge-emitting LED emits more highly collimated light.

Two electro-optic processes that produce loss in fibers are scattering and absorption. Scattering is caused by oscillating atomic dipoles being driven by the electric field of the incident wave that emits electromagnetic energy in directions not parallel to the incident beam. Electrons can absorb light and thereby be excited into a higher energy state. Electrons in the transition metals such as iron, copper, and chromium are particularly effective at doing this in the wavelength region of interest and, therefore, must be kept below the parts per billion range if the fiber is to be of high quality. Vibrating ions, called phonons, can also absorb light. The vibrations of the OH radical are particularly effective in the wavelength region of interest; therefore, great efforts are made to keep the water content below parts per million. The lattice vibration frequencies in the glass (SiO_2) set an upper wavelength limit to effective optical propagation in fibers.

There are also losses due to mechanical effects. These include bending, separation, displacement, and rotation at a joint. Loss due to bends is not great unless the radius of curvature of the bend is of the order of 100 times the fiber core.

The chapter concludes with sections on fiber manufacture and fiber jointing.

2-2 PLANAR DIELECTRIC

2-2.1 Total Internal Reflection

We begin the discussion of light propagation by considering propagation in the planar dielectric shown in Fig. 2-1. We chose this geometry because the math-

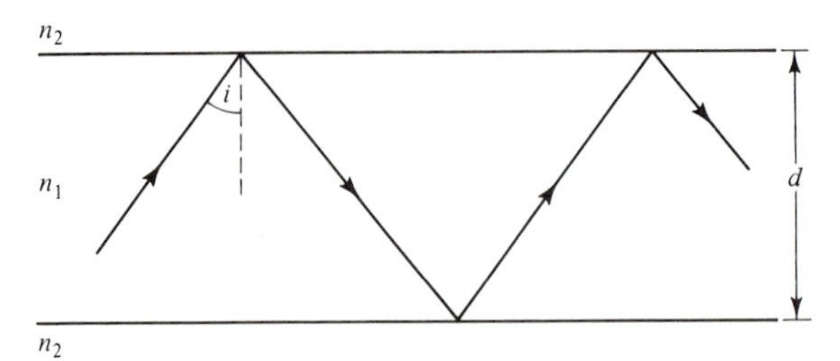

Figure 2-1 A ray in a planar dielectric that is totally internally reflected.

ematics is simpler than it is for a fiber, which has a circular cross section, yet the physical concepts are the same. The planar dielectric is two-dimensional and lies in the x-z plane. The inner core has an index of refraction that is larger than that of the outside cladding ($n_1 > n_2$), and a ray will be totally reflected back and forth if the angle of incidence is larger than the critical angle ($i > i_c$).

One can show that all of the light is reflected when $i > i_c$ by considering the Fresnel equations for reflection, which can be written

$$r_\perp = \frac{n_1 \cos i - n_2 \cos t}{n_1 \cos i + n_2 \cos t} \tag{2-1a}$$

and

$$r_\| = \frac{n_1 \cos t - n_2 \cos i}{n_1 \cos t + n_2 \cos i} \tag{2-2a}$$

when $n_1 > n_2$. Using Snell's law

$$\cos t = (1 - \sin^2 t)^{1/2} = \left[1 - \left(\frac{n_1}{n_2}\right)^2 \sin^2 i\right]^{1/2} = -jB \tag{2-3}$$

when $i > i_c$. We now choose $-j$ rather than $+j$. Later we will see that this predicts the true physical behavior as the beam attenuates in the cladding for $i > i_c$. Equation 2-1a can now be written

$$r_\perp = \frac{n_1/n_2 \cos i + jB}{n_1/n_2 \cos i - jB} = \frac{A + jB}{A - jB} \tag{2-1b}$$

Clearly, $r_\perp$ is complex with a magnitude of one. Thus, all of the light is reflected.

The numerator and denominator of Eq. 2-1b can be represented by the phasors in Fig. 2-2. Thus, Eq. 2-1b can also be written

$$r_\perp = \frac{\cos \psi_\perp + j \sin \psi_\perp}{\cos \psi_\perp - j \sin \psi_\perp} = \frac{e^{j\psi_\perp}}{e^{-j\psi_\perp}} = e^{2j\psi_\perp} \tag{2-1c}$$

where

$$\tan \psi_\perp = \frac{B}{A} \tag{2-4}$$

Physically, this equation means that all of the light is reflected and there is a phase change of 2ψ; the reflected ray leads the incident ray by 2ψ. Using a similar procedure, one can show that

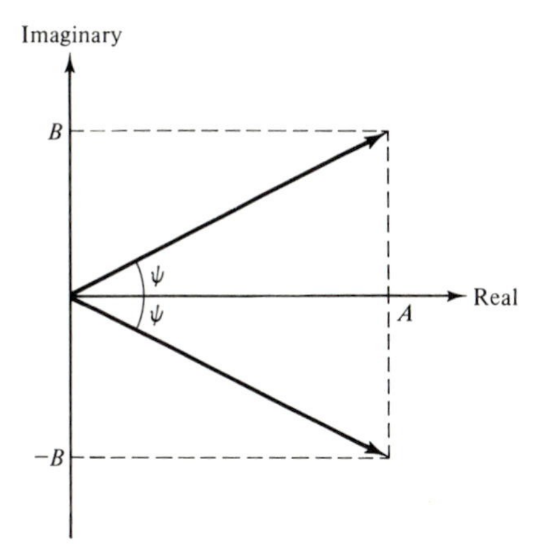

Figure 2-2 The phasor representation of $A + jB$ and $A - jB$.

$$r_{\parallel} = e^{2j\psi_{\parallel}} \tag{2-2b}$$

and

$$\tan \psi_{\parallel} = \left(\frac{n_1}{n_2}\right)^2 \tan \psi_{\perp} \tag{2-5}$$

Even though the beam is totally reflected, it does penetrate into the cladding. We now find what that penetration distance is. The plane wave propagating in the cladding in the x-z plane is described by the equation:

$$\xi = \xi_0 \exp[j(\omega t - \mathbf{k} \cdot \mathbf{r})] = \xi_0 \exp\left\{j\left[\omega t - \frac{2\pi n_2}{\lambda_0}(x \cos t + z \sin t)\right]\right\} \tag{2-6a}$$

Using the expression for $\cos t$ given in Eq. 2-3, Eq. 2-6a becomes

$$\xi = \xi_0 \exp\left[-\frac{2\pi n_2}{\lambda_0} Bx + j\left(\omega t - \frac{2\pi n_2}{\lambda_0} z \sin t\right)\right] \tag{2-6b}$$

The beam, therefore, attenuates as x increases. If we had chosen cost equals $+jB$ for Eq. 2-3, the amplitude of the beam would have grown as x increases; this is not a physically acceptable solution. Note that B decreases as i decreases toward i_c, and it equals 0 at $i = i_c$.

EXAMPLE 2-1

(a) For a planar dielectric with $n_1 = 1.50$ and $n_2 = 1.48$, find the phase changes, $2\psi_{\parallel}$ and $2\psi_{\perp}$, and the penetration depth when $\lambda_0 = 1.0\ \mu$m (the wavelength in free space) and $i = 1.05 i_c$ and when $i = 1.10 i_c$. **(b)** Repeat for a dielectric-air boundary with $n_1 = 1.5$.

(a) $\quad \sin i_c = \dfrac{n_2}{n_1} = \dfrac{1.48}{1.5} = 0.98666$

$$i_c = 80.60° \qquad 1.05 i_c = 84.63° \qquad 1.10 i_c = 88.66°$$

For $i = 1.05i_c = 84.63°$

$$A = \frac{n_1}{n_2}\cos i = \frac{1.50}{1.48}\cos(84.63°)$$

$$= \frac{1.50}{1.48}(0.09359) = 0.0949$$

$$B = \left[\left(\frac{n_1}{n_2}\right)^2 \sin^2 i - 1\right]^{1/2} = \left[\left(\frac{1.50}{1.48}\right)^2 \sin^2(84.63) - 1\right]^{1/2}$$

$$= \left[\left(\frac{1.50}{1.48}\right)^2 (0.9956)^2 - 1\right]^{1/2} = 0.135$$

$$\tan\psi_\perp = \frac{B}{A} = \frac{0.135}{0.0949} = 1.4228$$

$$\psi_\perp = 54.90° \qquad 2\psi_\perp = 109.80°$$

$$\tan\psi_\parallel = \left(\frac{n_1}{n_2}\right)^2 \tan\psi_\perp = \left(\frac{1.50}{1.48}\right)^2 1.4228 = 1.4615$$

$$\psi_\parallel = 55.62° \qquad 2\psi_\parallel = 111.24°$$

$$\text{penetration depth} = \frac{\lambda_0}{2\pi n_2 B} = \frac{1.0}{(2\pi)(1.48)(0.135)} = 0.797\ \mu\text{m}$$

For $i = 1.10i_c = 88.66°$

$$A = \frac{n_1}{n_2}\cos i = \frac{1.50}{148}(0.0234) = 0.0237$$

$$B = \left[\left(\frac{n_1}{n_2}\right)^2 \sin^2 i - 1\right]^{1/2} = \left[\left(\frac{1.50}{1.48}0.9997\right)^2 - 1\right]^{1/2} = 0.163$$

$$\tan\psi_\perp = \frac{B}{A} = \frac{0.163}{0.0237} = 6.887$$

$$\psi_\perp = 81.74° \qquad 2\psi_\perp = 163.48°$$

$$\tan\psi_\parallel = \left(\frac{n_1}{n_2}\right)^2 \tan\psi_\perp = \left(\frac{1.50}{1.48}\right)^2 6.887 = 7.075$$

$$\psi_\parallel = 81.96° \qquad 2\psi_\parallel = 163.92°$$

Note that the phase angle increases as the angle of incidence increases.

$$\text{penetration depth} = \frac{\lambda_0}{2\pi n_2 B} = \frac{1.0}{(2\pi)(1.48)(0.163)} = 0.660\ \mu\text{m}$$

Note that the penetration depth decreases as the angle of incidence increases.

(b) $$\sin i_c = \frac{n_2}{n_1} = \frac{1}{1.5} = 0.6667$$

$$i_c = 41.82^\circ \qquad 1.05 i_c = 43.91^\circ \qquad 1.10 i_c = 46.00^\circ$$

For $i = 1.05 i_c = 43.91^\circ$

$$A = \frac{n_1}{n_2} \cos i = (1.50)(0.7204) = 1.08$$

$$B = \left[\left(\frac{n_1}{n_2} \sin i\right)^2 - 1\right]^{1/2}$$

$$= [(1.50)^2(0.6935)^2 - 1]^{1/2} = 0.287$$

$$\tan \psi_\perp = \frac{B}{A} = \frac{0.287}{1.08} = 0.2653$$

$$\psi_\perp = 14.86^\circ \qquad 2\psi_\perp = 29.72^\circ$$

$$\tan \psi_\parallel = \left(\frac{n_1}{n_2}\right)^2 \tan \psi_\perp = (2.25)(0.2653) = 0.5970$$

$$\psi_\parallel = 30.84^\circ \qquad 2\psi_\parallel = 61.68^\circ$$

$$\text{penetration depth} = \frac{\lambda_0}{2\pi n_2 B} = \frac{1.0}{(2\pi)(1.0)(0.287)} = 0.555\ \mu\text{m}$$

For $i = 1.1 i_c = 46.00^\circ$

$$A = \frac{n_1}{n_2} \cos i = (1.50)(0.6947) = 1.042$$

$$B = \left[\left(\frac{n_1}{n_2} \sin i\right)^2 - 1\right]^{1/2}$$

$$= [(1.50)^2(0.7193)^2 - 1]^{1/2} = 0.4053$$

$$\tan \psi_\perp = \frac{B}{A} = \frac{0.4053}{1.042} = 0.4053$$

$$\psi_\perp = 22.16^\circ \qquad 2\psi_\perp = 44.32^\circ$$

$$\tan \psi_\parallel = \left(\frac{n_1}{n_2}\right)^2 \tan \psi_\parallel = 2.25(0.4053) = 0.9119$$

$$\psi_\parallel = 42.36^\circ \qquad 2\psi_\parallel = 84.72^\circ$$

$$\text{penetration depth} = \frac{\lambda_0}{2\pi n_2 B} = \frac{1.0}{(2\pi)(1)(0.4053)} = 0.393\ \mu\text{m}$$

2-2.2 Transverse Modes

A ray can be broken up into its z component parallel to the dielectric axis and its x component normal to the core-cladding interface. Only those rays whose x components form standing waves can propagate, since only they constructively interfere with each other. As a result, only a finite number of rays, or modes, are allowed to propagate. We now find that number, m.

To coincide with the nomenclature used in other textbooks, we will now and throughout this chapter define the incident angle to be ϕ. The path length traveled by the ray reflected up and back is $2d$, where d is the core width of the planar dielectric in Fig. 2-3, and the ray is reflected twice during this interval. The phase difference, δ, for this path length difference is

$$\delta = (2\pi)2\,\frac{d}{\lambda_0/(n_1 \cos \phi_m)} - 4\psi = 2m\pi \tag{2-7}$$

since the effective wavelength in the x direction is $\lambda_0/(n_1 \cos \phi_m)$ where λ_0 is the wavelength in a vacuum.

The phase change must be an integral multiple of 2π for there to be constructive interference of the x components, and 4ψ is subtracted because δ is a phase lag and 4ψ is a phase gain. Solving for the allowable values of ϕ, ϕ_m, we find that

$$\cos \phi_m = \frac{\lambda_0}{2\pi n_1 d}(m\pi + 2\psi) \tag{2-8}$$

Since

$$\cos \phi_m \leq \left[1 - \left(\frac{n_2}{n_1}\right)^2\right]^{1/2} \tag{2-9}$$

$$m \leq \frac{2n_1 d}{\lambda_0}\left[1 - \left(\frac{n_2}{n_1}\right)^2\right]^{1/2} - \frac{2\psi}{\pi} = \frac{V - 2\psi}{\pi} \tag{2-10}$$

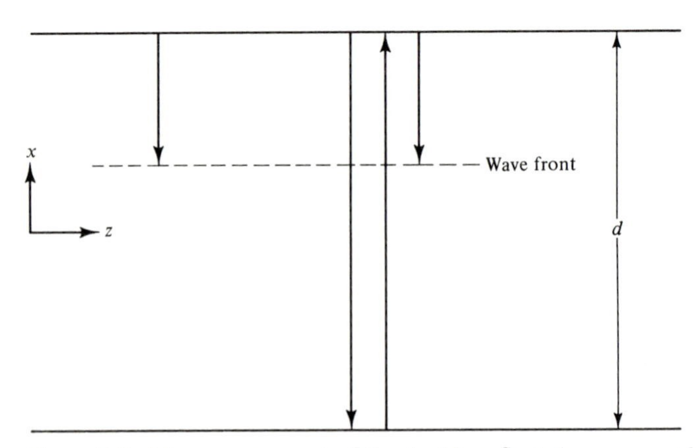

Figure 2-3 The x component of the doubly reflected wave must have a phase that is an integer multiple of 2π greater than that of the unreflected ray since the x components form standing waves.

The quantity, V, is often called the normalized film thickness, and it appears in a number of later equations. It can be shown that for a planar dielectric symmetric about the core axis, $m \geq 1$. Thus, at least one mode will always propagate.

EXAMPLE 2-2

Find the maximum diameter for which only one mode propagates in a planar dielectric when $\lambda_0 = 1\ \mu$m and **(a)** $n_1 = 1.50$ and $n_2 = 1.48$, and **(b)** $n_1 = 1.50$ and $n_2 = 1.00$. m in Eq. 2-10 must be less than 2

$$\therefore \frac{V - 2\psi}{\pi} < 2$$

ψ depends on i as well as n_1 and n_2 but, from Fig. 2.2, $2\psi \leq \pi$

$$\therefore V < 3\pi \quad \text{or} \quad d < \frac{3\pi\lambda_0}{2\pi(n_1^2 - n_2^2)^{1/2}}$$

(a)
$$d < \frac{(3)(1.0)}{2[(1.5)^2 - (1.48)^2]^{1/2}} = 6.14\ \mu\text{m}$$

(b)
$$d < \frac{(3)(1.0)}{2[(1.5)^2 - (1.00)^2]^{1/2}} = 1.34\ \mu\text{m}$$

Note that single mode planar dielectrics can be larger if n_2 is more nearly equal to n_1.

The higher order transverse modes have a smaller value of ϕ and, therefore, travel a longer distance before they reach the end of the fiber. This leads to the signal broadening called mode dispersion, which is shown in Fig. 2-4. The time, t_1, required for the central ray to travel a distance, L, in the z direction is $n_1 L/c_0$, and the time, t_2, required for the ray making the smallest angle with the normal to the cladding is $n_1 L/(c_0 \sin \phi_c)$. The time spread, Δt, for the signal is, therefore

$$\Delta t = \frac{n_1 L}{c_0(n_2/n_1)} - \frac{n_1 L}{c_0} = \frac{n_1 L}{n_2 c_0}(n_1 - n_2) \tag{2-11}$$

The larger Δt is, the lower the data transmission rate will be. The maximum data rate, DR_{max}, is the maximum rate at which individual signals can be trans-

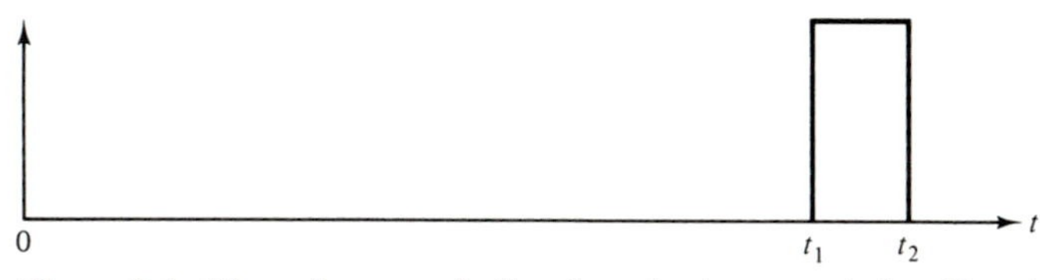

Figure 2-4 The pulse spread of an impulse in a step index fiber due to the radial wave arriving earlier (t_1) than the highest order mode (t_2).

mitted. If we choose the separation between the individual signals to be the pulse width, then

$$DR_{\max} = \tfrac{1}{2}\Delta t \tag{2-12}$$

Clearly, the data rate decreases as the length increases. It also decreases as $n_1 - n_2$ increases. This fact as well as the magnitude of the mode dispersion is illustrated in Example 2-3.

EXAMPLE 2-3

Find the pulse width broadening due to modal dispersion in a planar dielectric when **(a)** $n_1 = 1.50$ and $n_2 = 1.48$, and **(b)** $n_1 = 1.50$ and $n_2 = 1.00$. From Eq. 2-11

$$\Delta t = \frac{n_1 L}{n_2 c_0}(n_1 - n_2)$$

(a) $\Delta t = 1.5(1.50 - 1.48)/(1.48)(3.00 \times 10^5) = 67.6$ nsec/km

(b) $\Delta t = 1.5(1.5 - 1.0/(1.0)(3.00 \times 10^5) = 2500$ nsec/km

It is easy to see why n_2 must approach n_1 for high data rate fibers.

When Maxwell's equations are solved for a waveguide as they are in Appendix B, one finds that the magnitudes of the ξ and H fields are a function of the position and the dependence is different for the different modes. We now consider a similar situation by computing the amplitudes of the standing waves in the x direction. Only the interference pattern between a wave and a wave reflected one more time needs to be considered because, for a standing wave, the phase difference for one round-trip of the beam is an integral multiple of 2π (see Eq. 2-7).

The path length difference between the ray and the reflected ray at x is $2(d/2 - x)$, x being measured from the center of the dielectric. The path length difference is thus $d - 2x$, and the phase difference, Φ, is

$$\Phi = 2\pi \frac{d - 2x}{\lambda_0/n_1 \cos \phi_m} - 2\psi \tag{2-13a}$$

Again, $\lambda_0/(n_1 \cos \phi_m)$ is the effective wavelength in the x direction, and 2ψ is the phase gain on reflection. Substituting the equation for $\cos \phi_m$, Eq. 2-8, into Eq. 2-13a yields

$$\Phi = m\pi - \frac{2x}{d}(m\pi + 2\psi) \tag{2-13b}$$

The amplitude of the wave that is the sum of two waves of equal amplitude and angular frequency, but with one wave lagging the other by Φ, can be found from the phasor diagram in Fig. 2-5. The vertex angle of the isosceles triangle is $\pi - \Phi$. Hence, the side opposite the vertex is $2 \cos (\Phi/2)$. The amplitude of the standing wave, $\xi_0(x)$, is thus

$$\xi_0(x) = 2\xi_0 \cos \left[\frac{m\pi}{2 - \frac{x}{d}(m\pi + 2\psi)} \right] \tag{2-14}$$

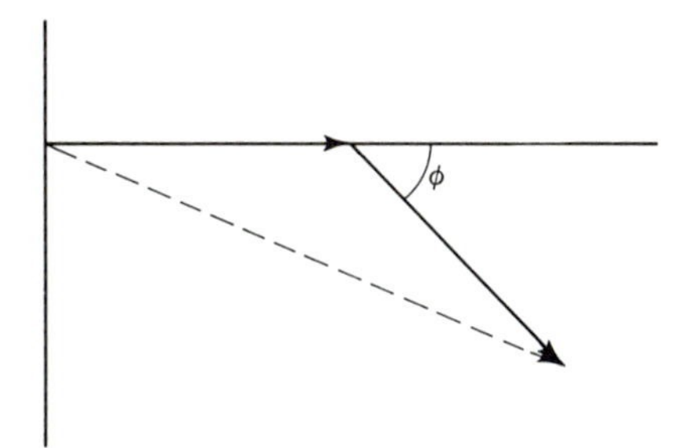

Figure 2-5 Phasor used to determine the amplitude of the resultant wave formed from two waves of equal amplitude and frequency, but one lags the other by the phase angle, Φ.

At $x = 0$, $\xi_0(0) = \pm 2\xi_0$ for m even, and $\xi_0(0) = 0$ for m odd. The first two modes are plotted in Fig. 2-6.

Recall that the modes for $\xi_\perp$ and $\xi_\parallel$ are different. When Maxwell's equations are solved, there is a set of solutions for which the z component of ξ is zero. These modes are called the TE, or transverse electric modes. The modes for which $H_z = 0$ are the TM, or transverse magnetic modes.

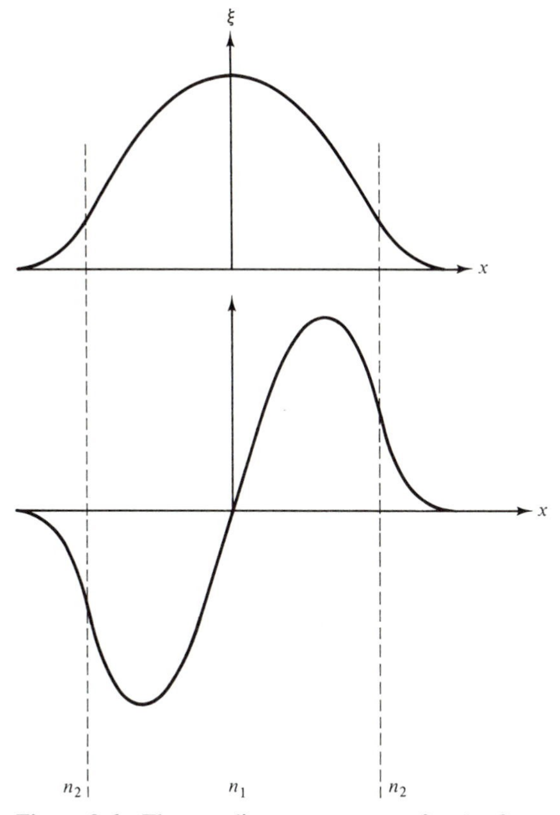

Figure 2-6 The standing wave pattern for the first two transverse modes in a planar dielectric.

2-3 WAVELENGTH DISPERSION

2-3.1 Physical Description

In Section 1-4.3 we mentioned that the index of refraction is a function of λ, and that in the visible and near infrared n decreases as λ increases. This is shown in Fig. 2-7 along with one of the ramifications of this effect—the breaking up of a light wave into its component colors as it passes through a prism. The red light, having the longest wavelength, has the smallest index of refraction and is bent the least; the blue light, having the shortest wavelength, is bent the most.

The index of refraction is a measure of how strongly the electromagnetic radiation interacts with the material and can be quantified by considering the polarization. **P** is the dipole moment per unit volume; thus, it is the vector sum of the individual dipoles, **p**, over a unit volume.

$$\mathbf{P} = \sum_{i=1}^{N} \mathbf{p}_i \tag{2-15a}$$

where N is the number of dipoles per unit volume. For one type of dipole all pointing in the same direction

$$\mathbf{P} = N\mathbf{p} \tag{2-15b}$$

where the magnitude of **p** is

$$p = Zqx \tag{2-16}$$

Z is the number of charges forming the dipole, q is the charge on a proton, and x is the separation between the center of positive charge and the center of negative charge. This is illustrated in Fig. 2-8. For an induced dipole, x is proportional to the applied field and the constant of proportionality is the polarizability, α. Thus

$$p = \alpha\xi \tag{2-17}$$

The electric field seen by an individual dipole is different from the external field because one must also consider the field created by all the other dipoles. This field is called the internal field. In this book the internal field will be neglected. This introduces essentially no error when gases are considered but does introduce

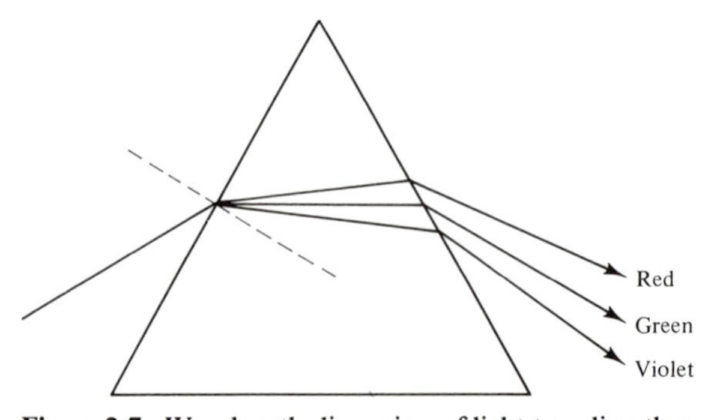

Figure 2-7 Wavelength dispersion of light traveling through a prism.

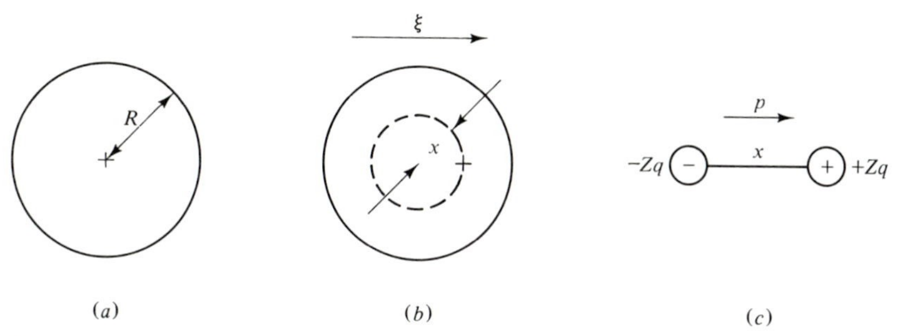

Figure 2-8 (*a*) A spherical atom in which the centers of positive and negative charge coincide. (*b*) A spherical atom in which an electric field with different centers of positive and negative charge. (*c*) The dipole moment created by the separation of the centers of positive and negative charge.

considerable error when solids are considered. We will, however, consider solids in a qualitative manner.

Assuming that the local field seen by an individual dipole is the same as the external field, P can be written

$$P = Np = N\alpha\xi = \chi\epsilon_0\xi = NZqx$$

using Eq. 2-15, 2-17, 1-33, or 2-16. Therefore

$$\chi = \frac{N\alpha}{\epsilon_0} \tag{2-18a}$$

or

$$\chi = \frac{NZqx}{\epsilon_0\xi} \tag{2-18b}$$

From Eqs. 1-56 and 1-32b, $n = \sqrt{\epsilon/\epsilon_0} = \sqrt{1 + \chi}$ when $\mu = \mu_0$. Thus, n is a $f(\lambda)$ if χ is $f(\lambda)$. We will show that the latter is the case by showing that $\chi \propto \xi$ and $\chi = f(\lambda)$. This will be done by finding the frequency response of χ in a simple harmonic oscillator in much the same way that one would find the frequency response of the charge on the capacitor in a simple *LRC* series circuit.

2-3.2 Susceptibility of an Atomic Gas

An atom can be approximately modeled as it is shown in Fig. 2-8. The positively charged nucleus with charge, Z_q, is surrounded by a spherical "pudding" of negative charge that is uniformly distributed throughout the sphere. When there is no applied electric field, the centers of positive and negative charge coincide and thus there is no dipole moment. When an external field is applied in the $+x$ direction, the positive nucleus is repelled to the right while the negatively charged cloud is attracted to the left. The equilibrium separation distance, x, between the centers of negative and positive charge is the distance for which the force of attraction between the negative cloud and positive nucleus is equal but opposite to the force on the atom generated by the external electric field. To find the force of attraction between the displaced negative cloud and positive nucleus, we must first find the electric field produced by the negative cloud at the point where the nucleus is located. From Gauss' law, Eq. 1-1b,

$$\int_s \epsilon_0 \xi \cdot d\mathbf{S} = 4\pi x^2 \epsilon_0 \xi = \int_V \rho \, dV = \frac{-Zq}{\frac{4}{3}\pi R^3} \frac{4}{3} \pi x^3 \tag{2-19a}$$

or

$$x = \frac{4\pi R^3 \epsilon_0}{Zq} \xi \tag{2-19b}$$

R is the radius of the negative cloud, and the minus sign is missing because the external field is equal, but opposite to that of the ξ field of the negative cloud at the nucleus. For a system at rest, the restoring force, $k_f x_{st}$, is equal and opposite to the applied force, $Zq\xi$.

The subscript, *st,* is used to emphasize that it is the static displacement. Thus, using Eq. 2-19b,

$$k_f x_{st} = \frac{(Zq)^2}{4\pi\epsilon_0 R^3} x_{st} \tag{2-20}$$

When ξ has a sinusoidal dependence, $\xi_0 \exp(j\omega t)$, the amplitude of vibration, x_0, and therefore, n, has a frequency dependence. We now find that dependence.

The forces acting on the atom are the external force, $Zq\xi$, the restoring force, $-k_f x$, and the frictional force $-\beta_1 \dot{x}$, which is proportional to the velocity and opposed to the direction of motion. From Newton's second law, the sum of the forces equals the mass times the acceleration,

$$Zq\xi - k_f x - \beta_1 \dot{x} = m\ddot{x} \tag{2-21a}$$

or

$$m\ddot{x} + \beta_1 \dot{x} + k_f x = Zq\xi \tag{2-21b}$$

This equation is similar to that of an *LRC* series circuit and can be solved using phasors. Thus

$$x = x_0 e^{j(\omega t - \phi)} \tag{2-22}$$

Substituting Eq. 2-22 into 2-21b to find the magnitude of x_0, yields

$$x_0 = \frac{Zq\xi_0}{k_f - m\omega^2 + j\omega\beta_1} \tag{2-23a}$$

Dividing through by k_f yields

$$x_0 = \frac{x_{st}}{1 - \omega^2/\omega_0^2 + j\omega\tau} \tag{2-23b}$$

x_{st} is found from Eq. 2-20, ω_0 is the natural frequency of the system where

$$\omega_0 = \sqrt{k_f/m} \tag{2-24}$$

and τ is the decay time for the system where

$$\tau = \frac{\beta_1}{k_f} \tag{2-25}$$

The magnitude of x_0 is given by

$$|x_0| = \frac{x_{st}}{\{[1 - (\omega/\omega_0)^2]^2 + (\omega\tau)^2\}^{1/2}} \tag{2-26}$$

the real part of x_0 is

$$x_0' = \frac{[1 - (\omega/\omega_0)^2]x_{st}}{[1 - (\omega/\omega_0)^2]^2 + (\omega\tau)^2} \tag{2-27}$$

the imaginary part of x_0 is

$$x_0'' = \frac{\omega\tau x_{st}}{[1 - (\omega/\omega_0)^2]^2 + (\omega\tau)^2} \tag{2-28}$$

and the lag angle is found from

$$\tan\phi = \frac{x_0''}{x_0'} = \frac{\omega\tau}{1 - (\omega/\omega_0)^2} \tag{2-29}$$

The student should be familiar with these equations through his experience with circuits. Here we highlight only the salient points with the assistance of Fig. 2-9 where $|x_0|/x_{st}$, x_0'/x_{st}, and x_0''/x_{st} are plotted versus the reduced angular frequency, ω/ω_0.

For electronic resonances ω_0 is in the ultraviolet spectrum so the visible region and the near infrared are in the longer wavelength region, $\omega \ll \omega_0$. In this

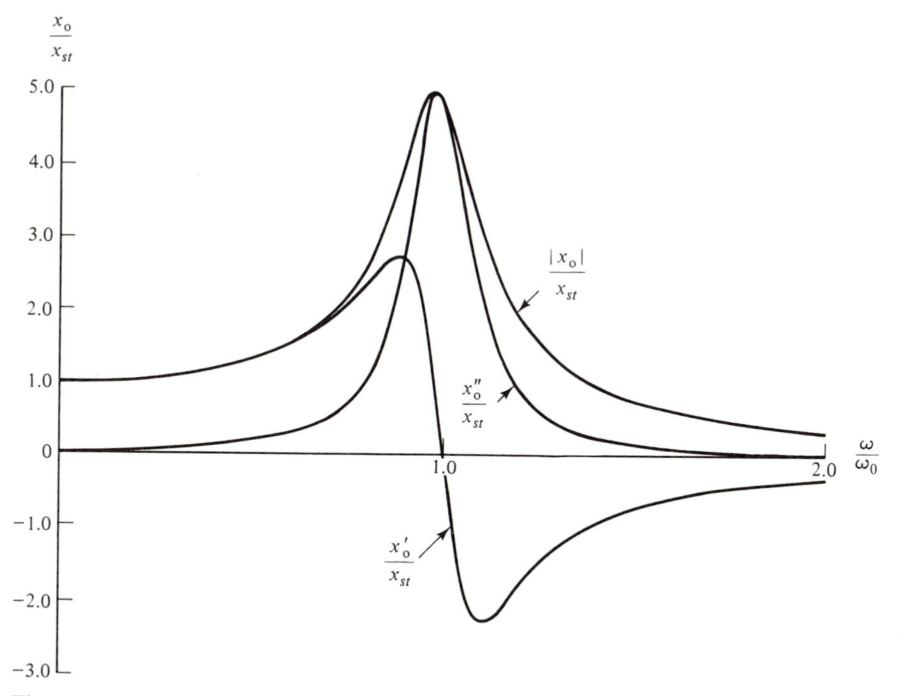

Figure 2-9 The magnitudes of x_o/x_{st}, x_0'/x_{st} and x_0''/x_{st} plotted as a function of ω/ω_0 when $m\omega_0/\beta_1 = 5.0$.

region $|x_0| \sim x_0'$ and $x_0'' \sim 0$. Furthermore, x_0 and therefore n increase with increasing ω as was predicted. The quantity x_0 can be greater than x_{st} because it is able to store energy in the system, and it reaches its largest value at $\omega \sim \omega_0$. The electrical analog is the charge on the capacitor plates in an *LRC* series circuit.

As is shown in Fig. 2-9, x_0'' is large only in the vicinity of $\omega = \omega_0$. x_0'' is associated with ϵ'' which, in turn, is associated with absorption. We will see later in this chapter and in more detail in Chapter 3 that absorption occurs primarily at resonant frequencies.

2-3.3 Wavelength Dispersion

As is true of modal dispersion, wavelength dispersion causes the signal to broaden thereby limiting the data rate. The longer wavelength light traveling the same path as the shorter wavelength light will travel faster and, therefore, will arrive at its destination sooner. The difference between the time of arrival of the slowest moving wave and the fastest moving wave is equal to the pulse broadening.

Values for $n(\lambda)$ can be computed from Cauchy's phenomenological equation

$$n = A + \frac{B}{\lambda^2} + \frac{C}{\lambda^4} \tag{2-30}$$

where the values of A, B, and C are found from the experimental values of n at different wavelengths like those shown in Table 2-1.

EXAMPLE 2-4

(a) Find the pulse broadening and associated maximum data rate due to wavelength dispersion when an LED light source with peak emission near 0.8 μm and a peak width of 300 Å is used. For this peak width at this wavelength $\Delta n = 6.0 \times 10^{-4}$. **(b)** Repeat part a if a laser diode light source with a peak width of 20 Å with an associated $\Delta n = 4 \times 10^{-5}$ is used.

TABLE 2-1 THE INDEX OF REFRACTION OF A NUMBER OF TRANSPARENT SOLIDS FOR DIFFERENT WAVELENGTHS IN THE VISIBLE

Substance	Violet (410 nm)	Blue (470 nm)	Green (550 nm)	Yellow (580 nm)	Orange (610 nm)	Red (660 nm)
Crown glass	1.5380	1.5310	1.5260	1.5225	1.5216	1.5200
Light flint	1.6040	1.5960	1.5910	1.5875	1.5867	1.5850
Dense flint	1.6980	1.6836	1.6738	1.6670	1.6650	1.6620
Quartz	1.5570	1.5510	1.5468	1.5438	1.5432	1.5420
Diamond	2.4580	2.4439	2.4260	2.4172	2.4150	2.4100
Ice	1.3170	1.3136	1.3110	1.3087	1.3080	1.3060
Strontium titanate ($SrTiO_3$)	2.6310	2.5106	2.4360	2.4170	2.3977	2.3740

(a)
$$\Delta t = \frac{L}{c_{\text{slow}}} - \frac{L}{c_{\text{fast}}} = \frac{L}{c_0}\,\Delta n$$

For LED $\Delta t = (6 \times 10^{-4})/3 \times 10^5 = 2$ ns/km
$DR_{\max} \sim \frac{1}{2}\,\Delta t = \frac{1}{4} \times 10^9 = 2.5 \times 10^8$ bits · km/sec

(b) For laser diode $\Delta t = (4 \times 10^5)/3 \times 10^5 = 0.133$ ns/km
$DR_{\max} \sim \frac{1}{2}\,\Delta t = 3.75 \times 10^9$ bits · km/sec

This example illustrates one of the dominant problems an engineer will face when he or she designs an optical communications system. Should the engineer use a laser diode or an LED as the light source. The laser diode system can be much faster, but it is more expensive than an LED. In addition, the laser diode frequently requires a feedback system to adapt to different temperatures, and its lifetime is quite a bit shorter than that of an LED.

As we stated previously, ray theory in a wave guide with small dimensions is only an approximation, and more accurate results can be obtained by using mode theory.

For a group of waves with different velocities the group or mode moves at the group velocity

$$v_g = \frac{d\omega}{dk} = -\frac{dv}{d\lambda}\,\lambda^2 \qquad (2\text{-}31)$$

The difference in the velocity of the different modes is then

$$\Delta v_g = \frac{dv_g}{d\lambda} = \frac{c_0\lambda}{n^2}\left[\frac{\partial^2 n}{\partial\lambda^2} - \frac{2}{n}\left(\frac{\partial n}{\partial\lambda}\right)^2\right]\Delta\lambda \qquad (2\text{-}32)$$

The pulse spread is thus

$$\Delta t = -L\,\frac{\Delta v_g}{v_g^2} = \frac{-L\lambda[\partial^2 n/\partial\lambda^2 - (2/n)(\partial n/\partial\lambda)^2]\Delta\lambda}{c_0\left(\dfrac{\lambda\partial n}{n\partial\lambda} + 1\right)^2} \qquad (2\text{-}33a)$$

$$\simeq -\frac{L\lambda}{c_0}\,\frac{\partial^2 n}{\partial\lambda^2}\,\Delta\lambda \qquad (2\text{-}33b)$$

The curvature of $n(\lambda)$ is plotted in Fig. 2-10, which shows that the curvature, and therefore the pulse spread, is zero near $\lambda = 1.3\ \mu$m. By using ray theory, one finds that the wavelength dispersion decreases as λ increases. This is one reason why people are interested in developing reliable light sources emitting near or at this wavelength.

2-4 OPTICAL FIBERS

2-4.1 Step Index Fibers

A step index fiber is one for which the index of the core, n_1, is constant and is larger than the index of the cladding, n_2, as shown in Fig. 2-11. Because $n_1 >$

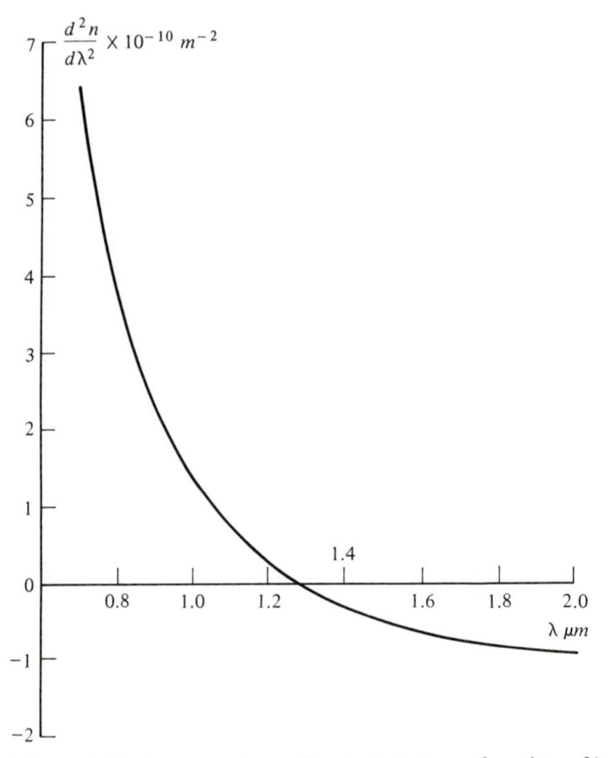

Figure 2-10 The curvature of $n(\lambda)$ plotted as a function of λ used to calculate the wavelength dispersion.

n_2, there is total internal reflection for $\phi > \phi_c$ just as there was for a planar dielectric. The step index fiber differs from the planar dielectric only in that it is two-dimensional with a circular cross section.

The meridional ray illustrated in Fig. 2-12*a* is one that passes through the center of the fiber and, as a result, follows the same saw-tooth planar path followed by the rays in the planar dielectric. The skew ray illustrated in Fig. 2-12*b* is one

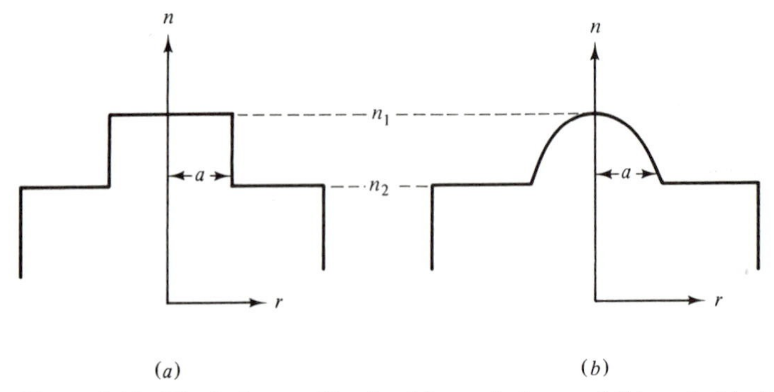

Figure 2-11 The index profiles for (*a*) step index, and (*b*) graded index fibers.

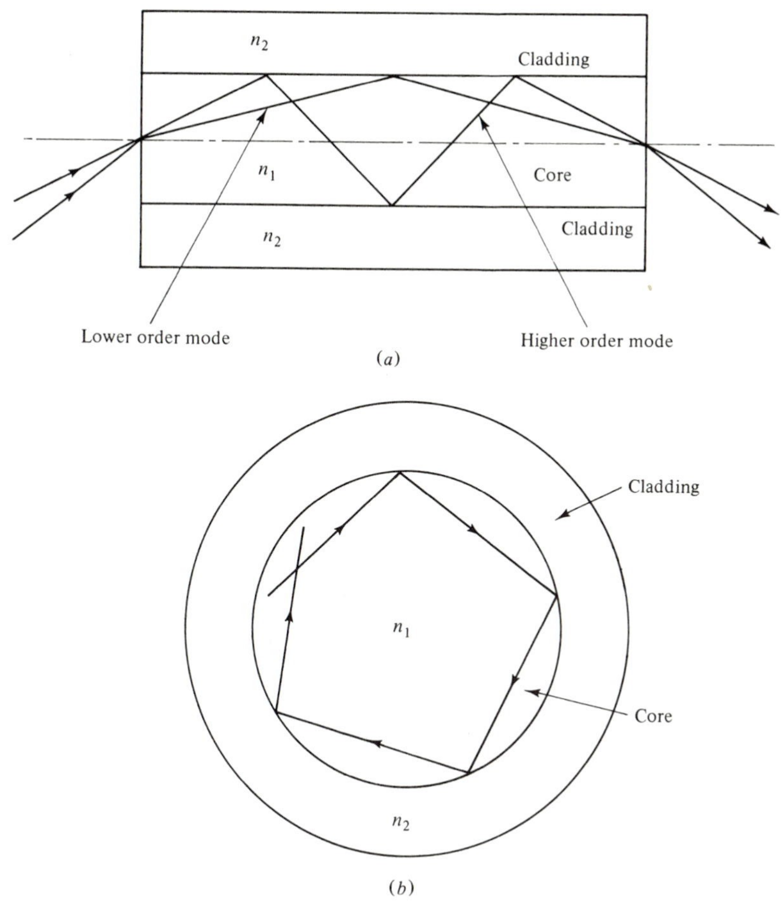

Figure 2-12 The (*a*) planar ray path of meridional rays, and (*b*) the three-dimensional path of a skew ray.

that does not pass through the center and, as a result, follows a three-dimensional path.

In terms of the mode description of the electromagnetic radiation in the fiber the meridional rays correspond to the TE_{1m} and TM_{1m} modes. Two subscripts are necessary, since they are two-dimensional. The skew rays correspond to what are called the HE_{1m} and EH_{1m}. The first term describes which field dominates and the subscript, 1, is a measure of how tight the helix is.

The modes in a fiber are, however, often described by the LP_{1m}—linearly polarized—modes. The subscript, 1, is one-half the number of local maxima around a circumference, and m is the number of local maxima along a radius. Two examples are illustrated in Fig. 2-13.

As was true for the planar dielectric, the number of allowed mode in the fiber is determined by the magnitude of V described in Eq. 2-10. The quantity for a fiber differs only in that the radius of the fiber, a, is used instead of the width, d, of the planar dielectric. Note that the smaller a and Δn are, the smaller

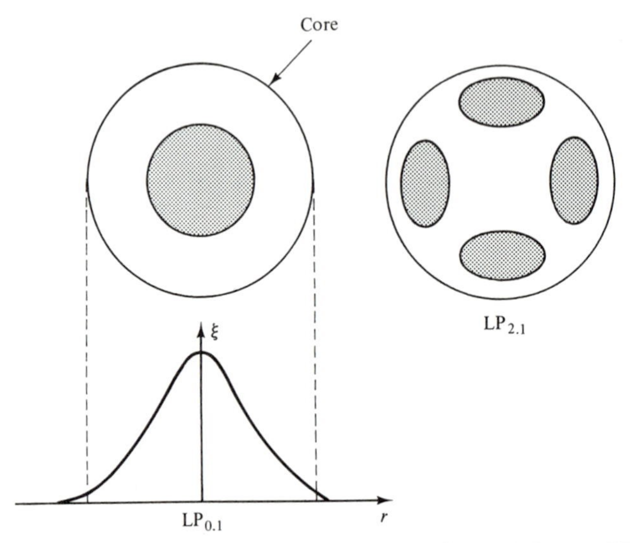

Figure 2-13 The electric field distribution for the $LP_{0,1}$, and $LP_{2,1}$ modes in a fiber.

V is. It can be shown that for $V \leq 2.405$, only one mode can propagate (see ref. 11, Section 5.7), and as V decreases below this value, the penetration into the cladding increases. For larger values the number of modes, N, is given approximately by (see ref. 11, Section 5-6)

$$N \simeq \tfrac{1}{2}V^2 \tag{2-34}$$

EXAMPLE 2-5

(a) What is the maximum value of a for which there is only one mode if $n_1 = 1.50$, $n_2 = 1.48$, and $\lambda_0 = 1\ \mu$m? Repeat for $n_2 = 1.45$. **(b)** How many modes can propagate if $a = 50\ \mu$m for the conditions listed in part a?

(a)
$$V = \frac{2\pi a}{\lambda_0}(n_1^2 - n_2^2)$$

$$\therefore a_{\max} = \frac{2.405\lambda_0}{2\pi(n_1^2 - n_2^2)^{1/2}}$$

$$a_{\max,1} = \frac{(2.405)(1)}{2\pi[(1.50)^2 - (1.48)^2]^{1/2}} = 1.57\ \mu\text{m}$$

$$a_{\max,2} = \frac{(2.405)(1)}{2\pi[(1.50)^2 - (1.45)^2]} = 1.00\ \mu\text{m}$$

(b)
$$V_1 = \frac{(2\pi)(50)}{1}[(1.50)^2 - (1.48)^2]^{1/2} = 76.7$$

$$N_1 \simeq \frac{V_1^2}{2} = 2941$$

$$V_2 = (2\pi)(50)[(1.50)^2 - (1.45)^2]^{1/2} = 121$$

$$N_2 \simeq 7279$$

Wavelength and modal dispersion in optical fibers is similar to that in the planar dielectric. The only major difference is that the modal dispersion in fibers greater than 1 km in length is less than that predicted by the simple theory. This is because there is a slow interchange of energy between the modes. Thus the originally fast-moving modes slow down and the slower moving modes speed up.

2-4.2 Graded Index Fibers

The index of the core in the graded index fiber illustrated in Fig. 2.11 has a maximum value at the center and decreases toward the edges. The advantage of using this type of fiber is that the modal dispersion can be greatly reduced. The axial ray travels the shortest distance, but it moves the slowest since it sees the largest average index of refraction. The off axis meridional rays move in sinusoidal paths like those shown in Fig. 2-14. The higher order modes have a larger amplitude and, therefore, a longer path length, but they spend more time in the lower index material. As a result, they travel farther but faster. Hence, it is possible for them to arrive at their destination at about the same time as the axial ray.

Physically, the rays follow a curved path because the vibrating electric field sees a different index of refraction above the direction of propagation than it does below. For a ray above the axis, the ray sees a smaller index of refraction above the direction of propagation than it does below. Thus it slows down when

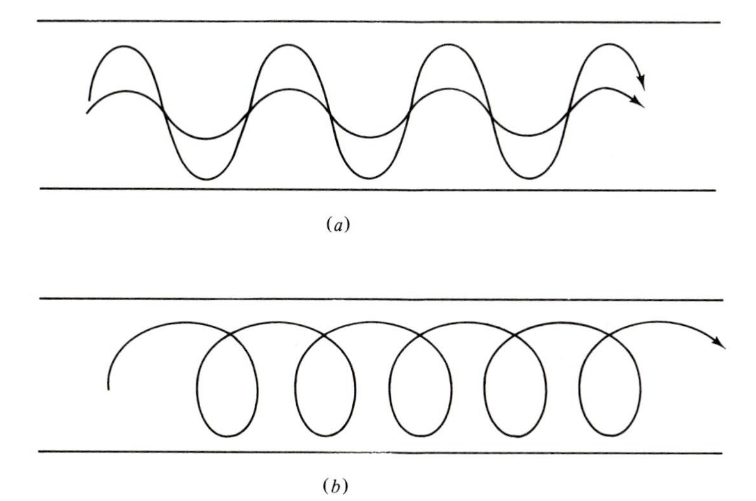

Figure 2-14 The (*a*) sinusoidal path of a meridional ray, and (*b*) helical path of a skew ray in a graded index fiber.

it is vibrating below the direction of propagation; this braking effect bends the ray down. For propagation below the axis the ray is continually bent upward.

The input angle of incidence determines the amplitude of the ray path and the mode order. The highest order mode that will propagate is the mode with an amplitude equal to the fiber radius. If the amplitude is greater than a, then the ray will not be totally internally reflected. Clearly, the larger the fiber radius, the larger the allowed amplitude, the larger the number of modes that can propagate.

To match the time of arrival of the different modes, the variation in $n(r)$ must be closely controlled. It is given by

$$n(r) = n_1\left[1 - 2\Delta\left(\frac{r}{a}\right)^\gamma\right]^{1/2} \tag{2-35a}$$

where

$$\Delta \simeq \frac{n_1 - n_2}{n_1} \tag{2-35b}$$

with n_2 being equal to $n(a)$. As is shown in Fig. 2-15, the pulse spread is extremely sensitive to the value of γ, the profile parameter. A detailed analysis shows that Δt is a minimum when $\gamma \simeq 2(1 - 1.2\ \Delta)$ and its magnitude is

$$\Delta t_{\min} = \frac{Ln_1^2}{8}\Delta^2 \tag{2-36}$$

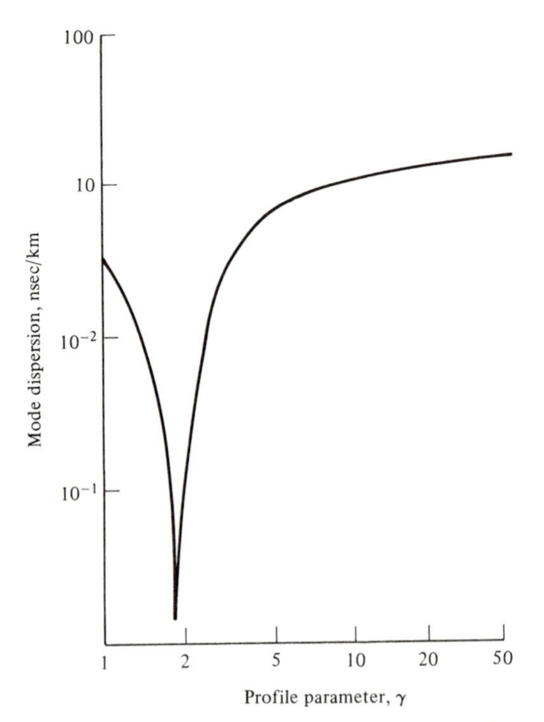

Figure 2-15 The mode dispersion in a graded index fiber plotted as a function of the profile parameter, $\gamma\upsilon$.

This value is $\Delta/8$ or $\sim$300 times smaller than it is for a step index fiber (see ref. 7, Section 5-6). (Note that a step index fiber is a graded index fiber for which $\gamma = \infty$).

The number of allowed modes is also a function of γ (see ref. 7, Section 5-6).

$$N = \left(\frac{2\pi n_1 a}{\lambda_0}\right)^2 \frac{\gamma\Delta}{\gamma + 2} \tag{2-37}$$

Thus, the graded index fiber with the optimum value of γ contains about one-half as many modes as the step index fiber with the same a and Δ values.

2-4.3 Light Insertion

Only light that is totally internally reflected travels very far in an optical fiber. Thus, only the rays that make an angle of $\theta \leq \theta_c$, the critical fiber axis in Fig. 2-16, propagate down the fiber. From Snell's law

$$\sin\theta_c = \frac{n_1}{n_0}\sin(\pi/2 - \phi_c) = \frac{n_1}{n_0}\cos\phi_c \tag{2-38a}$$

when the index of the material adjacent to the fiber is n_0. From Eq. 2-9, using the equality sign,

$$\sin\theta_c = \frac{(n_1^2 - n_2^2)^{1/2}}{n_0} \tag{2-38b}$$

The numerator is often called the numerical aperture, NA and, of course, it decreases as n_2 approaches n_1.

Semiconductor light emitters are used for fiber optic light sources because they can be made as small as the cross section of the optical fibers. This is a particularly stringent condition for single mode fibers, which have radii of only a few microns.

Not only should the diameter of the light source be smaller than the fiber diameter, it is preferable for it to emit collimated light. This is a particular problem

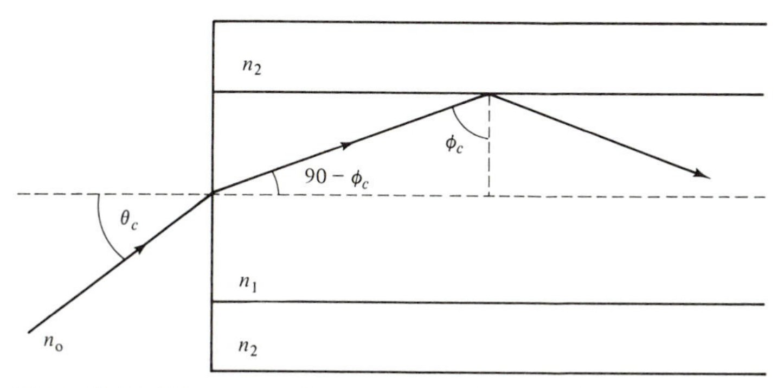

Figure 2-16 Diagram used to determine the critical input angle, θ_c, and the numerical aperture.

for surface emitting LEDs. The fraction of the emitted light that is accepted by the fiber, F, when reflection losses are ignored is

$$F = \frac{\int_0^{\theta_c} AI(\theta)2\pi \sin\theta\, d\theta}{\int_0^{\pi/2} AI(\theta)2\pi \sin\theta\, d\theta} \tag{2-39a}$$

where $I(\theta)$ is the light intensity between θ and $\theta + d\theta$, A is the area of the light emitter, and $2\pi \sin\theta\, d\theta$ is the fraction of the area of a sphere that lies between θ and $\theta + d\theta$. Radiation from surface emitting LEDs often has a Lambertian distribution. That is

$$I(\theta) = I_0 \cos\theta \tag{2-40}$$

Substituting Eq. 2-40 into Eq. 2-39a yields

$$F = \sin^2\theta_c \tag{2-39b}$$

EXAMPLE 2-6

(a) Find NA, θ_c, and F for a step index fiber for which $n_1 = 1.50$, $n_2 = 1.48$, and $n_0 = 1.00$. **(b)** Repeat part a for $n_0 = 1.50$.

(a)

$$\text{NA} = (n_1^2 - n_2^2)^{1/2} = (1.50^2 - 1.48^2)^{1/2} = 0.2441$$

$$\sin\theta_c = \frac{\text{NA}}{n_0} = 0.2441$$

$$\theta_c = 14.13°$$

$$F = \sin^2\theta_c = 5.96 \times 10^{-2}$$

(b)

$$\text{NA} = 0.2441$$

$$\sin\theta_c = \frac{\text{NA}}{n_0} = \frac{0.2441}{1.5} = 0.1627$$

$$\theta_c = 9.37°$$

(Note that when $n_0 = n_1$, $\theta_c = \dfrac{\pi}{2 - \phi_c}$, as it must.)

$$F = \sin^2\theta_c = 2.65 \times 10^{-2} = 2.65 \times 10^{-2}$$

F is reduced by index matching, but index matching also reduces the reflection losses.

The problem of reflection losses can be reduced by using the pigtail, or Burrus type, LED configuration shown in Fig. 2-17*a*. The fiber is centered over the light emitting area and then glued into place with index matching epoxy. This also eliminates alignment problems, but it has the disadvantage of reducing θ_c and, therefore, the fraction of the light that will propagate down the fiber. F can be increased by using a hemispherical diode that bends the rays toward the fiber axis. This is illustrated in Fig. 2-17*b*.

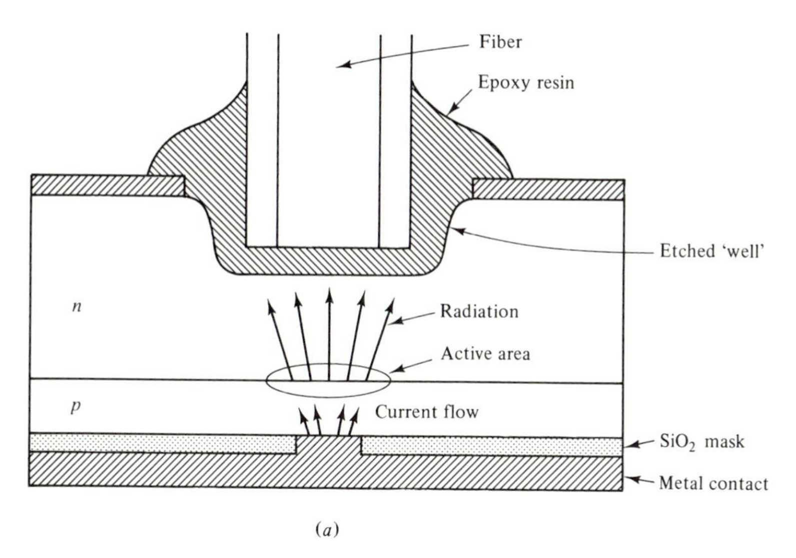

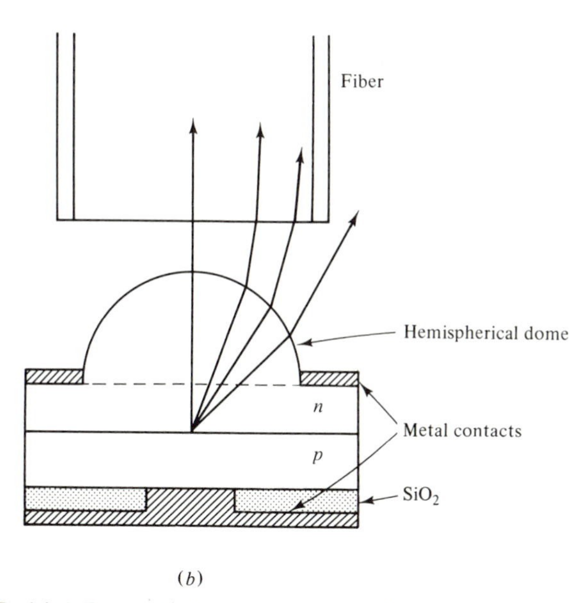

Figure 2-17 (*a*) A Burrus diode structure. (*b*) A diode with a hemispherical dome used to increase the fraction of the light injected into a fiber.

The edge emitting LED in Fig. 2-18*a* has the advantage of emitting a beam that is more highly collimated. In the plane parallel to the *pn* junction (horizontal) and perpendicular to the plane of the paper, $I(\theta)$ is Lambertian as it is for a surface emitter. However, in the plane perpendicular to the junction (parallel to

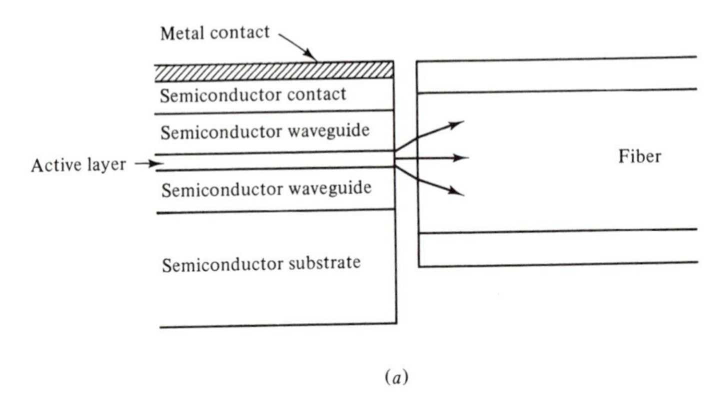

(*a*)

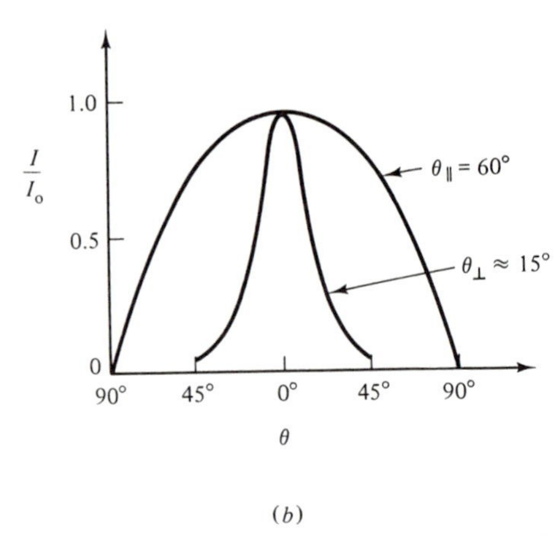

(*b*)

Figure 2-18 (*a*) Schematic of an edge-emitting LED. (*b*) Intensity distribution plotted as a function of θ in the plane parallel to the *pn* junction ($\theta_{\parallel}$), and perpendicular to the junction ($\theta_{\perp}$).

the plane of the paper) $I(\theta)$ decreases much more rapidly with increasing θ. The half-power point is at $\theta \cong 15°$ instead of $\theta = 60°$. This is shown in Fig. 2-18*b*. The beam is more collimated in this plane because the semiconductor materials above and below the active region behave as waveguides much in the same way as the cladding on a fiber does. We will explore this in much more detail when we discuss LEDs and laser diodes.

The disadvantage of using edge emitters is that they generate only one-half as much power as surface emitters. As a general rule of thumb, edge emitters are used when NAs are less than 0.4.

Laser diodes have the same configuration as edge emitting LEDs. They, however, can emit more power, their light is much more highly collimated, and their peak width is much narrower. They are the only reasonable light source for single mode fibers, and they are used when high data rates are required. As we mentioned previously, they are more expensive, less stable, and less reliable than LEDs.

2-5 LOSS

2-5.1 Scattering and Absorption

In discussing dispersion, we described the atoms as harmonic oscillators that were driven by the oscillating electric field of the wave. These oscillating dipoles are antennas, and they, therefore, emit radiation; they emit radiation in directions other than the direction of the primary beam. From the conservation of energy the beam energy is attenuated by the amount of energy not radiated in the direction of the primary beam. This loss is called scattering loss.

One can show that the average power emitted by an antenna, P_a, is (ref. 12, Section 11-2)

$$P_a = \frac{\omega^4}{12\pi\epsilon_0 c^3} p_0^2 \tag{2-41a}$$

where again p_0 is the amplitude of oscillation of the induced dipole. Substituting Eqs. 2-16 and 2-23b into Eq. 2-41a yields

$$P_a = \frac{\omega^4}{12\pi\epsilon_0 c^3}\left[\frac{(Zq)^2\xi_0/m}{\omega_0^2 - \omega^2 + j\omega\beta_1/m}\right]^2 \tag{2-41b}$$

For a dielectric in the visible and the near infrared, $\omega_0 \gg \omega$ and $\omega_0^2 \gg \omega\beta_1/m$. Thus Eq. 2-41b reduces to

$$P_a \simeq \frac{(Zq)^4\xi_0^2\omega^4}{12\pi\epsilon_0 c^3 m^2 \omega_0^4} \tag{2-41c}$$

The important thing to remember is that $P_a \propto \omega^4$ or $1/\lambda^4$. In solids, which are more complex than isolated atoms, since one must consider scattered waves in periodic structures, the scattering power has the same wavelength dependence and is called Rayleigh scattering. This is shown in Fig. 2-19*a* where the attenuation of the primary beam is plotted as a function of λ. One of the primary reasons that longer wavelength light emitters are preferred to those with shorter wavelengths is that there is less Rayleigh scattering. This is illustrated in Example 2-7.

EXAMPLE 2-7

If the Rayleigh scattering loss in a fiber is 1.5 dB/km at $\lambda = 0.84\ \mu$m, find the Rayleigh scattering loss at 1.3 μm.

The power loss in decibels is

$$\text{dB}(\lambda_1) = 10 \log\left[\frac{P_0}{P(\lambda_1)}\right] = 10 \log \frac{1}{1 - [(K/P_0)/\lambda_1^4]}$$

$$\therefore 1 - \frac{K/P_0}{\lambda_1^4} = 10^{-\text{dB}(\lambda_1)/10}$$

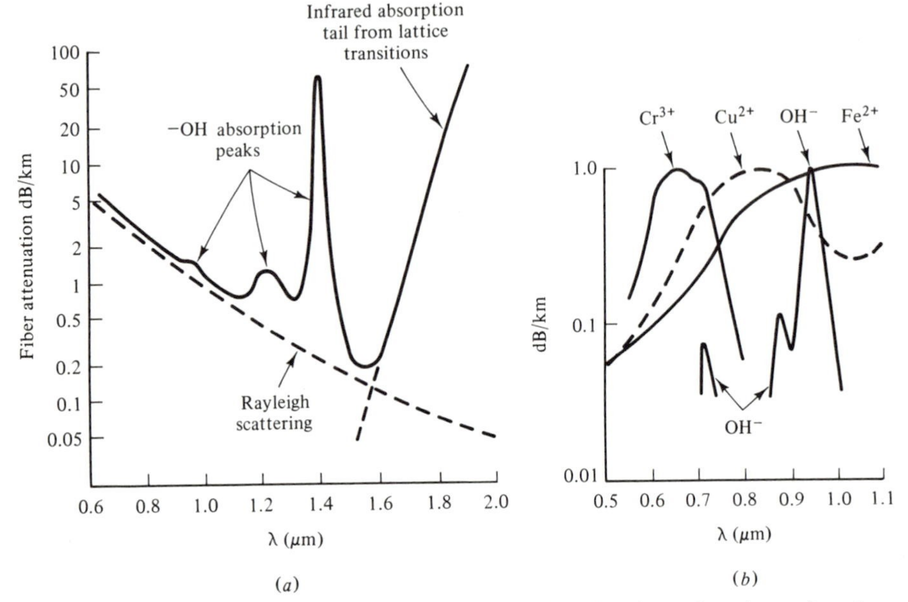

Figure 2-19 (*a*) The fiber attenuation in a high-quality silica fiber plotted as a function of the wavelength. (*b*) Fiber attenuation in silica fibers for 10 ppb of Fe^{+2} and Cr^{+3}, 2.5 ppb of Cu^{+3}, and 1.25 ppm of OH.

$$\text{or} \quad \frac{K}{P_0} = \lambda_1^4[1 - 10^{-\mathrm{dB}(\lambda_1)/10}] = 0.84^4(1 - 10^{-0.15}) = 0.1454$$

$$\mathrm{dB}(\lambda_2) = 10 \log \frac{1}{1 - [(K/P_0)/\lambda_2^4]}$$

$$= 10 \log \frac{1}{1 - \dfrac{0.1454}{(1.3)^4}} = 0.227 \text{ dB/km}$$

The attenuation curve in Fig. 2-19*a* does not monotonically decrease with increasing wavelength. Rather, there are local maxima that are absorption peaks.

The resonance peaks for electronic transitions in transition metal impurities are in the visible and not in the ultraviolet spectrum, as they are for pure dielectrics. As is shown in Fig. 2-19*b*, 1 ppb (parts per billion) of Fe^{+2} or Cr^{+3} ions or 2.5 ppb of Cu^{+2} ions can attenuate the signal by 1 dB/km at certain wavelengths.

Also shown in Fig. 2-19 are the absorption peaks associated with the OH^- radical of water. These peaks are not electronic resonance peaks; rather, they are ionic, or vibration, resonance peaks. The fundamental OH vibration peak is at 2.7 μm, but there are overtones at 1.39, 1.25, and 0.95 μm. If the fiber contains 1.25 ppm (parts per million) H_2O the attenuation at 0.95 μm will be 1 dB/km, but it will be 30 dB/km at 1.39 μm. Thus, high-quality fibers transmitting light at 1.3 μm should have no more than 10 ppb of H_2O.

The natural frequencies of the ionic resonance peaks are, in general, at longer wavelengths than they are for electronic resonance peaks because they have larger effective masses. The OH^- radical has one of the shortest wavelength resonance peaks because its effective mass is relatively small. The resonance peaks of the SiO_2 fiber are at longer wavelengths, but their tails extend down to ~1.6 μm. As a result, there is a minimum in the attenuation at ~1.55 μm. More will be said about vibration resonance peaks in Chapter 3.

Glass fibers have been emphasized in this discussion because they are the high-quality fibers. However, plastic fibers are sometimes used because they are cheaper and easier to handle. The primary reason they will not be extensively used is that they are much more absorbing. This is shown in the attenuation versus wavelength plot in Fig. 2-20. Note that the absorption minima occur at shorter wavelengths—near 0.67 and 0.79 μm.

2-5.2 Mechanical Effects

There is some loss at bends in the fiber, but as we shall see, it is small if the radius of curvature is much larger than the fiber radius. This is usually the case.

The bending loss can be calculated by using ray theory with the aid of Fig. 2-21. When the fiber is bent, the input beam making an angle θ_c with the fiber interface now makes an angle $\phi < \phi_c$ with the normal to the cladding so it is no longer totally internally reflected. The problem is, therefore, to find the angle ϕ' for which the ray in the fiber makes an angle of ϕ_c with the normal to the cladding. The loss is then the energy of the beams making an angle with the fiber interface between ϕ_c and ϕ'. The fraction of the energy loss at the bend is

$$\frac{\Delta E_{\text{loss}}}{E} = \frac{AI(\theta_c)2\pi \sin \theta_c \, \Delta\theta}{\int_0^{\theta_c} AI(\theta)2\pi \sin \theta \, d\theta} \tag{2-42a}$$

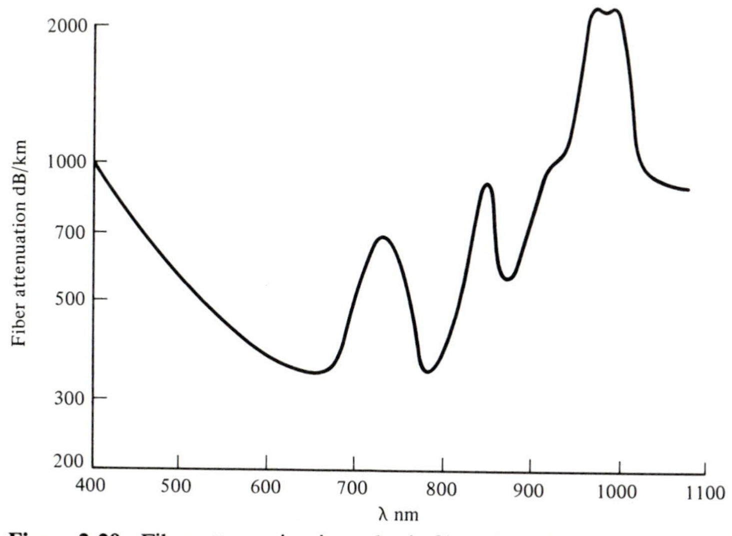

Figure 2-20 Fiber attenuation in a plastic fiber plotted as a function of the wavelength.

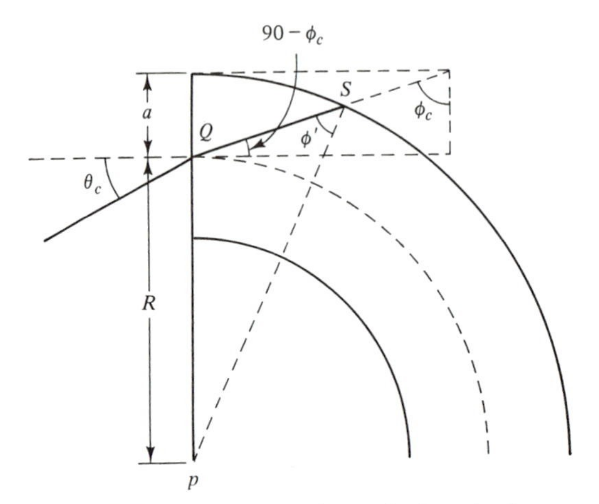

Figure 2-21 Diagram used to calculate the loss at a bend using ray theory. Note that $\phi' < \phi_c$.

where $\Delta\theta = \theta_c - \theta'$ is small. For a Lambertian source

$$\frac{\Delta E_{\text{loss}}}{E} = 2 \cot \theta_c \, \Delta\theta \tag{2-42b}$$

Taking the derivative of Eq. 2-38a and solving for $\Delta\theta$ when $n_0 = 1$ yields

$$\Delta\theta = \frac{n_1 \sin \phi_c}{\cos \theta_c} \Delta\phi \tag{2-43}$$

There is no negative sign because we are finding the decrease in ϕ created by the bend that is equal to the effective increase in ϕ_c.

From Fig. 2-21 and the law of sines for the triangle *PQS*

$$\frac{\sin \phi'}{R} = \frac{\sin (\pi/2 + \pi/2 - \phi_c)}{R + a} = \frac{\sin \phi_c}{R + a} \tag{2-44}$$

With $\Delta\phi = \phi_c - \phi'$, $\cos \Delta\phi \simeq 1$, $\sin \Delta\phi \simeq \Delta\phi$, and $R \gg a$, Eq. 2-44 reduces to

$$\Delta\phi = \frac{a}{R} \tan \phi_c \tag{2-45}$$

Finally, by substituting Eqs. 2-45 and 2-43 into Eq. 2-42b, we obtain

$$\frac{\Delta E_{\text{loss}}}{E} = \frac{2n_1 a}{R} \frac{\sin^2 \phi_c}{\sin \theta_c \cos \phi_c} \tag{2-42c}$$

EXAMPLE 2-8

(a) Find the fraction of the energy lost at a bend with a radius of curvature of 1 m if the core radius of the multimode step index fiber $a = 50\ \mu$m, $n_1 = 1.50$, and $n_2 = 1.48$. **(b)** What is the fraction lost when $R = 1$ cm?

(a)
$$\frac{\Delta E_{\text{loss}}}{E} = \frac{(2)(1.5)(5.0 \times 10^{-5})}{1} \times \frac{(1.48/1.50)^2}{[(1.50)^2 - (1.48)^2]^{1/2}[1 - (148/150)^2]^{1/2}}$$
$$= 3.68 \times 10^{-3}$$

(b) The loss is 100 times greater = 0.368

When solving Maxwell's equations, one finds that there is loss because, in order to keep up, the wave at the outside of the bend must travel faster than the one on the inside. A wave cannot continue to travel faster than the speed of light so it must radiate away some energy. The effective absorption coefficient, α_β, due to bending is (ref. 9, Section 2-3.1)

$$\alpha_B = Ce^{-R/R_c} \tag{2-46}$$

where C and R_c are constants characteristic of the material.

Other sources of mechanical losses occur at a joint. They are due to a displacement of the axes, a physical separation of the two ends, a rotation of the axes, and surface roughness at the joint. These effects are illustrated in Fig. 2-22.

When fibers are displaced from each other, the primary loss is the result of a ray missing the displaced fiber as it crosses the interface. When the displacement/diameter, d/D, ratio is 0.1, the loss is 1 dB. The loss that occurs when the fibers are separated is primarily due to the rays transmitted from the first fiber missing the second fiber and the rays reflected from the second fiber missing the first. Since θ_c is larger in large NA fibers, the loss for a given separation is larger. The loss for a fiber with NA = 0.5 is 3 to 4 times larger than it is for a fiber with NA = 0.15. The loss mechanism for misoriented fibers is similar to that for separated fibers. For a misorientation of 5° the loss in a fiber with NA = 0.15 is 1 dB whereas it is 1.5 dB for NA = 0.5. It is important for the fiber end to be smooth and perpendicular to the axis. This reduces the loss created by the highly refracting rays.

EXAMPLE 2-9

Determine the reflection loss in decibels for light reflected at normal incidence from an air-fiber interface if $\epsilon' = 2.25\ \epsilon_0$ and $\epsilon'' = 0$.

$$n = \sqrt{\epsilon/\epsilon_0} = \sqrt{2.25} = 1.50$$

$$R = r^2 = \left(\frac{n-1}{n+1}\right)^2 = \left(\frac{1.5-1}{1.5+1}\right)^2 = 0.04$$

The power lost by reflection is therefore RP_0 so that the input power

$$P = P_0 - RP_0 = P_0(1 - R)$$

$$\text{loss} = 10 \log \frac{P_0}{P} = 10 \log\left(\frac{1}{1-R}\right) = 10 \log\left(\frac{1}{1-0.04}\right) = 0.177 \text{ dB}$$

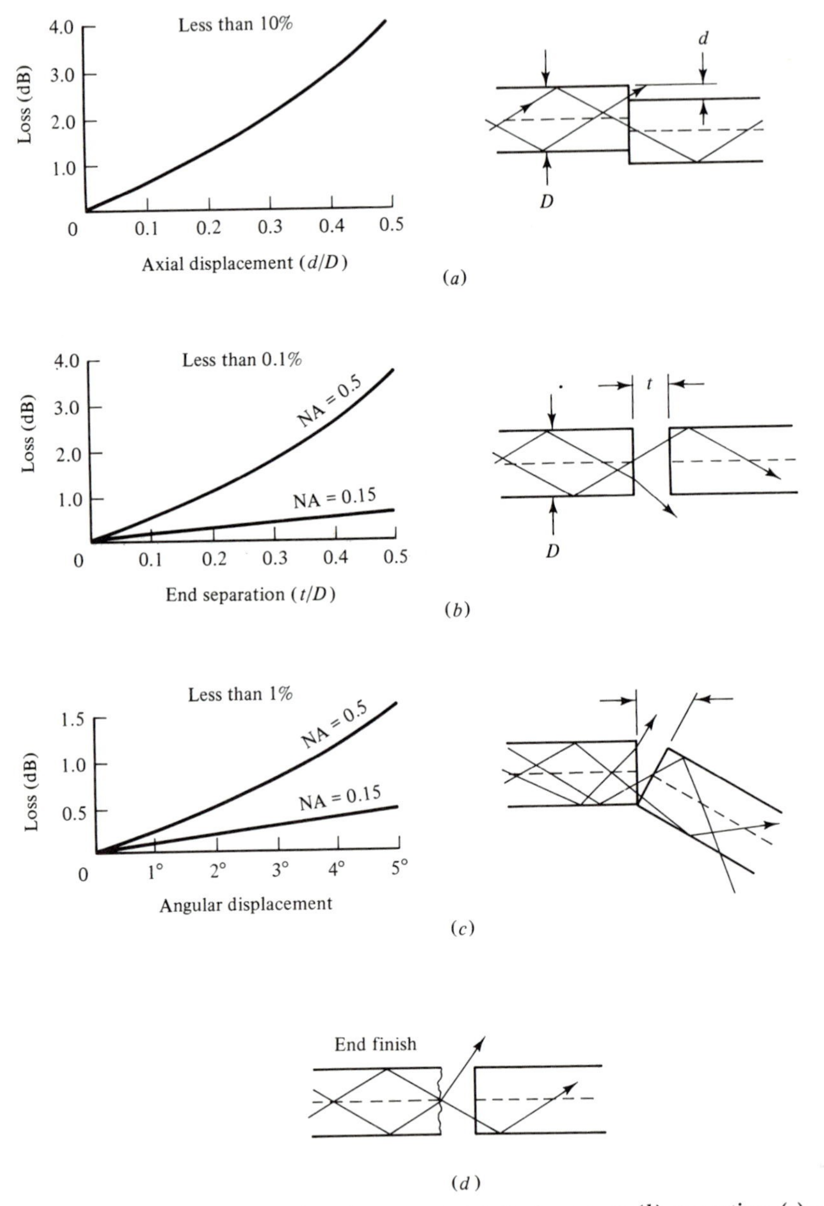

Figure 2-22 Losses at fiber junctions due to (*a*) displacement, (*b*) separation, (*c*) rotation, and (*d*) a rough surface.

2-6 FIBER CHOICE

The fiber choice is determined by both its physical and optical properties. Single mode fibers have cores with diameters of ~10 μm whereas their cladding diameters are ~100 μm so that they can be handled. Step and graded index multimode fibers have core diameters of ~50 μm and cladding diameters of ~125 μm. One of the factors that limits the use of single mode fibers is their small diameter, which makes it difficult to couple light into them and to join them.

The advantage of using single mode fibers is that they can transmit at the highest data rate. They, however, require a laser light source to couple enough energy into them, and they are more expensive than step index multimode fibers.

Graded index fibers have the advantage of having less modal dispersion than the step index multimode fibers. When the wavelength dispersion is reduced by using a laser source, the data rate in graded index fibers can be as much as 300 times larger than it is in step index fibers with the same dimensions. However, graded index fibers are the most expensive to manufacture, and only one-half as much power is coupled into them as is coupled into step index fibers with the same dimensions.

Step index multimode fibers are used when high data rates over long distances are not required, or when there is a desire to maximize the power transmission.

As an example, let us consider the case where we want to lay a long cable and wish to minimize the number of repeater stations. If the maximum allowable attenuation is 50 dB, then the maximum distance between repeater stations, L_{max}, is L_{max} = 50/loss per km. If the fiber loss is 2 dB/km, L_{max} = 25 km. If we require that the data be at least 20 Mbits/sec, then we will have to choose a fiber that has a bandwidth of at least 500 MHz/km.

2-7 FABRICATION AND JOINING

2-7.1 Fabrication

The primary fiber material, SiO_2, is formed by mixing the vapor of a silicon chloride, such as $SiCl_4$, or silicon oxychlorides, such as $SiOCl_2$, with oxygen at elevated temperatures. Chlorides are used instead of pure silicon because they have a much higher vapor pressure. The index of pure SiO_2, which is 1.45, can be increased by substituting germanium, which has more electrons than silicon, for silicon. It can be decreased by substituting the lighter element, boron.

In the CVD fabrication method a blank is made by depositing a SiO_2 soot on the inside of a SiO_2 tube called a bait. (The bait can also be a rod with the SiO_2 deposited on the surface.) The soot is deposited uniformly by moving the heating unit, which is at a temperature of 1600 to 2000° C, back and forth at a uniform velocity, as is shown in Fig. 2-23*a*. After the soot has been deposited, the glass unit is collapsed into a solid unit by heating it to ~2000° C. If a step index fiber is to be pulled from the blank, the outside layer has a uniform index of n_2 and the inner layer, which contains more germanium, will have a uniform index, n_1. If a graded index fiber is to be pulled from the blank, each layer of soot deposited will have a slightly higher index. This layered structure will be given a homogenizing anneal so that n will vary continuously. The annealing procedure is closely controlled to yield the desired profile parameter, γ, with

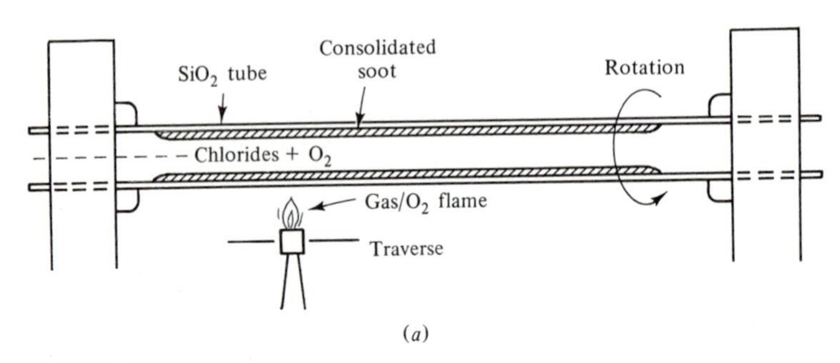

(*a*)

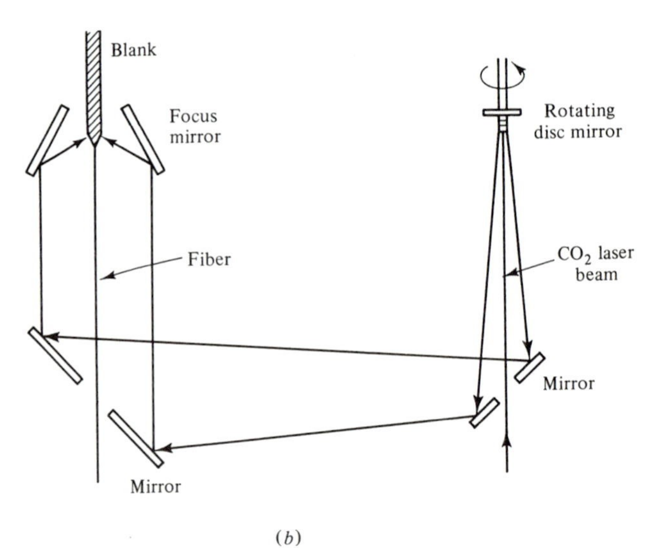

(*b*)

Figure 2-23 (*a*) Schematic of the chloride CVD system used to form the silica blank from which optical fibers are pulled. (*b*) An optical system used to heat the blank and pull a fiber from it.

great accuracy. Recall that the amount of modal dispersion is a very sensitive function of γ.

Fibers are formed by pulling them from the blank. One type of pulling assembly, which uses a CO_2 laser, is illustrated in Fig. 2-23*b*. The laser beam is incident on a rotating eccentric mirror that directs the light to the tip of the blank where it is absorbed and heats the area up to above the softening point. The thickness of the pulled fiber can be monitored optically, and this unit can be used as a feedback system to the power unit of the laser.

Another method is the double crucible method illustrated in Fig. 2-24. This method is used to make step index fibers with the material in the inside crucible forming the core and the material in the outside crucible forming the cladding. They are concentrically pulled through a small opening in the bottom of the two containers.

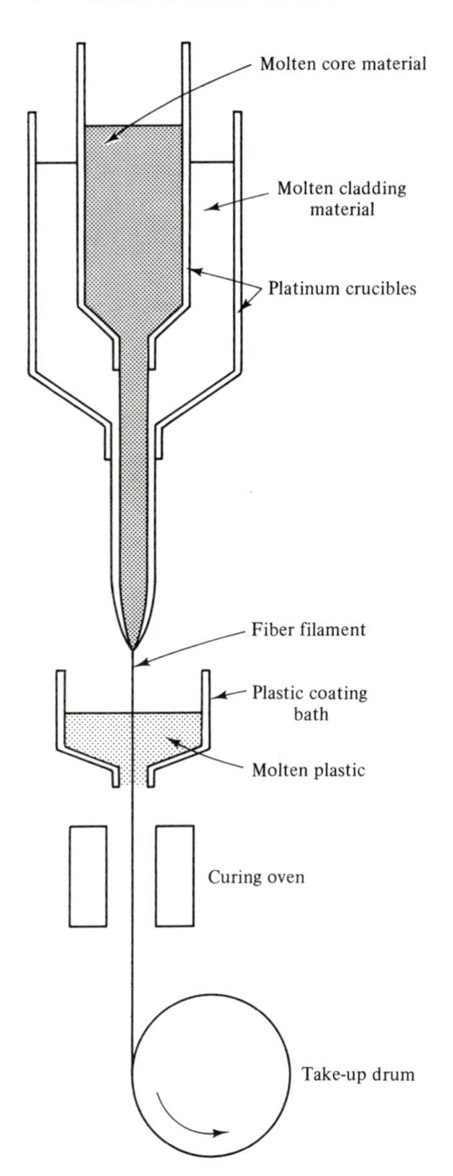

Figure 2-24 Schematic diagram of the double crucible method used to pull a fiber.

The fibers (see Fig. 2-25) are coated almost immediately after they are pulled to protect them from the environment, which can cause microcracks to form. They are first coated with a silicone oil and then are coated with an organic material. A number of outside coatings are applied by solution or extrusion techniques at a rate of 0.1 to 10 m/sec. They are promptly cured. These coatings must have a high softening temperature, be a barrier to water, act as a flame retardant, be stable for a long period of time, be strippable from the fiber, and have no detrimental effects on the fiber's optical properties.

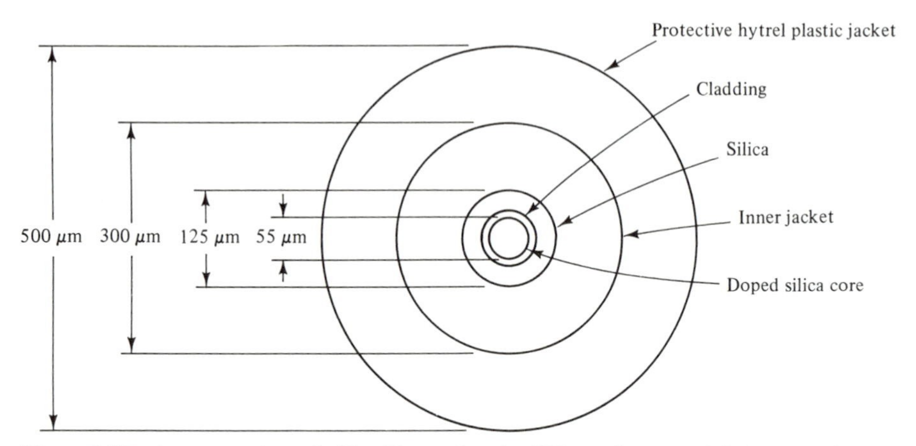

Figure 2-25 A cross section of a fiber illustrating the different layers and their approximate dimensions.

2-7.2 Joining

A permanent joint is made by splicing two fibers together. This is a much more complicated process than soldering two copper wires together; it is one of the more unattractive features of fiber-optic systems.

Before a splice is made, the ends should be carefully prepared. A clean cross section perpendicular to the fiber axis can be obtained by scoring the surface with a diamond scribe and then cleaving it at the scratch. The surface should then be polished with a fine powder, such as small grit Al_2O_3 powder.

The best splice is a fusion splice. The two ends are aligned, usually with the aid of a microscope, and then are melted together using an electric arc. The loss at a fused splice is typically 0.4 dB, but it can be as low as 0.2 dB.

Fused splices require heavy, sophisticated apparatus that cannot normally be used in the field. Thus, simpler, but more lossy, mechanical splices must be made. One type of splice is the loose tube splice shown in Fig. 2-26*a.* A tube a little larger than the fibers is filled with index matching epoxy and allowed to set after the fibers have been aligned. The precision pin splice, illustrated in Fig. 2-26*b,* is composed of three steel pins held together by a heat shrink tube. The fibers are aligned in the opening between the three fibers, heat is applied to the heat shrink tube to hold the pins in place, and an index matching epoxy is applied. When two cables containing a number of individual fibers are to be spliced, a groove alignment splice like that shown in Fig. 2-26*c* is often made. Grooves, which have been etched in a silicon chip, act as guides for the fibers. When the fibers are aligned in the grooves, a foam-rubber-coated plate is clamped on and index matching epoxy is applied.

There are a number of applications where fibers have to be repeatedly connected and separated. The connections are made with units appropriately called connectors. The loss at the connectors is typically 0.5 to 2 dB. This is larger than it is for splices. The primary reason for this is that there is always an

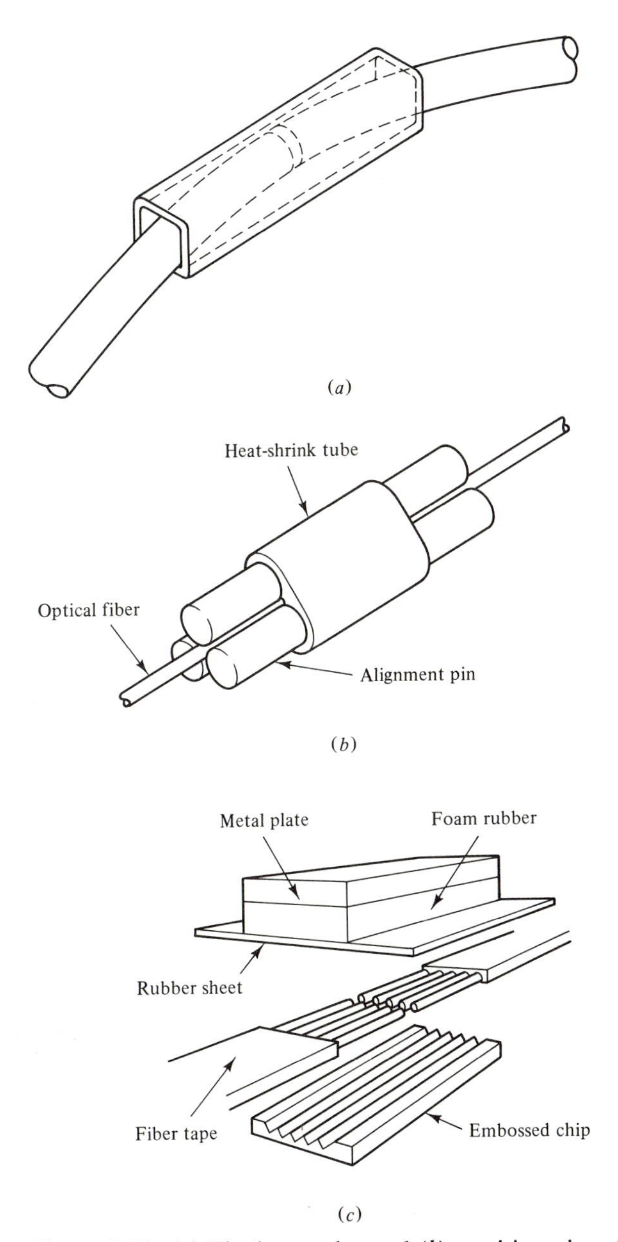

Figure 2-26 (*a*) The loose tube, and (*b*) precision pin splices used for splicing single fibers. (*c*) The groove splice used to splice fiber cables.

air gap between the fibers to keep them from scratching each other; hence, there will be some separation loss as well as reflection loss.

The simplest type of connector is the tube connector shown in Fig. 2-27*a*. It is composed of a metal jack and plug held together by a metal coupling (not

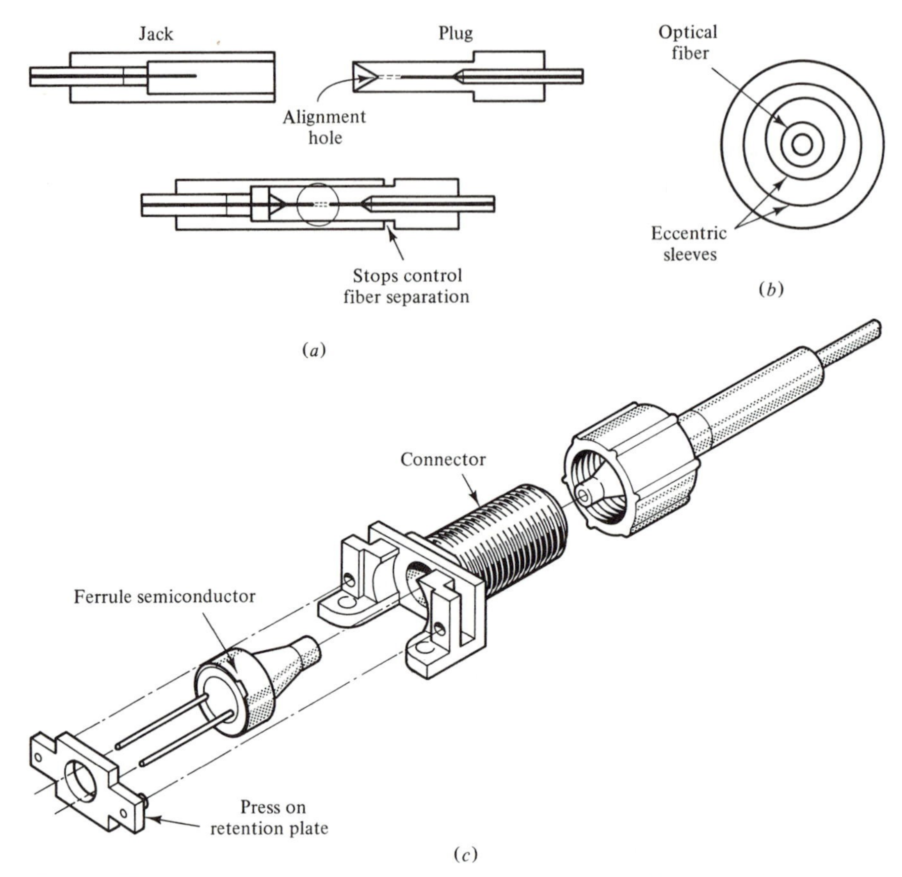

Figure 2-27 The (*a*) tube, (*b*) double eccentric, and (*c*) semiconductor ferrule connectors.

shown in the figure). The fiber in the jack is inserted into the tapered hole in the plug. The alignment is often not as precise as one would like because the tapered hole has to be able to accommodate fibers of slightly different sizes.

This problem can be partly overcome by using the double eccentric connector shown in Fig. 2-27*b*. After the fiber has been inserted, the connector is rotated about its axis until the signal through the connection is maximized. Then, the connectors are locked into place with a sleeve. However, this does require the use of test equipment to optimize the connection, and it can be time consuming.

Another type of connector is the resilient ferrule shown in Fig. 2-27*c*. The alignment is achieved by pressing the tapered ferrules into a tapered bushing. When the ferrule contains an emitter or detector, the ferrule is held in place with a press plate. When two optical fibers are coupled, a threaded cap is used.

A connector used to connect cables is the silicon chip multiple-fiber connector which is illustrated in Fig. 2-28. Grooves are etched in the inside and the outside of the positive chip, and fibers are placed in the grooves of two matched positive plates. The grooves on the outside of the positive plates are mated with

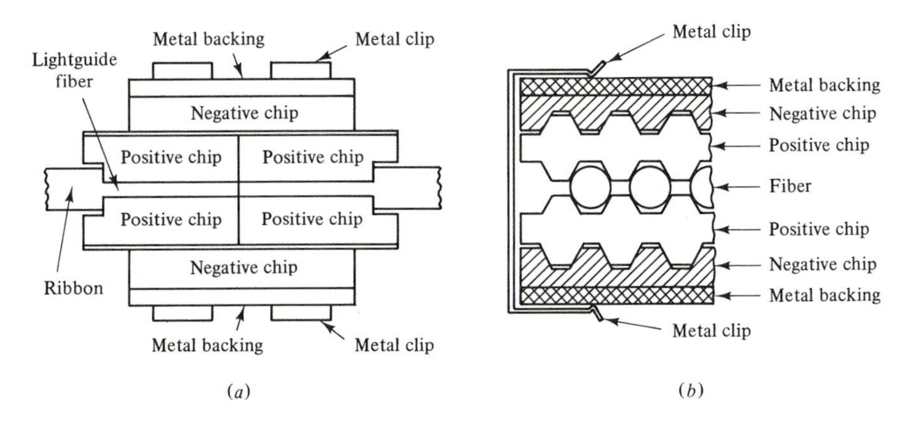

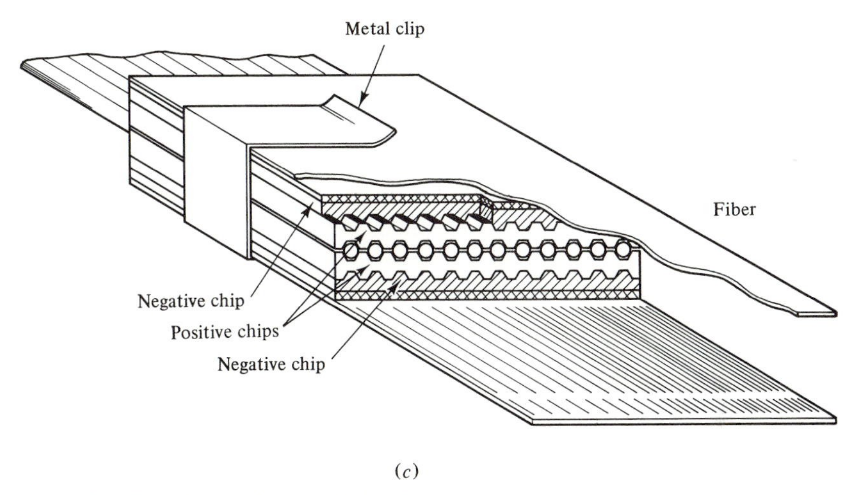

Figure 2-28 The silicon chip multiple fiber connector.

the grooves of the negative plates, which have a metal backing on them. This structure is held together with a metal clip.

REFERENCE LIST

1. J. S. Cook, "Communication by Optical Fiber," *Scientific American,* November 1973, pp. 28–34.
2. W. S. Boyle, "Light-Wave Communication," *Scientific American,* August 1977, pp. 40–48.
3. E. A. Lacy, *Fiber Optics.* Englewood Cliffs, NJ: Prentice-Hall, 1982, Chapters 1, 4, and 5.
4. J. Wilson, and J. F. B. Hawkes, *Optoelectronics: An Introduction.* Englewood Cliffs, NJ, Prentice-Hall, 1983, Chapter 8.
5. N. S. Kapany, and J. J. Burke, *Optical Waveguides.* New York: Academic Press, 1972.

6. R. D. Maurer, "Glass Fibers for Optical Communication," Proc. IEEE **61,** 452–63 (1973).
7. D. Gioge and E. A. J. Marcatili, "Multimode Theory of Graded Core Fibers," *Bell Syst. Tech. J.* **52,** 1563–1578 (1973).
8. J. E. Midwinter, *Optical Fibers for Transmission.* New York: Wiley-Interscience, 1979, Chapters 5 and 8.
9. M. K. Barnoski (Ed.), *Fundamentals of Optical Fiber Communications.* New York: Academic Press, 1976, Chapter 2.
10. C. K. Kao, *Optical Fiber Systems.* New York: McGraw-Hill, 1982.
11. A. H. Cherin, *An Introduction to Optical Fibers.* New York: McGraw-Hill, 1983.
12. N. H. Frank, *Introduction to Electricity and Optics,* 2d Ed. New York: McGraw-Hill, 1950, Chapters 11 and 17.
13. L. H. Van Vlack, *Materials Science for Engineers.* Reading, MA: Addison-Wesley, 1970, Chapter 13.
14. M. Born and E. Wolf, *Principles of Optics,* 6th Ed. Oxford: Pergamon Press, 1980, Chapter 1.
15. D. Marcuse, *Theory of Dielectric Optical Waveguides.* New York: Academic Press, 1974.

PROBLEMS

2-1. **(a)** What are the minimum and maximum values of $\psi_\perp$ if $i = 60°$? If $i = 75°$?
(b) What are the corresponding maximum and minimum values of $\psi_\parallel$?

2-2. **(a)** Find i_c for $n_1 = 1.50$ and $n_2 = 1.45$.
(b) Make a plot of $\psi_\perp$ versus i for $1_c \leq i \leq 90°$.
Is there a local maximum or minimum? Show this mathematically as well as in your plot.
(c) Repeat parts a and b for $n_1 = 1.50$ and $n_2 = 1.40$.

2-3. On the same graph as in Problem 2-2 make a plot of the penetration depth versus i for $n_1 = 1.50$ and $n_2 = 1.45$ and for $n_1 = 1.50$ and $n_2 = 1.40$. Assume $\lambda_0 = 1\ \mu m$.

2-4. Prove that

$$\tan \psi_\parallel = \left(\frac{n_1}{n_2}\right)^2 \tan \psi_\perp$$

2-5. For $n_1 = 1.50$ and $n_2 = 1.45$ make a plot of $\psi_\parallel - \psi_\perp$ versus i for $i_c \leq i \leq 90°$. Is there a local maximum or minimum? Show this mathematically as well as in your plot.

2-6. Consider a planar dielectric for which $n_1 = 1.50$ and $n_2 = 1.48$.
(a) What is the maximum value of $2\psi_\perp$?
(b) Estimate the maximum number of modes in the wave guide if $d = 50\ \mu m$ and $\lambda_0 = 1.0\ \mu m$.

2-7. Make a plot of $\psi_\perp$ versus n_2/n_1 for $i = 60°$ and $i = 75°$.

2-8. On the same graph as in Problem 2-7 make a plot of the penetration depth versus n_2/n_1 for $i = 60°$ and $75°$. Assume $\lambda_0 = 1.0\ \mu m$ and $n_1 = 1.50$.

2-9. Light of wavelength $0.85\ \mu m$ traveling in a medium of refractive index 1.50 is incident upon an interface with another medium of refractive index 1.48 at an angle of 82°. Calculate the relative amounts of power that penetrate distances of:

(a) 5 μm.
(b) 50 μm into the second medium.

2-10. Using ray theory, make a plot of the maximum data rate/km versus n_2/n_1 for a step index fiber.

2-11. **(a)** Make a plot of the standing wave pattern in a planar dielectric for $m = 3$, $\phi = 82°$, $n_1 = 1.50$, and $n_2 = 1.48$.
(b) On the same graph, plot the pattern when n_2 is changed to 1.45.

2-12. Under conditions of standard temperature and pressure (STP) the dielectric constant of neon is 1.000106.
(a) Find the polarizability and the radius of the atom.
(b) If an electric field of 1900 V/cm is applied, what is the polarization and the dipole moment of the atom?
(c) What would be the effects of increasing the pressure by an order of magnitude?

2-13. **(a)** Show that for hydrogen gas

$$n^2 = 1 + \frac{Nq^2}{m\epsilon_0(\omega_0^2 - \omega^2)}$$

and that

$$n - 1 \approx \frac{Nq^2}{2m\epsilon_0(\omega_0^2 - \omega^2)}$$

when $\omega_0 \gg \omega$. N is the number of atoms per unit volume.
(b) The observed values of $n - 1$ for hydrogen at a number of wavelengths are as follows:

λ, Å	5460	4080	3340	2890	2540	2300	1900
$(n - 1) \times 10^7$	1400	1426	1461	1499	1547	1594	1718

Construct a plot of $1/(n - 1)$ versus ω^2, and from two points on this curve compute the values of N and ω_0. How does this value of N compare with the number of hydrogen molecules per unit volume under standard conditions?

2-14. Blue light (λ = 470 nm) and red light (λ = 660 nm) strike the surface of a 60, 60, 60 triangular prism at an angle of 30°. Using the indices of refraction in Table 2-1, find the exit angle of each wave if the prism material is:
(a) Crown glass.
(b) Diamond.

2-15. The refractive indices of a piece of optical glass for the blue and green lines of the mercury spectrum, λ = 4358 Å and λ = 5461 Å, are 1.65250 and 1.62450, respectively. Using the two-constant Cauchy equation, calculate values for:
(a) The constants A and B.
(b) The refractive index for the sodium yellow lines at λ = 5898 Å.
(c) The dispersion at this same wavelength.

2-16. **(a)** Compute the values of A, B, and C for Cauchy's formula for crown glass using the magnitude for n at 410, 550, and 660 nm in Table 2-1.
(b) Compute n at 470, 580, and 610 nm and compare them with the values tabulated in Table 2-1.
(c) Compute the pulse spread and DR_{max} using ray theory for an LED source emitting at 900 nm if its bandwidth is 500 Å.
(d) Repeat part c for a laser emitting at 900 nm with a bandwidth of 20 Å.

2-17. **(a)** Using the values of A, B, and C calculated for crown glass in Problem 2-16, compute the pulse spread and the maximum data rate using Eq. 2-33b for the LED described in the previous problem.
(b) Repeat part a for the laser described in the previous problem.

2-18. By using the values of A, B, and C calculated for crown glass in Problem 2-16, show that Eq. 2-33b is a good approximation to Eq. 2-33a. Make your computations for $\lambda = 1.0\ \mu$m.

2-19. A laser operating at 0.9 μm with a linewidth of 2 nm is used in conjunction with a single mode silica-based fiber. What is the material dispersion if the fiber is made from crown glass?

2-20. It has been suggested that if a material could be found where the onset of the infrared lattice absorption bands is at a higher wavelength than in glass/silica fibers, then much lower ultimate fiber losses could be obtained. Estimate from Fig. 2-19*a* what the minimum absorption is likely to be at 4 μm assuming that Rayleigh scattering is the dominant loss mechanism.

2-21. The index of refraction of the core of a step index optical fiber is 1.5238 at $\lambda =$ 4000 Å, 1.5085 at 6000 Å, and 1.5040 at 7500 Å.
(a) What is the maximum data rate per km when a LED with an emission peak at 0.82 μm with a peak width of 350 Å is used as the light source?
(b) What is the maximum data rate per km when a laser with an emission peak centered at 0.82 μm with a width of 20 Å is used as the light source?
(c) Repeat parts a and b when the LED and laser peaks are centered at 1.06 μm. Use the formula

$$n = A + \frac{B}{\lambda^2} + \frac{C}{\lambda^4}$$

assume that

$$\frac{n_1 - n_2}{n_1} = 0.01$$

and the maximum data rate is given by $1/2\ \Delta t$.

2-22. Explain why the index of refraction in the X-ray region of the spectrum is independent of the wavelength. Use both physical and mathematical arguments.

2-23. Explain why the afternoon sky is blue and the evening sky is red.

2-24. At low concentrations the absorption coefficient is proportional to the concentration.
(a) How much is the attenuation increased in decibels at 0.95 μm if the OH^- concentration is increased from 1.25 ppm to 2.50 ppm?
(b) How much is the attenuation at 1.39 μm increased?

2-25. A hollow glass tube 35.0 cm long with end windows contains tiny smoke particles that produce Rayleigh scattering. Under these conditions it transmits 65.0 percent of the light. After precipitation of the smoke particles it transmits 88.0 percent of the light. Calculate the value of:
(a) The scattering coefficient.
(b) The absorption coefficient.

2-26. A solid plastic rod 60.0 cm long transmits 85.0 percent of the light entering it at one end. When it is subjected to a strong beam of radiation, tiny particles are produced in it that give rise to Rayleigh scattering. Under these modified conditions the rod transmits 55.0 percent of the light. Calculate:
(a) The absorption coefficient.
(b) The scattering coefficient.

2-27. A certain plastic rod 40.0 cm long has an absorption coefficient of 0.00249 cm^{-1}. If 50.0 percent of the light entering the end of this tube is transmitted, find
(a) The scattering coefficient.
(b) The total loss coefficient.

2-28. Make a plot of the number of allowed modes in a step index fiber versus n_2 if $n_1 = 1.50$, $\lambda_0 = 1.0$ μm, and $a = 2.5$ μm. Consider values of $1.5 < n_2 < 1.4$.

2-29. Make a plot of the maximum allowed radius for a single mode fiber versus n_2 if $n_1 = 1.50$ and $\lambda_0 = 1.0$ μm. Consider values of $1.5 < n_2 < 1.4$.

2-30. **(a)** Find NA, θ_c, and the fraction of the light inserted into the fiber, F, if $n_0 = 1.0$, $n_1 = 1.50$, and $n_2 = 1.40$, and the emission is Lambertian.
(b) Repeat part a if the fiber is placed in water so that $n_0 = 1.33$.

2-31. Repeat the preceding problem if the angular dependence of the emission is given by $I = I_0 (\cos \theta)^m$. Consider the cases $m = 2$, 4, and 8.

2-32. A point source of light emits uniformly over a hemisphere.
(a) What fraction of the light can be coupled into the fiber if the source abuts the fiber along the axis? For the fiber $n_1 = 1.50$ and $n_2 = 1.48$.
(b) A sphere with index of refraction, n, and a radius equal to the fiber radius is inserted between the light source and the fiber touching each along a diameter. If the ray makes an angle ω with the axis on the input side, find an expression for the deviation of the ray after it exits the sphere. Express your answer in terms of ω, and the transmitted angle, t.
(c) Find an expression for θ_c in terms of n and ω_{max} where ω_{max} is the maximum input angle into the sphere that will be captured by the fiber.
(d) Find an expression for the ratio of the fraction of the light captured with and without the sphere.

2-33. Explain physically why the maximum number of modes in a planar dielectric is proportional to the dielectric width, d, whereas it is proportional to the square of the radius, a^2, of a fiber.

2-34. A step index fiber has a numerical aperture of 0.16, a core refractive index of 1.450, and a core diameter of 90 μm. Calculate
(a) The acceptance angle, θ_c, of the fiber.
(b) The refractive index of the cladding.
(c) The approximate maximum number of modes with a wavelength of 0.9 μm that the fiber can carry.

2-35. Estimate the intermodal dispersion expected for the fiber in Problem 2-34.

2-36. Calculate the maximum core diameter needed for a circular dielectric wave guide (refractive index 1.50) to support a single mode of radiation with wavelength 1 μm when
(a) The wave guide has no cladding and
(b) The wave guide is of the step index type with $n_2 = 1.48$. Comment on the values obtained from the point of view of manufacturing single mode wave guides.

2-37. A LED source emits 1 mW from an area 100 μm in diameter. It is butt joined to a step index fiber with a core diameter of 50 μm and a core refractive index of 1.5. The cladding has a refractive index of 1.48. Assuming the LED source is a Lambertian emitter, estimate the energy coupled into the fiber.

2-38. The much narrower beam divergence of lasers enables them to couple substantially more power into a fiber than can LEDs. A convenient expression for their surface illuminance as a function of angle is given by $B(\theta) = I_0 \cos^n \theta$. A Lambertian source is characterized by $n = 1$, whereas lasers may be characterized by $n \sim 20$. Using this expression, calculate the coupling efficiency into a fiber as a function of n assuming NA $\ll 1$.

2-39. From mode theory, the phase velocity of the $1m$ mode, c_{1m}, is

$$c_{1m} = \frac{c_0 V}{(u_{1m}^2 n_2^2 + w_{1m} n_1^2)^{1/2}}$$

where u_{1m} and w_{1m} characterize the field in the cladding and the core, respectively, and are related to each other by the equation:

$$u_{1m}^2 + w_{1m}^2 = v^2$$

If

$$u_{01} = \frac{(1 + \sqrt{2})V}{1 + (4 + V^4)^{1/4}}$$

Find c_{01} if $n_1 = 1.50$, $n_2 = 1.48$, $\lambda_0 = 1.0\ \mu$m, and $a = 25\ \mu$m, and compare it with the ray velocity of the axial ray. Also consider the case $a = 10\ \mu$m and $a = 50\ \mu$m. Comment on the accuracy of the ray theory as the fiber radius decreases.

2-40. If all of the light reflected at normal incidence from a single fiber-air interface is lost, what is the attenuation in decibels if $n = 1.50$?

2-41. **(a)** If all of the rays cross the fiber axis at the interface as is shown in the figure, what is the maximum separation, S_{max}, in terms of the core diameter, D, between the two fibers for which all of the rays will be captured if $n_1 = 1.50$ and $n_2 = 1.48$?

(b) If $n_1 = 1.50$ and $n_2 = 1.40$?

(c) Answer parts a and b if the two fiber interfaces are parallel and make an angle $\theta \neq 90°$ with the axis. Find S_{max} if $\theta = 60°$.

(d) Answer parts a and b if the interface on the left-hand side makes an angle of θ with the axis and the interface on the right-hand side is normal to the axis. Find S_{max} if $\theta = 60°$.

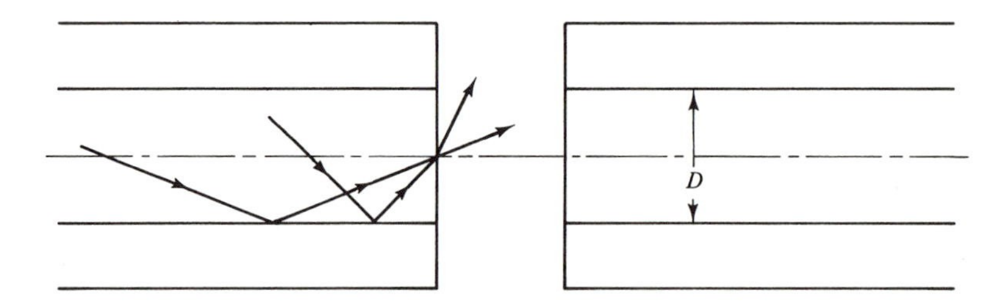

2-42. **(a)** If all the rays cross the fiber axis at the interface as is shown in Fig. 2-29 in Problem 2-41, what is the maximum, ρ_{max}, rotation between the two fibers for which all of the rays will be captured if $n_1 = 1.50$ and $n_2 = 1.48$?

(b) If $n_1 = 1.50$ and $n_2 = 1.40$?

(c) Answer parts a and b if the two fiber interfaces are parallel and make an angle $\theta \neq 90°$ with the axis. Find ρ_{max} for $\theta = 60°$.

(d) Answer parts a and b if the interface on the left-hand side makes an angle of θ with the axis and the interface on the right-hand side is normal to the axis. Find ρ_{max} for $\theta = 60°$.

2-43. **(a)** If light reflected from the second interface of two separated fibers is lost only when the reflected ray passes outside the diameter of the first fiber, what is the maximum fiber separation, S_{max}, in terms of the fiber core diameter, D, for which there is no loss if all of the rays cross the axis at the interface as is shown in the figure for Problem 2-41 if $n_1 = 1.50$ and $n_2 = 1.48$?

(b) If $n_1 = 1.50$ and $n_2 = 1.40$?

(c) Answer parts a and b if the two fiber interfaces are parallel and make an angle $\theta \neq 90°$ with the axis. Find S_{max} for $\theta = 60°$.

(d) Answer parts a and b if the interface on the left-hand side makes an angle of θ with the axis, and the interface on the right-hand side is normal to the axis. Find S_{max} for $\theta = 60°$.

2-44. Consider the two planar dielectrics shown in the figure and assume the probability that the ray exits the dielectric between x and $x + \Delta x$ is equal to $\Delta x/D$. If the emission from the light source is Lambertian, find an expression for the fraction of the energy lost and the attenuation expressed in terms of the core fiber diameter, D, and the separation distance, S, if $n_1 = 1.50$, $n_2 = 1.48$, and $S < D/\tan \theta_c$.

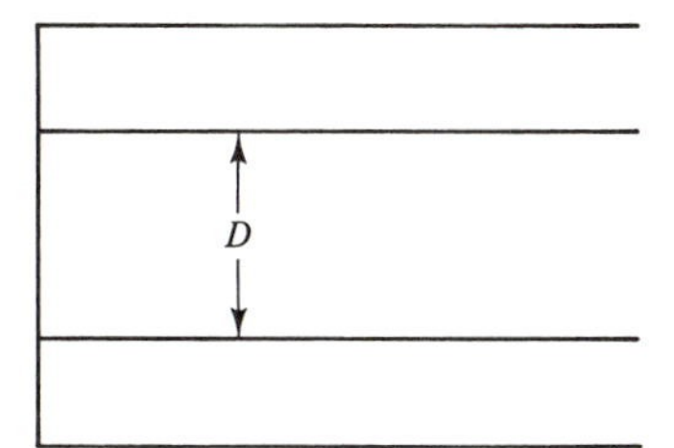

2-45. Consider the two planar dielectrics described in Problem 2-44.

(a) If the emission from the light source is Lambertian, find an expression for the fraction of the energy lost and the attenuation in terms of the core fiber diameter, D, and the displacement, d, if $n_1 = 1.50$ and $n_2 = 1.48$.

(b) If $n_1 = 1.50$ and $n_2 = 1.40$.

2-46. Consider the two planar dielectrics described in Problem 2-44.

(a) If the emission from the light source is Lambertian, find an expression for the fraction of the energy lost and the attenuation expressed in terms of the angle of rotation of one fiber relative to the other if $n_1 = 1.50$ and $n_2 = 1.48$.

(b) If $n_1 = 1.50$ and $n_2 = 1.40$. Assume the angle of rotation is less than $\pi/2 - \theta_c$.

2-47. Consider a wave guide bent into an arc of a circle of radius, R. Determine the point in the cladding at which the phase velocity of a guided mode equals the phase velocity of a plane wave in the cladding. Assume that the mode phase velocity is given by c/n_1. Hence, show that the rate of energy loss is proportional to $\exp(-R/R_0)$ where R_0 is a constant. Assume that the mode electric field intensity declines exponentially with distance into the cladding.

Chapter 3

Optical Spectra of Atoms, Molecules, and Solids

3-1 INTRODUCTION

Most introductory textbooks on electromagnetism show that radiation is generated by an oscillating dipole in an antenna. The frequency of the radiation emitted by these antennas is almost always less than a gigahertz, which is much much less than the 10^{14}–10^{15} Hz used in optical communication systems.

The light for these systems is generated by microscopic dipoles composed of ions or electrons or both. However, the behavior of these atomic dipoles cannot be adequately described by using only classical concepts and, therefore, some quantum mechanical concepts must be introduced. Also, instead of describing the motion of the charged particle, it is more convenient to determine the initial and final energies of the dipole, and then from the conservation of energy calculate the frequency of the emitted radiation.

The electronic energy levels in atoms determine their optical spectra. A characteristic of these spectra is that they are discrete; that is, the spectra are composed of a number of emission or absorption peaks with extremely narrow bandwidths. The peak frequencies for the hydrogen atom can be determined in a straightforward manner, but it is a much more complicated process for any other atom.

Molecular spectra are more complicated because, in addition to a more complex electronic structure, they have a vibrational and a rotational spectrum. A diatomic molecule can be modeled as two masses connected by a spring and vibrating at its natural frequency, or as a rigid rotator rotating at specified frequencies. As in an antenna, the frequencies of the emitted photons are equal to the vibrational and rotational frequencies.

Doping semiconductors and insulators with small amounts of impurities can profoundly change their optical and electrical properties. The impurity atoms do this by creating atomiclike states in the energy gap. Some specific examples are chromium in Al_2O_3 to form ruby, phosphorus in silicon to form *n*-type silicon, and boron in silicon to form *p*-type silicon.

3-2 THE HYDROGEN ATOM

3-2.1 Bohr Theory of the Hydrogen Atom

The atomic spectra of atoms, in general, and the hydrogen atom, in particular, led to the development of quantum mechanics because they could not be explained theoretically by using classical concepts. An atomic spectrum is the electromagnetic radiation emitted from an atom when energy is added to it by, for example, heating or passing an electrical current through its vapor. What puzzled the early theoreticians is that radiation with only very specific energies was emitted, that is, the spectrum was discrete and not continuous. According to classical concepts, if an electron were revolving about an atomic nucleus, it would radiate energy because it is being accelerated. To conserve energy the diameter of the orbit would have to decrease and the electron would spiral into the nucleus. In so doing, it would emit a continuous spectrum of radiation.

The experimentalists noticed that the discrete energies of the emitted light could be represented by simple mathematical series. One series known as the Lyman series has the relationship

$$E = C\left(1 - \frac{1}{n_1^2}\right) \qquad n_1 = 2, 3, 4 \ldots \tag{3-1}$$

where E is the energy and C is a constant. Another series, the Balmer series, is given by the relationship

$$E = C\left(\frac{1}{4} - \frac{1}{n_2^2}\right) \qquad n_2 = 3, 4, 5 \ldots \tag{3-2}$$

and the Ritz-Paschen series is given by the equation

$$E = C\left(\frac{1}{9} - \frac{1}{n_3^2}\right) \qquad n_3 = 4, 5, 6 \ldots \tag{3-3}$$

As is shown in Fig. 3-1, these series can be created from a discrete energy level diagram in which the energy levels are given by

$$E = -\frac{C}{n^2} \tag{3-4}$$

where n is any positive integer. The energy of the electron is negative because it is bound to the atom. As is shown in the figure, the energy of a free electron at rest is defined to be zero. When an electron drops from level E_1 to E_2, by the

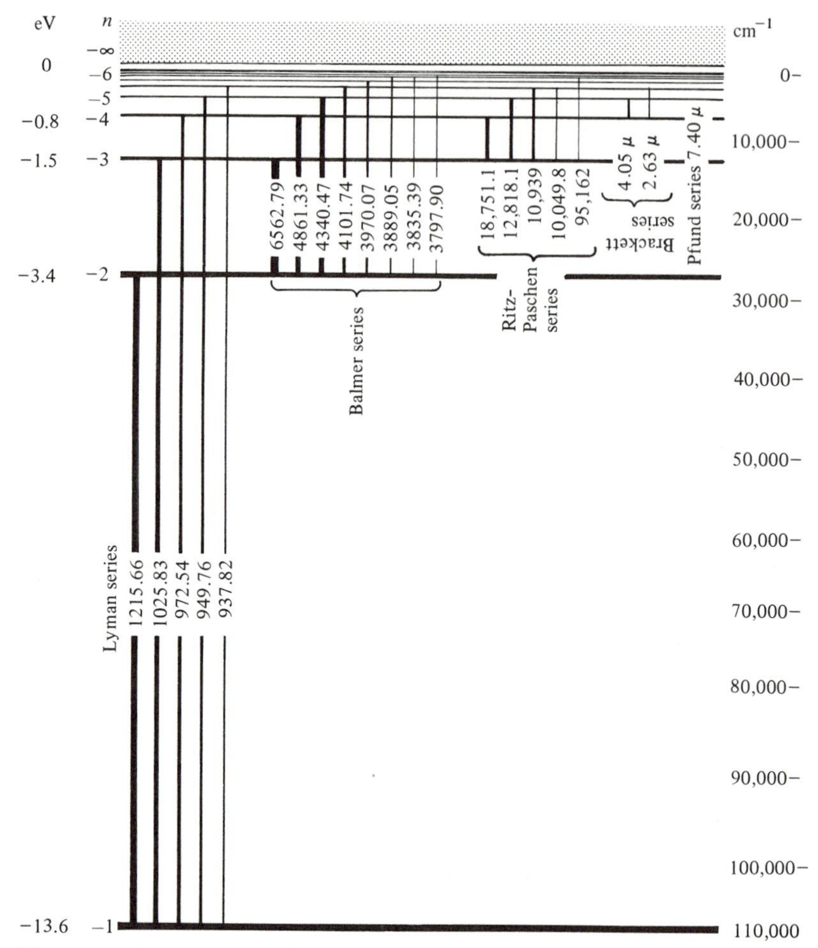

Figure 3-1 Energy levels of the hydrogen atom and the reciprocal wavelengths in cm^{-1} or microns associated with the electronic transitions.

conservation of energy the energy of the radiation is $E_1 - E_2$. It is found that the frequency of the radiation, ν, is

$$\nu = \frac{E_1 - E_2}{h} \tag{3-5}$$

where h is Planck's constant (6.625×10^{-34} J · sec).

Before we turn to Bohr's explanation of these experimental results, we must first look at the important contribution of de Broglie to wave mechanics. He hypothesized that, since it had been shown that electromagnetic waves have both wavelike and particlelike behavior, particles should also have wavelike behavior. He suggested that all particles have a wavelength, λ, associated with them and that the wavelength can be calculated from the equation

$$\lambda = \frac{h}{p} \tag{3-6}$$

where p is the particle momentum. Davisson and Germer verified this relationship in 1927 when they experimentally showed that electrons are diffracted by crystalline solids in the same manner that X rays are.

This relationship was not discovered until early in the twentieth century because all but the smallest momentum particles have wavelengths that are too short to experimentally detect. Because an electron has a very small mass ($m = 9.11 \times 10^{-31}$ kg), both its wavelike and particlelike behavior are important. This is especially true when the electron is confined to a volume that has dimensions similar to its wavelength. This is the domain where quantum mechanics is particularly important—small masses and/or small volumes.

Bohr theorized that the only stable electron orbits around the nucleus are those in which the orbit circumference is an integral number of wavelengths so that destructive interference will not occur (see Fig. 3-2). For a circular orbit this boundary condition can be expressed mathematically as

$$n\lambda = 2\pi r \tag{3-7}$$

He incorporated this equation with the classical equation for force equilibrium, the centrifugal force equals the force of attraction,

$$\frac{mv^2}{r} = \frac{kq^2}{r^2} \tag{3-8}$$

where q is the charge on an electron (1.6×10^{-19} C) and k is the constant of proportionality (8.99×10^9 N$\cdot$m^2/C^2) [k is often written as $\frac{1}{4}\pi\epsilon_0$ where ϵ_0 is the electric permittivity of free space (8.85×10^{-12} C^2/N$\cdot$m^2)]. By substituting Eqs. 3-6, 3-7, and 3-8 into the equation for the total energy of the electron, both kinetic and potential energy,

$$E = -\frac{kq^2}{r} + \tfrac{1}{2}mv^2 \tag{3-9}$$

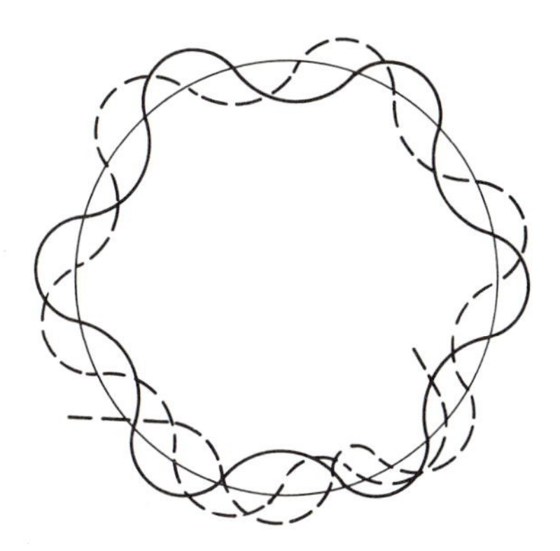

Figure 3-2 Schematic drawing of an electron wave constrained to move around the nucleus. The solid line represents a possible stationary wave. The dashed line shows how a wave of somewhat different wavelength would be destroyed by interference.

it can be shown that the radius of the nth circular electron orbit is

$$r_n = \frac{n^2\hbar^2}{mkq^2} \tag{3-10}$$

where $\hbar = h/2\pi$, and the energy of that orbit is

$$E_n = -\frac{mk^2q^4}{2\hbar^2}\frac{1}{n^2} \tag{3-11}$$

This is precisely the equation used to describe the discrete electron energy levels that were needed to describe the observed hydrogen spectrum. The value of the constant, C, is 2.17×10^{-18} J (13.6 eV) and this also agrees with the experimental results.

Spectroscopists frequently express spectra in terms of the reciprocal wavelengths. The transfer from one unit to the other can be made by using the simple relationship

$$\frac{1}{\lambda} = \frac{E}{hc} \tag{3-12}$$

which is a variation of Eq. 3-5. Again, c is the speed of light.

The energy that must be added to the electron in its ground state ($n = 1$) to remove it from the nucleus is called the ionization energy. Other energies that will be referred to are the excitation energy and the binding energy, and they are shown schematically in Fig. 3-3. The excited energy of an electron is the difference between the energy of the occupied state and the ground state, and the binding energy is the difference in the energy of a free electron at rest and the energy of the excited electron.

EXAMPLE 3-1

A hydrogen atom is in an excited state having a binding energy of 0.85 eV in this state. The atom makes a transition to a state whose excitation energy is 10.2 eV. **(a)** Calculate the wavelength of the photon emitted in this transition. **(b)** Find the value of n for each excited state.

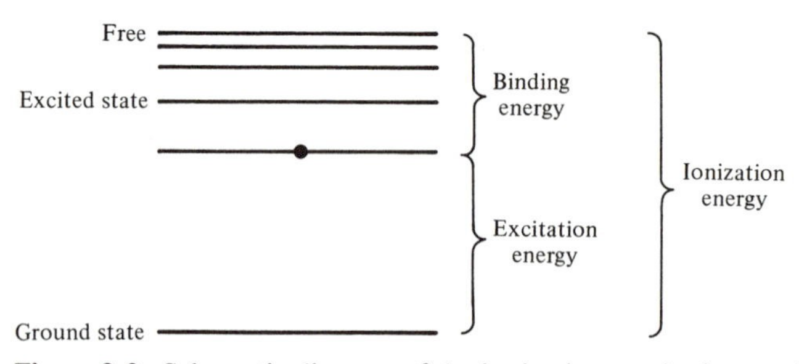

Figure 3-3 Schematic diagram of the ionization, excitation, and binding energies.

(a) $E = (13.6 - 10.2) - 0.85 = 2.55 \text{ eV}$

$$\lambda = \frac{hc}{E} = \frac{(6.63 \times 10^{-27} \text{ erg} \cdot \text{sec})(3.00 \times 10^{10} \text{ cm/sec})}{(2.55 \text{ eV})(1.60 \times 10^{-12} \text{ erg/eV})}$$

$$= 4.86 \times 10^{-5} \text{ cm} = 4860 \text{ Å}$$

(b) $$n = \left(\frac{-C}{E}\right)^{1/2} = \left[\frac{2.17 \times 10^{-18} \text{ J}}{(13.6 - 10.2)\text{eV} \cdot 1.60 \times 10^{-19} \text{ J/eV}}\right]^{1/2} = 2$$

$$n = \left[\frac{2.17 \times 10^{-18} \text{ J}}{0.85 \text{ eV} \cdot 1.60 \times 10^{-19} \text{ J/eV}}\right]^{1/2} = 4$$

3-2.2 A More Complete Description of the Hydrogen Atom

One way in which the Bohr theory is incomplete is that it considers only the energy of the electron and not the angular momentum, $\bar{L}$, associated with the orbit or the spin of the electron. A more complete quantum mechanical analysis shows that more insight into the behavior of the hydrogen atom can be obtained when these quantities are also considered.

They can be considered using the planetary model in Fig. 3-4 where elliptical orbits as well as a circular orbit are displayed. The nucleus is located at one of the two foci of the orbit. It can be shown that the energy of an electron in a specific orbit is determined only by the length of the major diameter; the ellipticity of the orbit does not affect the energy (see Problem 3-6). As was described in the previous section, the quantized circular orbits are designated by the quantum number, n, which is called the principal quantum number. The elliptical orbits with the same major axis as a specific circular orbit have the same principal quantum number.

The angular momentum, like the energy, is quantized, and it is given by the equation

$$L = [l(l + 1)]^{1/2}\hbar \tag{3-13}$$

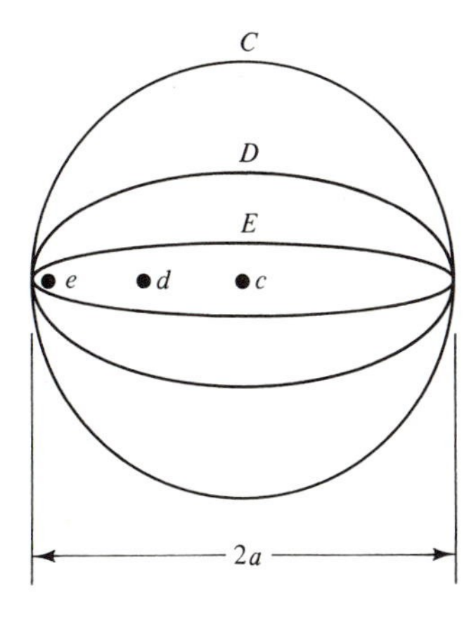

Figure 3-4 Elliptical orbits with the same major axis ($2a$); c, d, and e are the respective 1 of 2 foci for a particle moving in the orbits C, D, and E.

where l is the angular momentum quantum number and the restrictions placed on it are

$$l = 0, 1, 2 \ldots n - 1$$

Because l depends on n, there are $n - 1$ possible orbits for a given major diameter. The angular momentum vector in Fig. 3-5 passes through the nucleus and is perpendicular to the plane of the orbit. For a given orbit it is constant (not a function of time) because the torque ($d\bar{L}/dt$) through the nucleus must be zero. It can be shown that L is proportional to the area of the orbit so that the more elliptical the orbit, the smaller is the angular momentum.

The z component of the angular momentum, L_z, is also quantized, and this is shown schematically in Fig. 3-6. The z direction is defined to be the direction of the magnetic field that is used to measure the angular momentum. L_z can have only the values

$$L_z = m\hbar \tag{3-14}$$

and the restrictions placed on m are

$$m = -l, -l + 1 \ldots 0 \ldots l - 1, l$$

In terms of the Bohr model this means that the orbit can make only certain angles with the z direction.

The electron spin is also quantized, and it is designated by the spin quantum number, m_s, which can have the two values $\pm\frac{1}{2}$. If the value is $+\frac{1}{2}$ the electron is said to have spin up; if the value is $-\frac{1}{2}$ it is said to have spin down. Once again referring to the Bohr model, one can think of the spin resulting from the magnetic field formed by the rotation of the charged electron about its axis. If it rotates clockwise, the spin is up, and if it rotates counterclockwise, the spin is down. This is shown in Fig. 3-7.

Now, all the quantum states of the hydrogen atom can be described. A quantum state is defined by a set of quantum numbers, and some of these states are shown in the energy level diagram in Fig. 3-8. Each level in the figure actually has two quantum states associated with it because the two possible spins have not been taken into account.

In the ground state $n = 1$ so that l and m must be zero. Therefore, there are only 2 states that can have the ground state energy. There are 8 states that can have the energy defined by $n = 2$. For $l = 0$ there are 2 states and for $l = 1$

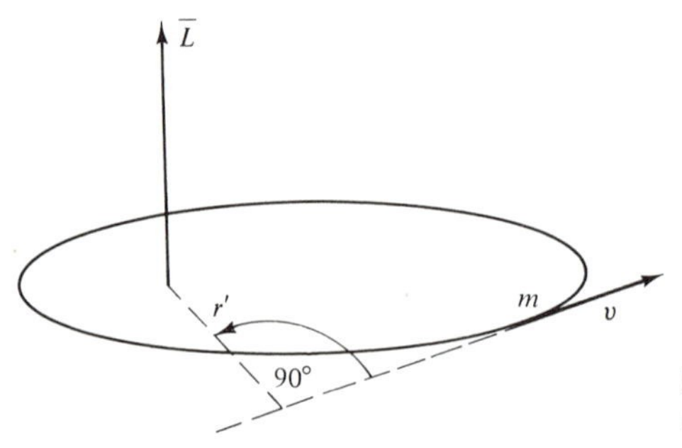

Figure 3-5 Angular momentum, $L = mvr'$, of a particle in an elliptical orbit.

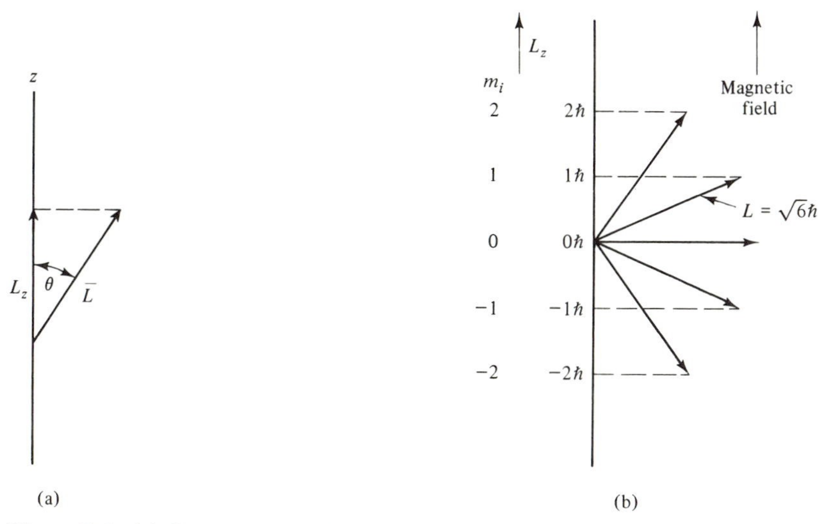

Figure 3-6 (*a*) Component of the angular momentum along the z direction. (*b*) Space quantization of the orbital angular momentum vector for $l = 2$.

there are 6 states. When $n = 3$, there are 18 states because l can be zero, 1, or 2. When $l = 0$, the quantum states are said to be s states; when $l = 1$, they are called p states; when $l = 2$, they are called d states; when $l = 3$, they are called f states.

3-2.3 Light Emission from a Hydrogen Atom

The wavelength(s) of the light emitted by the hydrogen atoms as the excited electron returns to the ground state depends on both the initial excited state and the path the electron takes as it cascades down the energy level diagram. What

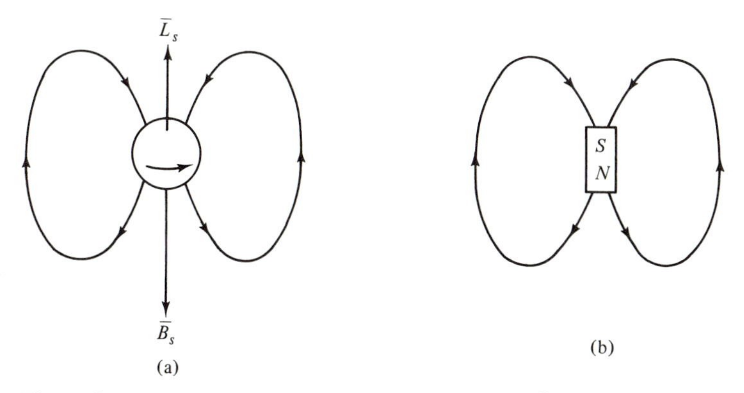

Figure 3-7 (*a*) Electron spin angular momentum, $\bar{L}_s$, with the associated magnetic induction vector, $\bar{B}_s$. (*b*) Equivalent permanent magnet.

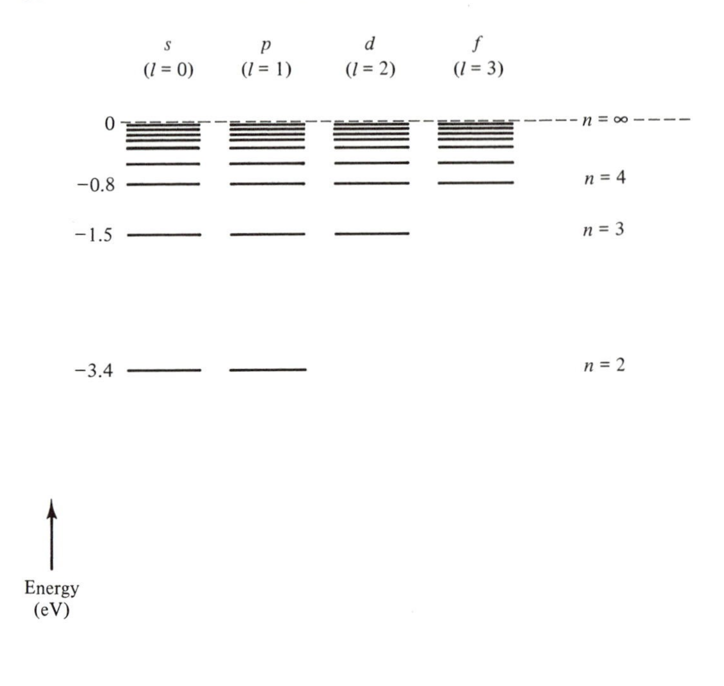

Figure 3-8 Energy level diagram for hydrogen showing the *s, p, d,* and *f* levels.

the initial excited state is depends on what the exciting mechanism is. An electron can be excited to a higher level thermally, but it is very unlikely that this will happen in a hydrogen atom at room temperature. The reason is the energy of the first excited state is much greater than the average thermal energy, which is $\frac{3}{2}kT$ per atom. k is Boltzmann's constant ($=1.38 \times 10^{-23}$ J/K) and T is the temperature in degrees kelvin. Therefore, at 300 K, $\frac{3}{2}kT = 0.039$ eV. Thus the probability that any one electron has an energy of 10.2 eV is so small it can essentially be ignored.

One common method of excitation is with an electric discharge. At large electric fields electrons are pulled off of some atoms and the resulting ions are accelerated by the field. They, in turn, can collide with hydrogen atoms, and some of their energy can be transmitted to the hydrogen atoms by exciting the hydrogenic electrons into higher energy states.

Energy can be transmitted from one atom with an excited electron to a hydrogen atom with its electron originally in the ground state when these two atoms collide. The probability that the energy will be transferred is much larger if the transition energy of the excited atom is within $\pm kT$ of the transition energy of the hydrogen atom. It is easy for the hydrogen atom to absorb from, or give up to, the surroundings kT units of energy, but it cannot easily accommodate larger changes. Thus, if the difference in the transition energies is larger than kT, it is unlikely that energy will be transferred.

The requirement of matched transition energies for good energy transfer is an important concept we will encounter in the discussion of both the helium-neon and the CO_2 laser. In the He-Ne laser, energy is transferred from an excited helium electron to a ground state neon electron. In the CO_2 laser an excited vibrating N_2 molecule transfers its energy to a CO_2 molecule in the ground vibrational state.

A hydrogen atom can also be excited optically. Of course, the energy of the absorbed photon must be equal to a hydrogen electron transition energy.

The cross section, σ, is a measure of the probability an exciting particle will strike an absorbing atom or molecule and be consumed by it. It is the cross-sectional area of the absorbing specie. This will cause the flow of exciting particles to be reduced by $A\,dI$ where A is the cross-sectional area of the beam and I is its intensity. The probability, P, that an exciting particle will strike an absorbing particle is

$$P = \frac{\sigma}{A} \tag{3-15}$$

where

$$\sigma = \pi b^2 \tag{3-16}$$

with b being the radius of the capture cross section. If we ignore the possibility that one absorbing specie "hides" behind another, the number of exciting particles being absorbed per unit time in a volume $A\,dz$ is proportional to the number of absorbing particles, N, in the volume element where

$$N = nA\,dz \tag{3-17}$$

and n is the number of absorbing particles per unit volume. Finally, the number of absorbing collisions per unit time is also proportional to the flux of particles in the exciting beam, AI. Thus

$$-A\,dI = \left(\frac{\sigma}{A}\right)(nA\,dz)(AI) \tag{3-18a}$$

The negative sign indicates an intensity decrease. Cross multiplying and integrating I between the limits I_0 and I and z between the limits 0 and z yields

$$I = I_0 e^{-n\sigma z} \tag{3-18b}$$

By comparing Eq. 3-18b with Eq. 1-43, we see that

$$\alpha = n\sigma \tag{3-19}$$

and learn the important physical fact that the absorption coefficient is proportional to the concentration of the absorbing specie as well as its cross section.

The cross section is a strong function of the wavelength. If the conditions described in the preceding paragraphs are not met σ and, therefore, α are virtually equal to zero.

The probability an electron will drop into a specific lower level is proportional to the density of states, ρ_i, in that level. For instance, $\rho = 2$ for the s states, $\rho = 6$ for the p states, $\rho = 10$ for the d states, and so forth.

A more important parameter is the spontaneous transition probability, A_{ij}, per unit time from one state to another. At optical frequencies $A_{ij} \simeq 10^8$ sec^{-1} when $\Delta l = \pm 1$ and is much smaller for any other Δl changes. This selection rule leads to the optical series illustrated in Fig. 3-9. The transitions from the p states to the ground state form the P (principal) series; the transitions from the s states to the $2p$ state form the S (sharp) series.

The transitions from the d states to the $2p$ state form the D (diffuse) series, and the transitions from the f state to the $3d$ state form the F (fundamental) series. The S and D series are essentially identical in the hydrogen atom, but in Section 3-3 we will learn that they are quite different in other single valence electron atoms.

An excited electron could cascade down along the path in Fig. 3-10 where one of the states it occupies is a $2s$ state; the electron would spend a long time in the $2s$ state, since the $2s$ to $1s$ transition is forbidden. In atoms other than hydrogen the $2p$ states are at higher energies; thus, the electron can, in fact, become locked into the $2s$ state. This state is said to be a metastable state.

Note that we said an electron in a state that has a low transition probability will stay in this state a long time. The average time it takes for the electron to fall is the decay time, τ, and it is equal to the reciprocal of the transition probability. Thus

$$\tau_{ij} = \frac{1}{A_{ij}} \tag{3-20a}$$

The average time, τ, an electron remains in the ith state is given by the equation:

$$\frac{1}{\tau} = \Sigma A_{ij} \tag{3-20b}$$

For most transitions in hydrogen $\tau \simeq 10^{-9}$ sec, but in a number of atoms there are metastable states for which $\tau \simeq 10^{-3}$ sec.

3-3 THE PERIODIC TABLE

The periodic table can be constructed by using qualitative arguments and our knowledge of the hydrogen atom. The arguments must be qualitative because

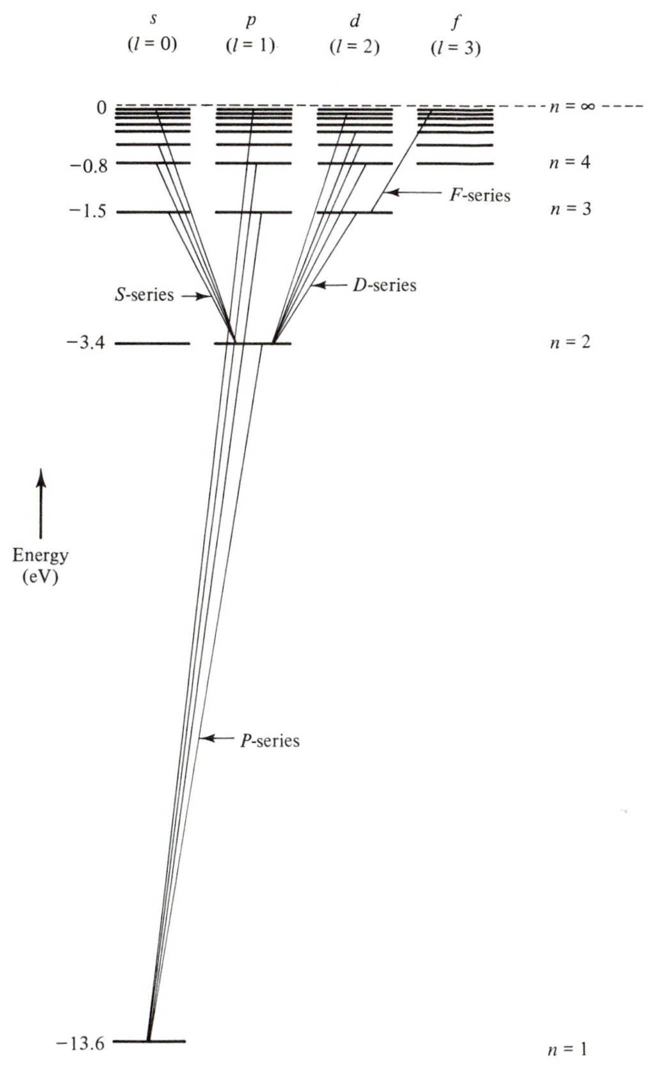

Figure 3-9 The *s, p, d,* and *f* series of the hydrogen atom.

the electron-electron repulsion term that must be added to the potential term in Eq. 3-9 presents enormous computational difficulties. The major differences between hydrogen and other atoms are that the binding energies for a given value of n are larger, and the energy of a quantum state now depends on l as well as n. The binding energy for a given n increases with the atomic number, Z, because the larger charge on the nucleus attracts the electron more strongly. The energy of a quantum state increases with l for a fixed n because the electron orbit becomes more circular as l increases. An electron with a small l, and hence a more elliptical orbit, spends more time near the nucleus than one with a larger l and, therefore, is more strongly attracted by it. Another way to describe this

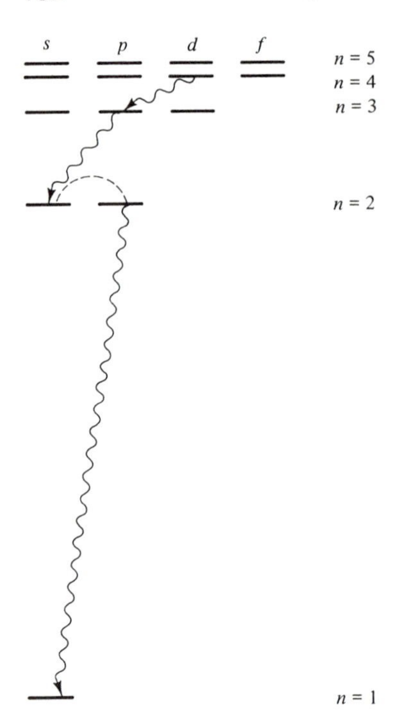

Figure 3-10 A cascading electron that becomes lodged in the metastable $2s$ state.

effect is that the electron with a more circular orbit is more effectively screened from the nuclear charge by the electrons in orbits inside of its orbit.

The periodic table is constructed by examining the ground state electronic configurations of the elements. The ground state configuration is the configuration that has the smallest overall energy. The initial suggestion is that all of the electrons should be put in the $n = 1$ state, but this is not acceptable because it violates the Pauli exclusion principle. The Pauli principle is that no two electrons can occupy the same quantum state.

The ground state electronic configurations for H($Z = 1$), He($Z = 2$), Li($Z = 3$), B($Z = 5$), F($Z = 9$), Ne($Z = 10$), and Ar($Z = 18$) and the nomenclature used to define them are shown in Fig. 3-11. The ground state for H is $1s$ because the lowest acceptable energy configuration occurs when the single electron is in the $n = 1$ state. The ground state for He is $1s^2$ because two electrons can be placed in the $n = 1$ state, one with spin up and one with spin down. He is a rare gas—an element that is chemically inert—so it is placed in the column of rare gases, column VIII, in the periodic table (Fig. 3-12). It is inert because this electronic configuration is very stable. The ground state of Li is $1s^2 2s^1$ because the $n = 1$ state can hold only two electrons. It is $2s^1$ rather than $2p^1$ because the $2s$ level is at a lower energy. B has a $1s^2 2s^2 2p^1$ configuration because the $2s$ state can hold only two electrons. F has the $1s^2 2s^2 2p^5$ configuration, and Ne has the $1s^2 2s^2 2p^6$ configuration because the $2p$ state can hold up to six electrons. Ne like He is a rare gas. It appears that an element is a rare gas if the element with the next higher atomic number begins to start filling in an energy level with a

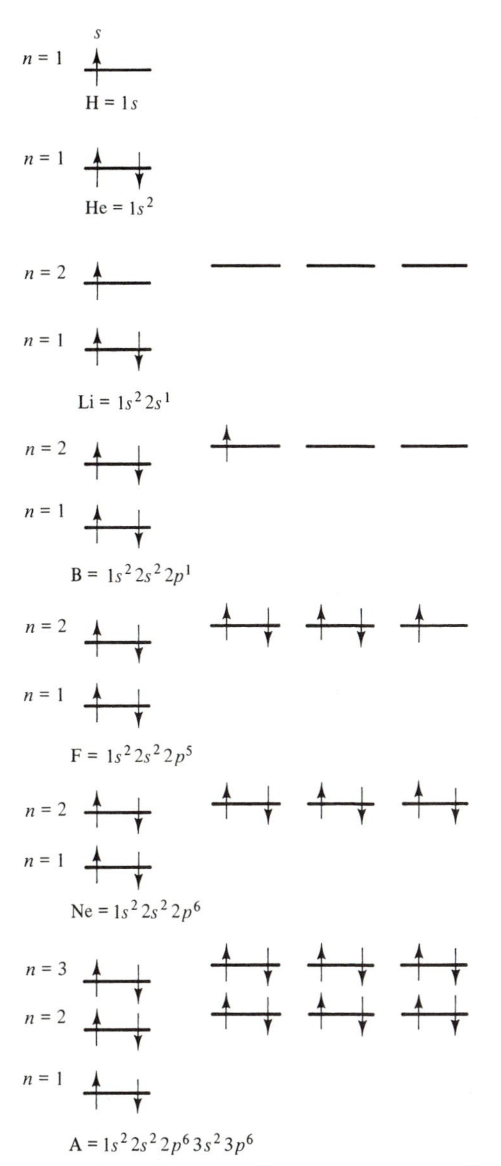

Figure 3-11 The ground state electronic configurations of H, He, Li, B, F, Ne, and Ar and the nomenclature used to describe them.

larger value of *n*. This, in fact, will always be true. But what about argon (Ar)? The 3*d* shell has not been filled. The explanation is the 4*s* energy level is lower than the 3*d*. Therefore, *K* has the configuration [Ar]$4s^1$, and Ca has the configuration [Ar]$4s^2$ where [Ar] represents the ground state configuration of argon. The 3*d* levels are lower than the 4*p* and, therefore, they fill up next. The 10 elements that are formed by filling the 3*d* levels (Sc $\rightarrow$ Zn) are called the first transition metal series. If we look further into the electron level filling sequence,

Metals | Nonmetals

1A	IIA	IIIB	IVB	VB	VIB	VIIB	VIII			IB	IIB	IIIA	IVA	VA	VIA	VIIA	0
1 H 1.00797																1 H 1.0079	2 He 4.0026
3 Li 6.939	4 Be 9.012											5 B 10.811	6 C 12.011	7 N 14.007	8 O 15.9994	9 F 18.998	10 Ne 20.183
11 Na 22.990	12 Mg 24.312											13 Al 26.98	14 Si 28.086	15 P 30.97	16 S 32.064	17 Cl 35.453	18 Ar 39.95
19 K 39.102	20 Ca 40.08	21 Sc 44.96	22 Ti 47.90	23 V 50.94	24 Cr 52.00	25 Mn 54.94	26 Fe 55.85	27 Co 58.93	28 Ni 58.71	29 Cu 63.54	30 Zn 65.37	31 Ga 69.72	32 Ge 72.59	33 As 74.92	34 Se 78.96	35 Br 79.91	36 Kr 83.80
37 Rb 85.47	38 Sr 87.62	39 Y 88.91	40 Zr 91.22	41 Nb 92.91	42 Mo 95.94	43 Te 99	44 Ru 101.07	45 Rh 102.91	46 Pd 106.4	47 Ag 107.87	48 Cd 112.40	49 In 114.82	50 Sn 118.69	51 Sb 121.75	52 Te 127.60	53 I 126.90	54 Xe 131.30
55 Cs 132.90	56 Ba 137.34	57–71 La series*	72 Hf 178.49	73 Ta 180.95	74 W 183.85	75 Re 186.2	76 Os 190.2	77 Ir 192.2	78 Pt 195.1	79 Au 196.97	80 Hg 200.59	81 Tl 204.37	82 Pb 207.19	83 Bi 208.98	84 Po 210	85 At 210	86 Rn 222
87 Fr 223	88 Ra 226	89– Ae series†															
				58 Ce 140.12	59 Pr 140.91	60 Nd 144.24	61 Pm 147	62 Sm 150.35	63 Eu 151.96	64 Gd 157.25	65 Tb 158.92	66 Dy 152.50	67 Ho 164.90	68 Er 167.26	69 Tm 168.93	70 Yb 173.01	71 Lu 174.97
				90 Th 232.04	91 Pa 231	92 U 238.03	93 Np 237	94 Pu 239	95 Am 241	96 Cm 242	97 Bk 249	98 Cf 252	99 Es 254	100 Em 253	101 Md	102 No	103 Lw

Figure 3-12 Periodic table of elements. The atomic number (upper) of each element equals the number of protons or electrons. The atomic weight (lower) of each element is based on C_{12} = 12.0000.

Metals ← → Nonmetals

1A	IIA	IIIB	IVB	VB	VIB	VIIB	VIII			IB	IIB	IIIA	IVA	VA	VIA	VIIA	0
1 H 1.00797																1 H 1.0079	2 He 4.0026
3 Li 6.939	4 Be 9.012											5 B 10.811	6 C 12.011	7 N 14.007	8 O 15.9994	9 F 18.998	10 Ne 20.183
11 Na 22.990	12 Mg 24.312											13 Al 26.98	14 Si 28.086	15 P 30.97	16 S 32.064	17 Cl 35.453	18 Ar 39.95
19 K 39.102	20 Ca 40.08	21 Sc 44.96	22 Ti 47.90	23 V 50.94	24 Cr 52.00	25 Mn 54.94	26 Fe 55.85	27 Co 58.93	28 Ni 58.71	29 Cu 63.54	30 Zn 65.37	31 Ga 69.72	32 Ge 72.59	33 As 74.92	34 Se 78.96	35 Br 79.91	36 Kr 83.80
37 Rb 85.47	38 Sr 87.62	39 Y 88.91	40 Zr 91.22	41 Nb 92.91	42 Mo 95.94	43 Te 99	44 Ru 101.07	45 Rh 102.91	46 Pd 106.4	47 Ag 107.87	48 Cd 112.40	49 In 114.82	50 Sn 118.69	51 Sb 121.75	52 Te 127.60	53 I 126.90	54 Xe 131.30
55 Cs 132.90	56 Ba 137.34	57–71 La series*	72 Hf 178.49	73 Ta 180.95	74 W 183.85	75 Re 186.2	76 Os 190.2	77 Ir 192.2	78 Pt 195.1	79 Au 196.97	80 Hg 200.59	81 Tl 204.37	82 Pb 207.19	83 Bi 208.98	84 Po 210	85 At 210	86 Rn 222
87 Fr 223	88 Ra 226	89– Ae series†															

58 Ce 140.12	59 Pr 140.91	60 Nd 144.24	61 Pm 147	62 Sm 150.35	63 Eu 151.96	64 Gd 157.25	65 Tb 158.92	66 Dy 152.50	67 Ho 164.90	68 Er 167.26	69 Tm 168.93	70 Yb 173.01	71 Lu 174.97
90 Th 232.04	91 Pa 231	92 U 238.03	93 Np 237	94 Pu 239	95 Am 241	96 Cm 242	97 Bk 249	98 Cf 252	99 Es 254	100 Em 253	101 Md	102 No	103 Lw

Figure 3-12 (*Continued*)

we will find that it fills up in the order, 1*s* 2*s* 2*p* 3*s* 3*p* 4*s* 3*d* 4*p* 5*s* 4*d* 5*p* 6*s* 4*f* 5*d* 6*p* 7*s* 5*f*. Figure 3.13 is a schematic representation of this order, which can be used to help you remember the proper sequence. There are a few minor exceptions to this rule, such as Cu([Ar]$4s^1 3d^{10}$), but we will not be concerned with them.

Some of the properties of the elements in the periodic table are reviewed in Example 3-2.

EXAMPLE 3-2

(a) What is the weight of an aluminum atom, and **(b)** how many aluminum atoms are there in a cubic centimeter if the density is 2.70 g/cm^3?

(a) One gram molecular weight has a weight in grams equal to the atomic weight. From the periodic table in Fig. 3-12, $W = 27.0$ g. There is an Avogadro's number, $N_0 = 6.03 \times 10^{23}$, of atoms in 1 g molecular weight.

$$\therefore\ W = \frac{27.0}{6.03} \times 10^{23} = 4.48 \times 10^{-23}\ \text{g}$$

(b) The weight of 1 cm^3 of material is the density, ρ. The number of atoms in 1 cm^3, n, then is $n = \rho/W = 6.03 \times 10^{22}\ \text{cm}^{-3}$.

3-4 EXCITED ATOMIC STATES

3-4.1 Single Valence Atoms

We have already learned in Section 3-3 that, for all atoms but the hydrogen atom, the energy of an electronic state increases with increasing l even though n remains constant. This is shown in Fig. 3-14 where the excited states of sodium are illustrated. The S series, which is generated by electrons falling from the *ns* states to the 3*p* state, and the D series, which is produced by electrons falling from the *nd* states to the 3*p* state, have different energies associated with them, since the *s* and *d* levels no longer have the same energy.

The different excited energy levels in single valence electron atoms, often called the alkali metals, can be calculated once a term known as the screening

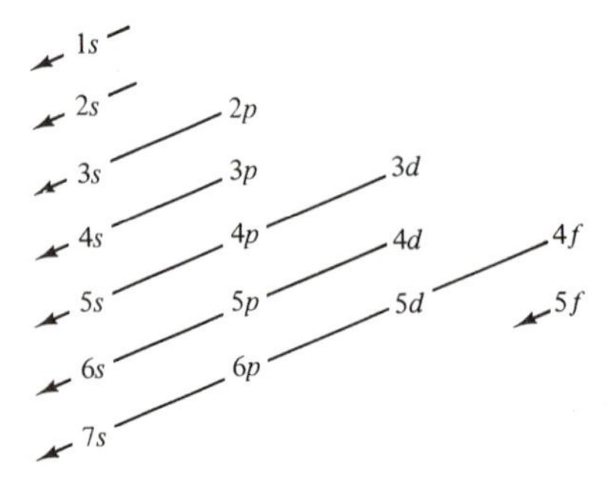

Figure 3-13 Schematic diagram showing the order in which electron energy levels are filled.

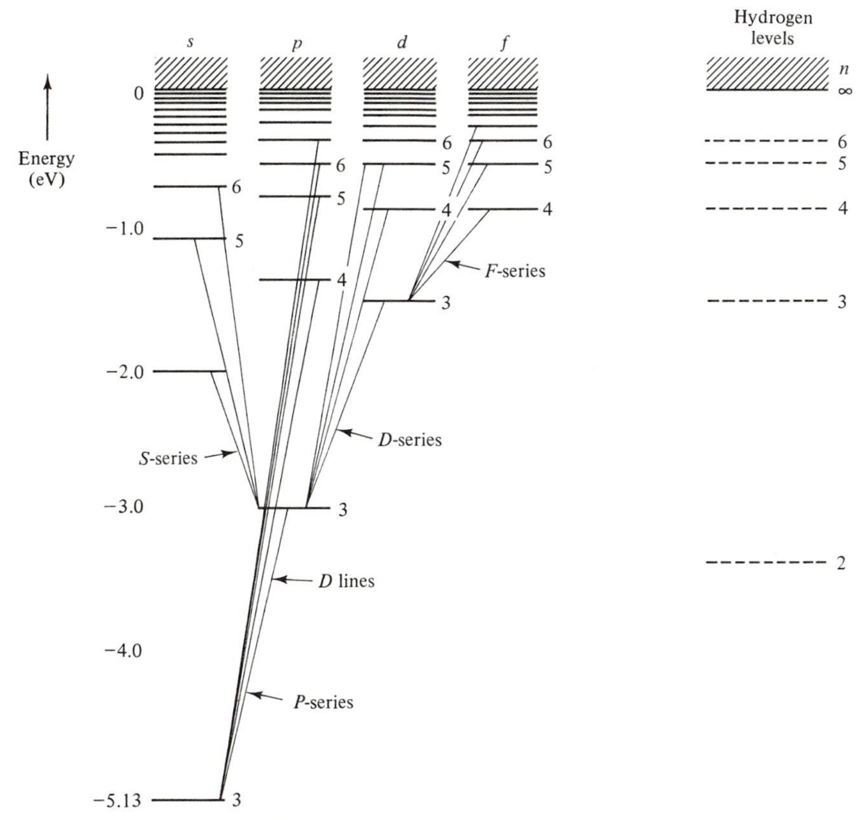

Figure 3-14 Energy level diagram for sodium. For comparison the hydrogen energy levels are shown on the right.

factor, Δ_i, has been determined. This factor is the same, for example, for all of the *s* levels in a given atom, but it is different from the screening factor for the *p* levels in the same atom, which is different from the factor for the *d* levels, and so forth. In addition, Δ_s, Δ_p, Δ_d, and Δ_f vary from atom to atom.

The excited energy levels for the alkali metals are given by the equation

$$E_n = -\frac{C}{(n - \Delta_i)^2} \tag{3-21}$$

where C is the same constant determined for the hydrogen atom. The value of Δ_i is always positive, $\Delta_s > \Delta_p > \Delta_d > \Delta_f$, and the magnitude of Δ with the same subscript increases as the atomic number increases. The Δ_i are positive because the inner core electrons do not completely screen the outer electron from the positively charged nucleus. The magnitude of the Δ_i decreases as the angular momentum increases because of the fact that the electrons in the more circular orbits are more effectively screened from the nucleus. Finally, the Δ_i for the heavier alkali metals are larger because the valence electron "sees" a larger effective charge on the nucleus.

EXAMPLE 3-3

The wavelengths for the transitions in potassium are as follows: $5s \rightarrow 4p$ = 12,523 Å, $6s \rightarrow 4p$ = 6939 Å, $4p \rightarrow 4s$ = 7645 Å, $5p \rightarrow 4s$ = 4044 Å, $3d \rightarrow 3p$ = 11,772 Å, and $4d \rightarrow 3p$ = 6965 Å. Find Δ_s, Δ_p, and Δ_d. The ground state of potassium is [Ar] $4s^1$.

$$\Delta E = hc\left(\frac{1}{\lambda_2} - \frac{1}{\lambda_1}\right) = 1.242 \times 10^{-6}\left(\frac{10^6}{0.6939} - \frac{10^6}{1.2523}\right) = 0.798 \text{ eV}$$

From Eq. 3-21

$$\Delta E = C\left[\frac{1}{(6 - \Delta_s)^2} - \frac{1}{(5 - \Delta_s)^2}\right] = 13.6\left[\frac{1}{(5 - \Delta_s)^2} - \frac{1}{(6 - \Delta_s)^2}\right] = 0.798$$

By using an iterative technique,

$$\Delta_s = 2.21$$

Finding Δ_p

$$\Delta E = 1.242\left(\frac{1}{0.4044} - \frac{1}{0.7645}\right) = 1.447 \text{ eV}$$

$$= 13.6\left[\frac{1}{(4 - \Delta_p)^2} - \frac{1}{(5 - \Delta_p)^2}\right]$$

$$\Delta_p = 1.78$$

Finding Δ_d

$$\Delta E = 1.242\left(\frac{1}{0.6965} - \frac{1}{1.1772}\right) = 0.7282 \text{ eV}$$

$$= 13.6\left[\frac{1}{(3 - \Delta_d)^2} - \frac{1}{(4 - \Delta_d)^2}\right]$$

$$\Delta_d = 0.11$$

The spectra of the alkali metals are more complicated than we have thus far described. What we discussed above is called the rough structure. We now turn to a discussion of the fine structure, which will be illustrated by the energy level diagram of potassium in Fig. 3-15. Notice that in this diagram the ground state, a $4s$ state by our convention, is designated as a $1S$ state. It is an unfortunate act of history that there is also a convention in which the ground state is a 1 state. Both conventions will be used in this book, and they will be differentiated by using capital letters and n' in the latter case.

The fine structure results from the fact that the electron spin and the angular momentum vectors are coupled. Thus, one must consider the total spin, M_s, the total orbital angular momentum, L, and the total angular momentum, J, which is a vector sum of M_s and L. The rules for finding them are as follows:

$$M_s = \frac{n}{2}, \frac{1}{2}(n - 2) \ldots \frac{1}{2}(n - 2m) \qquad (3\text{-}22)$$

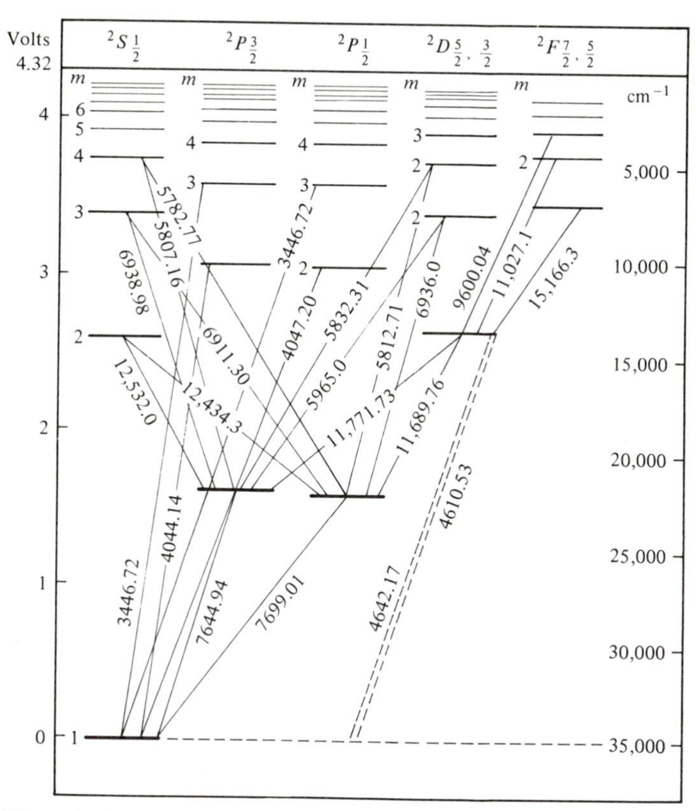

Figure 3-15 Energy level diagram for potassium.

where n is the total number of valence electrons and $2m = n$ if n is even and $2m = n - 1$ if n is odd. For two valence electrons the total orbital angular momentum quantum number can be

$$L = l_1 + l_2, l_1 + l_2 - 1, \ldots |l_1 - l_2| \tag{3-23}$$

By using the Bohr orbital model, this implies that the orbits can make only specific angles with each other. L for the atoms is like L for the single electron in that the magnitude of the total orbital angular momentum, L_{orbit}, is

$$L_{\text{orbit}} = \sqrt{L(L+1)}\hbar \tag{3-24}$$

The total angular momentum is also quantized and is found from the equation

$$\mathbf{J} = \mathbf{L} + \mathbf{M}_s, \mathbf{L} + \mathbf{M}_s - 1 \ldots |\mathbf{L} - \mathbf{M}_s| \tag{3-25}$$

The three possible J values, 2, 1, and 0, for $L = M_s = 1$ are shown pictorially in Fig. 3-16.

We are now in a position to gain an accountant's understanding of Fig. 3-15. The states are designated by the formula

$$n'^{(2M_s+1)}L_J \tag{3-26}$$

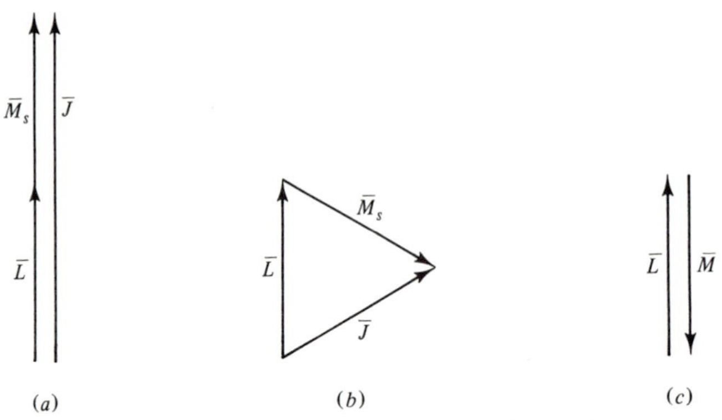

Figure 3-16 The three configurations for (*a*) $J = 2$, (*b*) $J = 1$ and (*c*) $J = 0$ when $M_s = L = 1$.

where $n' = 1$ for the valence electron in its ground state. In Example 3-4 we show how to find the different M_s, L, and J values.

EXAMPLE 3-4

Using Eqs. 3-22 to 3-26, analyze the energy level diagram of potassium in Fig. 3-15.

(a) There is only one value of $M_s = m_s = \frac{1}{2}$ because there is only one valence electron.

(b) For $l = 0$, $L = 0$.

(c) If $L = 0$ and $M_s = \frac{1}{2}$, the only allowed value of J is $\frac{1}{2}$.

(d) Thus, for the $L = 0$ states, the only states are $n'^2S_{1/2}$ states.

(e) For $l = 1$, $L = 1$.

(f) If $L = 1$ and $M_s = \frac{1}{2}$, the allowed values of J are $\frac{3}{2}$ and $\frac{1}{2}$.

(g) Thus, for the $L = 1$ states, there are two different types of states: the $n'^2P_{3/2}$ and $n'^2P_{1/2}$ states. States with the same n' values are said to be doublets.

(h) For $L = 2$, $J = \frac{5}{2}$ and $\frac{3}{2}$, and the two types of states are the $n'^2D_{5/2,3/2}$ states.

With the exception of the S states the states appear in pairs called doublets. This is designated by the superscript, $2M_s + 1$. The doublets appear as two distinct states in the P levels in Fig. 3.15, but the ^2D and ^2F doublets are shown as single states. This is because the state splitting for P states is greater than it is for D states which, in turn, is greater than it is for F states. Another general rule is that the magnitude of the splitting increases with the atomic number because the strength of the spin-orbit coupling increases.

The selection rules are also illustrated in Fig. 3-15. They are $\Delta L = \pm 1$ and $\Delta J = 0, \pm 1$. The forbidden $1^2D_{5/2,3/2} \rightarrow 1^2S_{1/2}$ transitions are also shown. This is done to show that forbidden transitions do occur, but they are much less probable.

3-4.2 Two Valence Electron Atoms

The description of excited electronic states in two valence electron atoms is more complex than that for the single valence electron atoms, as is shown in Fig. 3-17 where the energy level diagram for calcium is displayed. One of the more prominent differences between Figs. 3-15 and 3-17 is that the diagram for calcium is divided into two segments: singlets and triplets. States are singlet states when the two electron spins are antiparallel ($M_s = 0$). Another important difference is that now the angular momentum of two electrons must be considered. However, the valence electron that remains in the grown state has $l_2 = 0$. Hence, it does not greatly affect the results since, as before, $L = l_1$. Other details of the calcium energy level diagram are discussed in Example 3-5.

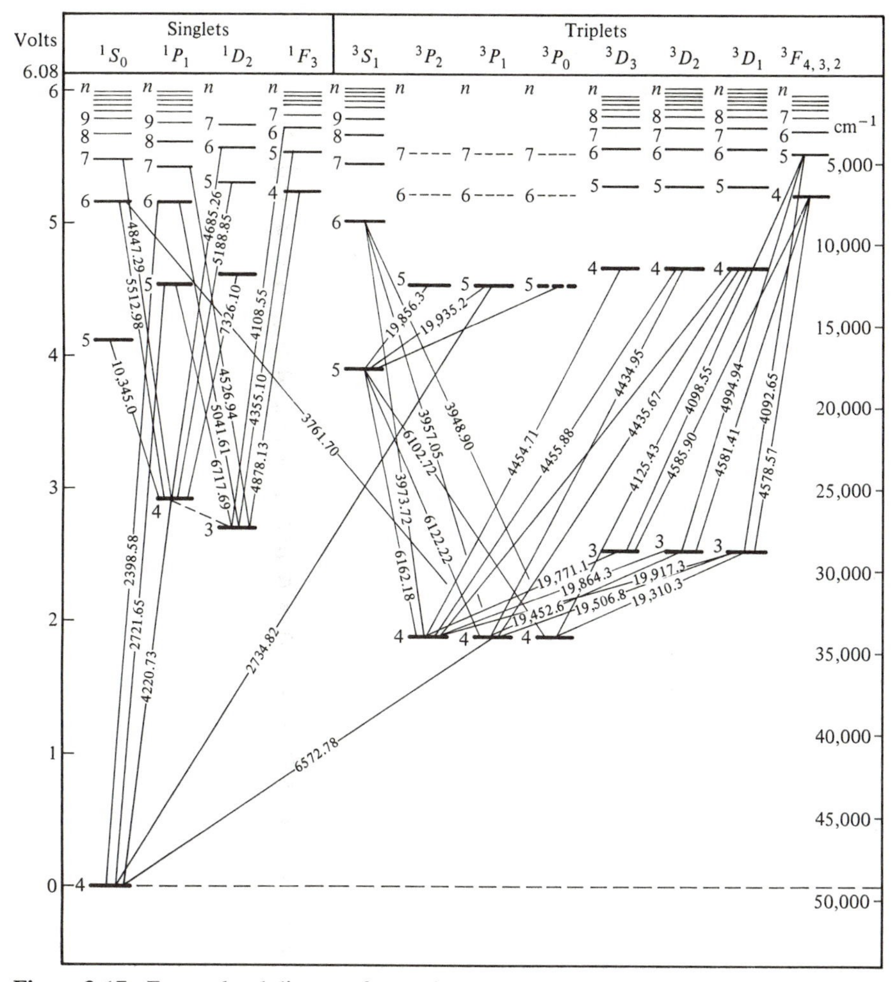

Figure 3-17 Energy level diagram for calcium.

EXAMPLE 3-5

Using Eqs. 3-22 to 3-26, analyze the energy level diagram of calcium in Fig. 3-17.

(a) Consider the singlet states first: $M_s = 0$.

1. For $l_1 = 0, L = 0$.
2. $J = L + M_s = 0$
3. Thus, the only S singlet states are n'^1S_0 states.
4. For $l_1 = 1, L = 1$.
5. $J = L + M_s \ldots (L - M_s) = 1$ only.
6. Thus, the only P singlet states are n'^1P_1 states.
7. $J = L$ also for the D and F singlet states, $^1D_2$ and $^1F_3$.

(b) For the triplet states, $M_s = 1$.

1. For $l_1 = 0, L = 0$.
2. $J = L + M_s \ldots (L - M_s) = 1$.
3. Thus, the only triplet S states are n'^3S_1. Note also that there is no ground state triplet state because it is disallowed by the Pauli exclusion principle.
4. For $l_1 = 1, L = 1$.
5. $J = L + M_s \ldots |L - M_s| = 2, 1, 0$.
6. Therefore, the three triplet P states are $^2P_{2,1,0}$.
7. $J = 3, 2, 1$ for $L = 2$, and $J = 4, 3, 2$ for $L = 3$ so that the three triplet D and F states are $^3D_{3,2,1}$ and $^3F_{4,3,2}$ states.

Another interesting feature of the calcium energy level diagram is that there are only a few transitions between the singlet and triplet states. These transitions are, in fact, forbidden and thus they have a low probability of occurring. Physically, the reason that transitions for which $\Delta M_s \neq 0$ are unlikely is that the electron must simultaneously alter its spin and jump to another state. Some other observations are that the triplet states are at a lower energy than the corresponding singlet states, and the splitting of the triplet states is greater for the P states than it is for the D states which, in turn, is greater than it is for the F states.

3-4.3 Rare Gas Atoms

The energy level diagram of helium shown in Fig. 3-18 is qualitatively similar to the one for calcium, since both atoms have two electrons in their outer shell. The only differences between the two diagrams are that no forbidden transitions are shown for helium, and the splitting in the triplet states is much less. At one time spectroscopists thought that there were two different types of helium, ortho and para helium, because electronic transitions between the two configurations are very difficult.

The ground state of the other rare gases, which have eight electrons in their outer core, is identical to that of helium. For every state with an electron with

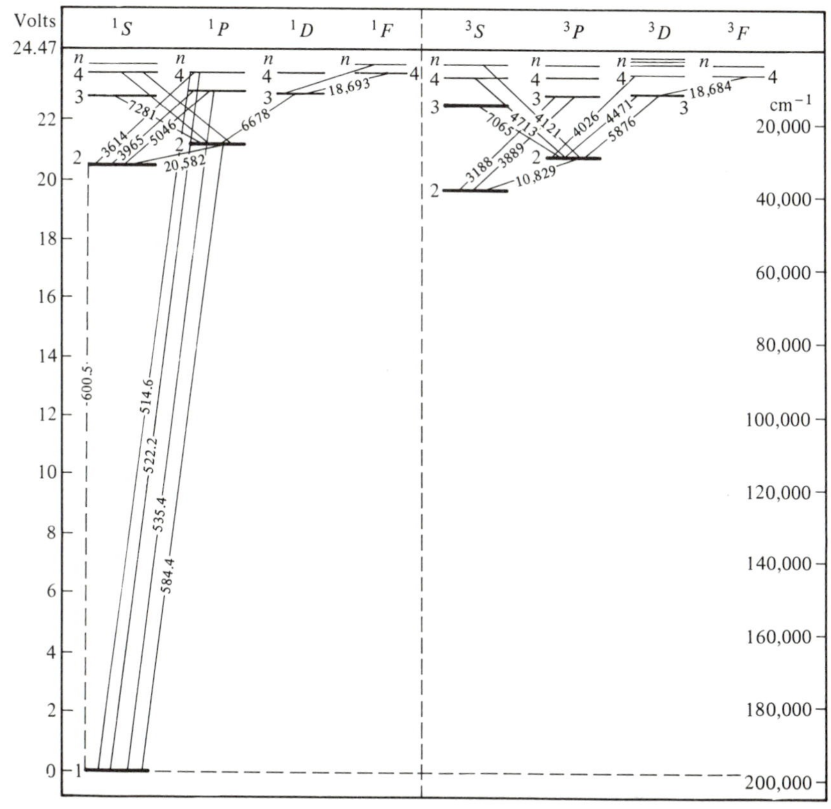

Figure 3-18 Energy level diagram for helium.

spin up there is one with spin down; hence, $M_s = 0$. Likewise, for every electron moving in a clockwise orbit, there is one moving in a counterclockwise orbit so that $L = 0$. Thus, the ground state must be a ${}^1S_0$ state.

When an electron is excited out of a p state in the outer core, the absence of the eighth electron causes the remaining seven electrons to have a net spin, $m_{s2} = \frac{1}{2}$, and a net angular momentum, $l_2 = 1$. Thus, for the excited states, $M_s = 1, 0$, and L can have many different values as is shown in Example 3-6 where the S and P excited levels are considered.

EXAMPLE 3-6

Find all of the states in the rare gases other than helium when the excited electron is in an s state ($l_1 = 0$) or in a p state ($l_1 = 1$).

Consider the excited s states first: $l_1 = 0$.

(a) For the singlet state, $M_s = 0$.

1. $L = l_2 = 1$.
2. $J = L + M_s = 1$.
3. There is one singlet state, ${}^1P_1$.

(b) For the triplet states, $M_s = 1$.
1. $L = l_2 = 1$.
2. $J = L + M_s, \ldots |L - M_s| = 2, 1, 0$.
3. There are three triplet states, $^3P_{2,1,0}$.

Next consider the excited p states $l_1 = 1$.

(a) For the singlet states, $M_s = 0$.
1. $L = l_1 + l_2 \ldots |l_1 - l_2| = 2, 1, 0$.
2. $J = L = 2, 1, 0$.
3. Therefore, the three singlet states are $^1D_2$, $^1P_1$, and $^1S_0$.

(b) For the triplet states, $M_s = 1$.
1. $L = l_1 + l_2 \ldots |l_1 - l_2| = 2, 1, 0$.
2. For $L = 2$, $J = L + M_s \ldots |L - M_s| = 3, 2, 1$.
3. The three triplet states for $L = 2$ are $^3D_{3,2,1}$.
4. For $L = 1$, $J = L + M_s \ldots |L - M_s| = 2, 1, 0$.
5. The three triplet states for $L = 1$ are $^3P_{2,1,0}$.
6. For $L = 0$, $J = L + M_s \ldots |L - M_s| = 1$.
7. The one triplet state for $L = 0$ is $^3S_1$.
8. Therefore, there are seven triplet states, the $^3D_{3,2,1}$, $^3P_{2,1,0}$, and $^3S_1$ states.

These $4s$ and $10p$ states are discussed in more detail in Chapter 9 when the He-Ne laser is examined.

3-5 MOLECULES

3-5.1 Atomic Orbitals

The chemical bond between atoms in a molecule is, in most instances, at least partially covalent in character. In binary molecules composed of like atoms such as in O_2, and in the group IV semiconducting elements in the periodic table the bond is 100 percent covalent. This bond is due to the quantum mechanical effect called the overlap of atomic orbitals.

Neither the mathematical description nor the concept of atomic orbitals can be obtained from the Bohr theory. Rather, this requires the concepts implicit in, and the mathematical solution of, the Schroedinger equation. One of the fundamental concepts of quantum mechanics, which until now has been neglected, is that the position of the electron cannot be known with absolute certainty. Instead of describing the exact position of the electron, one must be content with determining the probability of finding the electron in a given volume, V. This probability, P, is given by

$$P = \int_V \psi^*_{nlm} \psi_{nlm} \, d\tau \qquad (3\text{-}27)$$

where $d\tau$ is the differential volume element, and ψ^*_{nlm} is the complex conjugate of ψ_{nlm}, the one electron wave function associated with the nlm quantum state found by solving the Schroedinger equation.

TABLE 3-1 ORBITALS OF A HYDROGEN ATOM[a]

Wave function	General symbol	Mathematical description
ψ_{100}	$1s$	$\dfrac{e^{-\rho}}{a_0\sqrt{\pi a_0}}$
$\dfrac{(\psi_{211} + \psi_{21-1})}{\sqrt{2}}$	$2p_x$	$\dfrac{\sin\theta\cos\phi\,\rho e^{-\rho/2}}{4a_0\sqrt{2a_0\pi}}$
$\dfrac{(\psi_{211} - \psi_{21-1})}{\sqrt{2}j}$	$2p_y$	$\dfrac{\sin\theta\sin\phi\,\rho e^{-\rho/2}}{4a_0\sqrt{2a_0\pi}}$
ψ_{210}	$2p_z$	$\dfrac{\cos\theta\,\rho\,e^{-\rho/2}}{4a_0\sqrt{2a_0\pi}}$

[a] a_0 = Bohr radius, $\rho = r/a_0$.

It can be shown that the four wave functions for the 8-valence s and p electronic states are qualitatively similar to the ψ_{100}, ψ_{200}, $(\psi_{211} + \psi_{21-1})/\sqrt{2}$, and $j(\psi_{211} - \psi_{21-1})/\sqrt{2}$ hydrogenic wave functions. [A linear combination of the ψ_{21-1} and ψ_{211} wave functions is used because these two wave functions, which will be designated $2p_x(+)$ and $2p_y(-)$, are real, whereas ψ_{21-1} and ψ_{211} are complex.] The mathematical description of these wave functions is given in Table 3-1 ($\psi_{100} \equiv 1s$ and $\psi_{210} \equiv 2p_z$), and r, θ, and ϕ are defined in Fig. 3-19.

Two-dimensional constant ψ surfaces for the $1s$ and the $2p_z$ wave functions are plotted in Figs. 3-20*b* and 3-21*c*, and they are constructed with the aid of Figs. 3-20*a* and 3-21*b*. The constant $1s$ surfaces are circular (spherical), and their magnitudes are positive and decrease monotonically with r. The surfaces are spherical, since ψ is a function of r only, and the magnitude is positive and decreases monotonically with r because $\psi_{100} = Ae^{-\rho}$; ψ_{100} is plotted as a function of r in Fig. 3-20*a*.

The $2p_z$ wave function is directional since it is proportional to $\cos\theta$, and for $0 < \theta < 90°$ it initially increases with r, reaches a maximum at $r = 2a_0$, and decreases monotonically to zero for $r > 2a_0$. This is shown in Figs. 3-21*a* and

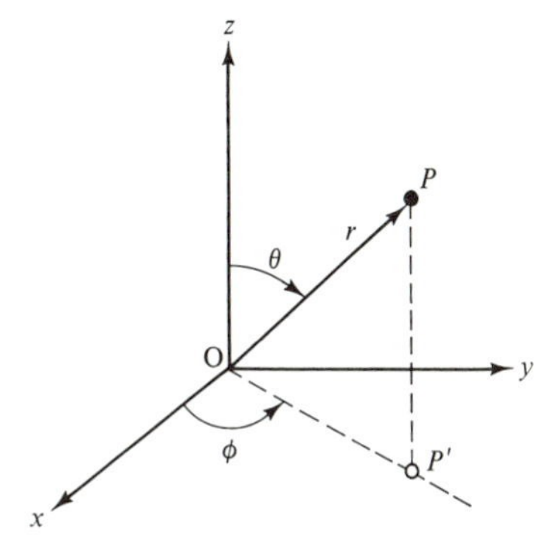

Figure 3-19 The relationship between the Cartesian coordinates x, y, and z and the spherical coordinates r, θ, and ϕ.

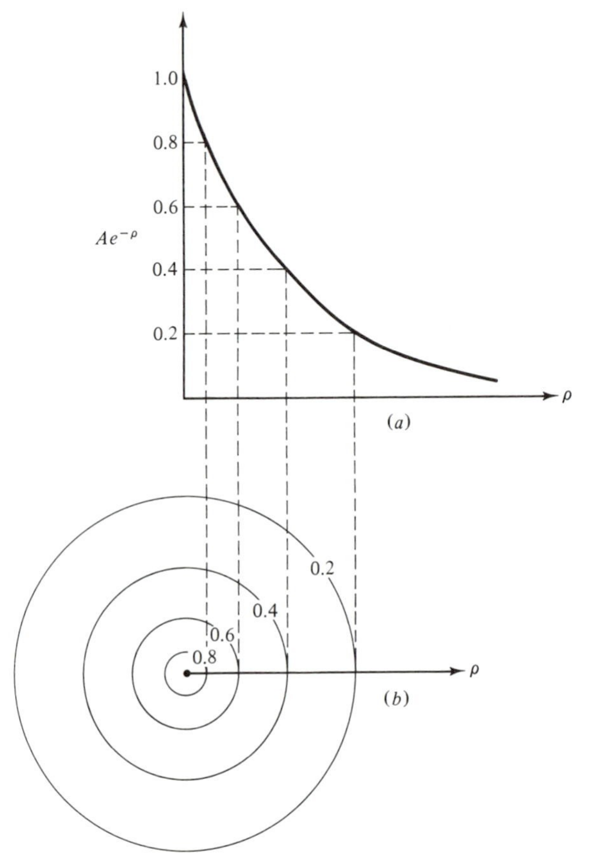

Figure 3-20 (*a*) The 1*s* wave function normalized to one at $\rho = 0$ plotted as a function of the normalized radius, ρ. (*b*) Some constant ψ "surfaces" for the 1*s* wave function.

b. Note that in Fig. 3-21*a* for $\theta = 180°$ the magnitude of $2p_z$ is the same as it is for $\theta = 0°$ for a given value of r, only the sign is different.

The constant $2p$ surfaces in Fig. 3-21*c* can be constructed with the aid of Fig. 3-21*b*. The maximum value occurs at A, which is located on the z axis ($\cos\theta = 1$) at $r = 2a_0$. This "surface" is a point in Fig. 3-21*c*. For $\theta = \theta_B$ the value of $2p_z$ at the points marked B along the z axis is the same as the value at $\theta = \theta_B$ and $r = 2a_0$. For $0 < \theta < \theta_B$ in a given direction there are two values of r for which the value of $2p_z$ is the same as it is for $\theta = \theta_B$ and $r = 2a_0$. The locus of these points in the yz plane is the line marked B in the positive lobe in Fig. 3-21*c*. The procedure can be repeated for the C, D, and E lines. (Note that the constant E line is also plotted in Fig. 3-21*b*.)

The same procedure can be used to construct the negative $2p_z$ lobe as well as the $2p_y$ and $2p_x$ constant ψ surfaces. $2p_y$ differs only in that it "points" in the y direction ($\sin\theta \sin\phi = 1$), and the $2p_x$ differs only in that it "points" in the x direction ($\sin\theta \cos\phi = 1$).

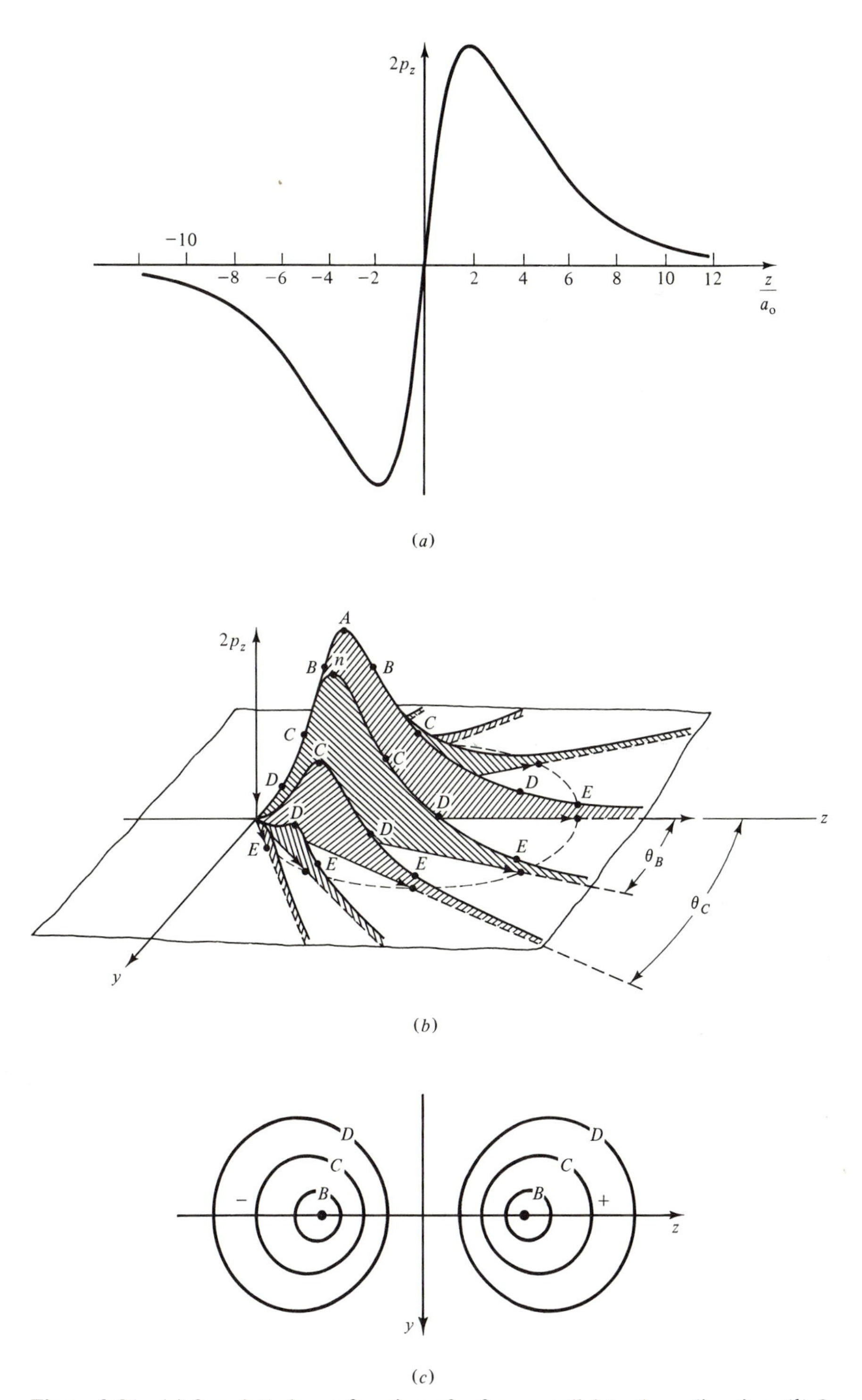

Figure 3-21 (*a*) $2p_z$ plotted as a function of ρ for r parallel to the z direction. (*b*) $2p_z$ plotted as a function of ρ for different directions of r. (*c*) Constant $2p_z$ "surfaces" in the yz plane constructed with the aid of part b.

A constant $2p_z$ surface in three dimensions can be created from the two-dimensional "surface" simply by rotating the latter about the z axis. This can be done since θ, the angle between $\bar{r}$ and the z axis, remains constant during the rotation. A constant $2p_z$ surface is shown in Fig. 3-22 along with constant $2p_y$, $2p_x$, and $1s$ surfaces.

The volumes enclosed by these constant ψ surfaces are called atomic orbitals. In an atomic orbital the absolute value of ψ must be greater than or equal to the value of ψ at the surface. For the $1s$ and $2p$ orbitals, $|\psi|$ inside of the enclosed surface is everywhere greater than $|\psi|$ at the surface and, therefore, they do not contain any internal surfaces. However, this is more the exception than the rule as there are a number of atomic orbitals that are hollow.

It is also important to emphasize that only the absolute magnitude of ψ on the surface needs to be constant. For the $2p$ orbitals each lobe has a different sign whereas in the $1s$ orbital the sign is everywhere positive.

The shape of the $2p_z$ orbital, which has been described qualitatively, is described quantitatively in the following example.

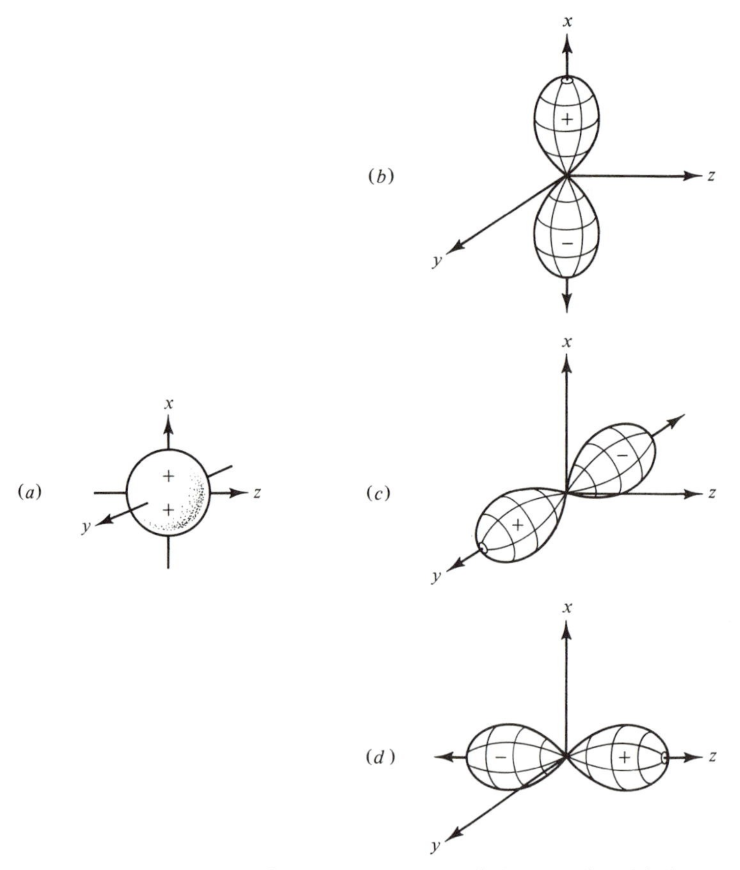

Figure 3-22 The (*a*) ls, (*b*) $2p_x$, (*c*) $2p_y$, and (*d*) $2p_z$ atomic orbitals.

EXAMPLE 3-7

(a) Show that for a given positive value of $\cos\theta$, $2p_z$ is a maximum for $r = 2a_0$. **(b)** What is the ratio of the values of $2p_z$ at $r = 2a_0$ for $\theta = 30°$ and $\theta = 0°$? **(c)** For what r values is $2p_z(\theta = 0) = 2p_z(\theta = 30°, r = 2a_0)$? **(d)** Repeat parts b and c for $2p_z(\theta, r) = 2p_z(45°, 2a_0)$.

(a) $\dfrac{\partial}{\partial r}(2p_z) = \dfrac{\partial}{\partial r}[A\cos\theta r\exp(-r/2a_0)] = A\cos\theta\exp(-r/2a_0)(1 - r/2a_0) = 0$ or $r = 2a_0$.

(b) Ratio $= \cos 30/\cos 0 = 0.866$.

(c) $A\cos(0°)\rho e^{-\rho/2} = A\cos(30°)2e^{-1} = 0.64$ Å

$\rho e^{-\rho/2} = 0.64$.

Solving by trial and error, $\rho = 1.12, 3.26$.

(d) Ratio $= 0.5$, $\rho e^{-\rho/2} = 0.52$, $\rho = 0.76, 0.16$; see Fig. 3-21.

3-5.2 Linear Combination of Atomic Orbitals

We now qualitatively examine the nature of the covalent bond in molecules. We do this by assuming that the wave functions of the valence electrons (electrons in the outermost s and p levels) are similar to the $1s$, $2p_x$, $2p_y$, and $2p_z$ hydrogenic wave functions, and then examine the changes in the atomic wave functions when two atoms are brought close together. The wave function changes because the potential valence electrons see changes.

The first order approximation to the molecular wave functions can be made by adding or subtracting the identical atomic wave functions that are centered about each atom. This has been done in Fig. 3-23. Notice that, if adding the atomic wave functions concentrates the electronic charge between the two nuclei, then subtracting the wave functions concentrates the electronic charge outside the region between the two nuclei, and vice versa. The molecular wave function that concentrates the electrons between the nuclei is called the bonding wave function, and the one that does not concentrate the electrons between the nuclei is called the antibonding wave function (*). They are so called because the energy associated with the bonding wave function is less (more negative) and that associated with the antibonding wave function is more (less negative) than the energy associated with the atomic wave function. The average of the bonding and antibonding energies is about the same as the energy associated with the atomic wave function.

Electrons, like anything else, prefer to be in the configuration that has the lowest energy associated with it. Therefore, molecules will be formed if the total energy of the electrons is lower when the atoms are combined. The total energy of the bonding wave functions is less than the total energy of the same number of atomic wave functions, but the total energy of the antibonding wave functions is greater. Therefore, a stable molecule will be formed if there are more bonding orbitals occupied by electrons than there are occupied antibonding orbitals.

We can now understand why O_2 (and also H_2, Li_2, N_2, and F_2) is a stable

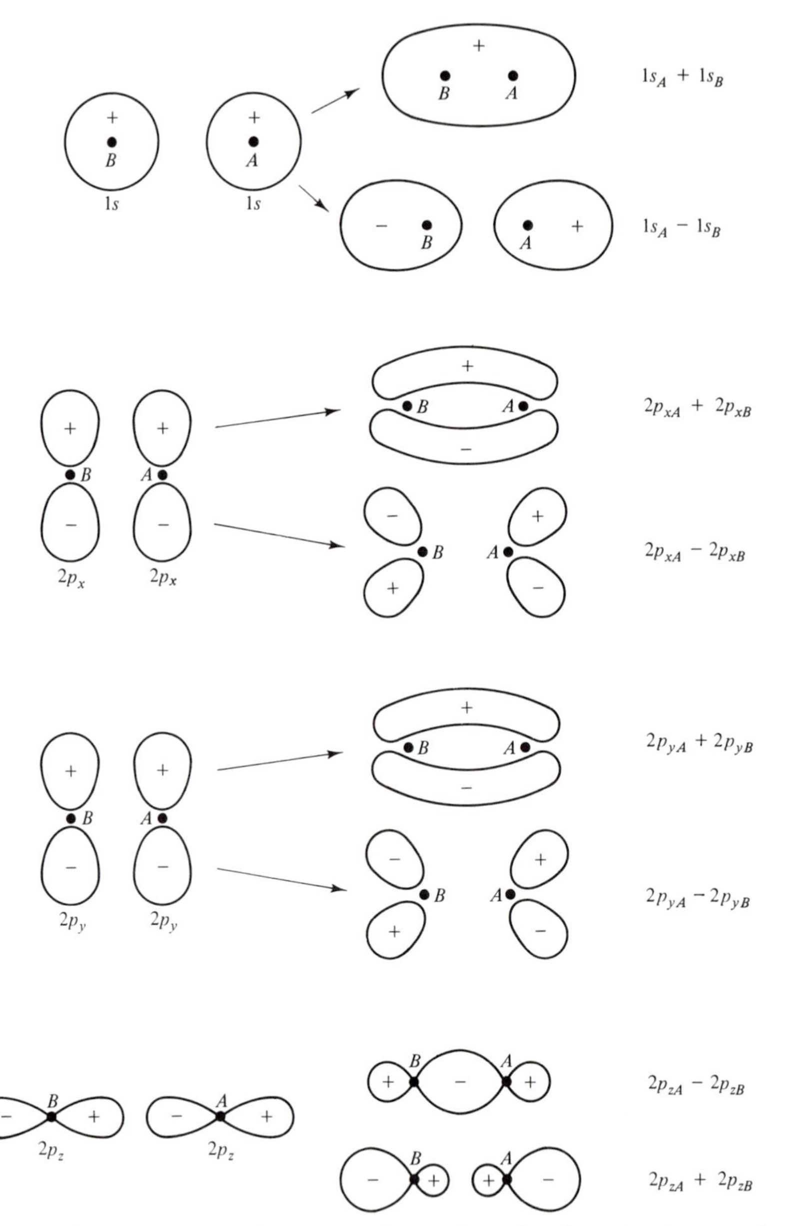

Figure 3-23 The four bonding and the four antibonding (*) molecular wave functions formed by adding or subtracting the atomic wave function of one atom with its identical counterpart in the other atom.

molecule. In the energy level diagram for the valence electrons in Fig. 3-24, one can see that the 12 valence electrons occupy 8 bonding and 4 antibonding energy levels. The energy of the O_2 molecule is less than the energy of the two separate O atoms; thus, O_2 is a stable molecule.

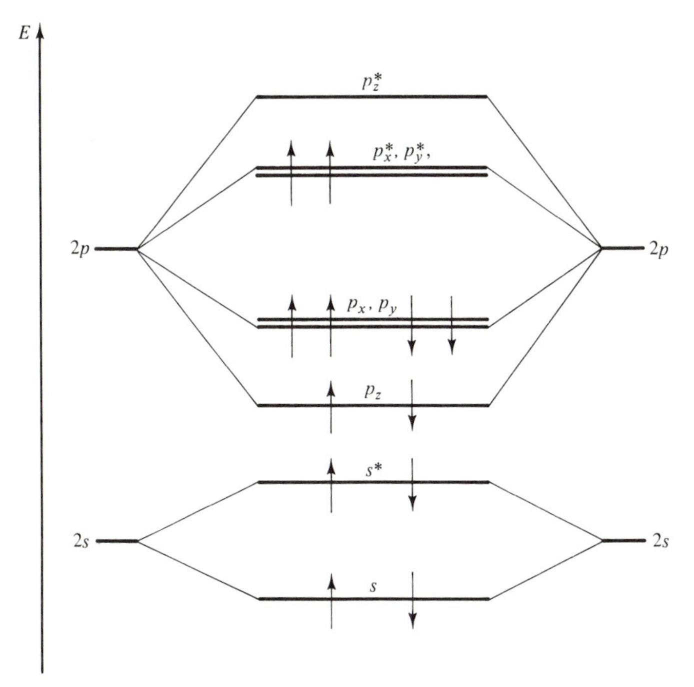

Figure 3-24 The molecular energy levels of the bonding and antibonding wave functions of the O_2 molecule.

The covalent bond is due to the quantum mechanical phenomenon called overlap. If the neighboring atomic orbitals positively overlap (as they do in bonding molecular orbitals), a covalent bond will be formed, and the strength of the bond increases with an increasing amount of overlap. The bond strength increases because the energy of the bonding orbitals decreases with increasing overlap. As the energy of the bonding orbital decreases with increasing overlap, the energy of the corresponding antibonding orbital increases. There is therefore said to be a greater splitting of the atomic energy level.

Like atoms the electronic states also have a fine structure. The designation of these states is found in a similar manner. For two atoms in which the atomic states have angular momentum quantum numbers L_1 and L_2, they combine to form a molecular angular momentum quantum number, Λ, where

$$\Lambda = L_1 + L_2, L_1 + L_2 - 1, \ldots |L_1 - L_2| \tag{3-28}$$

The equivalent Greek letters are also used to describe the states. For $\Lambda = 0$ the state is a Σ state; for $\Lambda = 1$, it is a Π state; for $\Lambda = 2$ it is a Δ state, and so on. The degeneracy is found by adding the atomic electron spins together. The degeneracy of the molecular state is $2S + 1$ where

$$S = S_1 + S_2, S_1 + S_2 - 1 \ldots |S_1 - S_2| \tag{3-29}$$

The degeneracy is again written as a superscript; a ${}^1\Sigma$ state is a singlet sigma state and a ${}^3\Pi$ is a triplet pi state.

3-6 BAND STRUCTURE OF SOLIDS

3-6.1 Pure Materials

It is now known that when two atoms are brought together to form a diatomic molecule each discrete atomic energy level breaks up into two discrete energy levels: the bonding and the antibonding levels. The question then is, what does the energy level diagram look like when a very large number (10^{23}) of atoms are brought together to form a solid? Let us see what happens qualitatively when the number of atoms increases from two to four. In Fig. 3-25*a** the number of bonding states has been increased from 8 to 16. The only difference is that the 8 new bonding states are at slightly different energies than the 8 old bonding states. They are at slightly different energies because the two original atoms perturb the energy levels of the two new atoms.

When the number of atoms is increased from four to N, the $N/2$ bonding levels associated with each bonding orbital (e.g., the s bonding orbital) are at slightly different energy levels. These energy levels are so close together that they can be considered to form a continuum, that is, they can be said to form an energy band. For reasons we will not discuss here, the s and p_z orbitals combine to form one band as do the p_x and p_y, the p_x^* and p_y^*, and the s^* and p_z^* orbitals. In Fig. 3-25*b* these levels are respectively referred to as the s, p, p^*, and s^* bands. The s band contains $N/2$ energy levels (and, therefore, N quantum states) from the s orbitals and $N/2$ energy levels from the p orbitals. Therefore, the s band contains $2N$ quantum states, that is, two quantum states per atom. The other three bands also contain two quantum states per atom.

In Fig. 3-25*b* the s and p bands overlap—the bottom of the p band is at a lower energy than the top of the s band. Because the s and p bands overlap, they are sometimes referred to collectively as the valence band. The s^* and p^* bands, which also overlap, are designated collectively as the conduction band. However, the p and p^* bands do not overlap. They are separated by an energy gap, E_g. This energy gap is the source of semiconductivity.

To understand how the energy gap is responsible for semiconductivity, let us see what happens when the valence electrons are in their lowest energy configuration. In Figure 3-25*b,* the s band is half filled when the material has one valence electron. The band is half filled because it has two quantum states per atom and, according to Pauli's exclusion principle, only one electron can occupy a quantum state.

This material is very conductive and, therefore, is a metal. It is conductive because the electrons at the top of the occupied energy levels can be excited to unoccupied energy levels with an infinitesimal amount of excitation energy. Once excited, the electron can move under the influence of a field because it can easily move to neighboring unoccupied energy levels. If all the neighboring levels were filled, the electron could not move because this would violate the

* Figure 3-25*a* differs from Fig. 3-24 in that the 2*s* energy levels are more similar to the 2*p* energy levels. Juggling acts like this often occur when you confine yourself to qualitative arguments.

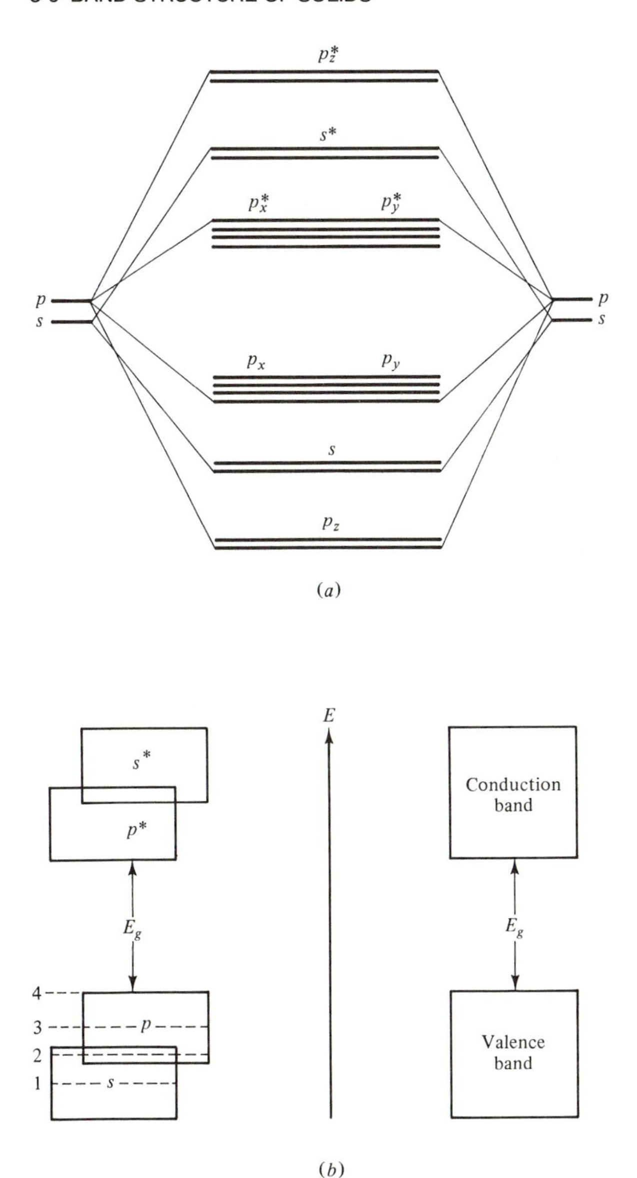

Figure 3-25 (*a*) The electronic energy levels formed when four similar atoms are brought together. (*b*) The electronic energy bands formed when N similar atoms are brought together to form a solid, and the ground state configuration (– – –) for solids having one, two, three, or four valence electrons per atom.

Pauli exclusion principle. These same arguments can be made for materials having two or three valence electrons.

However, if the material has four electrons per atom, the *s* and *p* bands are completely filled, and the p^* and the s^* bands are completely empty. Now the electrons at the top of the occupied levels cannot be excited into an unoccupied state with an infinitesimal amount of energy because there are *no* energy states in the energy gap. Rather, the electrons require enough energy to jump up into the p^* band. The number that can do this at a given temperature depends on the size of the energy gap; the larger the gap, the fewer that can make the jump. If the energy gap is very large (>3 eV), then essentially no electrons are able to jump over it, and the material is an electrical insulator.

In Section 3-5 we described how the splitting between the bonding and the antibonding levels increased as the amount of orbital overlap increased. It would, therefore, be logical to expect the energy gap to increase as the amount of overlap increases, since the energy gap results from the splitting between these two levels. As is shown in Table 3-2, the energy gap is largest for diamond and decreases as the amount of overlap decreases. One can make the same arguments for the III-V semiconductors. The relative size of the energy gaps of the three sequences shown in the table can also be explained by the amount of overlap if you remember that the smaller the atom, the greater is the overlap.

Another qualitative rule concerning the relative size of the energy gap is that E_g is larger in semiconductors containing the same average number of electrons per atom when the difference in the number of valence electrons between neighboring atoms is larger. For example, for the semiconductors from the Zn Ga Ge As Se sequence in the periodic table $E_g(\text{ZnSe}) > E_g(\text{GaAs}) > E_g(\text{Ge})$. Another way of saying this is that, everything else being equal, the more ionic the bond the larger the energy gap.

3-6.2 Doped Semiconductors and Insulators

Electronic states can be created in the energy gap by doping materials with chemical impurities. These impurity states can profoundly affect the electrical properties of semiconductors, and they can completely alter the optical properties of insulators.

In a semiconductor an electron in the conduction band is free to move; it is not bound to any one atom. One can, therefore, think of the energy difference between the bottom of the conduction band and the impurity energy state as the binding energy of an electron in that state.

An example of an n-type dopant is phosphorus in silicon. Phosphorus has one more valence electron than silicon; hence, there is one electron "left over" after four of them form bonds with the surrounding silicon atoms, as is shown in Fig. 3-26*a*. The fifth electron is bound to the phosphorus atom much like an electron is bound to a single valence atom, and this binding energy is given approximately by Eq. 3-11. Now, however, $k = \frac{1}{4}\pi\epsilon$ instead of $\frac{1}{4}\pi\epsilon_0$ so that the

TABLE 3-2 THE ENERGY GAPS (AT ABSOLUTE 0) OF THE GROUP IV AND SOME III-V SEMICONDUCTORS

C	7.0 eV	AlP	2.50 eV	GaP	2.40 eV	AlP	2.50 eV
Si	1.21 eV	GaAs	1.53 eV	GaAs	1.53 eV	GaP	2.40 eV
Ge	0.75 eV	InSb	0.23 eV	GaSb	0.78 eV	InP	1.41 eV

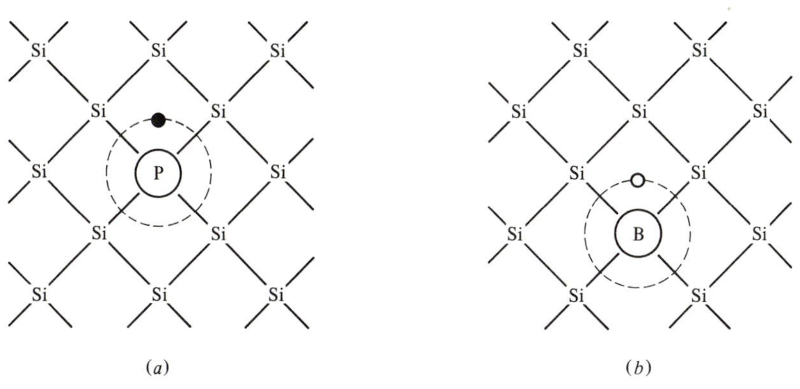

Figure 3-26 Schematic diagrams of (*a*) the *n*-type donor dopant, phosphorus, and (*b*) the *p*-type acceptor dopant, boron, in silicon.

binding energy is reduced by the factor $(\epsilon_0/\epsilon)^2$. For silicon, $\epsilon = 11.8\epsilon_0$, and for phosphorus $n = 3$ so that the binding energy of this donor electron, E_d, is $\sim$.03 eV. The donor level is illustrated in Fig. 3-27*a,* and the donor levels for a number of *n*-type dopants are listed in Table 3-3.

When silicon is doped with an impurity such as boron, which has only three valence electrons, one of the B—Si bonds is one electron short. The empty state associated with this bond is called an acceptor state, and it is located E_a above the valence band. The empty state bound to the boron atom is illustrated in Fig. 3-26*b,* and the acceptor level is illustrated in Fig. 3-27*b.* Semiconductors doped with acceptor impurities are said to be *p*-type, and a number of them are listed in Table 3-3.

The impurity states in the energy gap of insulators are distorted atomic states. They have atomiclike characteristics because the electrons in the states are not part of the chemical bond; they are the *d* electrons in transition metals and *f* electrons in the rare earths. The number of levels a single level can be split into depends on the crystal symmetry of the atomic site, and the magnitude of the splitting is determined by the strength of the interaction between the dopant and the crystal field. The symmetry of the site is designated by a letter further

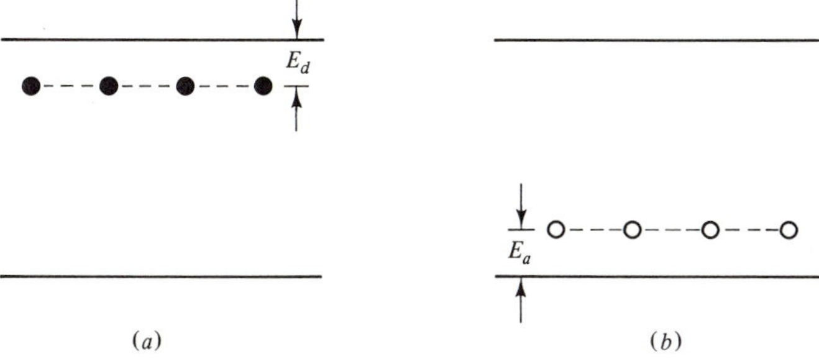

Figure 3-27 (*a*) Donor levels located E_d below the conduction band, and (*b*) acceptor levels located E_a above the valence band.

TABLE 3-3 THE DONOR AND ACCEPTOR LEVELS OF SEVERAL DOPANTS FOR THE SEMICONDUCTORS SILICON, GERMANIUM, AND GALLIUM ARSENIDE

Semiconductor	Dopant	E_d (eV)	E_a (eV)
Silicon	P	0.044	
	As	0.049	
	Sb	0.039	
	Bi	0.069	
	B		0.045
	Al		0.057
	Ga		0.065
	In		0.16
Germanium	P	0.0120	
	As	0.0127	
	Sb	0.0096	
	B		0.0104
	Al		0.0102
	Ga		0.0108
	In		0.0112
GaAs	O	Shallow	
	Se	0.005	
	Te	0.003	
	Si	0.002	
	Mg		0.012
	Zn		0.024
	Cd		0.021

complicating the nomenclature for the excited states. We will not discuss this problem.

The states are distorted by the crystal field of the matrix atoms. The crystal field can split one level with a number of states into a number of levels with fewer states; it can shift the levels; and it can spread a single level out into a band. The shifting of the levels by the matrix can be seen by looking at the chromium states in a ruby and in an emerald. Cr^{+3} ions are substituted for Al^{+3} ions in Al_2O_3 and are responsible for the red color of rubies. Cr^{+2} ions are substituted for Be^{+2} ions in BeO and are responsible for the green color of emeralds.

The rapidly changing electric fields of the crystal field can cause some atomic states to spread out into bands, since the atomic level depends on what the field strength is.

All of these phenomena are present in ruby and are responsible for the energy levels in Fig. 3-28. The ground state, which has the label ${}^4A_2$, originates from the atomic ground level, and has a degeneracy, $\rho_1 = 4$. The 2 levels, labeled ${}^4F_1$ and ${}^4F_2$, are bands ~ 1000 Å wide, and they, too, originate from the atomic ground state. The degeneracy of each band is 12. The 3 level is labeled ${}^2E$, and it has a degeneracy of 4. On closer scrutiny one can see that this level is split into two levels of degeneracy $2 \sim 15$ Å apart. It is the transition from the lower of these two levels to the ground state that is responsible for the room temperature

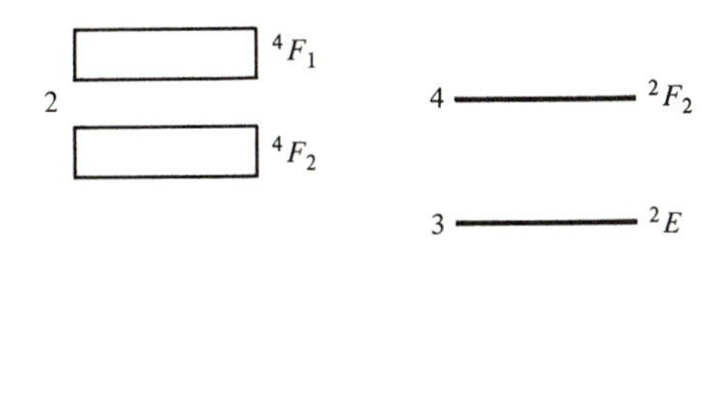

Figure 3-28 Schematic of a diatomic molecule. The "point masses" m_1 and m_2 are connected to each other by a spring with spring constant, k, and the center of mass is located a distance l from m_1.

6943 Å laser line. The fourth level, the $^2F_2$ level, has little effect on the optical properties.

3-7 VIBRATIONAL AND ROTATIONAL STATES

3-7.1 Vibrational States

The two atoms in a diatomic molecule are bound to each other and have an equilibrium separation distance, d. One must apply a force to either compress or extend the molecule along its axis and, to a good approximation, the displacement, x, is proportional to the applied force, that is, Hooke's law

$$F = k_f x \tag{3-30}$$

applies where k_f is the force, or spring, constant. Thus, the molecule can be represented by two masses, m_1 and m_2, connected together by a spring. This is illustrated in Fig. 3-29. According to classical mechanics, the energy of this harmonic oscillator is

$$E = \int_0^E dE = \int_0^{x_0} k_f x \, dx = \tfrac{1}{2} k_f x_0^2 \tag{3-31}$$

where x_0 is the original displacement. It is also the amplitude of vibration when the system is released and is allowed to vibrate freely at its natural frequency

$$\nu_0 = \frac{1}{2\pi} \sqrt{k_f/m_{\text{eff}}} \tag{3-32}$$

where m_{eff} is the effective mass.

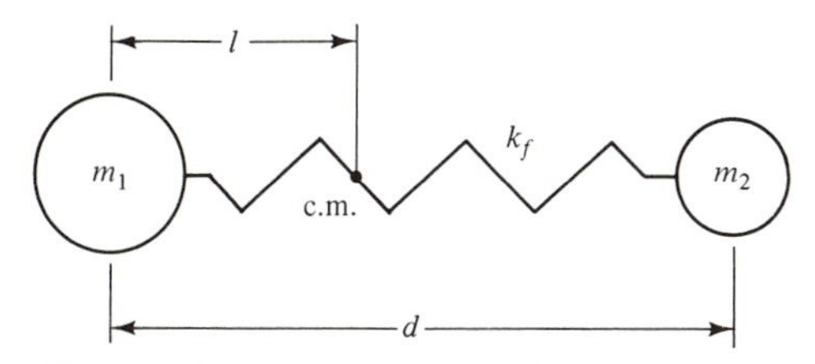

Figure 3-29 Schematic of a diatomic molecule. The "point masses" m_1 and m_2 are connected to each other by a spring constant, k_f, and the center of mass is located a distance, l, from m_1.

It is illustrative to derive an equation for the effective mass because it emphasizes that the center of mass located a distance, l, from m_1 in Fig. 3-29 remains fixed in any vibrating system. One can imagine cutting the spring at l and having the masses vibrating independently on springs of length l and $d - l$, respectively. Because the spring constant is inversely proportional to its length, the spring constant, k_{f1}, for mass, m_1, is

$$k_{f1} = k_f \frac{d}{l} \tag{3-33a}$$

From the center of mass equation

$$m_1 l = m_2(d - l) \tag{3-34a}$$

$$l = \frac{m_2}{m_1 + m_2} d \tag{3-34b}$$

so that

$$k_{f1} = \left(\frac{m_1 + m_2}{m_2}\right) k_f \tag{3-33b}$$

Thus, the effective mass in Eq. 3-32 is

$$m_{\text{eff}} = \frac{m_1 m_2}{m_1 + m_2} \tag{3-35}$$

Clearly, the same result would have been obtained if the mass, m_2, had been considered separately, since both atoms vibrate about the center of mass at the same frequency.

According to the classical equation for the energy, Eq. 3-31, the oscillator can have any energy. However, this is not allowed by quantum mechanics. The energy is again quantized and is given by the equation

$$E_v = (v + \tfrac{1}{2})h\nu_0 \tag{3-36}$$

where ν_0 is the natural vibration frequency and v is the vibration quantum number, which can have any integer value from 0 to infinity. These quantized vibrational states, which are frequently called phonons, along with the classical energy continuum are shown in Fig. 3-30.

Note that the ground state energy is $\frac{1}{2}h\nu_0$ and not 0, and that the energy levels are equally spaced. The ground state energy is not zero since, according to the Heisenberg uncertainty principle, we cannot precisely know the position and the momentum at the same time. If the ground state energy were zero, we would know what its precise position was at any time and we would know that its momentum would be zero. We will return to another implication of the uncertainty principle when we discuss emission peak widths in the next section.

The uniform spacing between the states allows for the easy transfer of energy between molecules in the higher energy state to molecules in the lower energy states. When one unit of energy is lost from one molecule, it can be absorbed easily by another, since the energy change for one transition is very

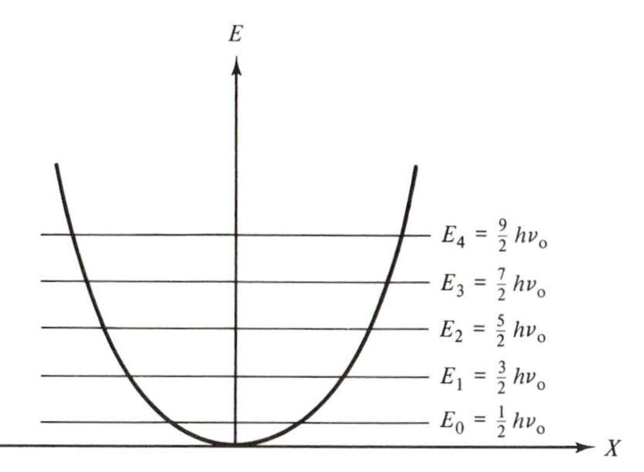

Figure 3-30 The potential energy, *E*, of a harmonic oscillator plotted as a function of the displacement, *x*, and the quantized energy levels.

nearly the same as the other. They will not be exactly the same because the molecule is not a perfect harmonic oscillator; some small deviations are present (see Problem 3-24). However, the differences in the transition energies are much less than the average thermal energy of the molecules, so that the small energy difference, depending on its sign, can easily be either absorbed from, or given up to, the translational kinetic energy of the gas.

In addition to the equal spacing of the energy states, quantum mechanical calculations show that the density of states at each energy level is 1, and the selection rule for the emission of radiation is $\Delta v = \pm 1$. The classical explanation for the selection rule is that the frequency of the emitted wave should be the same as that of the oscillating dipole.

For the oscillator to be an oscillating dipole, the molecule must have a dipole moment. The N_2 molecule in Fig. 3-31*a* does not form an oscillating dipole because, even as it vibrates, the center of negative charge of the electron cloud is at the same point as the center of positive charge of the nuclear cores. The situation is different, however, for the HCl molecule in Fig. 3-31*b* where the electron spends more time near the chlorine nucleus, thus moving the center of negative charge over toward the chlorine.

An important implication of the requirement that the oscillator must be an oscillating dipole to emit radiation is that the symmetric diatomic molecules such as N_2 only weakly absorb and emit radiation. As a result, an N_2 molecule excited into the first excited vibrational state by, for example, a collision with an accelerated ion will remain in an excited state for an extended period of time, since the molecule cannot easily return to the ground state by emitting a photon. We will return to this important point when the CO_2 laser is discussed.

Figure 3-31 (*a*) The superimposed center of positive charge for a symmetric molecule such as N_2, and (*b*) the separated center of positive and negative charge in an asymmetric molecule such as HCl.

The light emitted by an HCl molecule when it returns to the ground state from the first excited vibrational state has a wavelength, $\lambda = 3.47\ \mu m$, the frequency is $\nu_0 = 8.65 \times 10^{13}$ Hz, and the photon energy is $h\nu_0 = 0.36$ eV. This light is in the near infrared and represents the higher frequency end of typical first order phonon transitions. The HCl transition has a higher natural frequency both because its effective mass is relatively small and because its bond strength is relatively high. The large bond strength leads to a larger force constant.

Important vibrational transitions that we have already encountered in our discussion of absorption in optical fibers are the OH transitions. The fundamental transition is at $\lambda = 2.8\ \mu m$. The harmonics that appear in the wavelength region of interest between 0.8 and 1.6 μm are the first harmonic, $\Delta v = 2$, at 1.4 μm and the second harmonic, $\Delta v = 3$, at 0.93 μm. Classically, these higher order transitions occur because the potential seen by the vibrating molecule is not a perfect parabola (see Problem 3-24). However, the deviation is small so that the Fourier coefficients of the harmonics, and therefore the higher order transition probabilities, are small.

The only other molecule that will be discussed is the linear CO_2 molecule. As shown in Fig. 3-32, this molecule has three distinct natural frequencies. The first mode is a longitudinal mode with a natural frequency of $\nu_{01} = 4.18 \times 10^{13}$ Hz. The two oxygen atoms move in opposite directions, and the central carbon atom remains fixed. Thus, both the center of negative charge of the negatively charged oxygen atoms and the center of positive charge of the positively charged carbon atom are always located at the center; hence, no dipole moment is formed.

There actually are two modes with the natural frequency $\nu_{02} = 2.00 \times 10^{13}$ Hz—one vibrating in the plane of the paper and the other vibrating normal to it. A small dipole moment is generated by this mode since, when the positively charged carbon atom is moving up, the negatively charged oxygen atoms are moving down. The third mode has the highest natural frequency, $\nu_{03} = 7.06$

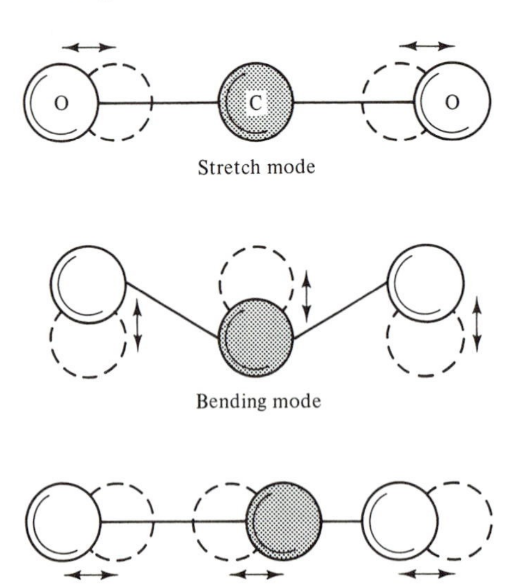

Figure 3-32 Three different modes of vibration for the linear CO_2 molecule.

$\times 10^{13}$ Hz and it, too, is a longitudinal mode. A substantial dipole is generated because the motion is asymmetric.

EXAMPLE 3-8

The prominent lasing transitions for the CO_2 laser are the (001) → (100) and (001) → (020) vibrational transitions in Fig. 3-33. What are the wavelengths of the photons emitted during these transitions? $\nu_2(020) = 3.935 \times 10^{13}$ Hz. For the (001) → (100) transition, $\nu = \nu_{03} - \nu_{01} = 2.88 \times 10^{13}$ Hz, $\lambda = c/\nu = 3 \times 10^8/2.88 \times 10^{13} = 10.4\ \mu m$; for the (001) → (020) transition, $\nu = \nu_{03} - \nu_{(020)} = 3.125 \times 10^{13}$ Hz, $\lambda = c/\nu = 3 \times 10^8/3.25 \times 10^{13} = 9.60\ \mu m$.

Transitions from the (001) state can be accomplished by an emission of a photon because an oscillating dipole is generated by this asymmetric mode. However, transitions from the (100) state do not produce photons, since an oscillating dipole is not generated by this mode, and it is unlikely that a photon will be emitted by a transition from the (020) state because the dipole is small. The most probable way that the molecule will be removed from these states is by colliding with a CO_2 molecule in the ground state transferring about one-half of the vibrational energy to it. This can be accomplished by the reactions:

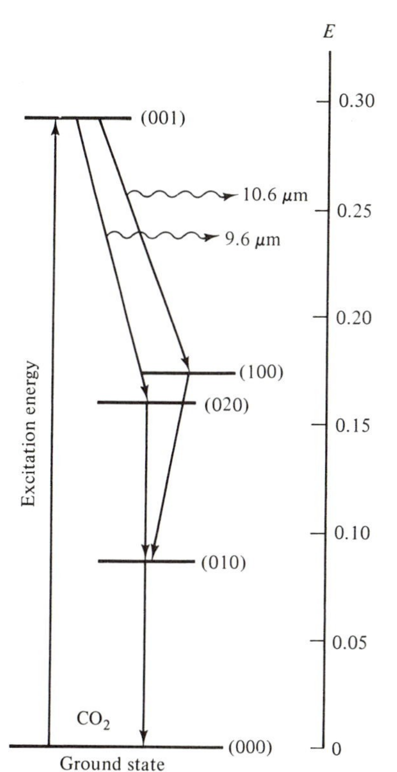

Figure 3-33 A partial energy level diagram for the low-lying stretching ($n_1$00), bending (0$n_2$0), and asymmetric (00n_3) modes.

$$CO_2(100) + CO_2(000) \rightarrow 2CO_2(010) + 7.45 \times 10^{-3}\ \text{eV} \quad (3\text{-}37a)$$

$$CO_2(020) + CO_2(000) \rightarrow 2CO_2(010) - 6.21 \times 10^{-3}\ \text{eV} \quad (3\text{-}37b)$$

The small amount of energy that must be emitted or absorbed by these transitions can easily be supplied to or absorbed from the translational energy of the gas, which is $\frac{3}{2}kT = 0.039$ eV. Because the transitions to the (010) state from (100) and (020) states can be made quickly, these upper states remain relatively empty. This makes it easier to achieve population inversion, a condition in which there are more molecules in the higher (001) state than there are in the lower (100) or (020) states. Population inversion is a condition that must be met for a system to lase. The molecules in the (010) state cannot easily return to the ground state because it is difficult for this mode to generate a photon or give its energy up to the translational energy, since it is substantially greater than $\frac{3}{2}kT$.

The transition energies for rotating molecules are similar to $\frac{3}{2}kT$, and we will discuss them in the next section.

3-7.2 Rotational States

The diatomic molecule in Fig. 3-34 acting as a rigid rotator rotating about its center of mass has a rotational kinetic energy of

$$E_r = \tfrac{1}{2}m_1v_1^2 + \tfrac{1}{2}m_2v_2^2 \quad (3\text{-}38a)$$

$$= \tfrac{1}{2}m_1(r_1\omega)^2 + \tfrac{1}{2}m_2(r_2\omega)^2 \quad (3\text{-}38b)$$

$$= \tfrac{1}{2}I\omega^2 \quad (3\text{-}38c)$$

where $I = \Sigma m_i r_i^2$ is the moment of inertia. Because the angular momentum is $L = I\omega$, Eq. 3-38c can also be written

$$E_r = \frac{L^2}{2I} \quad (3\text{-}39)$$

According to classical mechanics, L can have any value. However, it can be

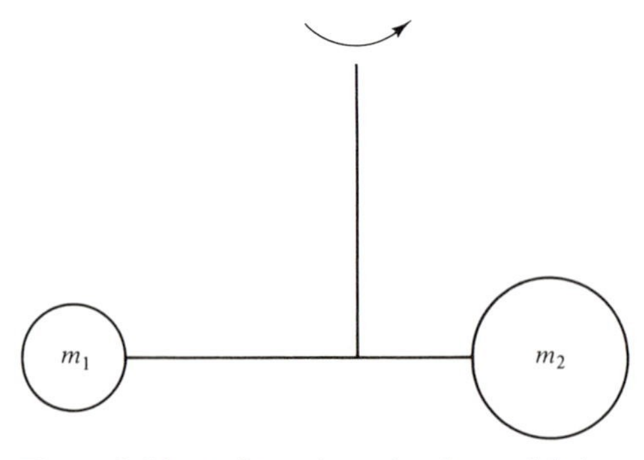

Figure 3-34 A diatomic molecule modeled as a rigid rotator.

shown by using quantum mechanics that L is quantized and is given by the familiar expression

$$L = \sqrt{J(J+1)}\hbar \tag{3-40}$$

where J in this equation is the molecular rotation quantum number and can have integer values from 0 to ∞. The z component of the angular momentum again can only have the values $m\hbar$ where $m = J, J-1 \ldots -J$ so that there are $2J+1$ orientations for a given L. Thus, the density of states for a given value of L is $2J+1$. The selection rule, $\Delta J = \pm 1$, is also similar, and for the molecule to be able to easily emit or absorb radiation, it must have a permanent dipole. Since $\Delta J = \pm 1$, the emitted photon will have the same frequency as the rotating molecule.

Substituting Eq. 3-40 into Eq. 3-39, for example, yields

$$E_r = \frac{\hbar^2}{2I} J(J+1) \tag{3-41}$$

One can see that the rotational energy levels are not evenly spaced; their separation increases as J increases. Also, rotational frequencies are lower than vibrational frequencies and the energy is accordingly lower. For HCl $\nu = 6.2 \times 10^{11}$ Hz, which is in the far infrared near the microwave region, and $\Delta E_r = 2.55 \times 10^{-3}$ eV, which is significantly less than the thermal kinetic energy at room temperature. Thus, the rotational states can quickly return to an equilibrium distribution after molecules have been perturbed by colliding with each other.

The number of molecules in each state is proportional to the density of states and to the probability that a state is occupied. This probability, in turn, is proportional to $e^{-E_r/kT}$. Therefore, the number of molecules in the Jth state, n_J, is

$$n_J \propto (2J+1) \exp\left[-\frac{(\hbar^2/2I)J(J+1)}{kT}\right] \tag{3-42}$$

At room temperature the pre-exponential factor, $2J+1$, dominates so that n_J initially increases with J. However, n_J soon reaches a maximum and then begins to decrease when the exponential term begins to dominate. This is shown in Fig. 3-35.

3-7.3 Multiple Transitions

A molecule can simultaneously undergo a combination of transitions. The dyes used in dye lasers have combined electronic, vibrational, and rotational transitions, and CO_2 has combined vibrational-rotational transitions. It is often assumed that the transitions are completely separate, but they are loosely coupled. For example, a vibrating atom has a slightly different moment of inertia from one that is not vibrating. However, the correction factors are usually less than 0.1%. Hence, they are frequently ignored.

We will only briefly consider the vibrational-rotational spectra of the CO_2 molecule. When the CO_2 molecule drops from the (001) vibrational state to the

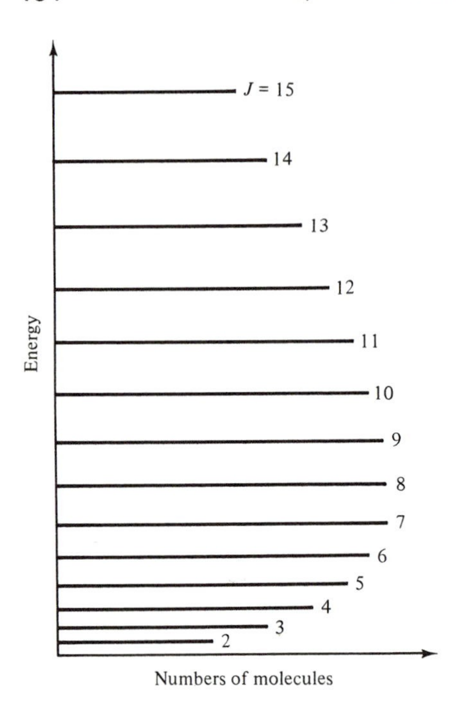

Figure 3-35 The rotational energy level diagram illustrating the increasing separation in the energy levels and that the population depends on the density of states as well as the energy.

(100) or (020) state, there can also be a change in the rotational energy level. If the value of J increases by 1 during the transition, it is called a P transition; if J is unchanged it is called a Q transition; and if J is reduced by 1 it is called an R transition. Because it is easier for a P transition to achieve the condition of population inversion, the lasing transitions are always P transitions. However, because the energy levels are not evenly spaced, the different P transitions will have different energies. Thus, the wavelength for the (001) $\rightarrow$ (100)P(12) transition is 10.5135 μm, where $J = 12$ in the (100) state, and the wavelength for the (001) $\rightarrow$ (100)P(38) transition is 10.7880 μm.

The vibrational and rotational levels for two different molecular electronic states are illustrated in Fig. 3-36. A transition can simultaneously involve changes in all three types of states, but the selection rules for each type of transition still apply.

3-8 EMISSION PEAK BANDWIDTHS

The light waves emitted by the transitions discussed in this chapter do not have a single frequency, as has been implied by the formulas that have been derived. Rather, they have a frequency distribution like that illustrated in Fig. 3-37. The formulas apply only to the peak frequencies.

The probability that the frequency lies between ν and $\nu + d\nu$ is $g(\nu)d\nu$, where $g(\nu)$ is called the line shape function. Because the probability that the wave has a frequency between $\pm\infty$ is one,

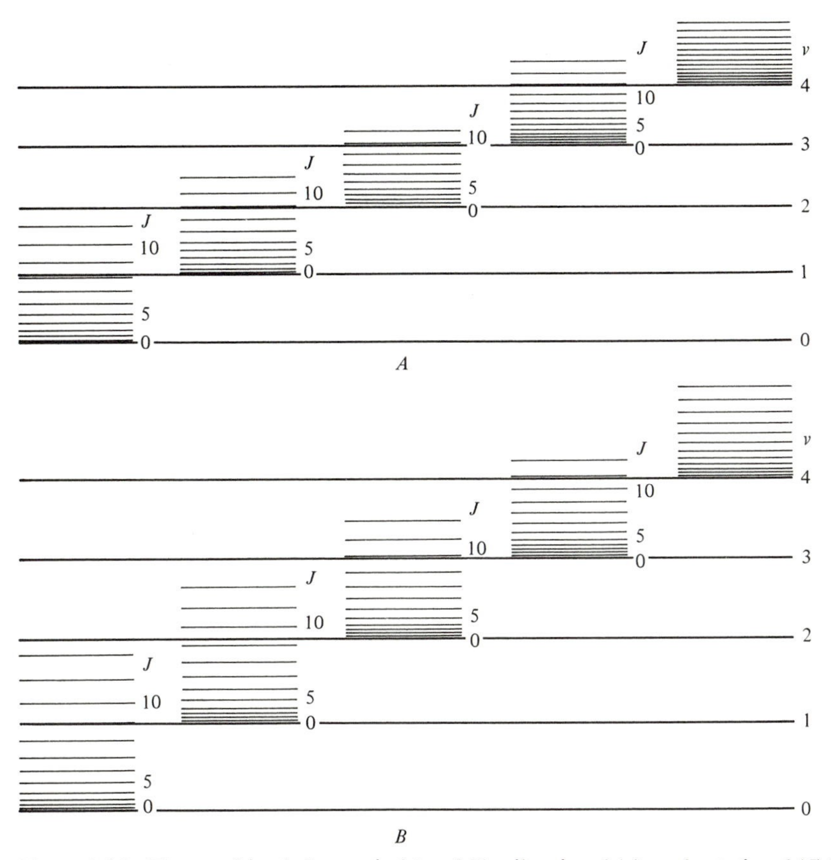

Figure 3-36 The combined electronic (A and B), vibrational (v), and rotational (J) levels of a molecule.

$$\int_{-\infty}^{\infty} g(\nu)d\nu = 1 \tag{3-43}$$

As is often the case, the bandwidth is defined to be the distance between the two half power points. For a number of calculations $g(\nu)$ is idealized by assuming that it is one between the half power points and zero elsewhere. This is illustrated in Fig. 3-37.

The quantum mechanical source of the frequency distribution is the Heisenberg uncertainty principle, which is that we cannot simultaneously know the exact energy of a state and the exact time an electron is in that state. Stated mathematically

$$\Delta E \, \Delta t \geq \hbar \tag{3-44}$$

where ΔE is the uncertainty in the energy of the state and Δt is the time an electron is in that state. Taking Δt to be the decay time, τ, and $\Delta E = h \, \Delta\nu$

$$\Delta\nu = \tfrac{1}{2}\pi\tau \tag{3-45}$$

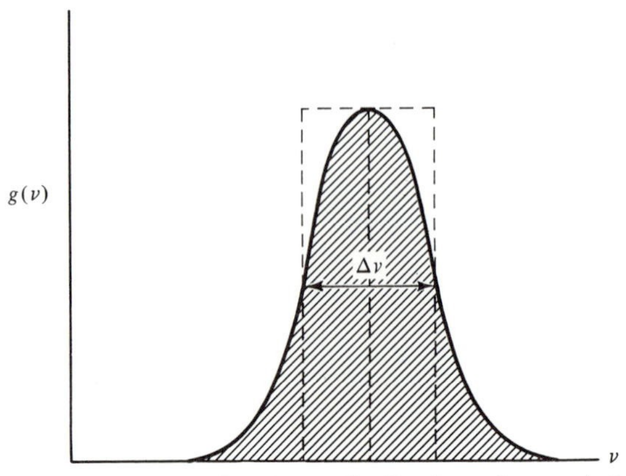

Figure 3-37 The frequency distribution of an emission peak ($\Delta\nu$), and the rectangular pulse approximation to it.

Thus, the faster the decay from the excited state, the wider the bandwidth. This is called homogeneous line broadening.

A classical analogy to this phenomenon is the decay of the charge on a capacitor plate in a simple high Q_l (quality factor) *LRC* series circuit when the system is short circuited at $t = 0$. In both the electrical and optical systems the systems are decaying from a higher to a lower energy state. The charge q is given by

$$q(t) = q_0 e^{-\alpha t} \cos \omega_d t \tag{3-46}$$

where

$$\alpha = \frac{R}{2L} \tag{3-47}$$

and

$$\omega_d = (\omega_0^2 - \alpha^2)^{1/2} \simeq \omega_0 \tag{3-48}$$

and ω_0 is the natural frequency of the system, $1/\sqrt{LC}$. The frequency distribution is

$$\begin{aligned} Q(\omega) &= \frac{q_0}{2}\int_0^\infty [e^{(-\alpha+j\omega_0)t} + e^{(-\alpha-j\omega_0)t}]e^{-j\omega t}\, dt \\ &= j\frac{q_0}{2}\left[\frac{1}{\omega_0 - \omega + j\alpha} - \frac{1}{\omega_0 + \omega - j\alpha}\right] \end{aligned} \tag{3-49}$$

The stored energy of the capacitor is

$$E(\omega) = \frac{1}{2C}|Q(\omega)|^2 \simeq \frac{q_0^2}{4C}\left[\frac{1}{(\omega_0 - \omega)^2 + \alpha^2}\right] \tag{3-50}$$

At the peak frequency, ω_0, $E(\omega) = q_0^2/4\alpha^2 C$; thus, the half power points ω_1 and ω_2 are

$$\omega_{1,2} = \omega_0 \pm \alpha \tag{3-51}$$

Thus, the bandwidth, $\Delta\nu$, is

$$\Delta\nu = \frac{\alpha}{\pi} = \frac{1}{\pi\tau} \tag{3-52}$$

$g(\nu)$ is proportional to $E(\nu)$, and the proportionality constant can be found from the version of Eq. 3-50

$$1 = \int_{-\infty}^{\infty} \frac{A\,d\nu}{(\nu_0 - \nu)^2 + (\Delta\nu/2)^2} \tag{3-53a}$$

so

$$g(\nu) = \frac{1}{2\pi(\nu_0 - \nu)^2 + (\Delta\nu/2)^2} \tag{3-53b}$$

Peaks that can be described by this equation are said to be Lorentzian.

Peaks that obey the relationship

$$g(\nu) = Ae^{-a(\nu_0 - \nu)^2} \tag{3-54}$$

are said to be Gaussian. This behavior is seen in gases because there is a frequency shift when a gas atom absorbs thermal energy from the surroundings or gives up energy to it. This shift is called the Doppler shift. The change in the frequency caused by a change in the velocity, Δv, of a gas atom is

$$\nu - \nu_0 = \frac{\Delta v}{c}\,\nu_0 \tag{3-55}$$

The average velocity change is equal to the rms thermal velocity in the direction of observation $\pm\sqrt{\langle v_x^2\rangle}$, which can be found from the equation for the average thermal energy

$$\tfrac{1}{2}m(\langle v_x^2\rangle + \langle v_y^2\rangle + \langle v_z^2\rangle) = \tfrac{3}{2}kT \tag{3-56}$$

or

$$\sqrt{\langle v_x^2\rangle} = \sqrt{\frac{kT}{m}} \tag{3-57}$$

On substitution into Eq. 3-55

$$\langle\Delta\nu\rangle = \frac{2\nu_0}{c}\sqrt{\frac{kT}{m}} \tag{3-58}$$

EXAMPLE 3-9

Find the line width of the neon line at 6328 Å at 300 K

$$\langle\Delta\nu\rangle = \frac{2}{c}\frac{c}{\lambda}\sqrt{\frac{kT}{m}}$$

$$= \frac{2}{6.328 \times 10^{-7}}\left[\frac{(1.38 \times 10^{-23})(3 \times 10^2)}{(20.183 \times 10^{-3})/(6.025 \times 10^{23})}\right]^{1/2}$$

$$= 1.11 \times 10^9 \text{ Hz}$$

The atomic weight of neon is 20.183.

The bandwidths of spectral lines in semiconductors produced by electrons falling from the valence band to the conduction band are broader than atomic line spectra because the electrons can fall from energy levels above the bottom of the conduction band. As the temperature increases, the average thermal energy of the conduction band electrons increases, and therefore the bandwidth increases.

In general, glasses have wider bandwidths than crystalline materials. This is because the environment of the optically active atoms is not the same in amorphous materials; hence, the emission peak of each atom will not be at the same frequency. The observed emission peak is the sum of all of the single emission peaks, and the envelope of these separate peaks clearly has a greater width than the individual peaks.

For much the same reason, inhomogeneously strained materials have broader emission peaks. A strain will shift a spectral line, and because the strain is different in different regions, the individual peaks will be shifted by a different amount.

READING LIST

1. A. Javan, "The Optical Properties of Materials," *Scientific American,* September 1968, pp. 239–248.
2. V. F. Weisskopf, "How Light Interacts with Matter," *Scientific American,* September 1968, pp. 60–71.
3. K. Nassau, "The Causes of Color," *Scientific American,* September 1980, pp. 124–154.
4. R. T. Weidner and R. L. Sells, *Elementary Modern Physics,* 2d Ed. Boston: Allyn and Bacon, 1973, Chapters 4 to 6.
5. R. M. Eisberg, *Fundamentals of Modern Physics,* New York: John Wiley, 1961, Chapters 5 to 8.
6. C. A. Wert and R. M. Thomson, *Physics of Solids,* 2d Ed. New York: McGraw-Hill, 1970, Chapters 7 and 8.
7. L. Pauling, *The Nature of the Chemical Bond,* 3d Ed. Ithaca, New York: Cornell University Press.
8. G. Herzberg, *Atomic Spectra and Atomic Structure,* 2d Ed. New York, Dover Publications, 1944.
9. G. Herzberg, *Molecular Spectra and Molecular Structure,* Vol. I. *Spectra of Diatomic Molecules,* 2d Ed. Princeton, NJ: D. Van Nostrand, 1950.

PROBLEMS

3-1. **(a)** How many copper atoms are there per gram?
(b) What is the volume of a grain of metal containing 10^{20} copper atoms?

3-2. **(a)** Calculate the velocity of an electron in the ground state of the H atom using the Bohr model.
(b) What is the de Broglie wavelength?

3-3. Electron microscopes utilize the fact that electrons have wavelike properties. What is the wavelength of an electron accelerated by a 50-kV potential?

3-4. Strictly speaking, the electron in the H atom and the proton revolve about the center of mass. This effect is accounted for by substituting the effective mass, $\mu = m_1 m_2 / m_1 + m_2$ for the electron mass into the equations developed by using the Bohr theory.

(a) Find an analytical expression for the difference in the wavelength of the photon emitted by a deuterium (an isotope of H with one neutron) atom and the corresponding photon emitted by an H atom.

(b) Calculate the difference in the wavelength of the first Balmer line in the H and D spectra.

3-5. A single electron moving in the field of a nucleus having a charge Z times the elementary charge has a total energy:

$$E = \frac{-13.6Z^2}{n^2}$$

The energy of the 1s levels in neutral free atoms of He and Li is about −24 and −65 eV, respectively. Calculate the energies for the single electron in singly ionized He and doubly ionized Li. What is a likely reason for the discrepancy between the 1s and the calculated values?

3-6. The energy of an electron revolving about a proton in an elliptical orbit is

$$E = \tfrac{1}{2}m\dot{r}^2 + \frac{L^2}{2mr^2} - \frac{kq^2}{r}$$

where $\dot{r}$ is the radial velocity and L is the value of the angular momentum.

(a) Derive this equation beginning with the equation $E = \frac{1}{2}mv^2 - (kq^2/r)$.

(b) Calculate the distance of closest approach, $r_{\min}$, and the distance of farthest approach, $r_{\max}$, and show that the energy associated with the orbit depends only on the length of the major axis of the ellipse, $r_{\min} + r_{\max}$.

(c) Show that the angular momentum associated with the orbit decreases as the orbit becomes more elliptical and the length of the major axis is held constant.

3-7. What is the ground state configuration for zinc? For cadmium?

3-8. Show that a maximum number of 14 electrons can be accommodated in an $f(l = 3)$ subshell. How many states are there in the $n = 4$ shell?

3-9. What is the maximum wavelength of a photon that can ionize an oxygen atom? The ionization energy is 13.61 eV.

3-10. An electron absorbs the energy from a photon of ultraviolet light (λ = 2768 Å). How many electron volts are absorbed?

3-11. Calculate the energy change (in kJ/mol) associated with the electronic transition in Hg atoms responsible for the emission of photons of wavelength 253.7 nm (1 nm = 10 Å).

3-12. In a sodium vapor lamp the visible light is produced with a wavelength equal to 5839 Å. How many photons are emitted per second if the lamp has a rating of 1 W?

3-13. **(a)** Find an expression for the probability of finding the electron of the hydrogen atom between r and $r + dr$ when it is in the 100 quantum state. (To do this you must integrate over all θ and all ϕ.)

(b) For what value of r is the probability the largest for an electron in this state?

(c) ψ_{100} decreases exponentially with r. Explain physically why the value of r for which the probability is a maximum is finite and not zero.

3-14. It can be shown that

$$\psi_{200} = \frac{1}{a_0\sqrt{2\pi a_0}}(2 - \rho)e^{-\rho/2}$$

(a) Sketch a plot of ψ_{200} versus ρ. Label the points where $\psi_{200} = 0$ and where it is a minimum.

(b) Construct a two-dimensional view of the ψ_{200} atomic orbital. Shade the area(s) [volume(s)] enclosed by the orbital and include the sign.

3-15. **(a)** Determine the values for Δ_s, Δ_p, and Δ_d for lithium using the energy level diagram in the figure.

(b) Using the results of part a, calculate the energy difference between the 5s and 2p levels. Check your answer with the wavelength of light emitted for this transition.

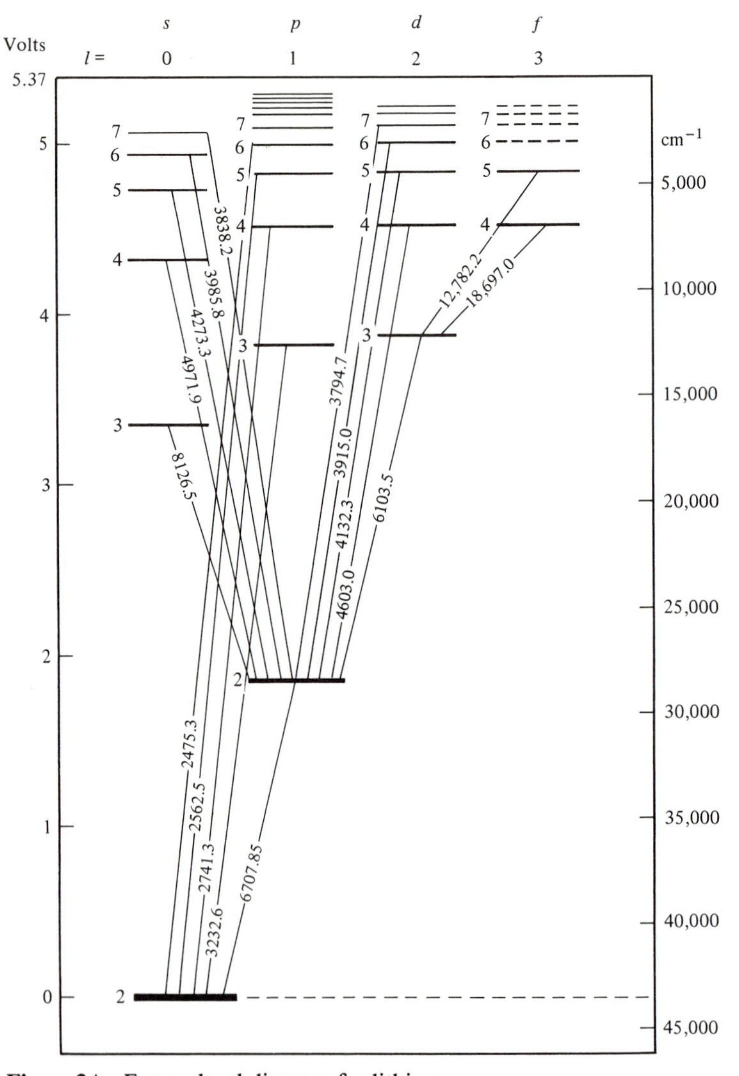

Figure 3A Energy level diagram for lithium.

3-16. Using a full sheet of millimeter graph paper, draw an energy level diagram so that it is as large as possible. Use the range 130,000 to 170,000 cm^{-1}. Use the energy levels listed below, which are given in wave numbers, and label the levels as given here. Take differences between levels to find which ones are involved in the lines at wavelengths:

(a) 6328 Å.
(b) 11,523 Å.
(c) 11,177 Å.
(d) What is
  (i) the smallest energy mismatch of the helium metastable levels with the levels of neon?
  (ii) What percentage of mismatch are these values?

Element	Electron configuration	Level designation	Energy, cm^{-1}
He	$1s^2$	$^1S_0$	0
He	$1s2s$	$^3S_1$ $^1S_0$	159,843 166,265
Ne	$2p^6$	$^1S_0$	0
Ne	$2p^53s$	$^3P_3$ $^3P_1$ $^3P_0$ $^1P_1$	134,042 134,460 134,820 135,889
Ne	$2p^53p$	*1*(1) *2*(3) *3*(2) *4*(1) *5*(2)	148,258 149.658 149,825 150,122 150,316

Element	Electron configuration	Level designation	Energy, cm^{-1}
Ne	$2p^53p$	*6*(0) *7*(1) *8*(2) *9*(0) *10*(0)	150,918 150,773 150,856 151,039 152,971
Ne	$2p^54s$	$^3P_2$ $^3P_1$ $^3P_0$ $^1P_1$	158,605 158,797 159,381 159,534
Ne	$2p^55s$	$^3P_2$ $^3P_1$ $^3P_0$ $^1P_1$	165,829 165,913 166,607 166,659

3-17. **(a)** The M—H bond energies in NH_3, PH_3, and AsH_3 are respectively 4.34, 3.31, and 2.54 eV. Explain this sequence.
(b) What are the wavelengths of the light that have the energy to break these bonds?

3-18. It takes approximately 10^{-19} cal to break the covalent bond between carbon and nitrogen. What wavelength would be required of a photon to supply this energy?

3-19. Using energy level diagrams, explain why the nitrogen molecule ion, N_2^+, has a smaller binding energy than N_2, whereas the oxygen molecule ion, O_2^+, has a larger binding energy than O_2.

3-20. **(a)** The ionization energies for the removal of the first electron are shown for the atom (I_1) and the binary molecule (I_2) in the table below along with the molecular binding energy (M.B.E.). From this data calculate the dissociation energy for the reaction $M_2^+ \rightarrow M + M^+$.
(b) Is the binding energy of N_2 greater than or less than that of N_2^+? Use an energy level diagram to explain your result.

Substance	I_1 (eV)	I_2 (eV)	M.B.E. (eV)
H	13.60	15.43	−4.48
N	14.53	15.58	−9.76
O	13.61	12.06	−5.11
F	17.42	15.7	−1.59

3-21. ZnO is transparent, CdS is a translucent yellow, and silicon is opaque. What does this say about the size of their energy gaps? Blue light has a wavelength $\lambda \cong 4000$ Å and red light has a wavelength $\lambda \cong 7000$ Å.

3-22. The energy gap of CdS is 2.42 eV.

(a) What is the maximum wavelength for a photon that has enough energy to excite an electron from the valence band into the conduction band?

(b) If light with this wavelength has a power density 0.1 W/cm^2, how many electrons are excited per centimeter squared per second?

3-23. If light of wavelength 2440 Å is incident on a platinum electrode, what retarding potential must be applied to the other electrode to guarantee that there is no photocurrent? The work function, which is the minimum energy needed to remove an electron from a metal, for platinum is 4.09 eV.

3-24. The potential energy curve of a diatomic molecule is composed of a core repulsion term, A/r^m, and an attraction term, $-B/r^n$. These two terms added together form a curve with a minimum energy, E_0, when the atoms are separated by the equilibrium distance, $r = d$.

(a) Find values for A in units of eV Å^m and B in units of eV Å^n for the CO molecule if the molecular binding energy is −9.144 eV, and the equilibrium separation distance is 1.1284 Å, $m = 12$ and $n = 6$.

(b) Draw an E(eV) versus r(Å) curve for the repulsion and attraction energies and for the sum of the two curves.

(c) If the reciprocal of the wavelength of the light emitted by the CO molecule when it jumps from its first excited state to the ground state is 2168 cm^{-1}, find the energy of the emitted photon.

(d) There are two values of r for which $E = E_0 + \Delta E$. They are $d - r_{min}$, the amplitude of vibration in the compression mode, and $r_{max} - d$, the amplitude of vibration in the stretching mode. If the molecule were a perfect harmonic oscillator, that is, the potential energy were a perfect parabola, these two amplitudes would be the same. Find these two amplitudes and compare them. (You should determine that they are, in fact, different. This is why the selection rule $\Delta v = \pm 1$ does not rigidly hold true, and why the vibration levels of the molecules are not precisely evenly spaced.)

(e) Find the two amplitudes when $E = E_0 + 10\Delta E$. Compare this result with the result in part d and comment on any trends that appear.

3-25. For the $J = 1$ to $J = 0$ transition for the HCl molecule, $1/\lambda = 20.6$ cm^{-1}. Calculate:

(a) The frequency and the energy of the photon emitted.

(b) The moment of inertia and the interatomic distance, d, for the molecule.

3-26. The wavelength of the light emitted by the (001) → (100)P(12) transition in CO_2 is 10.5135 μm, and the wavelength of the light emitted by the (001) → (100)P(38) transition is 10.788 μm.

(a) Calculate the frequency and the energy of the photons emitted.

(b) Calculate the moment of inertia and the $C = 0$ bond lengths.

(c) Determine the difference in the energy between the (001) and (100) states.

Chapter 4

Detectors

4-1 INTRODUCTION

There are a number of different types of detectors, but the discussion in this chapter is primarily limited to semiconductor detectors because they are the detectors used in fiber optical communication systems. A section is devoted to the photomultiplier, however, because of its universal applicability.

To understand how a semiconductor detector works, one must first learn about the electrical conduction process. This is done by first examining general conductivity, then examining conductivity in metals and in intrinsic and extrinsic semiconductors, and finally studying photoconductivity.

The photoconductivity cannot be quantified without first learning about generation and recombination. Electrons and holes are constantly being created and destroyed, but the concentrations remain constant at their equilibrium values, or at their steady state values when subjected to a constant intensity illumination. The steady state values are determined by both the light intensity and the recombination processes. The fact that the *pn* product is independent of the doping concentrations is also explained in terms of generation and recombination.

The constancy of the *pn* product is also explained by using the important concept of the Fermi energy. Understanding Fermi energies is absolutely crucial to understanding the *pn* junction.

The *pn* junction and its *i-V* curve are briefly discussed by examining the two types of current at the junction—the diffusion current and the electric field current. The diffusion current is the dominant current in an electrical diode, whereas the electric field current is the dominant current in a photodiode.

The general *pn* photodiode and the diffusion times, the transit times, and the junction capacitance, which limit the frequency response, are discussed in some detail. The *pin* diode, which eliminates the problem of the relatively slow

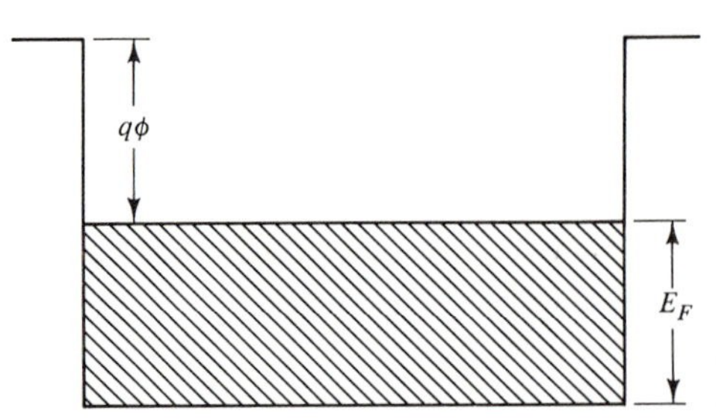

Figure 4-1 The work function, ϕ, and the Fermi energy, E_F, of a metal.

diffusion of the photocarriers to the junction, and the avalanche photodiode, which has internal gain, are also examined.

4-2 PHOTOMULTIPLIERS

Photomultiplier tubes are used to detect ultraviolet, visible, and near infrared light. They are very sensitive because they have high current gain and low noise, but they are not used in optical communication systems because they are too big, too expensive, and too fragile.

Their operation is quite straightforward. Recall that the valence electrons in a metal form an energy band. The difference in the energy of a free electron and an electron at the top of the band in Fig. 4-1 is q times the work function, $q\phi$, and a photon must have an energy greater than or equal to $q\phi$ to remove an electron from the metal. The photons are counted by collecting the electrons removed by the photons and measuring their current.

To detect the longer wavelength photons, one must have a small work function material. Materials often used for this purpose are Ag-O-Cs and Sb-Cs alloys, which have $\phi = 1.5$ V compared with 4–4.5 V for the average metal.

The electrons ejected from the cathode in Fig. 4-2 are collected and magnified by electrodes called dynodes, which are at progressively higher potentials. Frequently, the potential difference between adjacent dynodes is 100 V. For every electron colliding with a dynode, δ are emitted; δ is called the secondary

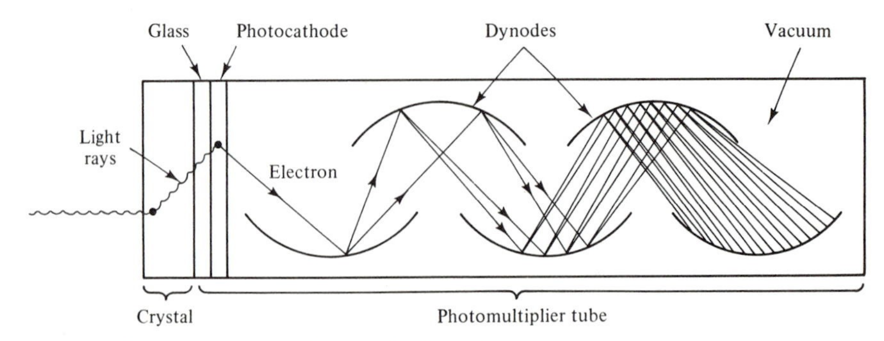

Figure 4-2 Schematic of a photomultiplier tube.

electron multiplication factor. There are often nine dynodes inside of the vacuum tube and $\delta \simeq 5$. Thus, a typical current gain, G, is $G = 5^9 \simeq 2 \times 10^6$.

For an individual photocathode there are two figures of merit. They are the internal quantum efficiency, η_i, and the responsivity, $\mathcal{R}$. The internal quantum efficiency is the fraction of the photons that create a photoelectron; in the ideal case $\eta_i = 1$, but in reality it is <1 and is a function of the wavelength. This is shown in Fig. 4-3. The responsivity is the current generated per unit of incident power. Thus $\mathcal{R}$ increases with the wavelength when η_i is constant because the energy per photon decreases.

Data for different photocathodes are represented in the responsivity versus wavelength diagram in Fig. 4-3. The current is given by

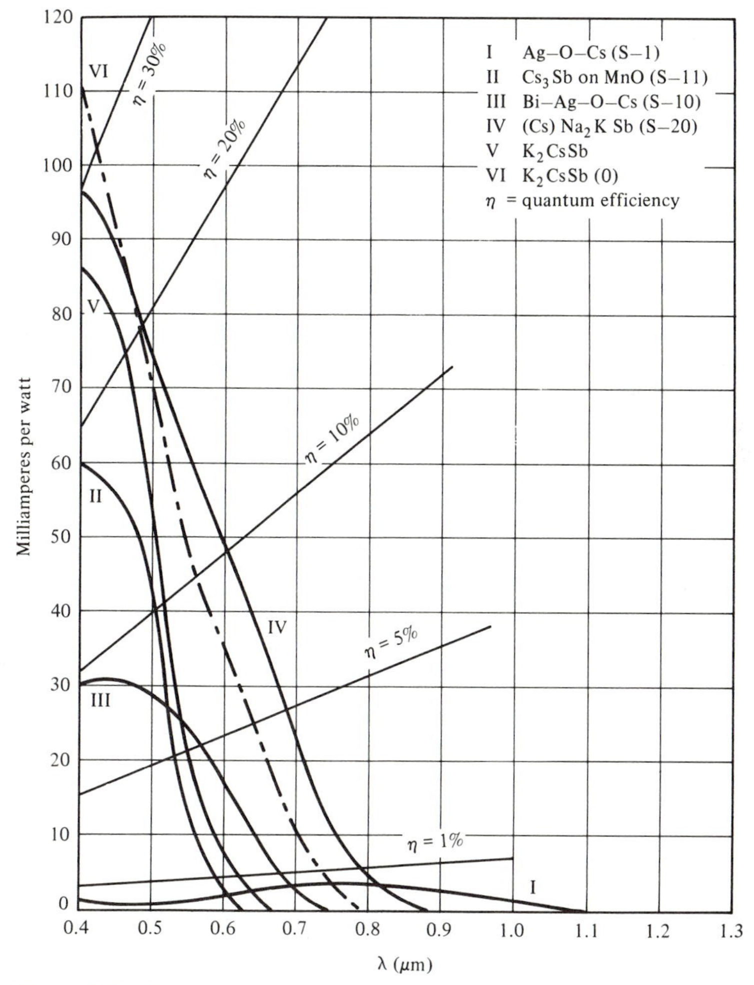

Figure 4-3 Photoresponse versus wavelength characteristics and quantum efficiency of a number of photocathodes. (From A. H. Sommer, *Photo-Emissive Materials,* Copyright © 1968 by John Wiley & Sons, New York.)

$$i = \frac{\eta_i q I A}{E_p} \tag{4-1}$$

where q is the charge on an electron, I is the incident intensity, A is the area, and E_p is the energy of the photon.

EXAMPLE 4-1

Using Fig. 4-3, determine the internal quantum efficiency of the $(C_s)Na_2KSb$(S-20) photocathode at $\lambda = 0.7\ \mu$m. From Fig. 4-3, $i/IA = 24 \times 10^{-3}$ A/W. From Eq. 4-1, $i/IA\eta_i = q\lambda/hc$.

$$\frac{i}{\eta_i IA} = \frac{q\lambda}{hc} = \frac{(1.6 \times 10^{-19})(7 \times 10^{7})}{(6.62 \times 10^{-34})(3 \times 10^{8})} = 5.64 \times 10^{-1}$$

Therefore

$$\eta_i = 24 \times \frac{10^{-3}}{5.64} \times 10^{-1} = 4.26 \times 10^{-2}$$

4-3 ELECTRICAL CONDUCTIVITY

4-3.1 General Conductivity

First, we consider charge carriers with no activation energy or quantum mechanical restrictions on their motion. An example of this situation is ions in a solution. When an electric field, ξ, is applied in the positive x direction as it is in Fig. 4-4, there is a net motion of positive charge to the right and of negative charge to the left. The net charge crossing unit area per unit time is the current density, J.

The fact that it is the *net* charge crossing a cross-sectional area and not the total charge should be emphasized. This is because the charge carriers are in constant thermal motion. There is no current when no field is applied, since

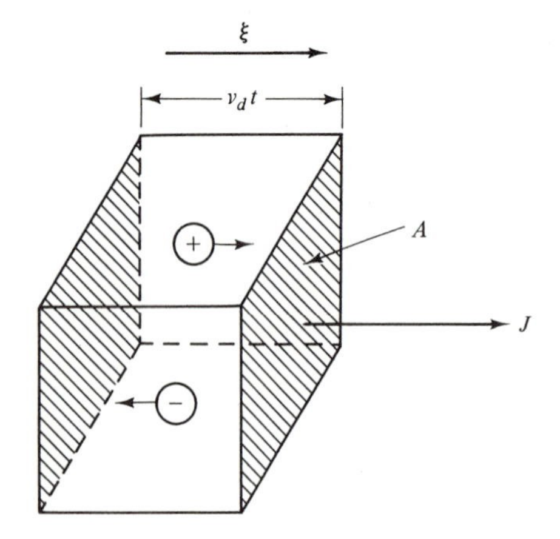

Figure 4-4 The net motion of charge carriers across the cross-sectional area, A, when an electric field is applied.

there are as many positive ions moving to the right as there are to the left; this is also true for the negative ions. There is a current when a field is applied because the positive ions have additional velocity components in the positive x direction, and the negative ions have additional velocity components in the negative x direction. This average additional velocity component is called the drift velocity, v_d, and only this velocity determines the current density.

In Fig. 4-4 it is seen that all of the positive ions within $v_d t$ of the cross-sectional area, A, cross A in time, t. Thus, the total net positive charge, Q^+, crossing the area in time, t, is

$$Q^+ = Z^+ q p v_{dp} t A \tag{4-2a}$$

where Z^+ is the positive ion valence, q is the electronic charge, and p is the positive ion density. One must also consider the charge Q^- of the negative ions moving to the left. By inspection

$$Q^- = Z^- q n v_{dn} t A \tag{4-2b}$$

Q^- like Q^+ is a positive quantity, since both Z^- and v_{dn} are negative. The total net charge crossing A in time, t, is then

$$Q = Atq(Z^+ p v_{dp} + Z^- n v_{dn}) \tag{4-2c}$$

Since J is the net charge crossing unit area per unit time,

$$J = Z^+ q p v_{dp} + Z^- q n v_{dn} \tag{4-3}$$

The current density is proportional to the electric field, since the drift velocity is proportional to ξ. The constant of proportionality is the conductivity, σ. Thus

$$J = \sigma \xi \tag{4-4}$$

The drift velocity is also proportional to the field, and the constant of proportionality is the mobility, μ, which is a characteristic of the carrier. Thus

$$v_d = \mu \xi \tag{4-5}$$

Combining Eqs. 4-3, 4-4, and 4-5 yields

$$\sigma = Z^+ q p \mu_p + Z^- q n \mu_n \tag{4-6}$$

The ionic charges, ion densities, and ion mobilities are all characteristics of the solution; hence, the conductivity must also be a solution characteristic.

The transference number, $T^{\pm}$, is the fraction of the current carried by the positive or negative ions. Therefore

$$T^+ = \frac{J_p}{J_p + J_n} = \frac{Z^+ q p \mu_p \xi}{Z^+ q p \mu_p \xi + Z^- q n \mu_n \xi} = \frac{\mu_p}{\mu_p + \mu_n} \tag{4-7}$$

for an electrically neutral solution.

The mobility can be further described by the carrier charge and mass and the average time between ion collisions, 2τ. This is done with the aid of Fig. 4-5. If one assumes that upon collision the drift velocity drops to zero and between

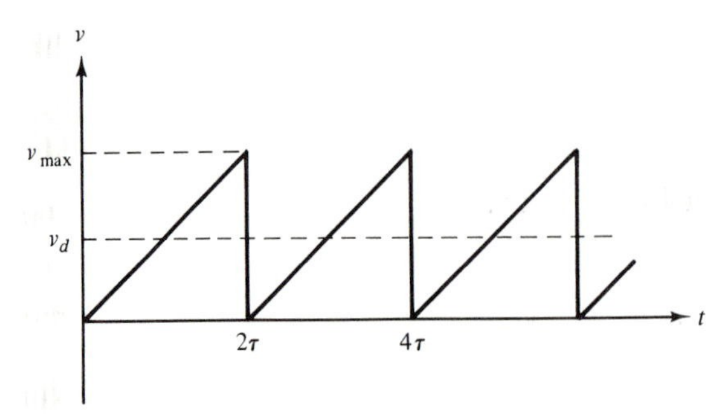

Figure 4-5 The maximum drift velocity obtained between collisions by a positively charged particle in an electric field, and the average drift velocity.

collisions the particle acceleration, a, is constant, then the maximum velocity attained is

$$v_{\max} = a(2\tau) = \frac{Zq\xi}{m}(2\tau) \tag{4-8}$$

since the force producing the acceleration is $Zq\xi$ and m is the mass of the ion. The drift velocity is the average electric field induced velocity. Thus

$$v_d = \tfrac{1}{2}v_{\max} = \frac{Zq\tau\xi}{m} \tag{4-9a}$$

or

$$\mu = \frac{Zq\tau}{m} \tag{4-9b}$$

The average distance traveled by the ions between collisions is the mean free path, l, and it is given by

$$l = (v_d + v_T)2\tau \tag{4-10}$$

where v_T is the thermal velocity.

The parameters that are usually measured are the current, i, the voltage, V, and the resistance, R. These terms are not characteristic of the material, since they also depend on the geometry. They are related to the more fundamental parameters by the following relationships. The current is simply

$$i = JA \tag{4-11}$$

The potential difference, or voltage between points a and b, is

$$V = \int_a^b \xi\, dx \tag{4-12a}$$

which becomes

$$V = \xi(b - a) = \xi L \tag{4-12b}$$

when ξ is constant.

$$R = \frac{V}{i} = \frac{\xi L}{JA} = \frac{L}{\sigma A} \tag{4-13}$$

EXAMPLE 4-2

A $0.1N$ K_2SO_4 solution in a tube with a cross-sectional area of 0.230 cm^2 and electrodes 5.00 cm apart produces a current of 14.1 ma when a voltage of 10.0 V is applied. The voltage is applied for 1 h. The K_2SO_4 is constantly replenished so the concentration remains constant; the K_2SO_4 completely dissociates into K^{+1} and SO_4^{-2} ions; the mobility of the K^{+1} ion is 0.762 $\times 10^{-3}$ $cm^2/V \cdot sec$, and the volume of the solution is 1.00 L. Calculate R, Q, J, ξ, σ, p, n, μ_n, v_{dp}, v_{dn}, t_p, t_n, the average time it takes each ion to move from one electrode to the other, and T^+.

$$R = \frac{V}{i} = \frac{10.0}{14.1 \times 10^{-3}} = 709\ \Omega$$

$$Q = it = (14.1 \times 10^{-3})(3.6 \times 10^3) = 50.8\ \mathrm{C}$$

$$J = \frac{i}{A} = \frac{14.1}{0.230} = 61.3\ \mathrm{mA/cm^2}$$

$$\xi = \frac{V}{L} = \frac{10.0}{5.00} = 2.00\ \mathrm{V/cm}$$

$$\sigma = \frac{J}{\xi} = 61.3 \times \frac{10^{-3}}{2.00} = 3.07 \times 10^{-2}\ \mathrm{S/cm}$$

$$p = \frac{2(0.1)N_0}{10^3} = (2 \times 10^{-4})(6.03 \times 10^{23}) = 12.1 \times 10^{19}\ \mathrm{cm^{-3}}$$

$$n = \frac{p}{2} = 6.03 \times 10^{19}\ \mathrm{cm^{-3}}$$

$$\mu_n = \frac{\sigma - Z^+ q p \mu_p}{Z^- q n}$$

$$= \frac{[3.07 \times 10^{-2} - (1.6 \times 10^{-19})(12.1 \times 10^{19})(0.762 \times 10^{-3})]}{2(1.6 \times 10^{-19})(6.03 \times 10^{19})}$$

$$= 0.829 \times 10^{-3}\ \mathrm{cm^2/V \cdot sec}$$

$$v_{dp} = \mu_p \xi = (0.762 \times 10^{-3})2.00 = 1.52 \times 10^{-3}\ \mathrm{cm/sec}$$

$$v_{dn} = \mu_n \xi = (0.829 \times 10^{-3})2.00 = 1.66 \times 10^{-3}\ \mathrm{cm/sec}$$

$$\tau_p = \frac{m_p \mu_p}{q} = \frac{(39.1)(0.762 \times 10^{-3})}{(6.03 \times 10^{23})(1.6 \times 10^{-19})} = 3.09 \times 10^{-7}\ \mathrm{sec}$$

$$\tau_n = \frac{m_n \mu_n}{q} = \frac{[32.1 + 4(16)](0.829 \times 10^{-3})}{(6.03 \times 10^{23})(1.6 \times 10^{-19})} = 8.26 \times 10^{-7}\ \mathrm{sec}$$

$$t_p = L/v_{dp} = \frac{5.00}{1.52 \times 10^{-3}} = 3.29 \times 10^3\ \mathrm{sec}$$

$$t_n = L/v_{dn} = \frac{5.00}{1.66 \times 10^{-3}} = 3.01 \times 10^3 \text{ sec}$$

$$T^+ = \frac{\mu_p}{\mu_p + \mu_n} = \frac{0.762 \times 10^{-3}}{(0.762 + 0.829) \times 10^{-3}} = 0.479$$

In concluding this section, we should point out that it has implicitly been assumed that the solution was an isotropic medium. In an isotropic medium the conductivity will be the same in all directions, and the current density will everywhere be parallel to the applied electric field. Examples of electrically isotropic materials are amorphous materials, polycrystalline materials in which the grains are randomly oriented, and single crystals with cubic crystal structures. Single crystals with crystal structures belonging to the other six crystal systems are electrically, as well as optically, anisotropic.

4-3.2 Metallic Conductivity

In Section 3-6.2 it was shown that the valence electrons are not bound to the individual atoms. Rather, they form an electron cloud that is bound to the metal by a flat potential energy well. In agreement with Pauli's exclusion principle, no two electrons occupy the same quantum state so that the electrons occupy successively higher kinetic energy quantum states. In so doing, they form a partially occupied energy band in which most of the electrons have a substantial kinetic energy. The uppermost occupied level has an energy, E_F (the Fermi energy), greater than the bottom of the potential well and an energy $q\phi$ less than that of a free electron.

All of the valence electrons participate in the conduction process even though their motion is constrained by the Pauli principle because the process is a cooperative phenomenon. This can be seen with the assistance of Fig. 4-6. First, the electrons are divided into two groups—those with a velocity component in the + direction and those with an equal but opposite velocity component. The conduction process can now be described by the following sequence of

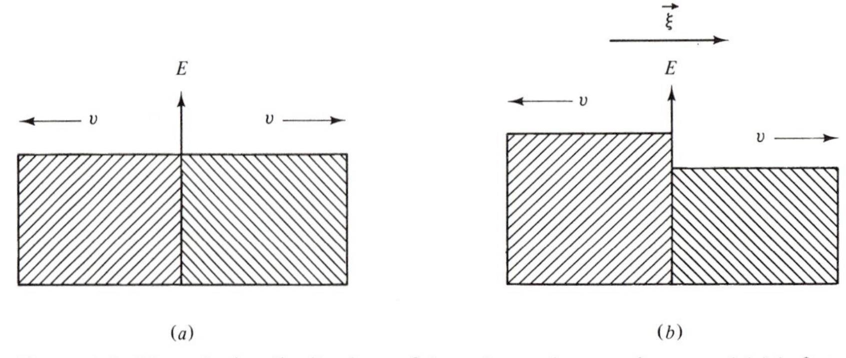

Figure 4-6 The velocity distributions of the valence electrons in a metal (*a*) before and (*b*) after an electric field has been applied.

events. These events occur simultaneously, but for clarity we will discuss them individually.

When an electric field is applied, the electron moving to the left with a kinetic energy, E_F, is accelerated and given an additional average component of velocity, v_d. The electron is now in a higher energy state so the state originally occupied by it is now empty. The electron with a kinetic energy slightly less than E_F can now jump to the state with an energy, E_F, leaving an empty state behind. This process continues with the net result of all those electrons originally moving to the left being given an additional velocity component equal to v_d, and those originally moving to the right having their velocity reduced by an amount v_d, that is, all of the electrons are given a drift velocity component to the left.

Until now we have assumed that the electrons do not interact with the lattice. They do, in fact, interact, but rather than consider these complicated interactions one can simply account for them by multiplying the mass of the electron by an experimentally determined constant. This new mass is called the effective mass and henceforth it will be used in all of the calculations involving electronic conduction. Some effective mass to electron mass ratios, m_n^*/m_n, along with the room temperature conductivities of some typical metals, are listed in Table 4-1.

Be certain to note that the conductivities do not have MKS units because the unit of length used is the centimeter and not the meter.

EXAMPLE 4-3

For sodium, which has a density of two atoms in a cubic volume 4.29 Å on a side, $\sigma = 2.17 \times 10^5$ S/cm at 300 K, $m^*/m = 1.2$, $E_F = 2.5$ eV. Calculate τ, v_d, and l when the applied field is 1.0 V/cm.

$$\tau = \frac{m^*\sigma}{q^2 n} = \frac{(1.2)(9.11 \times 10^{-31})(2.17 \times 10^7)}{(1.6 \times 10^{-19})^2} \cdot \frac{(4.29 \times 10^{-10})^3}{2}$$

$$= 3.63 \times 10^{-14} \text{ sec}$$

$$v_d = \mu\xi = \frac{\sigma\xi}{qn} = \frac{(2.17 \times 10^7)(10^2)}{(1.6 \times 10^{-19})} \cdot \frac{(4.29 \times 10^{-10})^3}{2} = 0.531 \text{ m/sec}$$

$$l = (2\tau)(v_d + v_T)$$

$$v_T = \left(\frac{2E_F}{m^*}\right)^{1/2} = \left[\frac{(2)(2.5)(1.6 \times 10^{-19})}{(1.2)(9.11 \times 10^{-31})}\right]^{1/2}$$

$$= 8.55 \times 10^5 \text{ m/sec}$$

$$l \simeq 2\tau v_T = (2)(3.63 \times 10^{-14})(8.55 \times 10^5) = 6.21 \times 10^{-8} \text{ m} = 621 \text{ Å}$$

TABLE 4-1 THE ROOM TEMPERATURE CONDUCTIVITIES AND EFFECTIVE MASS RATIOS FOR SOME REPRESENTATIVE METALS

Metal	Ag	Cu	Au	Al	Na	W	Cd	Ni	Fe	Pb
$\sigma \times 10^{+5}$ s/cm	6.15	5.82	4.09	3.55	2.17	1.80	1.30	1.28	1.00	0.45
m^*/m	—	1.0	—	0.97	1.2	—	—	—	—	—

4-3.3 Semiconductivity

We learned in Section 3-5.1 that a semiconductor such as Si, GaAs, or CdS has an average of four electrons per atom, and that at absolute zero the valence band of a pure semiconductor is completely filled and the conduction band is completely empty. All of the valence electrons cannot now take part in the conduction process because the conduction band is separated from the valence band by an energy gap, E_g. The electrons at the top of the valence band cannot absorb a small amount of energy and leave an unoccupied site behind because an electron cannot have an energy that lies in the energy gap.

Electrons now conduct electricity by being thermally excited into the conduction band. Also, when an electron is excited out of the valence band, it leaves behind an empty state which electrons in the valence band can jump to. Instead of attempting to keep track of the electrons that jump to this empty state, it is much simpler to follow the motion of the empty state. This empty state is called a hole, and it can be treated as if it has a positive charge. The effective mass of a hole, m_p^*, is usually larger than the effective mass of a conduction band electron, m_n^* (see Table 4-2), and this is reflected, in part, in the smaller hole mobilities listed in Table 4-3. Also, during the rest of this chapter the simple word, "electron," will be used interchangeably with conduction band electron.

Electron-hole pairs (EHPs) are constantly being created and destroyed. An EHP is thermally created, the hole and electron diffuse an average distance, L_n or L_p, which is called a diffusion length, and then they are annihilated. This is shown schematically in Fig. 4-7. At equilibrium the same number of electrons and holes are created as are destroyed.

For a pure, or intrinsic, semiconductor Eq. 4-6 becomes

$$\sigma_i = n_i q(\mu_p + \mu_n) \tag{4-14}$$

since $p = n$. σ_i is the intrinsic conductivity and n_i is the intrinsic carrier concentration. The most temperature sensitive component is n_i as

$$n_i \mathrel{\tilde{\propto}} e^{-E_g/2kT} \tag{4-15a}$$

The component n_i has a $-E_g/2kT$ dependence because the probability an electron has sufficient energy to jump across the energy gap is $\propto \exp(-E_g/2kT)$. The factor of two is needed because one-half of the bandgap energy is used to create

TABLE 4-2 VALUES FOR A NUMBER OF PARAMETERS FOR THE FREQUENTLY USED SEMICONDUCTORS, GERMANIUM, SILICON, AND GALLIUM ARSENIDE

Parameter	Ge	Si	GaAs
m_n^*/m	0.55	1.1	0.068
m_p^*/m	0.37	0.56	0.5
n_i (300° K) (cm^{-3})	2.5×10^{13}	1.2×10^{10}	1.1×10^{7}
D_n (cm^2/sec)	100	35	220
D_p (cm^2/sec)	50	12.5	10
B (cm^3/sec)	5.25×10^{-14}	1.79×10^{-15}	7.21×10^{-11}

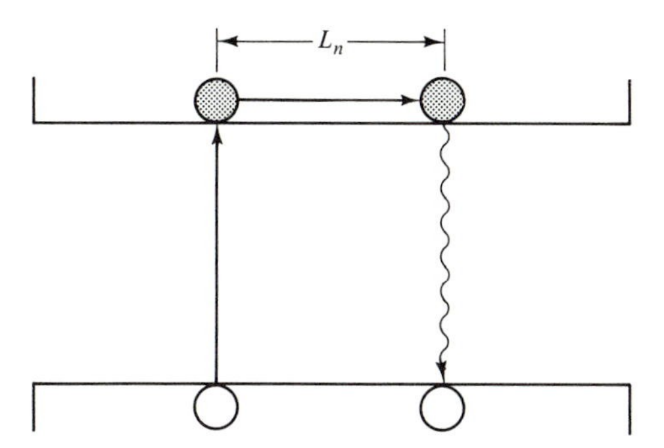

Figure 4-7 The creation, diffusion, and annihilation of an electron-hole pair.

an electron while the other one-half is used to create a hole; n_i is only approximately proportional to $\exp(-E_g/2kT)$ because $m_n^* \neq m_p^*$. The preexponential factor, n_c (see Appendix C), is

$$n_c = \frac{2}{h^3}(2\pi m_n^* kT)^{3/2} \tag{4-15b*}$$

so that Eq. 4-15a can be written

$$n_i \cong n_c e^{-E_g/2kT} = \frac{2}{h^3}(2\pi m_n^* kT)^{3/2} e^{-E_g/2kT} \tag{4-15c}$$

For silicon, $n_c = 2.5 \times 10^{19}\ \text{cm}^{-3}$ at room temperature. Likewise, for holes

$$p_i \cong n_v e^{-E_g/2kT} = \frac{2}{h^3}(2\pi m_p^* kT)^{3/2} e^{-E_g/2kT} \tag{4-15d}$$

By comparing Eqs. 4-15c and 4-15d, we can see that n_i and p_i are not equal as they must be for an intrinsic semiconductor, since every time a conduction band electron is created, a valence band hole must also be created. More accurate equations for n and p are

$$n = n_c e^{-(E_g - E_F)/kT} \tag{4-16a}$$

and

$$p = n_v e^{-E_F/kT} \tag{4-16b}$$

where E_F is again the Fermi energy. More will be said about this important parameter in Section 4-4.1. Equating Eqs. 4-16a and 4-16b to find the intrinsic carrier concentration and defining the Fermi energy for an intrinsic semiconductor to be E_i, we find that

$$E_i = \tfrac{1}{2}E_g + \tfrac{3}{4}kT \ln (m_p^*/m_n^*) \tag{4-17}$$

Clearly, when $m_n^* = m_p^*$, Eqs. 4-15c and 4-15d are identical to Eqs. 4-16a and 4-16b.

* The effective mass used in Eq. 4-15a is the density of states effective mass. It differs slightly from the conductivity effective mass used in Eqs. 4-8 and 4-9 because they are average values obtained by using different averaging techniques.

Values for n_i can be computed directly for a given temperature if m_n^*, m_p^*, and E_g are known. E_g is by far the most important parameter. This is shown in Table 4-2 where n_i for Ge, Si, and GaAs at 300 K are given. $n_i(\text{Ge}) \gg n_i(\text{Si}) \gg n_i(\text{GaAs})$, and this is directly attributable to the fact that $E_g(\text{Ge}) = 0.67$, $E_g(\text{Si}) = 1.11$, and $E_g(\text{GaAs}) = 1.43$ eV.

One can think of the energy $E_g - E_F$ as the amount of energy needed to create an electron and E_F as the amount of energy needed to create a hole. For an intrinsic material it is almost as difficult to thermally create an electron as it is a hole so that the two energies are essentially equal. This is why for an intrinsic semiconductor $E_F \approx E_g/2$. In an extrinsic semiconductor it can be easier or more difficult to create an electron; hence, the Fermi energy can be above or below the middle of the bandgap.

The electron and hole mobilities in semiconductors are approximately proportional to $T^{-3/2}$. They decrease with increasing temperature because the lattice vibrates more violently thereby decreasing the time between carrier collisions with it. Incorporating this fact and those implicit in Eqs. 4-15 and 4-16, Eq. 4-14 can be expressed as

$$\sigma_i = qAT^{3/2}e^{-E_g/2kT}(BT^{-3/2} + CT^{-3/2}) = \sigma_0 e^{-E_g/2kT} \tag{4-18}$$

Thus, a ln σ_i versus $1/T$ curve is a straight line with a slope of $-E_g/2k$. This is shown in the intrinsic conductivity curve for germanium in Fig. 4-8. The curve is not quite a straight line because E_g is a slowly varying function of T—it usually decreases. This is shown in Table 4-2 where E_g at 0 K and room temperature are listed.

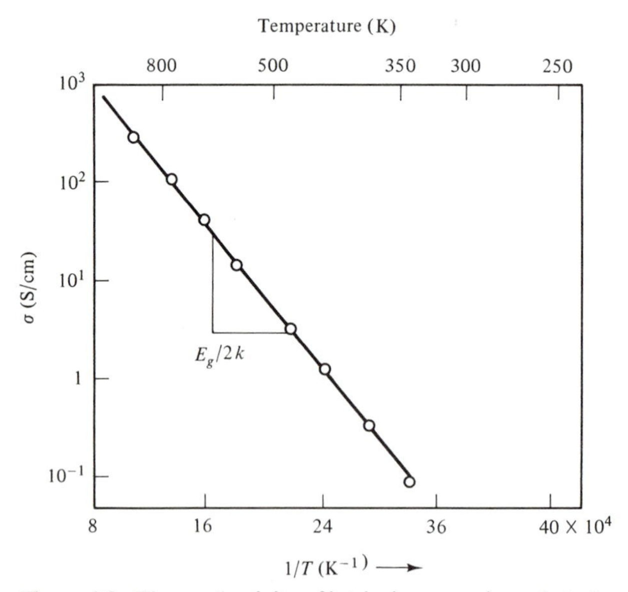

Figure 4-8 The conductivity of intrinsic germanium plotted as a function of the reciprocal temperature.

The dominant factor in determining the intrinsic conductivity is the energy gap—the larger E_g the smaller σ_i. Recall that in Section 3-5.1 it was shown that the greater the orbital overlap, the larger E_g is. This is shown to be true in Table 4-2 for both the decreasing cation and anion size sequences. Also, for semiconductors having the same average atomic number as they do in the sequence ZnSe, GaAs, and Ge, the more ionic the bond, the larger is E_g.

EXAMPLE 4-4

Calculate the intrinsic resistivity of silicon at 300 and 600 K.

$$\sigma = nq(\mu_n + \mu_p) \simeq n_c q(\mu_n + \mu_p)e^{-E_g/2kT}$$

$$= (2.50 \times 10^{19})(1.6 \times 10^{-19})(1.35 + 0.48)$$

$$\times 10^3 \exp\left[\frac{-(1.6 \times 10^{-19})(1.11)}{(2)(1.38 \times 10^{-23})(3 \times 10^2)}\right]$$

$$= 3.54 \times 10^{-6} \text{ S/cm}$$

$$\rho = \frac{1}{\sigma} = 2.82 \times 10^5 \ \Omega \cdot \text{cm}$$

$$\frac{\rho_2}{\rho_1} = \frac{\sigma_1}{\sigma_2} = \exp\left[-\frac{E_g}{2k}\left(\frac{1}{T_1} - \frac{1}{T_2}\right)\right]$$

$$\rho_2 = 2.82 \times 10^5 \exp\left[\frac{-(1.6 \times 10^{-19})(1.11)}{2(1.38 \times 10^{-23})}\left(\frac{1}{300} - \frac{1}{600}\right)\right] = 6.20 \ \Omega \cdot \text{cm}$$

The room temperature conductivity of semiconductors can be greatly increased when they are doped *p*- or *n*-type (see Section 3-6.2) with a concentration in the parts per million range. The sources of these "extra" electrons or holes are impurity electronic states which for electrons are filled states just below the conduction band and for holes are empty states just above the valence band.

The room temperature conductivity in doped semiconductors is much larger than it is in intrinsic semiconductors because at room temperature essentially all of the dopant levels of concentration n_d for donors or n_a for acceptors have been thermally ionized. That is, the donor electrons have been excited into the conduction band leaving a positively charged dopant ion behind, or valence band electrons have jumped into the acceptor sites creating negatively charged dopant ions. Instead of saying that an electron has jumped to an acceptor site, one could equally well say that a hole has jumped to the valence band where it is free to conduct.

EXAMPLE 4-5

Silicon, which has a density of eight atoms in a cubic volume 5.43 Å on a side, is doped with 1 ppm of phosphorus. **(a)** What is its room temperature conductivity? **(b)** What is the ratio of this conductivity to the intrinsic conductivity at room temperature?

(a) $n \simeq n_d = \dfrac{8 \times 10^{-6}}{(5.43 \times 10^{-8})^3} = 5.00 \times 10^{16}\ \text{cm}^{-3}$

$$\sigma = nq\mu = (5.00 \times 10^{16})(1.6 \times 10^{-19})(1.35 \times 10^3) = 10.8\ \text{S/cm}$$

(b) From Example 4-4, $\sigma_{\text{in}} = 3.54 \times 10^{-6}$ S/cm

$$\frac{\sigma_{\text{ex}}}{\sigma_{\text{in}}} = \frac{10.8}{3.54 \times 10^{-6}} = 3.05 \times 10^6$$

At very low temperatures (<100 K) not all of the impurity states are ionized. As a result, as the temperature increases the number of electrons or holes, and therefore the conductivity, will increase. The slope of the ln σ versus $1/T$ curve for an n-type material in this temperature regime is $-E_d/k$. This is shown in Fig. 4-9 for large values of $1/T$. Above about 150 K almost all of the donors have been ionized so that the carrier concentration cannot continue to increase—the carrier concentration is said to be in the exhaustion region. In this region the conductivity actually decreases because the mobility decreases with increasing temperature whereas the carrier concentration remains essentially constant. Eventually, the intrinsic region is reached where σ increases exponentially with the temperature. The temperature at which the intrinsic conductivity begins to dominate is called the intrinsic temperature, T_i.

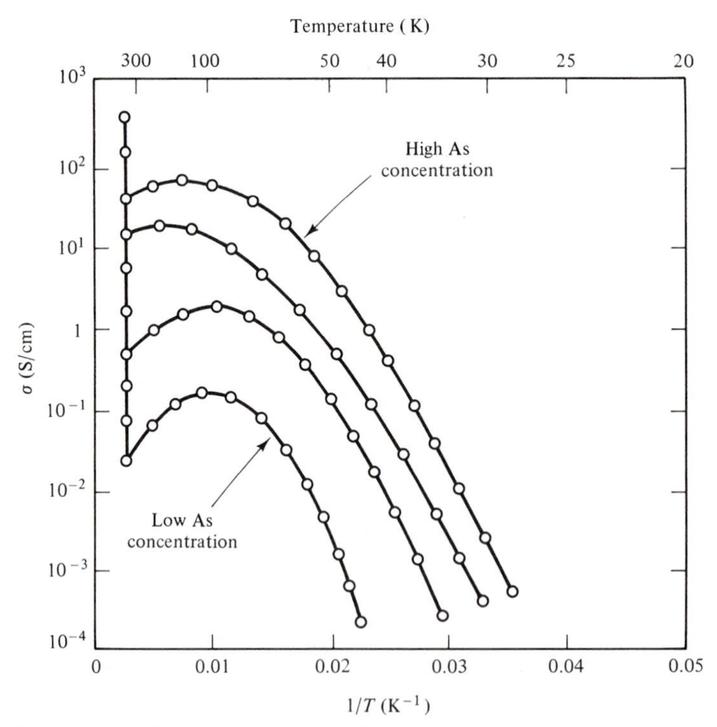

Figure 4-9 The conductivity of silicon doped with different amounts of arsenic plotted as a function of reciprocal temperature.

A temperature is eventually reached where σ_i dominates because there is a vast reservoir of electrons in the valence band. At lower temperatures this reservoir is essentially untapped whereas all of the donor electrons have been "mined." Since the number of electrons from the valence band increases exponentially with T, there is a temperature, T_i, at which the concentration of electrons from the valence band equals the donor concentration. When the donor concentration is higher, T_i is clearly higher, and this is also illustrated in Fig. 4-9. Also, T_i is higher when E_g is larger.

It is also interesting for those of us who have vaporized a semiconductor device to note why these devices are so easily vaporized. When current passes through a semiconductor, the i^2R heating warms it up. The resistance decreases, since more carriers are created, and the lower resistance allows more current to pass through. The net result is that i^2R increases so the semiconductor warms up more, so the i^2R increases . . . , and so it goes.

4-3.4 Photoconductivity

Light intensities can be measured using a photocell that has a semiconductor as the active element; the major difference between it and a photomultiplier tube is that an electron is excited into the conduction band rather than being removed. Thus, the minimum photon energy constraint is changed from $h\nu \geq \phi$ to $h\nu \geq E_g$ for an intrinsic photoconductor. The photoconductor that is frequently used for measuring the intensity of visible light is CdS. CdS is used because its dark resistivity can be made very large, and its bandgap of 2.42 eV corresponds to the energy of green light. The photons with energies greater than E_g are detected by the photocurrent they produce. When a photon is absorbed, electron-hole pairs are created which, in turn, increase the current flowing through the detector. This is illustrated in Fig. 4-10*a*. The amount of the current increase depends on the generation-recombination mechanisms involved. Thus, we must put off this calculation until after this subject is discussed in the next section.

Infrared radiation can be detected by narrow bandgap semiconductors, such as InSb, or by extrinsic semiconductors, such as mercury or copper doped germanium. Mercury has an acceptor state 0.087 eV above the valence band and copper has an acceptor state 0.04 eV above it. In the extrinsic photodetector the photocurrent is produced by ionizing the acceptors. This is illustrated in Fig. 4-10*b*. To be an effective detector, the photocurrent must be larger than the thermally induced current. Thus, in the longer wavelength region (>2 μm) the detector is often cooled to liquid nitrogen temperatures (77 K).

4-4 GENERATION–RECOMBINATION AND ABSORPTION PROCESSES

4-4.1 Fermi Energy of Extrinsic Semiconductors

The number of holes present at a given time depends on the number of electrons present since, if the number of electrons is increased, the probability that a hole

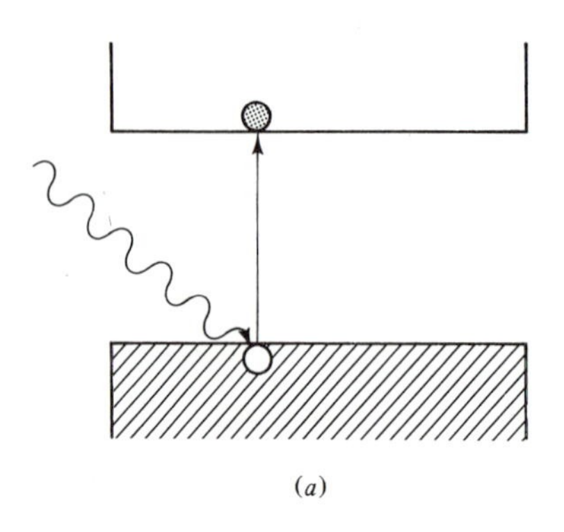

(a)

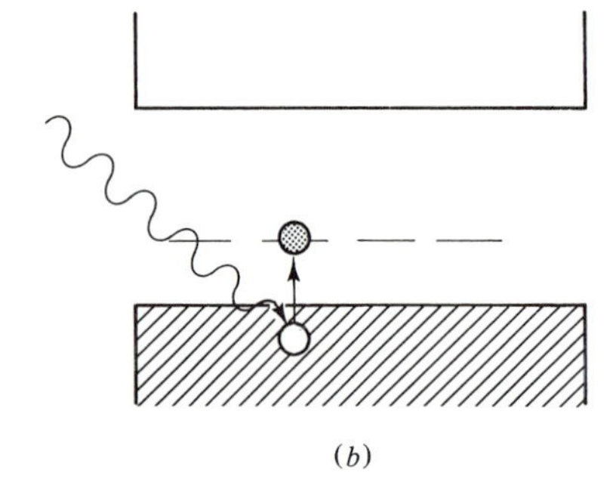

(b)

Figure 4-10 The creation of photoinduced charge carriers by (*a*) the formation of electron-hole pairs by bandgap radiation, and (*b*) ionizing an acceptor state.

will collide with an electron increases. Thus, if the equilibrium electron concentration is increased by doping the semiconductor *n*-type, the equilibrium hole concentration will decrease.

This fact is reflected in the change in the Fermi level produced by the *n*- or *p*-type dopant. Equation 4-16a can be written

$$n = n_c e^{-(E_g - E_F)/kT} = n_i e^{(E_F - E_i)/kT} \tag{4-16c}$$

For an *n*-type material at room temperature it can also be rewritten

$$n \simeq n_d = n_i e^{(E_F - E_i)/kT} \tag{4-16d}$$

Likewise for a *p*-type material at room temperature

$$p \simeq n_a = n_i e^{(E_i - E_F)/kT} \tag{4-16e}$$

Because n for an *n*-type material is larger than n_i, E_F must necessarily be larger than E_i. How much larger it is depends on both n_d and T. In Fig. 4-11 we see that E_F is larger for higher doping levels and lower temperatures, and that it asymptotically approaches E_i at high temperatures. Mathematically

$$E_F = E_i + kT \ln\left(\frac{n_d}{n_i}\right) \tag{4-19}$$

E_F decreases with increasing temperature because n_i increases exponentially with the temperature.

We can understand the physical reason for the decrease of E_F in n-type materials with increasing temperature by recalling that the average energy needed to create a conduction band electron is $E_g - E_F$. At very low temperatures essentially all of the electrons come from the donor levels. Thus, $E_g - E_F \simeq E_d$. As the temperature increases, the percentage of the electrons originating in the valence band increases so that the average energy needed to create an electron increases. At very high temperatures essentially all of the electrons come from the valence band, since the donor levels have been exhausted. Thus $E_g - E_F \cong E_g/2 \cong E_i$. Also, E_F decreases more slowly with increasing temperature for more heavily doped materials because at a given temperature a greater percentage of the electrons come from the donor levels.

EXAMPLE 4-6

Determine the Fermi energy in silicon (a) at 300 K for $n_d = 10^{15}$, 10^{16}, and 10^{17} cm^{-3}, and (b) at 200, 300, and 400 K for $n_d = 10^{16}$ cm^{-3}.

$$E_F = E_i + kT \ln\left(\frac{n_d}{n_i}\right) - E_g/2kT$$

$$n_i = n_c e$$

at 300 K

$$n_c = 2.50 \times 10^{19}$$

$$\therefore n_i = (2.50 \times 10^{19}) \exp\left[-\frac{(1.11)(1.6 \times 10^{-19})}{(2)(1.38 \times 10^{-23})(300)}\right]$$

$$= 1.21 \times 10^{10} \text{ cm}^{-3}$$

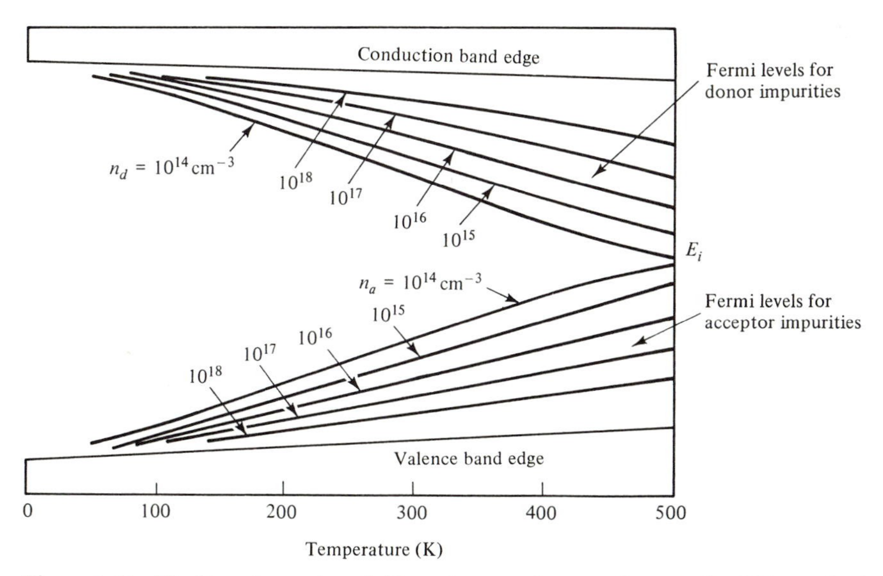

Figure 4-11 The Fermi energies of silicon for n- and p-type dopants for different doping concentrations plotted as a function of the temperature.

$$E_F(10^{15}) = 0.555 + \frac{(1.38 \times 10^{-23})(300)}{1.6 \times 10^{-19}} \ln \frac{10^{15}}{1.21 \times 10^{10}} = 0.848 \text{ eV}$$

$$E_F(10^{16}) = E_F(10^{15}) + kT \ln 10 = 0.908 \text{ eV}$$

$$E_F(10^{17}) = E_F(10^{16}) + kT \ln 10 = 0.968 \text{ eV}$$

At 200 K

$$n_i = (2.5 \times 10^{19})(\tfrac{2}{3})^{3/2} \exp - \left[\frac{(1.11)(1.6 \times 10^{-19})}{(2)(1.38 \times 10^{-23})(200)}\right]$$

$$= 1.45 \times 10^5 \text{ cm}^{-3}$$

$$E_F = 0.555 + \tfrac{2}{3}(0.0259) \ln \left(\frac{10^{16}}{1.45 \times 10^5}\right) = 0.986 \text{ eV}$$

At 400 K

$$n_i = (2.5 \times 10^{19})(\tfrac{4}{3})^{3/2} \exp - [(1.11)(1.6 \times 10^{-19})(2)(1.38 \times 10^{-23})(400)]$$

$$= 3.97 \times 10^{12} \text{ cm}^{-3}$$

$$E_F = 0.555 + \tfrac{4}{3}(0.0259) \ln \left(\frac{10^{16}}{3.97 \times 10^{12}}\right) = 0.825 \text{ eV}$$

The Fermi energy of a p-type material is $\sim E_a$ at very low temperatures, and it increases toward E_i as the temperature increases. This is also shown in Fig. 4-11. As one would expect, there are many more holes in a p-type material.

From Eq. 4-16d it is clear that $n > n_i$ when $E_F > E_i$ and from Eq. 4-16e it is clear that $p > n_i$ when $E_F < E_i$. These equations also show that the minority carrier concentrations, holes in n-type material and electrons in p-type material, are less than n_i. In fact, the product of Eqs. 4-16d and 4-16e

$$np = [n_i e^{(E_F - E_i)/kT}][n_i e^{(E_i - E_F)/kT}] = n_i^2 \tag{4-20}$$

is a constant. To understand why increasing the electron concentration decreases the hole concentration, and vice versa, one must understand generation–recombination.

4-4.2 Generation–Recombination

The density of holes present depends on the density of electrons present, since increasing the number of electrons per unit volume increases the number of electron-hole collisions per unit time. The probability that a single hole will collide with an electron is proportional to n; hence, the probability p holes will collide with n electrons in a unit volume is proportional to pn. Because electrons and holes recombine when they collide, the recombination rate, r, must be proportional to pn. Stated mathematically

$$r = Bpn \tag{4-21}$$

where Bp is the probability per unit time an electron will collide with a hole, and Bn is the probability per unit time a hole will collide with an electron. At equilibrium the thermal generation rate of charge carriers, g, must equal their rate of recombination. Thus

$$g = Bp_0n_0 = Bn_i^2 \tag{4-22}$$

The zero subscripts are used to designate the equilibrium concentrations. Note that the thermal generation rate decreases as E_g increases. This is as it should be, since the probability a valence band electron can be thermally excited into the conduction band decreases as E_g increases.

We now turn to the problem of what happens when excess charge carriers are created optically or injected by an electrical current. Consider first the situation that is shown in Fig. 4-12*a* where a *p*-type material is exposed to bandgap radiation. The photons produce an excess hole, Δp, and an excess electron, Δn, concentration which results in an increased recombination rate. We will assume that the photons are uniformly absorbed so that the excess charge carrier concentrations are everywhere the same. The thermal generation rate is unchanged,

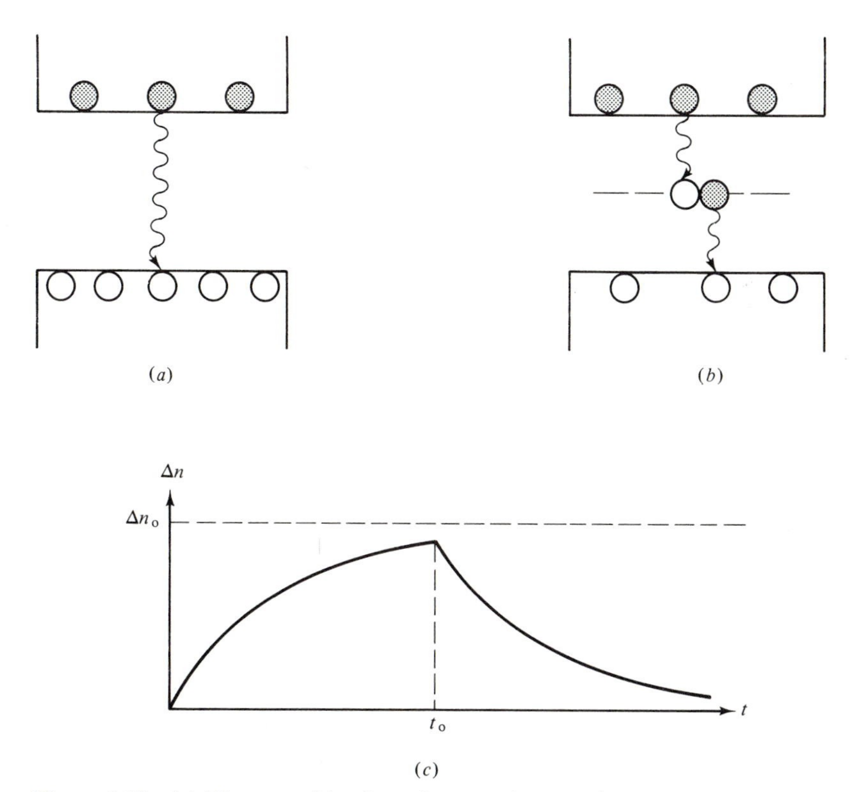

Figure 4-12 (*a*) The recombination of excess electrons in a *p*-type material via direct recombination with holes. (*b*) The excess electron concentration produced by a uniform photon flux for $0 < t < t_0$ plotted as a function of time. The photon flux is turned off at $t = t_0$. (*c*) The recombination of excess electrons and holes via recombination centers.

but there is an optical generation rate, g_{op}, which is equal to the number of photoinduced electron-hole pairs generated per unit volume per second. Thus, the change in the excess electron concentration per unit time is

$$\frac{\partial \Delta n}{\partial t} = Bn_0p_0 - B(n_0 + \Delta n)(p_0 + \Delta n) + g_{op}$$

$$= -B\,\Delta n(n_0 + p_0 + \Delta n) + g_{op} \tag{4-23}$$

For the case $p_0 \gg n_0$ and Δn, Eq. 4-23 reduces to

$$\frac{\partial \Delta n}{\partial t} \simeq -B\,\Delta n p_0 + g_{op} = -\frac{\Delta n}{\tau_n} + g_{op} \tag{4-24}$$

with $\tau_n = 1/Bp_0$ where τ_n is the electron recombination time—the average time it takes for an electron to recombine. The steady state value of $\Delta n = \Delta n_0$ for a step function input is found by setting the derivative equal to zero. Thus

$$\Delta n_0 = g_{op}\tau_n \tag{4-25}$$

The excess electron concentration does not reach its steady state concentration instantaneously. Rather, Δn grows exponentially. Rearranging terms in Eq. 4-24, using Eq. 4-25, and realizing that $\Delta n = 0$ at $t = 0$

$$\int_0^{\Delta n} \frac{d\,\Delta n}{\Delta n_0 - \Delta n} = \int_0^t \frac{dt}{\tau_n} \tag{4-26a}$$

Integrating

$$-\ln \frac{\Delta n_0 - \Delta n}{\Delta n_0} = \frac{t}{\tau_n} \tag{4-26b}$$

or

$$\Delta n = \Delta n_0(1 - e^{-t/\tau_n}) \tag{4-26c}$$

This increase in Δn with time is illustrated in Fig. 4-12*c*.

If the light, which was turned on at $t = 0$, is turned off at $t = t_0$, then for $t > t_0$ Eq. 4-24 becomes

$$\frac{\partial\,\Delta n}{\partial t} = -\frac{\Delta n}{\tau} \tag{4-27a}$$

Rearranging and integrating

$$\Delta n = \Delta n(t_0)e^{-t/\tau_n} \tag{4-27b}$$

where $\Delta n(t_0)$ is found from Eq. 4-26c. This exponential decaying concentration is also illustrated in Fig. 4-12*b*.

Next, consider a semiconductor that has recombination centers in the energy gap. For an electron a recombination center is an empty state in the energy gap into which an electron falls and, then, very soon after that, falls into the valence band by recombining with a hole. This recombination process is illustrated in Fig. 4-12*c*.

Assuming that the concentration of electron recombination centers, n_r, is $\gg p_0$ and n_0, and that they empty almost instantaneously so that their concentration remains essentially constant, Eq. 4-24 becomes

$$\frac{\partial \Delta n}{\partial t} = -B_r \, \Delta n n_r + g_{\text{op}} = -\frac{\Delta n}{\tau_n} + g_{\text{op}} \tag{4-28}$$

Clearly, the Δn time dependence for this case is identical to that of the previous one so that its behavior can be described by Fig. 4-12*b*.

Equation 4-28 can be used to describe photoconductivity in CdS. Electron-hole pairs are created in intrinsic CdS by bandgap radiation. The holes are quickly removed by defect centers called hole traps so that almost all of the photoconductivity is due to conduction by electrons, and the steady state value of Δn is given by Eq. 4-25. Thus, the photocurrent is directly proportional to the optical generation rate which, in turn, is proportional to the photon intensity.

The photocurrent is also directly proportional to the electron recombination time. However, the rate at which Δn grows or decays also depends on τ_n. Thus, the most sensitive detector also has the slowest response time. A typical value of τ_n for a CdS photoconductor is 10^{-3} sec. This electronically slow time is the primary reason that photoconductors see only limited use in optical detection systems.

Photodetectors do, however, have an advantage over the faster photodiode in that they have a photoconductive gain, G. It is defined as the ratio of the net number of photocarriers crossing the cross-sectional area, Zd, per unit time to the number of photons absorbed (see Fig. 4-13). For a photoconductor such as CdS for which $\Delta n \ll n_r$ and where essentially all of the charge is carried by electrons

$$G = \frac{\Delta n_0 v_d Z d}{g_{\text{op}} Z d L} = \frac{\Delta n_0 v_d}{g_{\text{op}} L} \tag{4-29}$$

Recalling that the drift velocity is given by the equation

$$v_d = \mu_n \frac{V}{L} \tag{4-5}$$

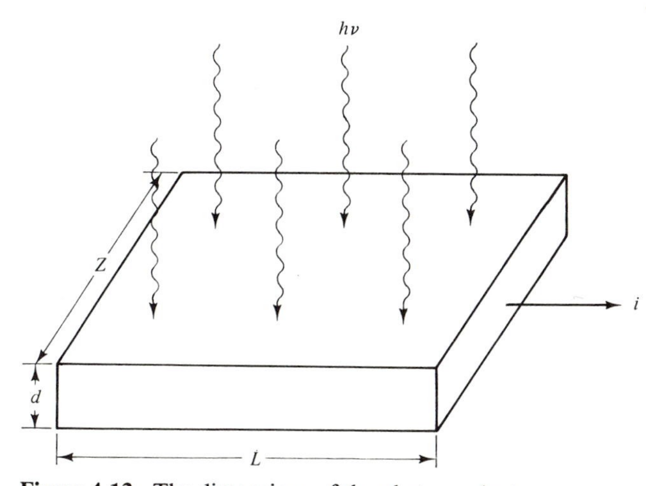

Figure 4-13 The dimensions of the photoconductor.

Eq. 4-29 becomes

$$G = \frac{\tau_n \mu_n V}{L^2} \tag{4-30}$$

Notice that the gain is determined by the material properties τ_n and μ_n as well as the external parameters, the voltage, and the length of the device. However, the gain cannot be increased indefinitely by increasing the voltage or decreasing the length because at the larger electric fields the drift velocity saturates, the current becomes space charge limited, or impact ionization or dielectric breakdown occurs. The largest practical value of G is about 10^5.

The result obtained in Eq. 4-30 can be derived by comparing the recombination time to the time of transit. We will do this because it gives insight into the physical mechanisms involved. When a photoelectron is created, it will, on the average, be swept out of the semiconductor if the transit time, t_{st}, is less than τ_n. However, this electron will be replaced by an electron injected from the opposite electrode so that electrical neutrality can be maintained. The process of electron ejection at one electrode and injection at the other electrode will continue for a time, τ_n, before the photocarrier is lost. The photoconductive gain is effectively the number of times a photoelectron crosses a cross section of the semiconductor; therefore, it is given by

$$G = \frac{\tau_n}{t_{st}} \tag{4-31}$$

Since the transit time is given by

$$t_{st} = \frac{L}{v_d} \tag{4-32}$$

it can be seen that Eq. 4-30 and Eq. 4-31 are equivalent. Physically, the reason G increases as V increases is that the photocarriers move faster, and the reason G increases as L decreases is that the effective path the electron travels is shorter and, therefore, it can be counted more times.

EXAMPLE 4-7

A CdS photoconductor has $E_g = 2.42$ eV, $\tau_n = 10^{-3}$ sec, the holes are trapped (i.e., they do not contribute to the conduction process), and $\mu_n = 100\ \text{cm}^2/\text{V}\cdot\text{sec}$. The photocell is 1 mm long, 1 mm wide, and 0.1 mm thick, with electric contacts at the end. Assume low-level excitation ($\Delta n \ll n_r$), that each photon produces a photoelectron, and that the photoelectrons are uniformly distributed. The cell is irradiated by violet light ($\lambda = 4096$ Å) of 1 mW/cm² intensity. Calculate **(a)** the long-wavelength cutoff for absorption, **(b)** the number of EHPs generated per second, **(c)** the increase in the number of conduction electrons in the sample, **(d)** the change, Δg, in the conductance of the sample, and **(e)** the photocurrent produced if 50 V are applied to the sample.

(a) $E_g = h\nu = \dfrac{hc}{\lambda}$

$$\lambda = \frac{hc}{E_g} = \frac{(6.63 \times 10^{-34})(3 \times 10^8)}{(1.6 \times 10^{-19})(2.42)} = 5.14 \times 10^{-7}\ \text{M} = 5140\ \text{Å}$$

(b) Energy per photon $= \dfrac{hc}{\lambda} = \dfrac{1.99 \times 10^{-26}}{4.096 \times 10^{-7}} = 4.85 \times 10^{-19}$ J

1 mW $= 10^{-3}$ J/sec

area of photodetector $= 1$ mm^2 so that photon energy striking crystal $= 10^{-5}$ J/sec

$\therefore$ no of photons striking cell per second $= \dfrac{10^{-5}}{4.85 \times 10^{-19}} = 2.07 \times 10^{13}$

since conversion efficiency $= 1$, EHP $= 2.07 \times 10^{13}$

(c) $\Delta n_0 = g_{op}\tau_n$

g_{op} = photon density per unit thickness $= \dfrac{2.07 \times 10^{13}}{0.1} = 2.07 \times 10^{14}$ mm^{-3} $\Delta n_0 = (2.07 \times 10^{14})10^{-3} = 2.07 \times 10^{11}$ electrons/mm^3 volume of sample $= 0.1$ mm^3

$\therefore$ total number of electrons created $= (2.07 \times 10^{11})0.1 = 2.07 \times 10^{10}$

(d) $\Delta g = \dfrac{\Delta\sigma A}{L} = \dfrac{\Delta n_0 q\mu A}{L} = (2.07 \times 10^{20})(1.6 \times 10^{-19})(10^{-2})\dfrac{10^{-7}}{10^{-3}}$

$= 3.32 \times 10^{-5}$ S

(e) $\Delta i = \Delta g V = (3.32 \times 10^{-5})(5 \times 10^1) = 1.66 \times 10^{-3}\text{A} = 1.66$ mA.

The final case we will consider is the injection of minority carriers at a *pn* junction. This is done by determining the steady state minority carrier concentration, Δp, as a function of x when the concentration is held constant at Δp_0 at $x = 0$. The concentration depends on x because, as the excess minority carriers diffuse in, they will recombine.

When the excess electrons or holes are inhomogeneously distributed, there will be a net motion of charge carriers in the direction that is opposite to the concentration gradient. The flux is proportional to the gradient so in one dimension the diffusion current densities are

$$J_{Dp} = -q\, D_p \frac{\partial \Delta p}{\partial x} \tag{4-33a}$$

$$J_{Dn} = q\, D_n \frac{\partial \Delta n}{\partial x} \tag{4-33b}$$

where D is the diffusion coefficient. The difference in the sign is due to the difference in the sign of the carriers. Since the current density is the net charge crossing unit area per unit time and the concentration is the number of particles per unit volume, D has the dimensions of length^2/second.

We now consider the net change in the concentration in the infinitesimal volume, $\Delta x\, \Delta y\, \Delta z$, shown in Fig. 4-14 by examining the net charge flow into and out of the volume. In one dimension the net hole flow into the volume is $J_{Dp}(x)\Delta y\, \Delta z$ and the net flow out is $J_{Dp}(x + \Delta x)\Delta y\, \Delta z$. There is also a loss of $\Delta p/\tau_p$ holes per unit volume per unit time due to recombination when $\Delta p \ll n$. Therefore, the rate of change in the number of holes in the volume $\Delta x\, \Delta y\, \Delta z$ per unit time is

$$\frac{\partial \Delta p}{\partial t}\Delta x\, \Delta y\, \Delta z = \frac{1}{q}\,[J_{Dp}(x) - J_{Dp}(x + \Delta x)]\, \Delta y\, \Delta z - \frac{\Delta p}{\tau_p}\Delta x\, \Delta y\, \Delta z \tag{4-34a}$$

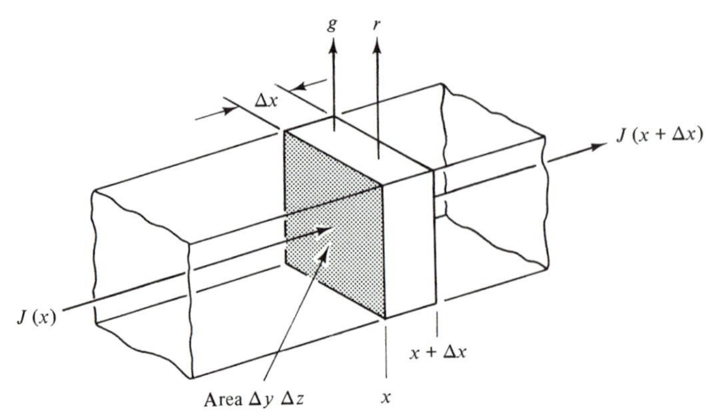

Figure 4-14 The flux in the x direction diffusing into $[J(x)]$ and out of $[J(x + \Delta x)]$ the volume element $\Delta x\, \Delta y\, \Delta z$.

Since as $\Delta x \rightarrow 0$

$$J_{Dp}(x + \Delta x) = J_{Dp}(x) + \frac{\partial J_p(x)}{\partial x}\Delta x \tag{4-34b}$$

$$\frac{\partial \Delta p}{\partial t} = -\frac{1}{q}\frac{\partial J_{Dp}}{\partial x} - \frac{\Delta p}{\tau_p} \tag{4-34c}$$

By substituting Eq. 4-33a into Eq. 4-34c and assuming that D_p is not a function of the concentration

$$\frac{\partial \Delta p}{\partial t} = D_p \frac{\partial^2 \Delta p}{\partial x^2} - \frac{\Delta p}{\tau_p} \tag{4-35}$$

The steady state solution to this partial differential equation has the form

$$\Delta p = Ae^{-kx} + Be^{kx} \tag{4-36}$$

The constants A and B are found from the boundary conditions $\Delta p = 0$ at $x = \infty$ and $\Delta p = \Delta p_0$ at $x = 0$. From the first boundary condition

$$\Delta p = Ae^{-\infty} + Be^{\infty} = 0 \tag{4-37a}$$

Thus B = 0. From the second boundary condition

$$\Delta p = Ae^{-0} = \Delta p_0 = \text{A} \tag{4-37b}$$

The constant, k, can be found by substituting the solution $\Delta p = \Delta p_0 e^{-kx}$ into Eq. 4-35. Thus

$$k^2 D_p \Delta p - \frac{\Delta p}{\tau} = 0 \tag{4-38a}$$

or

$$k = \frac{1}{\sqrt{D_p \tau_p}} = \frac{1}{L_p} \tag{4-38b}$$

where L_p is the hole diffusion length. The excess minority hole concentration is therefore given by

$$\Delta p = \Delta p_0 e^{-x/L_p} \tag{4-39}$$

and this profile is plotted in Fig. 4-15.

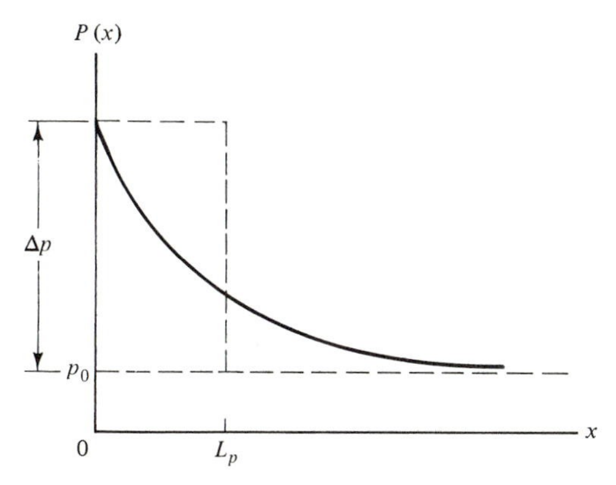

Figure 4-15 The steady state distribution of excess minority holes injected into an n-type material.

4-4.3 Diffusion Length, Diffusion Coefficient, and Decay Time

We have seen in Eq. 4-38b that the diffusion length, diffusion coefficient, and decay time are related to each other. If the diffusion coefficient is large, the carrier moves relatively rapidly, and if the decay time is long, the carrier moves a long time before it recombines. Thus, it is intuitive that the diffusion length is long if both D and τ are large.

The diffusion length is the average distance a carrier diffuses before it recombines; some move a longer distance and some move a shorter distance. The excess minority carrier distribution in Fig. 4-15 can also be thought of as a probability distribution curve. The fraction of, or probability that, carriers recombine between x and $x + dx$ is the ratio of the area, dA, between x and $x + dx$ and the total area. Since for holes

$$dA = \Delta p_0 e^{-x/L_p}\, dx \tag{4-40}$$

the probability, $P(x)$ is

$$P(x) = \frac{\Delta p_0 e^{-x/L_p}\, dx}{\Delta p_0 \int_0^\infty e^{-x/L_p}\, dx} = e^{-x/L_p}\frac{dx}{L_p} \tag{4-41}$$

Note also that if $P(x)$ is integrated between 0 and ∞, the integral equals one as it must, since a hole must recombine somewhere between 0 and ∞.

The average distance, $\langle x \rangle$, a hole diffuses is simply

$$\langle x \rangle = \sum xP(x)dx = \int_0^\infty xe^{-x/L_p}\frac{dx}{L_p} = L_p \tag{4-42}$$

For a number of applications one can treat all of the excess carriers as if they were the average carrier. An example of this is assuming that none of the carriers recombine in the region between 0 and L_p, and that they all recombine at $x = L_p$. This is illustrated in Fig. 4-15. Note that the area under each curve is the same.

We must be careful when using this approximation. For example, when finding an average product, we should realize that

$$\langle P\rangle\langle Q\rangle \neq \langle PQ\rangle \tag{4-43}$$

The diffusion coefficient is a measure of how fast a carrier moves out from its original position. Thus, one should expect it to be related to the mobility. We now show that they are directly proportional by deriving the Einstein relation.

When a uniform electric field is applied to a semiconductor, holes are attracted to the right. If the carriers are open circuited, they will build up on the right-hand side forming a gradient which, in turn, will create a hole diffusion current in the opposite direction. This is illustrated in Fig. 4-16. When the steady state is reached,

$$J_p = J_{\xi_p} + J_{D_p} = \Delta p q \mu_p \xi - q\, D_p \frac{\partial\, \Delta p}{\partial x} = 0 \tag{4-44a}$$

Solving for Δp, we find that

$$\Delta p = \Delta p(0) e^{\mu_p \xi x / D_p} \tag{4-44b}$$

One can arrive at the same hole distribution by realizing that more holes will move to where the energy is lower. The energy is lower by an amount

$$\Delta E = -q\xi x \tag{4-45}$$

Recalling that the particle distribution as determined by their energy is

$$\Delta p = \Delta p(0) e^{-\Delta E/kT} \tag{4-46}$$

Substituting Eq. 4-45 into Eq. 4-46 and equating the result with Eq. 4-44b, one finds that

$$D_p = \frac{kT}{q}\, \mu_p \tag{4-47}$$

The diffusion coefficients are relatively independent of the doping concentration. This is illustrated in Fig. 4-17 where the mobility is plotted as a function of the doping concentration. Note that it is relatively constant for n_d or $n_a < 10^{17}$, which is a relatively high doping level. Room temperature values for D_p and D_n can be computed directly from the mobilities in Table 4-3; values for the frequently used semiconductors, Ge, Si, and GaAs, are listed in Table 4-2.

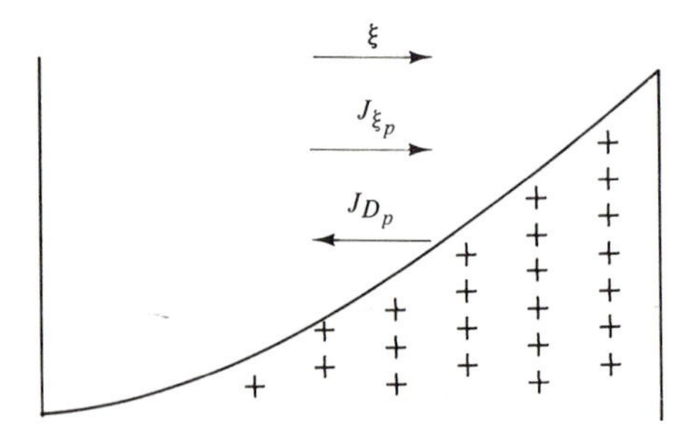

Figure 4-16 The hole distribution and the associated electric field and diffusion currents under open circuit conditions.

We have already seen from Eq. 4-24 that the low-level band-to-band minority carrier recombination time is inversely proportional to the majority doping concentration:

$$\tau_p = \frac{1}{Bn_0} \simeq \frac{1}{Bn_d} \tag{4-24}$$

This should be expected since the more donors there are the greater is the probability there will be a collision so that the time between collisions is less. Again, τ, like L, is an average value. Some carriers decay in a time less than τ while others decay at a later time.

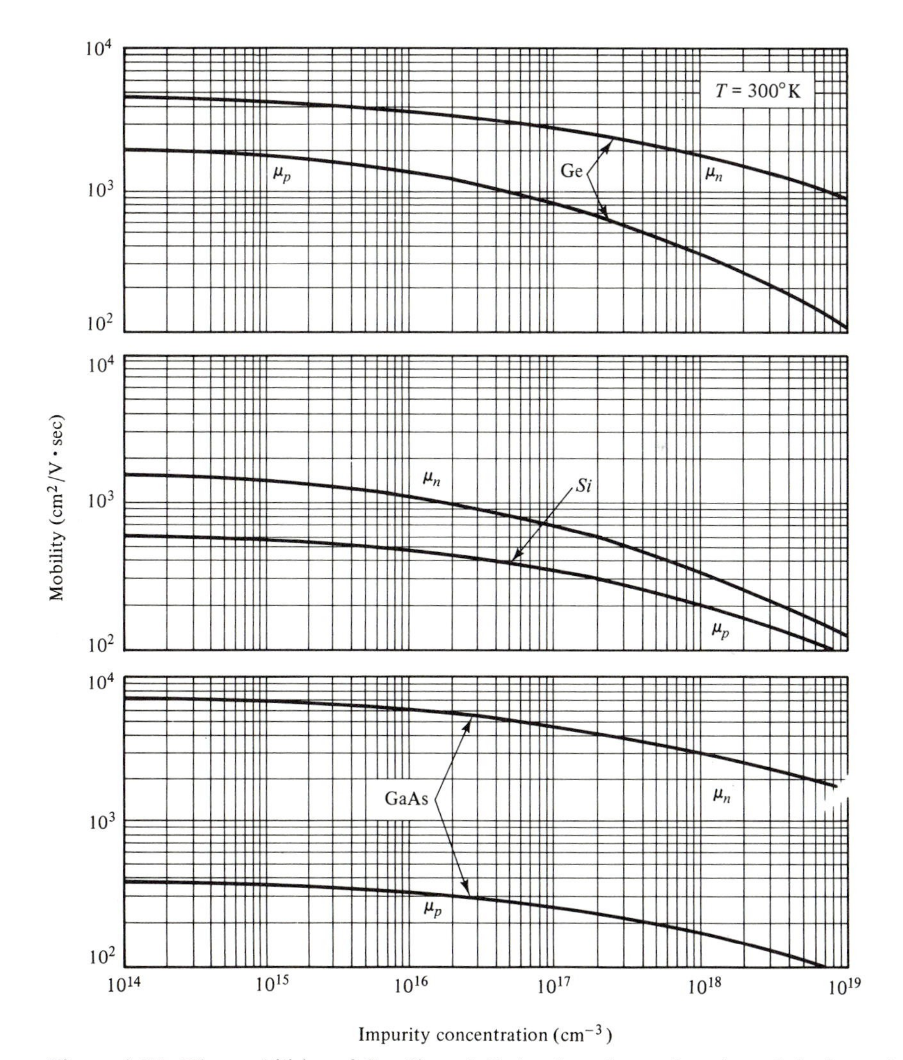

Figure 4-17 The mobilities of Ge, Si, and GaAs plotted as a function of the impurity concentration.

In Table 4-2 we see that B for GaAs is much larger than B for Si and Ge—about four orders of magnitude more. This is very important as it is the reason GaAs responds faster and is a much better light emitter than Si and Ge. For GaAs τ is much shorter because it is a direct bandgap semiconductor; Ge and Si have indirect bandgaps. To understand this we must use the E versus k curves in Fig. 4-18.

An electron is at rest when it is at the bottom of the conduction band, and all of the energy above the bottom is kinetic energy. Thus, the electron energy is given by

$$E_n = E_g + \frac{p_n^2}{2m_n^*} \tag{4-48a}$$

where p_n is the electron momentum. From Eq. 3-6

$$E_n = E_g + \frac{(h/\lambda)^2}{2m_n^*} = E_g + \frac{\hbar^2 k^2}{2m_n^*} \tag{4-48b}$$

where again $\hbar = h/2\pi$ and k is the wave vector, $2\pi/\lambda$. Likewise, the energy of a hole is given by

$$E_p = \frac{\hbar^2 k^2}{2m_p^*} \tag{4-48c}$$

These energies are plotted in the E versus k curves in Fig. 4-18*a*. There it is seen that the minimum in the conduction band is directly above the maximum in the valence band; they have the same k values. The location of the minimum in the conduction band is important because electrons excited into the conduction band fall to this point much more rapidly than they fall from the conduction to the valence bands. For the same reason the holes move very quickly to the

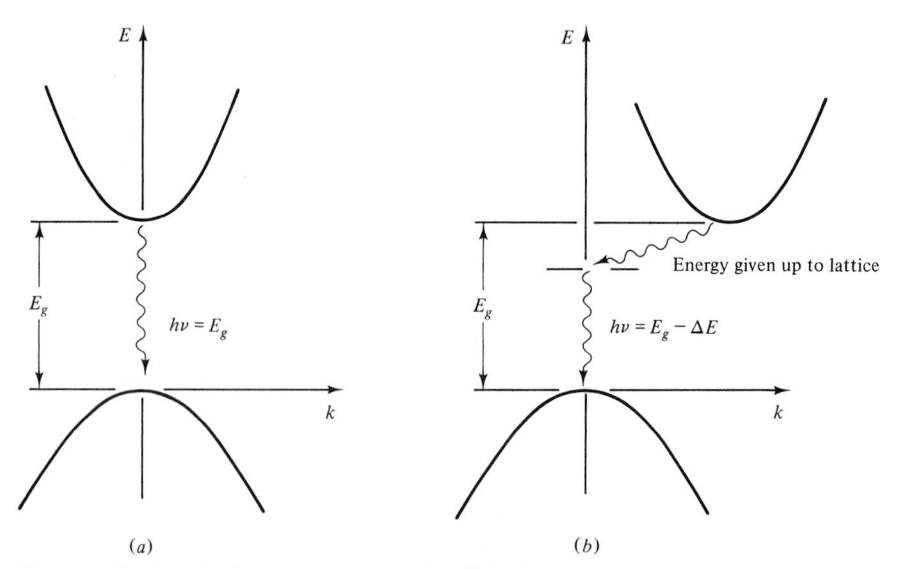

Figure 4-18 An (*a*) direct energy gap and a (*b*) indirect energy gap.

maximum (minimum hole energy) in the valence band. Thus, most band-to-band transitions are made between these two points.

In germanium and silicon the conduction band minimum is shifted from $k = 0$ to k_0. Thus Eq. 4-48b becomes

$$E_n \text{ (indirect)} = E_g + \frac{\hbar^2(k - k_0)^2}{2m_n^*} \tag{4-49}$$

Now, when a transition is made, there is a change in k, but k must be conserved since $\hbar k$ is the momentum. k can be conserved only if it can be absorbed from, or given up to, the lattice. This requires a simultaneous collision between an electron, a hole, and the lattice. A three-body collision is required for an indirect transition, but only a two-body collision is required for a direct transition. Two-body collisions are much more probable than three-body collisions so direct transitions are much more rapid. Whether a semiconductor has a direct or indirect energy gap is indicated in Table 4-3.

There is a way to circumvent the slow decay time in indirect bandgap materials. If silicon is doped with gold, the gold will create states in the middle of the energy gap. An electron can move much more rapidly by the two-step process illustrated in Fig. 4-12*c* by first dropping from the conduction band to a defect state and then from a defect state to the valence band, than it can drop from the conduction band directly to the valence band. These defect states are thus called recombination centers.

TABLE 4-3 THE MOBILITIES AND THE ENERGY GAP (DIRECT OR INDIRECT) AT ROOM TEMPERATURE AND 0° K ALONG WITH THE RELATIVE DIELECTRIC PERMITTIVITY FOR A NUMBER OF ELEMENTAL AND BINARY SEMICONDUCTORS

Material	Direct, *D* or indirect, *I* energy gap	μ_n ($cm^2/V \cdot sec$)	μ_p ($cm^2/V \cdot sec$)	E_g (eV)	$E_g(0)$ (eV)	ϵ_r
Si	*I*	1350	480	1.11	1.16	11.8
Ge	*I*	3900	1900	0.67	0.75	16
AlP	*I*	80	—	2.45	—	—
AlAs	*I*	180	—	2.16	—	10.9
AlSb	*I*	200	300	1.65	1.75	11
GaP	*I*	300	150	2.26	2.40	11.1
GaAs	*D*	8500	400	1.43	1.52	13.2
GaSb	*D*	5000	1000	0.73	0.80	15.7
InP	*D*	4000	100	1.35	1.40	12.4
InAs	*D*	22600	200	0.36	0.46	14.6
InSb	*D*	10^5	1700	0.18	0.26	17.7
ZnS	*D*	110	—	3.6	3.7	8.9
ZnSe	*D*	600	—	2.7	—	9.2
ZnTe	*D*	—	100	2.25	—	10.4
CdS	*D*	250	15	2.42	2.56	8.9
CdSe	*D*	650	—	1.73	1.85	10.2
CdTe	*D*	1050	100	1.58	—	10.2

In a similar way the emission of a photon from GaP, which is an indirect bandgap semiconductor, is speeded up by doping it with nitrogen. Nitrogen creates a state 8 meV below the conduction band. It is the transition from the N level to the valence band that produces the green light.

In concluding this section we should point out that E versus k curves are extremely important curves and are fundamental to the understanding of the physics of semiconductors. In more advanced texts one derives or experimentally determines these curves, and finds the effective masses from their curvature.

The decay times and diffusion lengths appear in many equations. It is therefore important to have a feeling for their magnitudes. Thus, we will compute them in Example 4-8.

EXAMPLE 4-8

Compute τ, L_p, and L_n for Ge, Si, and GaAs for a doping level of 10^{18} cm^{-3}.

$$\tau_{\mathrm{Ge}} = \frac{1}{Bn_d} = \frac{1}{(10^{18})(5.25 \times 10^{-14})} = 1.90 \times 10^{-5} \text{ sec}$$

$$\tau_{\mathrm{Si}} = \frac{1}{Bn_d} = \frac{1}{(10^{18})(1.79 \times 10^{-15})} = 5.59 \times 10^{-4} \text{ sec}$$

$$\tau_{\mathrm{GaAs}} = \frac{1}{Bn_d} = \frac{1}{(10^{18})(7.21 \times 10^{-10})} = 1.39 \times 10^{-9} \text{ sec}$$

$$L_p(\mathrm{Ge}) = \sqrt{D_p\tau} = \left[\frac{\tau kT\mu_p}{q}\right]^{1/2} = [1.90 \times 10^{-5} \times 0.0259)(375)]^{1/2} = 136\,\mu\text{m}$$

$$L_n(\mathrm{Ge}) = \sqrt{D_n\tau} = [(1.90 \times 10^{-5})(0.0259)(1100)]^{1/2} = 233\,\mu\text{m}$$

$$L_p(\mathrm{Si}) = \sqrt{D_p\tau} = [(5.59 \times 10^{-4})(0.0259)(200)]^{1/2} = 538\,\mu\text{m}$$

$$L_n(\mathrm{Si}) = \sqrt{D_n\tau} = [(5.59 \times 10^{-4})(0.0259)(350)]^{1/2} = 712\,\mu\text{m}$$

$$L_p(\mathrm{GaAs}) = \sqrt{D_p\tau} = [(1.39 \times 10^{-9})(0.0259)(180)]^{1/2} = 0.805\,\mu\text{m}$$

$$L_n(\mathrm{GaAs}) = \sqrt{D_n\tau} = [(1.39 \times 10^{-9})(0.0259)(3000)]^{1/2} = 3.28\,\mu\text{m}$$

The experimental values of τ, L_n, and L_p for Ge and Si are much shorter than the calculated values because the nonradiative recombination time, τ_{nr}, is much shorter than the recombination time, τ_r. The recombination time calculated above is τ_r.

4-4.4 Absorption

We discussed absorption in Chapter 1 where it was treated on a macroscopic basis when the penetration depth was studied. Here we consider the absorption in a semiconductor on a microscopic basis.

As one can see in Fig. 4-19, the absorption coefficient, α, increases abruptly

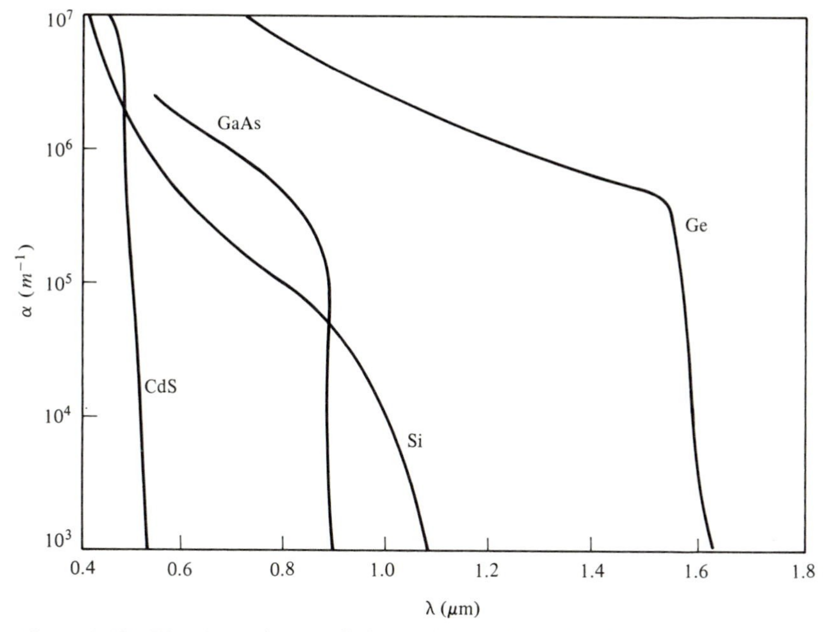

Figure 4-19 The absorption coefficients of the indirect bandgap semiconductors, Si and Ge, and the direct bandgap semiconductors, GaAs and CdS.

by three to four orders of magnitude for a direct bandgap semiconductor at $\lambda = hc/E_g$. This, of course, is because the photon has enough energy to create an EHP.

For $\lambda < hc/E_g$, α increases as λ decreases. The reason this occurs is that the density of electronic states in the conduction band increases as λ decreases. The larger the number of states there are for the electron to jump to, the more probable it is that the electron will absorb the photon, that is, the larger is α.

The density of states can be computed knowing that only standing electron waves are allowed. The maximum wavelength is

$$\lambda_{max} = 2L \tag{4-50a}$$

and the allowed wavelengths are

$$\lambda = \frac{2L}{m} \tag{4-50b}$$

where m is an integer. Thus

$$k_{min} = \frac{2\pi}{\lambda_{max}} = \frac{\pi}{L} \tag{4-51a}$$

and

$$k = \frac{\pi}{L} m \tag{4-51b}$$

Thus, each allowed k in three dimensions can be represented by a cubic lattice with a distance π/L between points when the semiconductor is a cube L on a side. The volume associated with each electronic state then is

$$V_{e.s.} = \frac{1}{2}\left(\frac{\pi}{L}\right)^3 \tag{4-52}$$

The source of the factor $\frac{1}{2}$ is the electron spin; for each k value there is a spin up and a spin down.

The number of states, N, for which $k \leq k^*$, where k^* is an arbitrary value, is found by dividing the volume for which $k \leq k^*$ by the volume per state. The volume for which $k \leq k^*$ is one-eighth (k must be positive) the volume of a sphere of radius k^*. Thus

$$N = \frac{(\frac{1}{8})(\frac{4}{3})\pi k^{*3}}{\frac{1}{2}(\pi/L)^3} = \frac{L^3 k^{*3}}{3\pi^2} \tag{4-53a}$$

Substituting Eq. 4-48b into Eq. 4-53a yields

$$\frac{N}{L^3} = \frac{8\pi}{3h^3}(2m_n^*)^{3/2}(E - E_g)^{3/2} \tag{4-53b}$$

Thus, the density of states per unit volume, $\rho(E)$, is

$$\rho(E) = \frac{\partial(N/L^3)}{\partial E} = \frac{4\pi}{h^3}(2m_n^*)^{3/2}(E - E_g)^{1/2} \tag{4-54}$$

and, as you can see, the density of states increases with the energy.

The small amount of absorption for $\lambda > hc/E_g$ can be attributed to band tailing, defect absorption, and free carrier absorption. By using a sophisticated quantum mechanical analysis, one can show that a random distribution of donors produces a continuum of states immediately below the conduction band, and their density increases with the doping concentration. Thus, a photon can be absorbed even when $h\nu < E_g$, but this is not a probable occurrence.

When there are defect states in the energy gap, an electron from the valence band can be excited into an empty state by the absorption of a photon, and an electron can be excited from a filled state into the conduction band. The defect absorption coefficient is therefore proportional to the density of defect states.

An electron in the conduction band can also absorb a photon. This is called free carrier absorption. This type of absorption is much less than band-to-band absorption simply because there are many more valence band electrons. Again, $\alpha_{f.c.}$ is proportional to the density of free carriers, and for GaAs

$$\alpha_{f.c.}(\text{GaAs}) = 5 \times 10^{-18} n \tag{4-55}$$

Absorption in indirect bandgap semiconductors is much less for $\lambda \sim hc/E_g$, since if the transition going one way is difficult, it is equally difficult to go the other way. Looking at the absorption coefficient for germanium, one sees that α increases gradually with decreasing λ until $\lambda \simeq 1.5$ μm at which point there is an abrupt increase. This increase occurs at the point where the photon has sufficient energy to excite an electron from the valence band to the conduction band without an accompanying change in k, that is, from Eq. 4-49

$$\lambda_{\text{abrupt}} = \frac{hc}{E_g + \hbar^2 k_0^2/2m_n^*} \tag{4-56}$$

4-5 PHOTODETECTOR OPERATION

In the schematic for a photodetector that is shown in Fig. 4-20, it can be seen that the output signal is the voltage, V_s, across the load resistor, R_L, in which the dc component has been filtered out by a capacitor. The source of the output signal is the increase in the potential drop across R_L when the resistance, R_c, of the photoconductor is decreased when it absorbs radiation. For the small signal case (i is constant) the voltage increase, ΔV, is given by

$$\Delta V = -i\,\Delta R_c \tag{4-57}$$

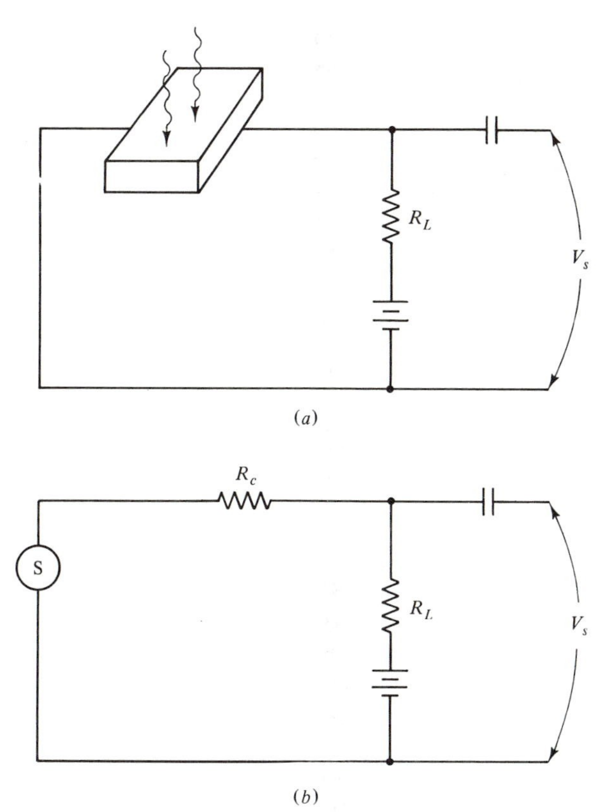

Figure 4-20 (*a*) The schematic representation of a photodetector, and (*b*) its equivalent circuit.

where for a photoconductor for which the primary photocarriers are electrons

$$\Delta R_c = \Delta\left(\frac{L}{nq\mu A}\right) = -\frac{\Delta n}{n} R_c \tag{4-58}$$

For photocurrents that rise and fall exponentially when the light is turned on and off, the steady state photoelectron concentration is $\Delta n_0 = \tau_n g_{op}$ where g_{op} is given by

$$g_{op} = \frac{\eta_i (I/E_p)(1 - R)}{d} \tag{4-59}$$

when the photon flux, I/E_p, is uniformly absorbed over the thickness d. $(1 - R)$ is the fraction of the light transmitted and, again, η_i is the internal quantum efficiency; I is the incident light intensity, and E_p is the photon energy. It is also assumed that none of the light is transmitted through the detector.

When the photon flux is turned off at $t = 0$, Δn is given by the equation

$$\Delta n = \Delta n_0 e^{-t/\tau_n} \tag{4-27b}$$

when $n_r \gg \Delta n$. By using the previous four equations, it can easily be shown that

$$\Delta V = \frac{(1 - R)\eta_i I i R_c e^{-t/\tau_n}}{n d E_p} \tau_n \tag{4-60}$$

when the photon flux is removed at $t = 0$. This would also be the decay voltage for a square wave light pulse if the pulse frequency $\ll 1/\tau_n$.

From Eq. 4-60 it can be seen that, for a given photon flux, ΔV is large when the resistance and the relaxation time are large. The resistance is largest when the semiconductor is intrinsic. The resistance can also be increased by lowering the operating temperature. Lowering the temperature of the wide band-gap semiconductors is not useful, but it is absolutely essential to operate the narrow gap materials at low temperatures. The extrinsic germanium detectors should be operated at temperatures as low as 20 K.

There is also a peak response when the response is measured as a function of the exciting wavelength, and the peak occurs at the absorption edge. This is shown in Fig. 4-21. The peak occurs at the absorption edge because the absorption coefficient increases with decreasing wavelength for wavelengths shorter than the edge wavelength. The more strongly absorbed photons produce a smaller photoconductive signal because a greater percentage of photocarriers are produced near the surface where they are lost via surface recombination (see Problem 4-53).

From Eq. 4-60 the voltage change is similar to the voltage across a capacitor in an RC series circuit. The frequency response is thus

$$\Delta n = \frac{\Delta n_0 \sin \omega t}{1 + (\omega \tau_n)^2} \tag{4-61}$$

The output signal $V_s \simeq \Delta V$ when the time constant for the circuit is much shorter than the response time. As is seen in Table 4-4, the response times for the photodetectors used for detecting visible light are on the order of milliseconds.

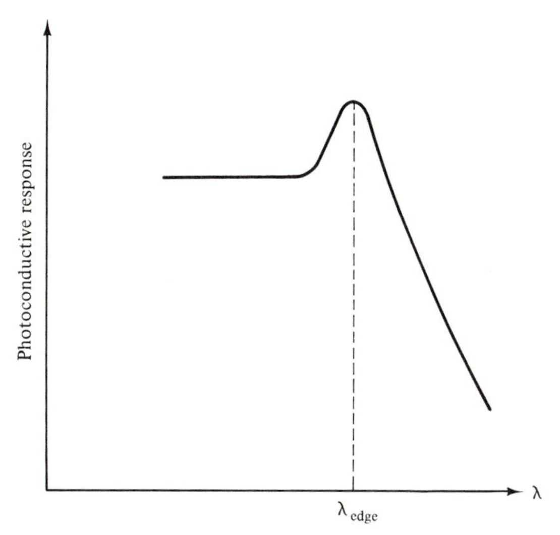

Figure 4-21 The photoconductive response plotted as a function of the wavelength of the incident radiation.

This is much longer than the time constants of the circuit. However, the infrared detector response times can be less than a microsecond, so the bandwidth in this case can be limited by the circuit time constant.

Some of the more common photodetectors, their optimum detection wavelengths, and their response times are shown in Table 4-4. Visible photodetectors are made out of thin polycrystalline CdS or CdSe films. CdS is used when it is desired to have maximum sensitivity in the green, and CdSe is used when maximum sensitivity in the red is desired. Most photographic light meters are made out of CdS because the wavelength of the most intense radiation from the sun is in the green region.

The lead chalcogenide photodetectors are also polycrystalline films, and they are used to detect radiation in the near infrared. These materials are unlike most other semiconductors in that there are more than an average of four valence electrons per atom if lead is considered to have four valence electrons.

TABLE 4-4 COMMON PHOTODETECTORS AND SOME OF THEIR IMPORTANT PARAMETERS

Material	Crystallinity	Peak sensitivity (λ in μm)	E_g or E_a (ev)	Response time (sec)
CdS	Poly	0.515	2.42	10^{-3}
CdSe	Poly	0.715	1.73	10^{-3}
PbS	Poly	2.4	0.31	10^{-3}
PbSe	Poly	4.0	0.17	10^{-5}
InSb	Single	6	0.19	10^{-6}
Hg doped Ge	Single	11	0.087	10^{-7}
Cu doped Ge	Single	24	0.04	3×10^{-7}

InSb photodetectors are single crystal detectors that are also used to detect radiation in the near infrared. One of the primary reasons that InSb is used is that it has a very large electron mobility (70,000 $\text{cm}^2/\text{V} \cdot \text{sec}$) and, therefore, has a large gain even though its response time is short.

The most popular photodetector material for $\lambda > 7\ \mu\text{m}$ is extrinsic germanium. Germanium is particularly well suited because, unlike silicon, its lattice does not absorb this infrared radiation. This is because its quantized vibrational energy states, or phonons, do not absorb radiation in this region (see Section 3-7.1). For the 8 to 14 μm region the first mercury acceptor state, $E_a = 0.087$ eV, is used. Having a detector in this region is particularly useful because in this region the atmospheric radiation absorption is small.

4-6 ELECTRICAL DIODES

4-6.1 *pn* Junction

When an n- and a p-type semiconductor are brought into contact with each other, a contact potential, V_0, is created. It forms because the Fermi energies of the separated n- and p-type material are different, but when they are brought into contact, they must equilibrate. This is illustrated in Fig. 4-22 where the charge transfer, electric field, and voltage drop in an abrupt junction are shown. An abrupt junction is one for which the doping on the p and n sides is uniform. The Fermi energies equilibrate by a transfer of negative charge from the n-type to the p-type material.

To a good approximation, the charge transfer is accomplished by all of the donor electrons within x_n of the boundary flowing over and filling all of the acceptor states within x_p of the boundary. This is shown in Fig. 4-22*a*. Also shown in Fig. 4-22 are the electric field pointing in the negative x direction, and the associated potential drop across this region, often called the depletion region. The contact potential is

$$V_0 = \frac{E_{Fn} - E_{Fp}}{q} \tag{4-62a}$$

EXAMPLE 4-9

The p-type material in a silicon pn abrupt junction is doped with $10^{15}\ \text{cm}^{-3}$ boron, and the n-type material is doped with $10^{16}\ \text{cm}^{-3}$ phosphorus. Calculate the charge transfer per unit area of the junction, the depletion layer width, W, and x_n and x_p at 300 K. For silicon $\epsilon = 10.45 \times 10^{-13}$ F/cm. The electric field for $-x_p < x < 0$ is

$$\xi = \int_{-x_p}^{x} \frac{\rho\, dx}{\epsilon} = -\frac{qn_a}{\epsilon}(x + x_p) \tag{4-63a}$$

where $\rho = -qn_a$ is the charge density, and the absolute magnitude of ξ at $x = 0$, $\xi_{\max}$, is

$$\xi_{\max} = \frac{qn_a x_p}{\epsilon} \tag{4-63b}$$

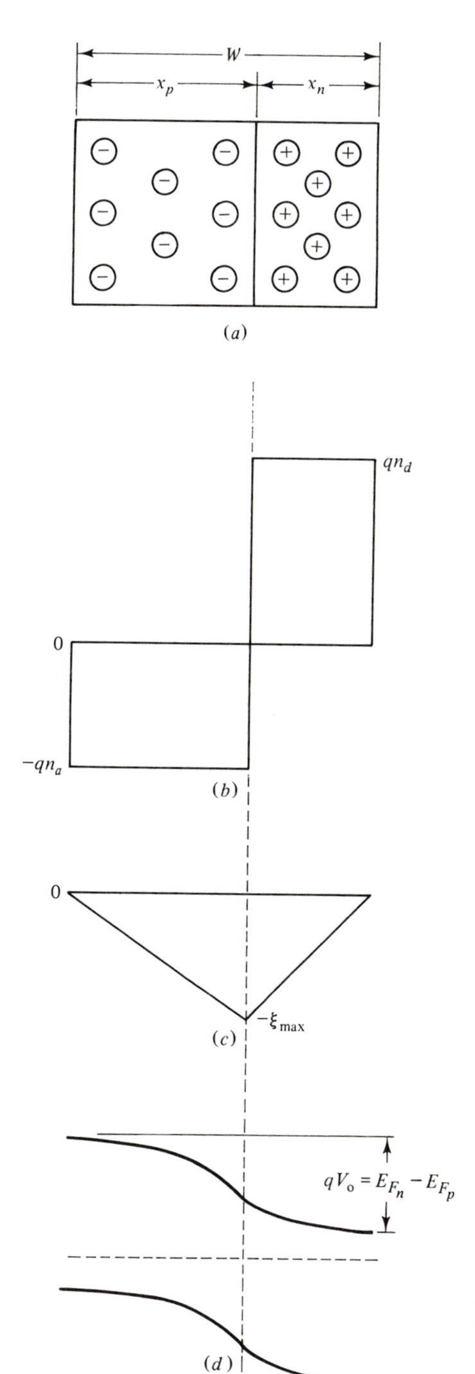

Figure 4-22 (*a*) Ionized donors (+) and acceptors (−), (*b*) charge distribution, and (*c*) electric field in the depletion region. (*d*) Band bending across the depletion region to equilibrate the Fermi energies.

The contact potential, V_0, is the area under the ξ versus x curve. Thus

$$V_0 = \tfrac{1}{2}\xi_{\max} W = \frac{1}{2}\frac{qn_a x_p}{\epsilon} W \tag{4-62b}$$

From charge conservation

$$n_a x_p = n_d x_n \tag{4-64}$$

and

$$x_n = W - x_p \tag{4-65}$$

$$x_p = \frac{n_d}{n_d + n_a} W \tag{4-66}$$

Substituting Eq. 4-66 into Eq. 4-63 yields

$$V_0 = \frac{q}{2\epsilon}\frac{n_a n_d}{n_d + n_a} W^2 \tag{4.67a}$$

or

$$W = \left(\frac{2\epsilon}{q} V_0 \frac{n_d + n_a}{n_d n_a}\right)^{1/2} \tag{4-67b}$$

Using the results of Example 4-6 and noting that for the same doping level E_{Fp} lies approximately the same distance below E_i as E_{Fn} lies above it,

$$V_0 = \frac{1}{q}(E_{Fn} - E_{Fp}) = (0.908 - 0.555) + (0.848 - 0.555) = 0.646 \text{ V}$$

$$W_\mu = \left[\frac{(2)(10.45 \times 10^{-13})(0.646)}{(1.6 \times 10^{-19})}\frac{10^{15} + 10^{16}}{(10^{15})(10^{16})}\right]^{1/2} = 9.63 \times 10^{-5} \text{ cm}$$

$$= 0.963\ \mu\text{m}$$

$$x_p = \frac{10^{16}}{10^{15} + 10^{16}}(0.963) = 0.876\ \mu\text{m}$$

$$x_n = \frac{n_0}{n_d} x_p = \frac{0.837}{10} = 0.0876\ \mu\text{m}$$

$$\frac{Q}{A} = qn_d x_n = (1.6 \times 10^{-19})(10^{16})(8.76 \times 10^{-6}) = 1.40 \times 10^{-8} \text{ C/cm}^2$$

4-6.2 Diodes

At the junction there are four current densities: the electron, $J_{D,n}$, and hole $J_{D,p}$ diffusion current densities and the electron, $J_{\xi,n}$, and hole, $J_{\xi,p}$, electric field current densities. The electron diffusion current is produced by the diffusion of electrons to the left from the n side where there are many electrons toward the p side where there are only a few. For similar reasons the hole diffusion current is produced by holes diffusing from the p to the n side. The currents are not as large as one would guess because the carriers have to diffuse up over the potential barrier. Note in Fig. 4-23 that the diffusion current is to the right; electrons with a negative charge moving to the left produce a current moving to the right. The electron electric field current is generated by electrons in the *p*-type material diffusing to the depletion region and being swept to the *n* side by the electric field in the depletion layer. Likewise the hole electric field current is generated by holes in the *n*-type region diffusing to the depletion region and being swept out by the electric field. This current is not usually very large, since there are only a few electrons in the *p*-type region and only a few holes in the *n*-type region. Also, the electric field current moves to the left. When no external voltage is applied, the net current must necessarily be zero or we could solve the energy problem merely by forming *pn* junctions.

The hole diffusion current density at $x = x_n$ is computed from the gradient in Δp shown in Fig. 4-15. Note, however, when we consider a *pn* junction, x is shifted x_n to the left. From Fick's first law (Eq. 4-33a)

$$J_{Dp}(x_n) = -q\, D_p\left(\frac{\partial \Delta p(x)}{\partial x}\right)_{x=x_n} \tag{4-33a}$$

When $x = x - x_n$ is substituted into the expression for Δp in Eq. 4-39, and then Δp is substituted into Eq. 4-33a, it becomes

$$J_{Dp}(x_n) = \frac{q\, D_p}{L_p}\, \Delta p(x_n) \tag{4-68a}$$

To find $\Delta p(x_n)$, we must take into account the fact that the holes must diffuse up the potential barrier of height $E_{Fn} - E_{Fp}$ shown in Fig. 4.23*a*. In the *p*-type material the equilibrium hole concentration is p_{p0} so that the concentration of holes that have an energy $\geq E_{Fn} - E_{Fp}$, so that they can diffuse over the potential barrier, is

$$\Delta p(x_n) = p_{p0}e^{-(E_{Fn}-E_{Fp})/kT} \tag{4-69a}$$

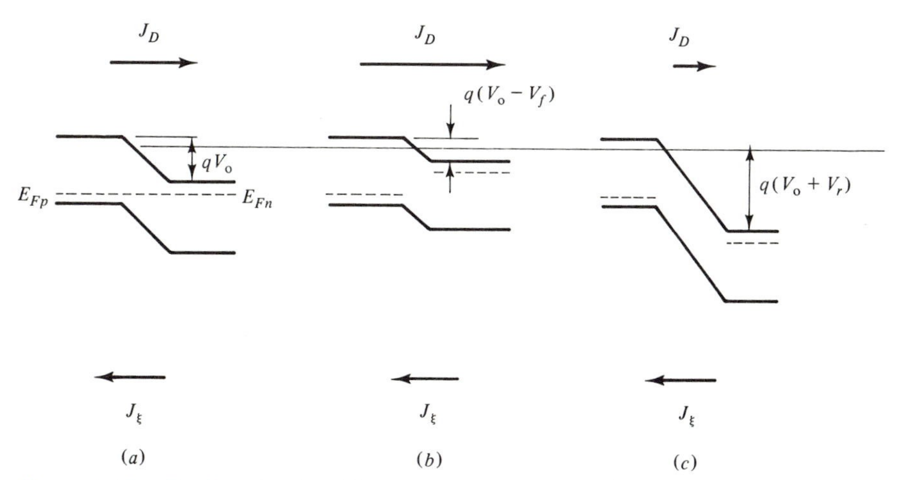

Figure 4-23 The bent bands, diffusion current, and electric field current for (*a*) zero biased, (*b*) forward biased, and (*c*) reverse biased diodes.

Inserting the expression for p_{p0} from Eq. 4-16e gives

$$\Delta p(x_n) = n_i e^{(E_i - E_{Fp} - E_{Fn} + E_{Fp})/kT} = n_i e^{(E_i - E_{Fn})/kT} = p_{n0} \tag{4-69b}$$

where p_{n0} is the equilibrium hole concentration in the n-type material. Thus, when no bias is applied to the junction, the hole diffusion current density is

$$J_{Dp} = \frac{q\,D_p}{L_p}\,p_{n0} \tag{4-68b}$$

The calculation of the hole electric field current density, $J_{\xi p}$, is less detailed. Recall in Section 4-4.3 we showed that the average minority hole diffuses a distance, L_p, and then recombines. Thus, we can say that all of the holes in the n-type material within L_p of the depletion layer will diffuse to it and be swept away to the left by the electric field. On the other hand, none of the holes further than L_p away from the depletion layer will reach it, since they recombine before reaching it. The swept out charges within L_p of the depletion layer are replaced every τ_p seconds so that

$$J_{\xi p} = -q p_{n0}\,\frac{L_p}{\tau_p} \tag{4-70a}$$

Recalling that $L_p^2 = D_p \tau_p$, Eq. 4-70a becomes

$$J_{\xi p} = -q p_{n0}\,\frac{D_p}{L_p} \tag{4-70b}$$

which is equal and opposite to the hole diffusion current density, as it must be. This result is shown schematically in Figure 4-23*a* along with the bent band structure.

When a forward bias is applied—that is, a negative voltage is applied to the n-type side or a positive voltage is applied to the p-type side—the potential barrier will be reduced. This is because a positive potential will attract electrons on the p-type side thereby lowering the p-type bands. Likewise, a negative potential would raise the n-type side. Now that a potential is being applied across the junction, the Fermi levels will not equilibrate. They will be separated by a value, qV, where V is the applied voltage, and this is shown in Fig. 4-16*b*. In the ideal diode we assume that all of the applied voltage drops across the junction.

The only effect the applied voltage has is to increase the concentration of charge carriers that can diffuse over the potential barrier, since now the barrier height is smaller. The excess minority hole concentration, $\Delta p(x_n)$, now becomes

$$\Delta p(x_n) = p_{p0} e^{(E_{Fn} + E_{Fp} + qV)/kT} = p_{n0} e^{qV/kT} \tag{4-69c}$$

or

$$J_{Dp} = \frac{q\,D_p}{L_p}\,p_{n0} e^{qV/kT} \tag{4-68c}$$

Both equations, of course, reduce to Eqs. 4-69b and 4-68b for $V = 0$. The hole electric field current is unchanged by the applied voltage so that the total hole current density is given by

$$J_p = J_{Dp} + J_{\xi p} = \frac{q\,D_p}{L_p}\,p_{n0}(e^{qV/kT} - 1) \tag{4-71}$$

Thus, the forward biased current increases exponentially with V and this fact is shown in Fig. 4-24.

The reverse condition is illustrated in Fig. 4-23c. For this bias the barrier height is raised so that the diffusion current is reduced. Equations 4-69c, 4-68c, and 4-71 can be used to describe the excess hole concentration and the hole current densities by defining a reverse biased voltage to be a negative voltage. As the reverse bias is increased, a point is soon reached where essentially no holes can diffuse over the barrier so that, as shown in Fig. 4-24, the current density saturates at the value of $J_{\xi p}$. This current density is often called the hole reverse saturation current density and it is given by

$$J_{ps} = -\frac{q\,D_p}{L_p}\,p_{n0} \tag{4-72}$$

The electron current density can be derived in a similar fashion. Thus, the total current density is

$$J = q\left(\frac{D_p}{L_p}\,p_{n0} + \frac{D_n}{L_n}\,n_{p0}\right)(e^{qV/kT} - 1) = J_s(e^{qV/kT} - 1) \tag{4-73}$$

EXAMPLE 4-10

For the p-type side of a silicon diode $n_a = 10^{16}$ cm^{-3} and for the n-type side $n_d = 10^{15}$ cm^{-3}. If $\tau_p = \tau_n = 10^{-5}$ sec, the cross-sectional area is 10 mm^2, and the applied voltage is 0.39 V, what are the hole, electron, and the total current at $T = 300$ K?

$$i_p = \frac{q\,D_p}{L_p}\,p_{n0}A\,e^{qV/kT}$$

$$p_{n0} = \frac{n_i^2}{n_d} \quad \text{(from Example 4-6)} \quad = \frac{(1.21 \times 10^{10})^2}{10^{15}} = 1.46 \times 10^5 \text{ cm}^{-3}$$

$$D_p = \frac{kT}{q}\,\mu_p = (0.026)480 = 12.5 \text{ cm}^2/\text{sec} \qquad (\mu_p \text{ from Fig. 4.17})$$

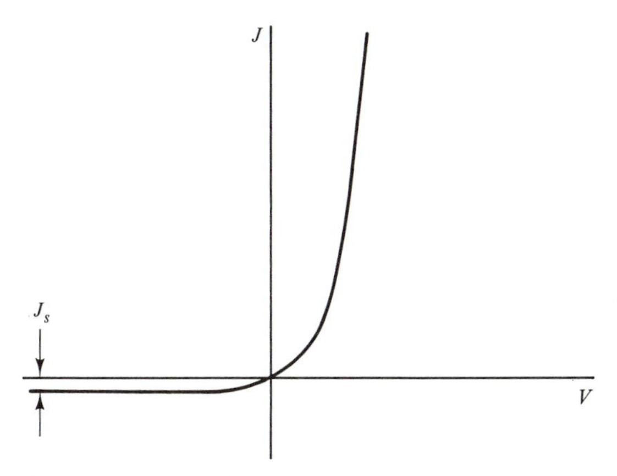

Figure 4-24 The I-V characteristic for a diode in the dark.

$$L_p = \sqrt{D_p\tau_p} = (12.5 \times 10^{-5})^{1/2} = 1.12 \times 10^{-2} \text{ cm}$$

$$i_p = \left[\frac{(1.6 \times 10^{-19})(12.5)(1.46 \times 10^5)(0.1)}{1.12 \times 10^{-2}}\right] e^{0.39/0.026} = 2.61 \times 10^{-12} e^{15}$$

$$= 8.52 \times 10^{-6} \text{ A}$$

$$i_n = \frac{q\, D_n}{L_n} n_{p0} A e^{qV/kT}$$

$$n_{p0} = \frac{n_i^2}{n_a} = 1.46 \times 10^4 \text{ cm}^{-3}$$

$$D_n = \frac{kT}{q}\mu_n = (0.026)(1350) = 35.1 \text{ cm}^2/\text{sec} \qquad \text{(Also see Fig. 4.17.)}$$

$$L_n = (D_n\tau_n)^{1/2} = (3.51 \times 10^{-4})^{1/2} = 1.87 \times 10^{-2} \text{ cm}$$

$$i_n = [(1.6 \times 10^{-19})(35.1)(1.46 \times 10^4)(0.1)/(1.87 \times 10^{-2})]e^{15}$$

$$= 4.38 \times 10^{-13} e^{15}$$

$$= 1.43 \times 10^{-6} \text{ A}$$

$$i = i_p + i_n = (8.52 + 1.43)10^{-6} = 9.95 \times 10^{-6} \text{ A}$$

4-7 PHOTODIODES

4-7.1 *pn* Photodiodes

Photodiodes are efficient light detectors that can be made in small sizes, have good linearity over as much as six decades, have a high response speed, have simple biasing requirements, and are relatively low-cost items. They are faster than photoconductors because they have a *pn* junction to collect the carriers, but they do not have the internal gain photoconductors have because the carriers are counted only once. This problem can be overcome to some extent in avalanche photodiodes where a different internal gain mechanism is exploited.

The photodetection mechanism in photodiodes is similar to that in photoconductors in that an EHP is created by a photon for which $h\nu \geq E_g$, but it differs in that only the photocarriers that diffuse to the depletion layer are counted, and only the minority carriers are collected. This is because the light intensity is usually low enough that the excess carrier concentration generated is such that $p_{p0} \gg \Delta p > p_{n0}$; a similar relationship exists for electrons. Thus, the diffusion current is essentially unaffected; the change in the current is due only to changes in the electric field current.

The optical current density, J_{op}, is the excess current density generated by the photocarriers. Because it represents an electric field current, it is a negative current. Thus, Eq. 4-73 becomes

$$J = J_s(e^{qV/kT} - 1) - J_{op} \tag{4-74}$$

Note that now there is a current at $V = 0$, and this is illustrated in Fig. 4-25*b*. Also illustrated is the fact that a photodiode is operated in reverse bias. Therefore, the current density is independent of the applied voltage and is equal to $-J_s - J_{op}$.

The magnitude of J_{op} is proportional to the incident photon flux I_0/E_p, since the minority photocarrier concentration is proportional to I_0/E_p. J_{op} is also proportional to the fraction of the light that is transmitted, the internal quantum efficiency, and the fraction of the minority photocarriers, F, that are collected. Thus

$$J_{op} = q \frac{I_0}{E_p} (1 - R)\eta_i F \tag{4-75}$$

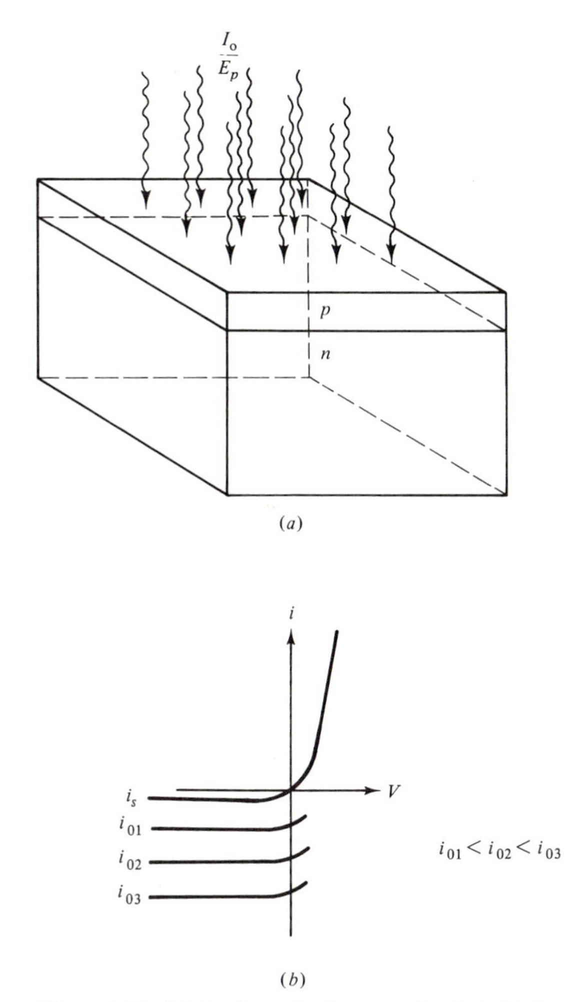

Figure 4-25 (*a*) A schematic diagram of a photodiode exposed to a uniform light intensity. (*b*) The reverse current of a photodiode for different intensities.

where R again is the fraction of the light refected. The external quantum efficiency, η_{ex}, is

$$\eta_{ex} = \frac{J_{op}/q}{I_0/E_p} = (1 - R)\eta_i F \tag{4-76}$$

Most of the photocarriers are absorbed in the n layer in Fig. 4-25a because the thicknesses of the top p layer and the depletion layer are usually much less than $1/\alpha$. The fraction of the carriers collected depends on α and L_p. If $\alpha > L_p$, then more of the photocarriers will be created close enough to the depletion layer so that they can diffuse to it before they recombine.

The ideal optical current density, J^i_{op}, is the current density when $R = 0$ and $\eta_i = 1$, and it can be found by first solving for the minority photocarrier distribution generated by a monochromatic beam of intensity, I_0. Combining Eqs. 4-24 and 4-35 to account for both carrier diffusion and generation

$$\frac{\partial \Delta p}{\partial t} = D_p \frac{\partial^2 \Delta p}{\partial x^2} - \frac{\Delta p}{\tau_p} + g_{op} = 0 \tag{4-77}$$

under steady state conditions. The optical generation rate is equal to the decrease in the photon intensity. Thus

$$g_{op} = -\frac{1}{E_p}\frac{\partial I}{\partial x} = \frac{I_0}{E_p}\alpha e^{-\alpha(x+d)} \tag{4-78}$$

where d is the thickness of the top p layer. The solution to Eq. 4-77, when the boundary conditions are $\Delta p = -p_{n0}$ at $x = x_n$ and $\Delta p = 0$ at $x = \infty$, is

$$\Delta p = \frac{I_0}{E_p D_p}\left[\frac{\alpha L_p^2 e^{-\alpha(x+d)}}{1 - \alpha^2 L_p^2}\right][1 - e^{(1/L_p - \alpha)(x_n - x)}] - p_{n0} e^{(x_n - x)/L_p} \tag{4-79}$$

The boundary condition, $\Delta p = -p_{n0}$ at $x = x_n$, results from the fact that all of the holes that reach the edge of the depletion layer are swept out by the electric field in it. This includes the thermally generated as well as the optically generated holes.

The electric field current at the edge of the depletion layer is equal to the rate at which holes diffuse to it. Thus

$$J^i_\xi = -J^i_{op} - J_{sp} = q\, D_p \frac{\partial \Delta p}{\Delta x}\bigg|_{x=x_n}$$

$$= -q\frac{I_0}{E_p}\frac{\alpha L_p}{1 + \alpha L_p} e^{-\alpha(d+x_n)} - q\frac{D_p}{L_p} P_{n0} \tag{4-80a}$$

Therefore

$$J^i_{op} = q\frac{I_0}{E_p}\frac{\alpha L_p}{1 + \alpha L_p} e^{-\alpha(d+x_n)} \tag{4-80b}$$

The term, $\exp[-\alpha(d + x_n)]$, is the fraction of photons that pass through the top layer and the depletion region and is $\simeq 1$, especially if the photodiode is an indirect bandgap material. If $L_p \gg 1/\alpha$, $J^i_{op} \cong qI_0/E_p$. This is because all of the photons are absorbed within a distance of the depletion region that is much less than the diffusion length. Hence, the holes diffuse to the depletion layer before they recombine. If $L_p \ll 1/\alpha$, $J^i_{\text{op}} \cong 0$.

The wavelength response for a silicon, germanium, and a back wall InP/$In_{0.53}Ga_{0.47}As$ photodiode are shown in Fig. 4-26. The external quantum efficiency for the silicon photodiode varies substantially with the wavelength with the maximum occurring at $\lambda \simeq 0.6$ μm. This corresponds to the direct energy gap of silicon. For $\lambda > 0.6$ μm the absorption coefficient decreases because light can be absorbed only by an indirect transition. As a result, carriers are created further from the depletion layer so that a smaller fraction of them is captured. For $\lambda < 0.6$ μm, α increases because the density of electron states in the conduction band increases with the energy above the bottom of the band. Now α is so large that carriers are absorbed near enough to the surface to be lost by surface recombination.

The germanium photodiode behaves in a qualitatively similar fashion. The major differences are that the direct and indirect energy gaps are smaller.

The InP/$In_{0.53}Ga_{0.47}As$ backwall photodiode differs in that the wavelength response is flat and drops off abruptly at the short and long wavelength limits. The abrupt drop off at the long wavelength limit is due to InGaAs being a direct gap semiconductor. The flat response and the abrupt drop off at the short-wavelength cutoff is due to the InP layer being on top with a thickness greater than the diffusion length. The response is flat, since now there is no surface recombination; the photocarriers in the InGaAs diffuse to the heterojunction

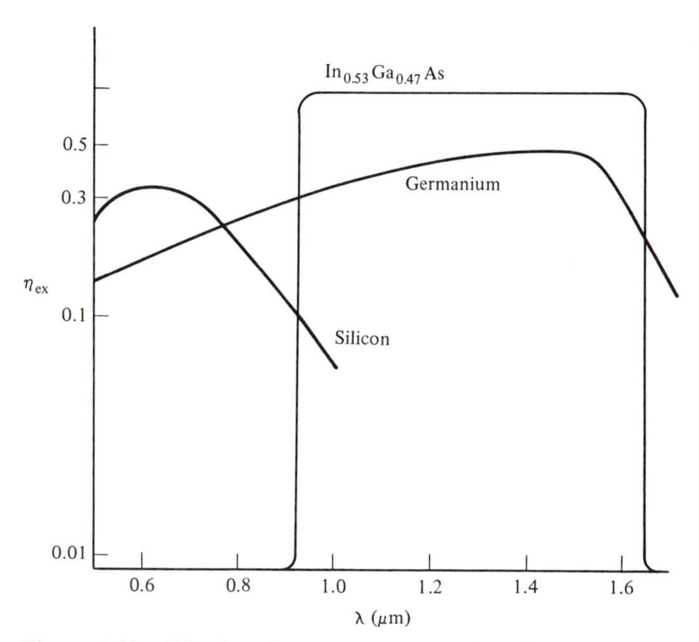

Figure 4-26 Wavelength response curves for silicon, germanium, and a backwall $In_{0.53}Ga_{0.47}As$ *pn* photodiode when $R = 0$.

instead of a free surface. The short wavelength cutoff occurs at the wavelength that can be absorbed by the direct gap InP layer.

The InGaAs must have the precise composition of 53 percent indium and 47 percent gallium. At this composition the InGaAs is lattice matched to the InP. This statement can be better understood by referring to Fig. 4-27. The repeat distance in the InAs crystalline lattice, called the lattice parameter, a_0, is

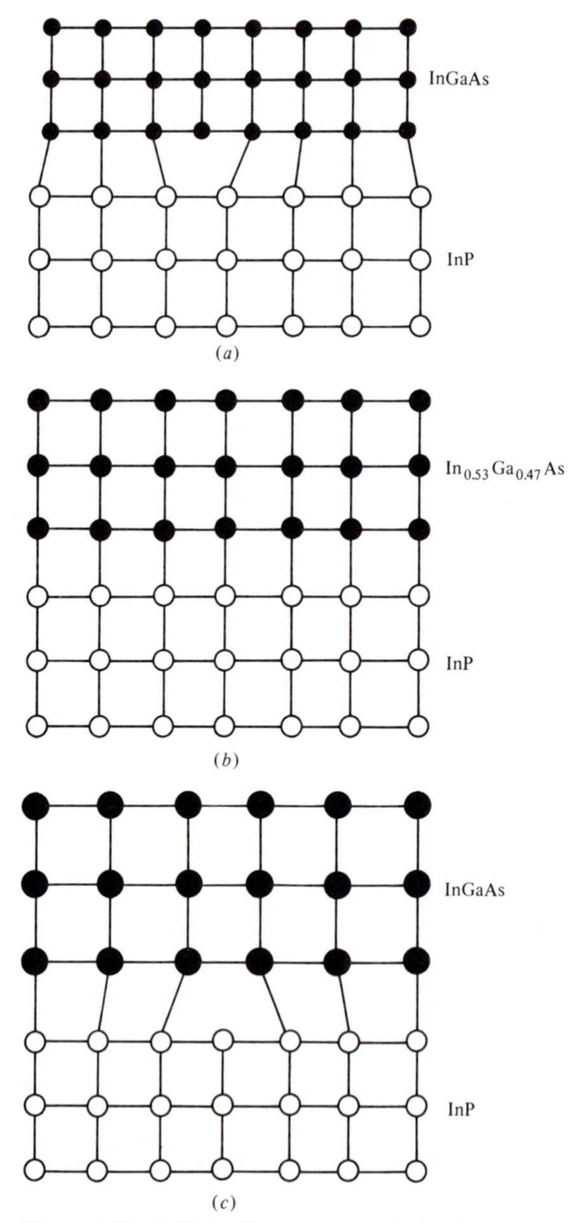

Figure 4-27 InGaAs films grown on InP substrates when the indium cation fraction is (*a*) <0.53, (*b*) =0.53, and (*c*) >0.53.

larger than the lattice parameter of InP because the arsenic atom is larger than the phosphorus atom. As a result, the atoms at the interface cannot line up; some atoms in the InP lattice are not bonded across the interface. These "dangling bonds" are electrically active and often act as recombination centers. They severely degrade the device properties so that every effort is made to eliminate them. The GaAs lattice constant is smaller than the lattice constant for InP so that now the dangling bonds emanate from the GaAs lattice. Replacing some of the gallium with indium increases the lattice parameter of GaAs, and at 53 percent indium the match is perfect. Thus, there are no dangling bonds, and the heterojunction device properties are much improved.

Another ternary heterojunction photodiode that is beginning to receive more attention, but at this date is less developed than the $InP/In_{0.53}Ga_{0.47}As$ photodiode, is the CdTe/HgCdTe photodiode. This system has the advantage that HgTe and CdTe have almost the same lattice parameter; hence, the ternary is essentially lattice matched at all compositions. This has the advantage of being able to vary the energy gap from 1.58 eV, E_g of CdTe, to zero. Thus, they can be tuned to detect the longer wavelengths. This material has been slow to develop because the mercury presents a number of difficult materials problems.

Being able to tune the energy gap to the wavelength being detected is an advantage because this allows one to use the largest possible E_g. This is desirable because increasing E_g reduces the reverse bias saturation current. With a smaller J_s the signal to noise ratio is larger. This is why it is advantageous to cool the narrow bandgap detectors.

The InGaAs photodiode cannot be usefully tuned because it is necessary for it to be lattice matched to the InP substrate. With the additional degree of freedom presented by the quaternary, $In_{1-x}Ga_xAs_{1-y}P_y$, the quaternary film can be both lattice matched and tuned to a desired E_g. This is also true for the quaternary $Ga_{1-x}Al_xAs_ySb_{1-y}$ on GaSb. More will be said about lattice matching and tuning E_g when we discuss LEDs in Chapter 6 and laser diodes in Chapter 7.

The wavelength response can be increased by reducing the amount of light reflected at the surface. As we will show in Chapter 8 (see Problem 8-19), R for a monochromatic beam at normal incidence can be reduced to zero if

$$d = \frac{\lambda_0}{4n_1} \tag{4-81}$$

and

$$n_1 = \sqrt{n_2} \tag{4-82}$$

where d is the antireflecting (AR) film thickness, λ_0 is the wavelength in air, n_1 is the index of refraction of the AR coating, and n_2 is the index of refraction of the semiconductor. An SiO_2 film $\sim$ 1000 Å comes close to meeting these conditions.

The frequency response is limited by the time it takes a carrier to diffuse to the depletion region, the time it takes to transit across the depletion layer, and the RC time constant of the circuit. The delay time produced by the diffusion and the transit times is illustrated in Fig. 4-28. When photocarriers are produced by a light impulse, the signal is not totally collected until the carriers have diffused

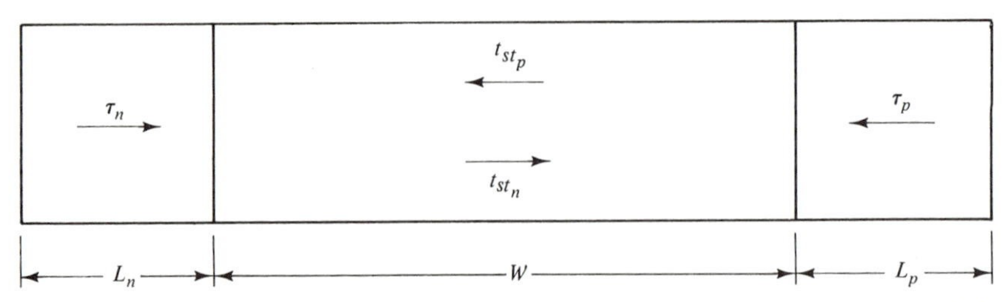

Figure 4-28 The hole and electron diffusion and transit times in a reverse biased diode.

to the edge of the depletion layer and have been transported across it. Thus, the delay time is the sum of the diffusion time, t_{diff}, and the transit time, t_{st}.

To obtain the approximate diffusion time, let us assume that the optical generation rate in Eq. 4-77 is constant. The solution is then

$$\Delta p = g_{\text{op}}\tau_p[1 - e^{(x_n - x)/L_p}] - p_{n0}e^{(x_n - x)/L_p} \tag{4-83}$$

and

$$J^i_{\text{op}} = q(g_{\text{op}}\tau_p)\left(\frac{L_p}{\tau_p}\right) \tag{4-84}$$

$g_{\text{op}}\tau_p$ is the excess carrier concentration away from the depletion region and L_p/τ_p is the average velocity at which it diffuses; the average photohole moves a distance L_p in τ_p sec. In this situation, then, the average diffusion time is τ_p. However, if $1/\alpha < L_p$, the holes do not have as long a distance to diffuse. Thus

$$t_{\text{diff},p} \simeq \frac{\tau_p}{\alpha L_p} \tag{4-85a}$$

and

$$t_{\text{diff},n} \frac{\tau_n}{\alpha L_n} \tag{4-85b}$$

In general, $L_n > L_p$ because $\mu_n > \mu_p$ so that $t_{\text{diff}_n} < t_{\text{diff}_p}$. This is why the absorbing layer is usually the *p*-layer.

Clearly, t_{diff} is shorter for direct gap materials, since their recombination times are much faster—as much as three orders of magnitude faster. We will learn in the next section that this deficiency in silicon and germanium photodiodes can be overcome by using a *pin* structure.

The transit time is simply

$$t_{st} = \frac{W}{v_{\text{sat}}} \tag{4-86}$$

where v_{sat} is the saturation velocity. Carriers reach a saturated, or terminal, velocity when they are subjected to large ξ fields. The ξ fields at which saturation is reached varies for the type of carrier and semiconductor as is seen in Fig. 4-29. For silicon and GaAs electrons $v_{\text{sat}} = 10^7$ cm/sec and v_{sat} is reached for $\xi \geq 2 \times 10^4$ V/cm. This velocity is reached very rapidly in a silicon photodiode so that the average velocity is $\sim v_{\text{sat}}$ (see Problem 4-55). For a *pn* photodiode $t_{\text{diff}} \gg t_{st}$ so that t_{st} is significant only in *pin* structures.

Carriers reach a saturated velocity for much the same reasons a free falling object reaches a terminal velocity. When a carrier increases its velocity it, in effect, increases its effective temperature. As its temperature increases, it gives up a larger percentage of its energy to the lattice. It finally reaches a point where any increase in the energy is completely given up to the lattice.

The time constant associated with the junction capacitance at the photodiode, C_j, is found by using the equivalent circuit in Fig. 4-30. The current source is the photocurrent; R_r is the reverse bias resistance of the junction; R_s is the series resistance of the semiconductor away from the junction; and $R_r \gg R_L > R_s$ so that the equivalent resistance is $\sim R_L$. Thus, the time constant is

$$\tau \simeq R_L C_j \tag{4-87}$$

and the upper frequency limit is

$$f = \tfrac{1}{2}\pi\tau \tag{4-88}$$

The source of the junction capacitance is the charge transfer across the junction as the bias is changed. When the forward bias is increased, the depletion layer shortens as electrons in some of the ionized acceptors are forced back across the junction. When the diode is reverse biased more electrons move across the junction to ionize the acceptors.

Mathematically

$$C_j = \frac{d|Q|}{d(V_0 - V)} \tag{4-89a}$$

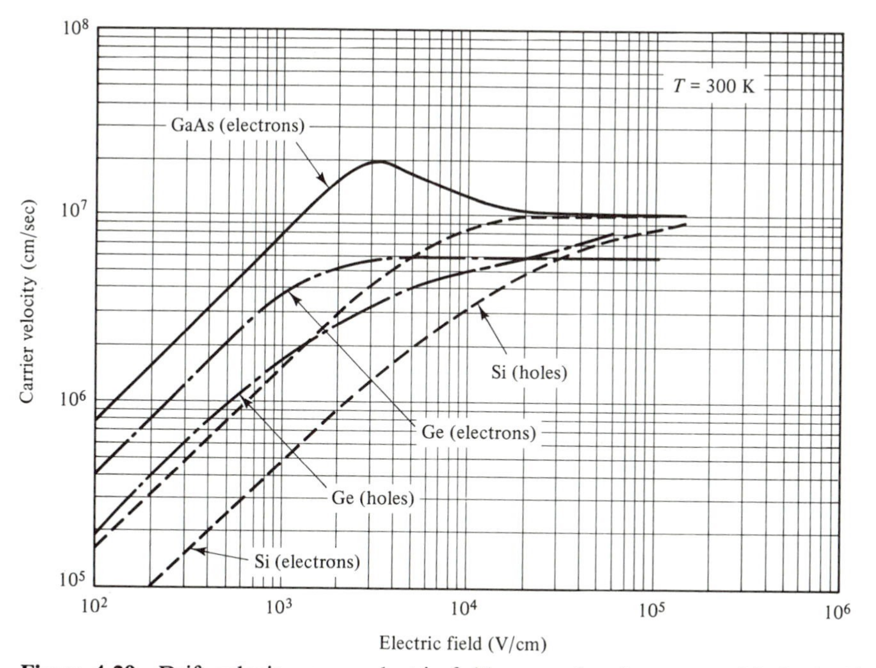

Figure 4-29 Drift velocity versus electric field curves for electrons and holes in high purity germanium, silicon, and GaAs. (From S. M. Sze, *Physics of Semiconductor Devices*, 2d Ed., Copyright © 1981 by John Wiley & Sons, New York.)

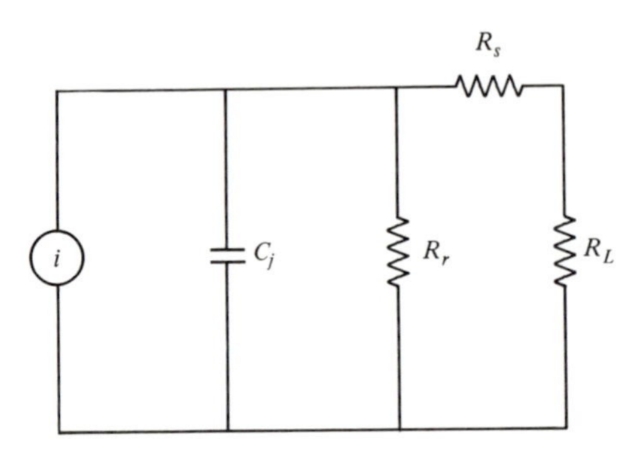

Figure 4-30 Approximate equivalent circuit for a photodiode.

Using Eqs. 4-66 and 4-67b,

$$|Q| = qAn_a x_p = A\left[2q\epsilon(V_0 - V)\frac{n_a n_d}{n_a + n_d}\right]^{1/2} \tag{4-90}$$

Substituting Eq. 4-90 into Eq. 4-89a yields

$$C_j = \frac{\epsilon A}{\left[\frac{2\epsilon}{q}(V_0 - V)\frac{n_a + n_d}{n_a n_d}\right]^{1/2}} = \frac{\epsilon A}{W} \tag{4-89b}$$

For high-speed detectors $C_j \approx 1$ pF so that for a load of 150 Ω, $f \approx 10^9$ Hz.

EXAMPLE 4-11

The light from a GaAlAs LED, which has an energy gap of 1.50 eV, strikes the surface of a silicon photodiode of area, A = 1 mm² with an incident intensity, $I_0 = 1$ mW. The diode quantum efficiency is 0.4, $n_a = 10^{16}$ cm^{-3}, $n_d = 10^{15}$ cm^{-3}, and the reverse bias on the diode is $V_r = 20$ V. The load of the external circuit is $R_L = 150$ Ω. Find the wavelength, λ, of the light emitted by the LED, J_{op}, ξ_{max}, W, x_n, x_p, $t_{st,n}$, $t_{st,p}$, C_j, and τ, the circuit RC time constant.

$$\lambda = \frac{hc}{E_g} = \frac{1.242}{1.5} = 0.828\ \mu\text{m}$$

$$J_{op} = \frac{\eta q I_0}{E_p} = \frac{(0.4)(1)}{1.5} = 26.7\ \text{mA/cm}^2$$

From Example 4-9, $\epsilon = 10.45 \times 10^{-13}$ F/cm and $V_0 = 0.646$ V

$$\xi_{max} = \left[\frac{2q}{\epsilon}(V_0 - V)\frac{n_a n_d}{n_a + n_d}\right]^{1/2} = \left[\frac{(2)(1.6 \times 10^{-19})(20.646)(10^{16})(10^{15})}{(10.45) \times 10^{-13})(10^{16} + 10^{15})}\right]$$

$$= 7.58 \times 10^4\ \text{V/cm}$$

$$W = \frac{2(V_0 - V)}{\xi_{max}} = \frac{2(20.646)}{7.93 \times 10^4} = 5.45\ \mu\text{m}$$

$$x_n = \frac{n_a}{n_a + n_d} W = \frac{10^{16}}{1.1 \times 10^{16}} (5.45)$$

$$= 4.95\ \mu\text{m}$$

$$x_p = \frac{n_d}{n_a} W = \frac{4.95}{10} = 0.495\ \mu\text{m}$$

From Fig. 4-29 $v_{sat,n} \approx v_{sat,p} = 10^7$ cm/sec

$$t_{sat,n} = t_{sat,p} = \frac{W}{v_{sat,p}} \approx \frac{5 \times 10^{-4}}{10^7} = 50 \text{ psec}$$

$$C_j = \frac{\epsilon A}{W} = \frac{(9.54 \times 10^{-13})(0.01)}{0.521 \times 10^{-3}} = 18.3 \text{ pF}$$

$$\tau = R_L C_j = (1.5 \times 10^2)(1.83 \times 10^{-11}) = 2.75 \text{ nsec}$$

4-7.2 *pin* Photodiodes

A *pin* diode is made by sandwiching an intrinsic layer between the p and n layers thereby greatly extending the depletion layer width. One reason for doing this is that now most of the photocarriers are created in the depletion layer where they are collected much faster by the electric field than they would be outside the depletion layer by diffusion. This is particularly true for the indirect bandgap silicon and germanium photodiodes. Another advantage is that the charge separation is larger and, therefore, the junction capacitance is smaller. The primary disadvantage is that *pin* diodes are more difficult, and therefore more expensive, to make. Also, ξ_{max} is smaller and t_{st} is longer. The former problem is easily overcome by applying a larger reverse bias, and the latter problem turns out to be one of the primary limiting factors of the frequency response.

These statements can be quantified with the assistance of Fig. 4-31. The charge is now separated by x_i, the i layer thickness, so that x_n and x_p are smaller than they are in a *pn* diode under the same conditions. ξ_{max} is still given by Eq. 4-62b and now $V_0 - V$ is given by

$$V_0 - V = \tfrac{1}{2}(W + x_i)\xi_{max} + \tfrac{1}{2}(W + x_i)\frac{q n_a x_p}{\epsilon} \tag{4-91a}$$

Noting that

$$W - x_i = x_n + x_p = \left(\frac{n_a}{n_d} + 1\right)x_p \tag{4-92}$$

Eq. 4-91a becomes

$$V_0 - V = \frac{q}{2\epsilon}\frac{n_a n_d}{n_a + n_d}(W^2 - x_i^2) \tag{4-91b}$$

or

$$W = \left[\frac{2\epsilon}{q}(V_0 - V)\frac{n_a + n_d}{n_a n_d} + x_i^2\right]^{1/2} \tag{4-93}$$

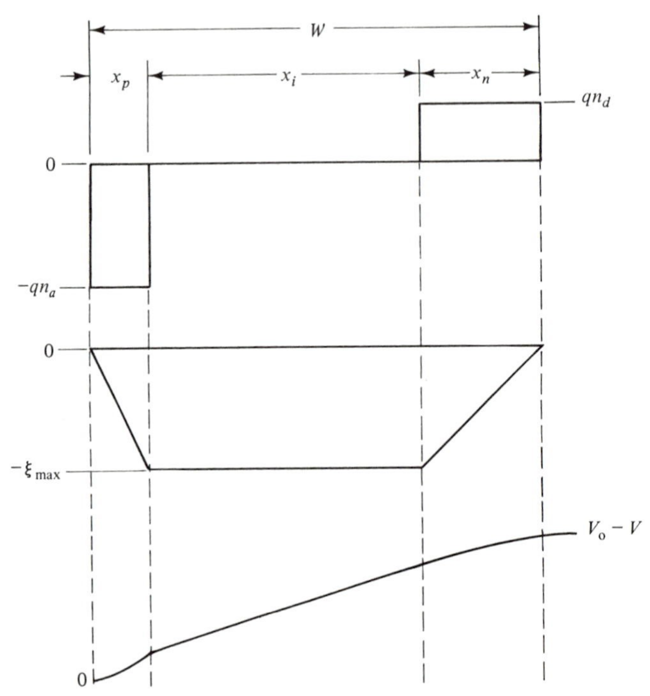

Figure 4-31 (*a*) Charge distribution, (*b*) electric field, and (*c*) potential in the depletion region of a *pin* photodiode.

The diffusion component of J_{op} is given by Eq. 4-80 except that the exponential term is now $\exp[-\alpha(d + W - x_p)]$. With W being relatively large, this component is quite small. The more dominant drift current component can be found from Eq. 4-75 once the fraction, F, of photons absorbed is known. The fraction of photons remaining in the beam at the front end of the depletion layer is $\exp[-\alpha(d - x_p)]$, and the fraction of photons remaining in the beam at the far end of the depletion layer is $\exp[-\alpha(d + W - x_p)]$. Thus

$$F = e^{-\alpha(d-x_p)}(1 - e^{-\alpha W}) \tag{4-94a}$$

so that

$$J_{op}(\text{drift}) = \frac{I_0}{qE_p}(1 - R)\eta_i e^{-\alpha(d-x_p)}(1 - e^{-\alpha W}) \tag{4-94b}$$

It should also be pointed out that one electron hole pair is equivalent to one carrier; the hole carries the charge a distance x to the left, and the electron carries the charge a distance $W - x$ to the right.

EXAMPLE 4-12

Find W, x_n, x_p, ξ_{max}, $t_{st,n}$, $t_{st,p}$, C_j, and τ for the *pin* diode, which has the same parameters as the *pn* photodiode in Example 4-11, when $x_i = 10\ \mu$m.

$$W = \left[\frac{2\epsilon}{q}(V_0 - V)\frac{n_a + n_d}{n_a n_d} + x_i^2\right]^{1/2} = [(5.45)^2 + (10)^2]^{1/2} = 11.39\ \mu\text{m}$$

$$x_n = \frac{n_a}{n_a + n_d}(W - x_i) = \frac{10^{16}}{10^{16} + 10^{15}}(11.39 - 10.0) = 1.26\ \mu\text{m}$$

$$x_p = \frac{n_d}{n_a} x_n = 0.126\ \mu\text{m}$$

$$\xi_{\max} = \frac{2(V_0 - V)}{W + x_i} = \frac{(2)(20.646)}{(11.39 + 10)10^{-4}} = 1.93 \times 10^4\ \text{V/cm}$$

$$t_{st_n} \approx t_{st_p} = \frac{W}{v_{\text{sat}}} = \frac{1.14 \times 10^{-3}}{10^7} = 0.114\ \text{nsec}$$

$$C_j = \frac{\epsilon A}{W} = \frac{(10.45 \times 10^{-13})0.01}{11.39 \times 10^{-4}} = 9.17\ \text{pF}$$

$$\tau = R_L C_j = (1.5 \times 10^2)(0.917 \times 10^{-11}) = 1.38\ \text{nsec}$$

Note that there is a trade-off between the transit times and the junction capacitance. Increasing W increases t_{st}, but it reduces C_j. However, C_j can also be reduced by reducing the area. For an optical fiber detector a typical diameter is 300 μm, which would reduce A, and therefore, C_j, by a factor of ten in Example 4-12. A more important trade-off is the one between the transit times and the collection efficiency. The usual compromise is $W \simeq 1/\alpha$.

Again, the wavelength response varies significantly with λ because $\alpha = f(\lambda)$ for indirect gap materials. This is illustrated in Fig. 4-32. There it is seen that for a silicon *pin* diode the collection efficiency increases with increasing W and decreases with increasing λ.

The frequency response for the drift component is computed for a sinusoidal signal by again recognizing that it takes a finite time to collect the photocarriers. Thus

$$J_{\text{op}} \propto \frac{1}{W}\int_0^W e^{j\omega(t - x/v_{\text{sat}})}\,dx = \frac{e^{j\omega t}}{j\omega t_{st}}(1 - e^{-j\omega t_{st}}) \tag{4-95}$$

using $t_{st} = W/v_{\text{sat}}$. The magnitude of J_{op} is then

$$|J_{\text{op}}|^2 \propto \frac{2}{\omega^2 t_{st}^2}(1 - \cos \omega t_{st}) \tag{4-96}$$

The half power, or -3-dB, point is reached at $\omega t_{st} \simeq 2.4$. Therefore, for the condition $W = 1/\alpha$

$$f_{1/2} \simeq 0.4\alpha v_{\text{sat}} \tag{4-97}$$

In general, α is larger in direct gap materials so that they are faster. For InGaAs detectors $f_{1/2}$ can be as high as 10 to 30 GHz.

4-7.3 Avalanche Photodiodes (APD)

One of the primary disadvantages of a photodiode is that its gain is less than one. When it is operated in the breakdown mode, however, the gain can be increased to greater than 100.

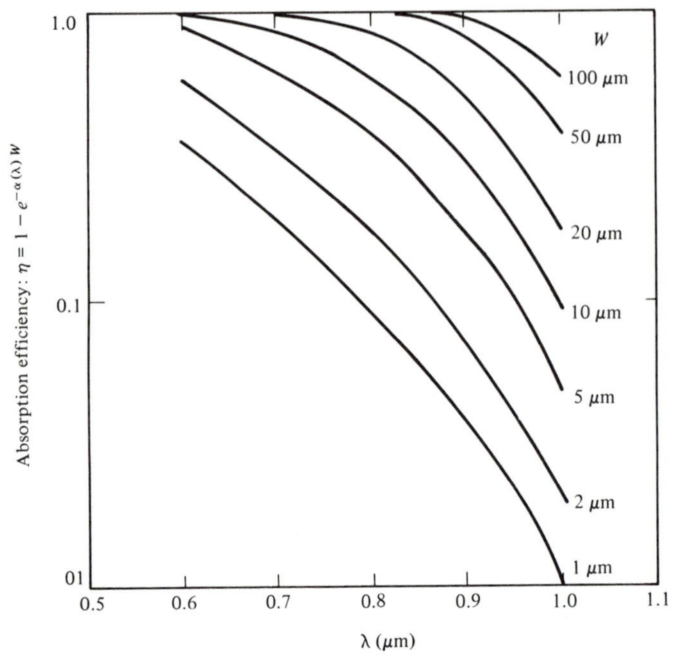

Figure 4-32 The fraction of the light absorbed in the depletion region for different depletion layer widths plotted as a function of the wavelength.

The gain mechanism is the creation of an EHP when a carrier collides with the lattice in the depletion region, and this is illustrated in Fig. 4-33*a*. The secondary hole created by the primary electron accelerates in the opposite direction of the electron, the tertiary electron accelerates in the opposite direction of the hole, and so forth. For the ideal case when the probability, P, that an accelerating

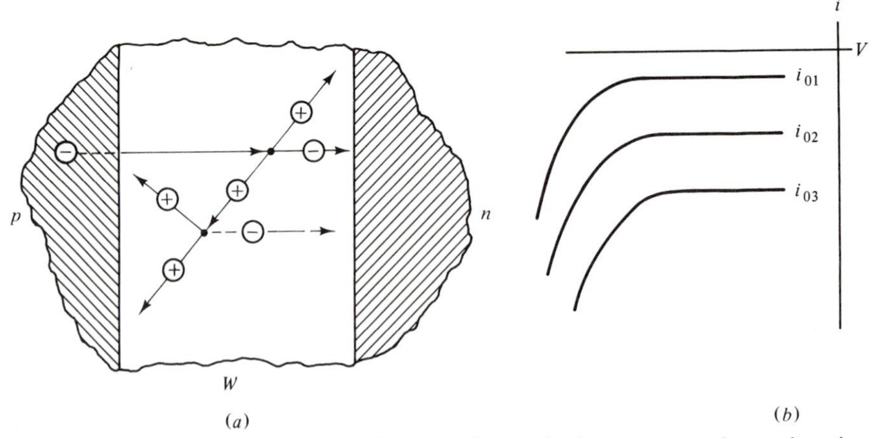

Figure 4-33 (*a*) Schematic representation of the impact ionization process in a *pn* junction. (*b*) Breakdown in an avalanche photodiode for different photon fluxes.

electron will create an EHP is the same as an accelerating hole creating an EHP, the number of carriers, M, generated by the initial photocarrier is

$$M = 1 + P + P^2 \cdots P^\infty = \frac{1}{1 - P} \tag{4-98}$$

The first term (one) is the injected carrier; the second term is the probability the injected carrier generates another carrier; the third term is the probability the injected carrier generates a second carrier and the second carrier generates a third, and so forth. The reverse saturation current now increases quite rapidly near the breakdown voltage, as is shown in Fig. 4-33*b*.

P is a very sensitive function of the voltage and in the dark it is given by

$$P = \left(\frac{V}{V_{\text{br}}}\right)^n \tag{4-99}$$

where V_{br} is the breakdown voltage and n is an experimentally determined constant.

The breakdown voltage is a function of the temperature, energy gap, and doping level. It increases with the temperature because the probability of collisions with the lattice increases. Thus, the charge carriers are accelerated for a shorter period of time and, therefore, do not accumulate as much kinetic energy; they must be accelerated by a larger voltage to obtain enough kinetic energy for breakdown. The energy gap and doping dependence are illustrated in Fig. 4-34. The larger is E_g then the larger is V_{br}, since the avalanche carriers must jump over

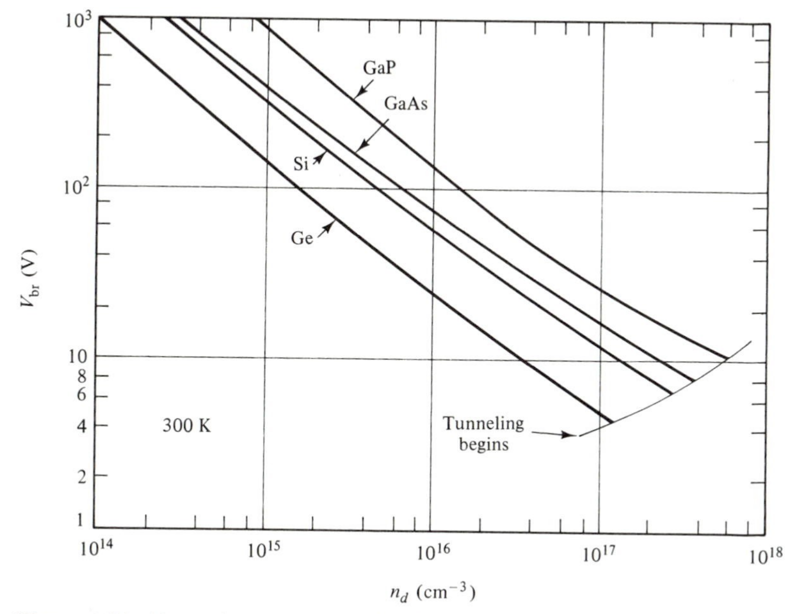

Figure 4-34 Dependence of the avalanche breakdown voltage on the doping level in a p^+n diode. p^+ indicates that the p side is heavily doped ($>10^{18}$ cm^{-3}).

a larger energy gap. V_{br} decreases with increasing doping levels because increasing the doping levels increases the accelerating field in the depletion region.

In reality electrons are often formed more easily than holes. The probability an electron has an ionizing collision in a distance, dx, is $\alpha\, dx$, and the probability a hole has an ionizing collision is $\beta\, dx$. In silicon $\alpha/\beta \simeq 50$. It is not important whether $\alpha > \beta$ or $\beta > \alpha$, but it is beneficial for them to have much different values. If they are the same, the gain bandwidth product is essentially constant. This is because the more avalanche carriers there are produced, the longer it takes to collect them. However, the greater the difference between α and β, the less sensitive M is to the frequency, and the gain bandwidth product does increase with increasing reverse bias.

The gain is a function of the light intensity as well as is the frequency; it decreases at higher intensities because there is more heating due to absorption, and there are more losses due to series resistance effects.

An APD often has the Read, or "reach through," structure shown in Fig. 4-35. Avalanche takes place in the p layer of the n^+pip^+ structure because the

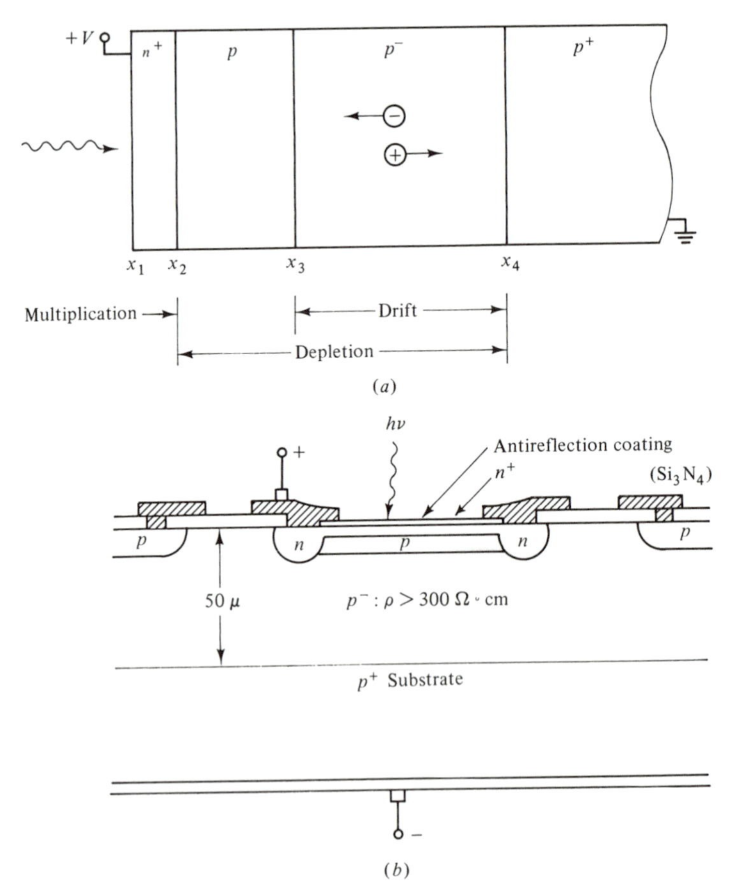

Figure 4-35 Schematic of the read reach through diode illustrating (*a*) the multiplication and drift regions, and (*b*) an antireflection coating and guard ring.

electric field is much larger there than it is in the i layer. For the device to operate at high speeds, a large enough voltage must be applied to it so that the ξ field punches through the p layer and extends to the p^+ layer. The ξ field in this wider depletion region more quickly collects the photocarriers. Voltages as large as 400 V are applied. This is why some APDs are hermetically sealed.

A silicon APD is constructed so that most of the photocarriers are generated in the i layer. Photoelectrons drift to the p layer where they are accelerated more rapidly and create avalanche breakdown. Primary electrons are used instead of primary holes because $\alpha/\beta = 50$. For $\beta > \alpha$ one would use a p^+nin^+ structure. Also, to prevent premature breakdown at the edges where larger ξ fields can be created, a guard ring is diffused around the avalanche region. This is illustrated in Fig. 4-35*b*.

The optimum operating magnification is to some extent determined by the noise sources. At room temperature the dominant noise is the thermal noise, which is independent of the gain, whereas the signal power is proportional to M^2. Other types of noise, however, are proportional to M^n where $n > 2$, so that they dominate at large values of M and reduce the signal/noise ratio. There is, therefore, an optimum value of M and this optimum value often lies between 30 and 100. This topic will be discussed in more detail in Section 5-4.3.

The dark reverse saturation current can be a primary contributor to the noise. This is one of the major disadvantages of germanium APDs. With its small (0.67 eV) energy gap, I_s at room temperature is 50 to 100 nA. The easiest, but also one of the more expensive, ways to solve this problem is to cool the device. $In_{0.53}Ga_{0.47}As$ APDs have a somewhat smaller reverse current, but it is even smaller in InGaAsP and GaAlAsSb APDs, since they have larger energy gaps. A structure that can greatly reduce I_s and yet still be sensitive to the longer wavelength radiation is a heterostructure consisting of an $In_{0.53}Ga_{0.47}As$ absorbing i layer and an InP avalanche layer. Recall that InP has $E_g = 1.35$ eV and $In_{0.53}Ga_{0.47}As$ has $E_g = 0.75$ eV.

4-8 REAL DIODE CHARACTERISTICS

Up to this point we have focused on the ideal diode characteristics. In this section the nonideal properties, what causes them, and where their effects are dominant will be discussed. These include the reverse bias generation–recombination, forward bias generation–recombination, high injection, and series resistance regions; they are illustrated in Fig. 4-36.

The generation–recombination currents are created in the depletion region by two step processes illustrated in Fig. 4-37. An electron-hole pair is formed when a valence band electron is excited into an empty recombination center (process d), and then is further excited into the conduction band (process b). An EHP can be destroyed by a conduction band electron falling into an empty recombination center (process a), and then falling further to the valence band by recombining with a hole (process c). The former process is the source of the

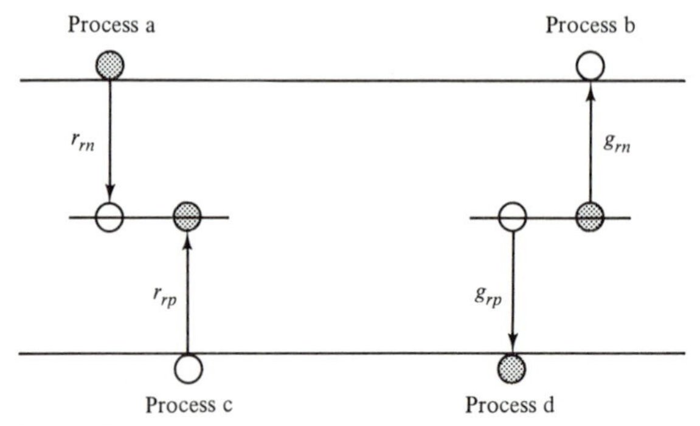

Figure 4-36 Schematic representations of (*a*) an electron recombining with an empty recombination center, (*b*) an electron being excited out of a recombination center into the conduction band, (*c*) a hole recombining with a filled recombination center, and (*d*) a hole being excited out of a recombination center into the valence band.

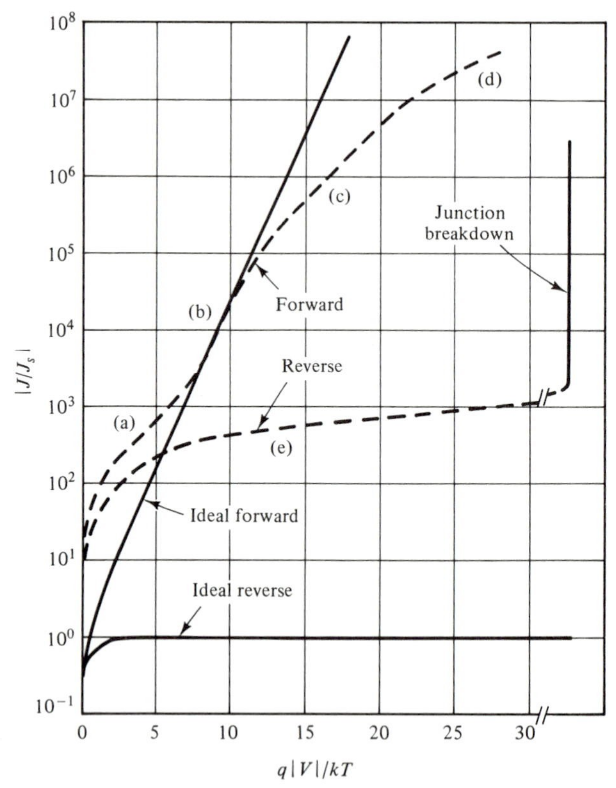

Figure 4-37 The real characteristics of a diode illustrating the (*a*) forward generation–recombination, (*b*) ideal, (*c*) high injection, (*d*) series resistance, and (*e*) reverse generation–recombination regions. (From S. M. Sze, *Physics of Semiconductor Devices,* 2d Ed., Copyright © 1981 by John Wiley & Sons, New York.)

reverse bias generation–recombination current, and the latter process is the source of the forward bias generation–recombination current.

Before finding these currents, we must first mathematically describe the four processes. For process a the rate at which an electron falls into a recombination center, r_{rn}, is the product of the probability per unit time an electron collides with a recombination center, the probability, $1 - f$, the recombination center is empty, and the density of electrons. From Eq. 4-28, the probability of a collision per unit time is $B_{rn}n_r$ so that

$$r_{rn} = B_{rn}n_r(1 - f)_n \tag{4-100}$$

The electron in the recombination center can follow one of two paths; it can be emitted into the conduction band, or it can fall into the valence band. The electron generation rate, g_{rn}, process b is the product of the probability per unit time of a collision of a vacant site, a hole, in the conduction band with a recombination center, the probability, f, the center is occupied, the density of recombination centers, and the probability the electron has a large enough thermal energy to jump out. n_c in Eq. 4-16a is the effective density of states in the conduction band so that the probability of a collision of a vacant site in the conduction band with a recombination center is $B_{rn}n_c$. For $\Delta E \gg kT$, the probability an electron has an energy greater than or equal to ΔE is $\exp(-\Delta E/kT)$. Thus, the probability an electron can jump from a recombination center into the conduction band is $\exp[-(E_g - E_r)/kT]$ where E_r is the energy level for the recombination center. Thus

$$g_{rn} = B_{rn}n_c f n_r \exp\left(\frac{E_r - E_g}{kT}\right) \tag{4-101a}$$

which can also be written using Eq. 4-16c

$$g_{rn} = B_{rn} f n_r n_i \exp\left(\frac{E_r - E_i}{kT}\right) \tag{4-101b}$$

The electron falling from a recombination center into the valence band, c, can also be viewed as a hole falling from the valence band into a recombination center. The rate at which holes fall into the recombination centers, r_{rp}, is the product of the probability per unit time a hole will collide with a recombination center, the probability the recombination center is occupied, and the density of holes. Thus

$$r_{rp} = B_{rp}n_r f p \tag{4-102}$$

We have already considered the rate at which the holes will fall into the conduction band when we found the rate at which electrons fall into the recombination centers; they are the same thing. The rate at which holes jump from the recombination centers back into the valence band is the product of the probability per unit time of a collision between an electron in the valence band with a recombination center, the probability a recombination site is empty, the density of recombination centers, and the probability that a hole has enough thermal energy to jump into the valence band. By analogy with Eq. 4-101a

$$g_{rp} = B_{rp}n_v(1 - f)n_r \exp\left(\frac{-E_r}{kT}\right) \tag{4-103a}$$

which can also be written

$$g_{rp} = B_{rp}(1 - f)n_r n_i \exp\left(\frac{E_i - E_r}{kT}\right) \tag{4-103b}$$

For a reverse biased diode both n and p are small so the important rates are g_{rp} (an electron jumping from the valence band into an empty recombination site) and g_{rn} (an electron jumping from a filled recombination site into the conduction band). The electrons and holes created by this process are then swept out by the electric field in the depletion region.

Under steady state conditions these two rates are equal. By using the simplifying assumption that $B_{rp} = B_{rn}$,

$$f \exp\left(\frac{E_r - E_i}{kT}\right) = (1 - f) \exp\left(\frac{E_i - E_r}{kT}\right) \tag{4-104}$$

under steady state conditions. These terms are maximized when $f = \frac{1}{2}$. Thus, the most effective recombination centers are those near the center of the energy gap where $E_r = E_i$.

The reverse bias generation–recombination current density, J_{gr}, is

$$J_{gr} = -qg_r W \tag{4-105a}$$

and for recombination centers near the mid-gap it is

$$J_{gr} = -\tfrac{1}{2}qB_r n_i W = \frac{-qn_i W}{2\tau_r} \tag{4-105b}$$

where τ_r is the recombination time (see Eq. 4-28). The negative sign simply indicates that the current is a reverse biased current. An important point to note is that the current increases in magnitude as the reverse bias increases because the depletion layer width increases. This is undesirable because the noise level increases with the magnitude of the current. To operate at low noise levels, one must operate at small reverse bias, but this has the undesirable effect of increasing the junction capacitance. The latter effect can be mitigated by using a low doping concentration that will reduce the capacitance by increasing the depletion layer width.

The reverse currents for a germanium, silicon, and GaAs diode at different temperatures are shown in Fig. 4-38. The current is the smallest through the GaAs diode, and the voltage dependence of the current is the most obvious. This is due to n_i being the smallest in the widest bandgap semiconductor. To see this we examine the ratio of the electric field current density, J_ξ, with J_{gr} for the case that $J_{\xi_p} = J_{\xi_n}$. The ratio is

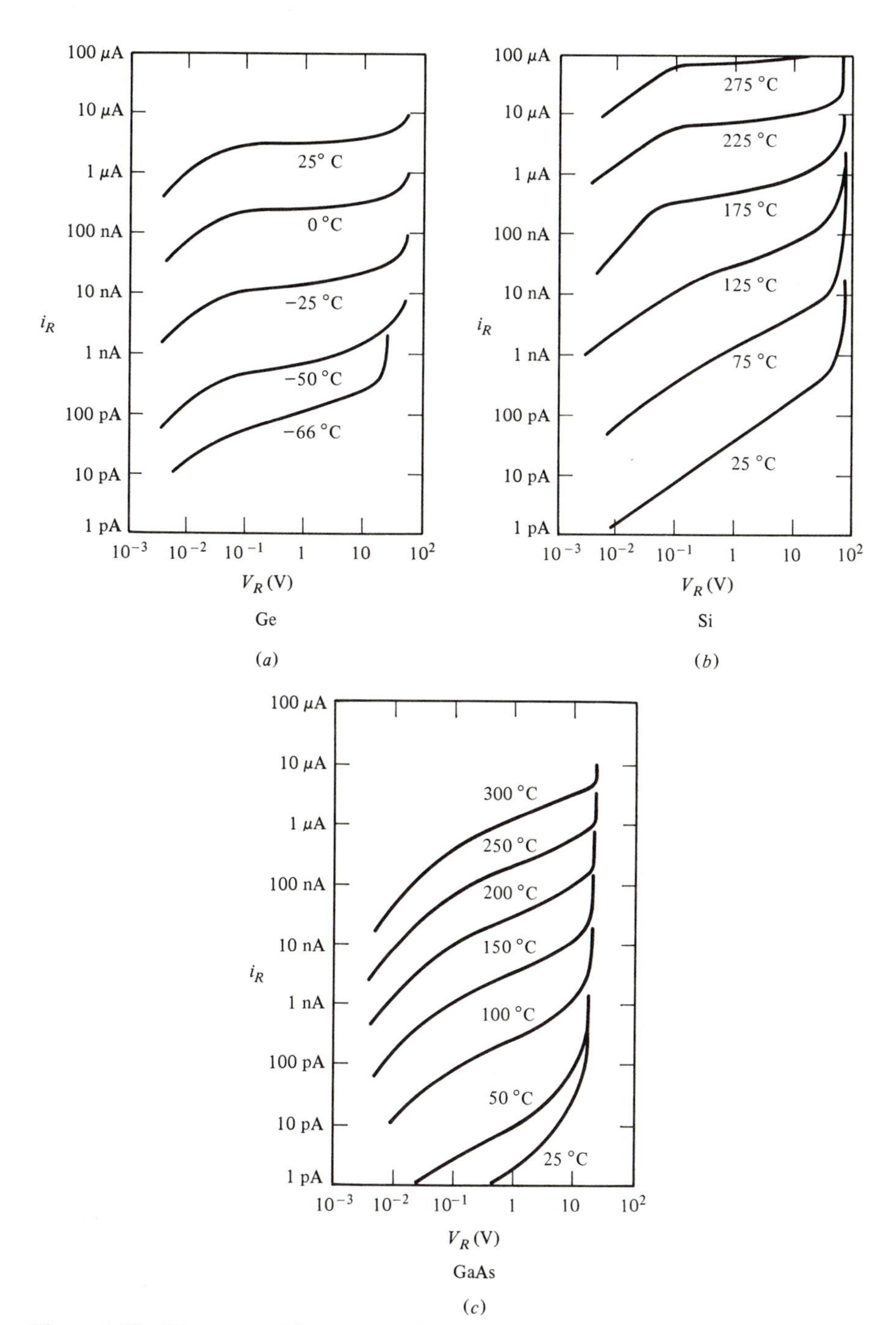

Figure 4-38 The reverse bias current characteristics for a Ge, Si, and GaAs diode at different temperatures. (From A. S. Grove, *Physics and Technology of Semiconductor Devices,* Copyright © 1967 by John Wiley & Sons, New York.)

$$\frac{J_\xi}{J_{gr}} = \frac{-2qp_{n0}L_p/\tau_p}{-qn_iW/2\tau_r} = \frac{4\tau_r L_p n_i}{\tau_p W n_d} \tag{4-106a}$$

Thus, for small n_i values J_{gr} will dominate.

Under forward bias n and p are quite large so that r_{rn} and r_{rp} dominate the generation–recombination process. Again, under steady state conditions $r_{rn} = r_{rp}$. Assuming $B_{rn} = B_{rp}$ the forward bias generation–recombination current density is

$$J_{gr} = qr_rW = fqB_rn_rW_p = \frac{fqW_p}{\tau_r} \tag{4-107}$$

The pn product under forward bias is

$$pn = n_ie^{(E_i-E_{F_p})/kT} \cdot n_ie^{(E_{F_n}-E_i)/kT} = n_i^2e^{qV/kT} \tag{4-108a}$$

J_{gr} is a maximum when $p = n$, and this implies that $f = \frac{1}{2}$, since it is equally hard to create an electron or a hole. Thus, J_{gr} becomes

$$J_{gr} = qWn_ie^{qV/2kT}/2\tau_r \tag{4-106b}$$

Note that there is now a $\frac{1}{2}$ in the exponent that is not present in the diffusion current. Thus, the general term for the exponent is qV/nkT where n is sometimes called the ideality factor. When the generation–recombination current dominates, $n = 2$; when the diffusion current dominates, $n = 1$; and in the transition region $1 < n < 2$.

As is illustrated in Fig. 4-39, the generation–recombination current dominates at low voltages and small n_i. This can be seen by examining the J_D/J_{gr} ratio, which is

$$\frac{J_D}{J_{gr}} \simeq \frac{2qL_pp_{n0}e^{qV/kT}/\tau_p}{qWn_ie^{qV/2kT}/2\tau_r} = \frac{4\tau_rL_pn_i}{\tau_pWn_d}\,e^{qV/2kT} \tag{4-109}$$

when the hole and electron currents are equal.

We now will briefly examine the high-injection region. In the high-injection region the forward bias lowers the potential barrier to a small enough value such that the minority carrier concentration approaches the majority carrier concentration. On the p side, for example, the excess electron concentration, Δn, is of the order of the hole concentration p_p. The holes lost through recombination with the excess electrons can no longer be ignored. Thus, there are fewer holes attempting to diffuse to the right over the potential barrier. Likewise, electrons are lost on the n side through recombination with the excess holes so that there are fewer of them diffusing to the left. From the fact that

$$pn = n_i^2e^{qV/kT} \tag{4-108b}$$

$$\Delta n \simeq p \simeq n_ie^{qV/2kT} \tag{4-109}$$

By using this value for Δn and Δp in Eqs. 4-68a and 4-73, one can see that the current does not increase in this region as fast as is predicted by the ideal equations since now there is a $\frac{1}{2}$ in the exponent.

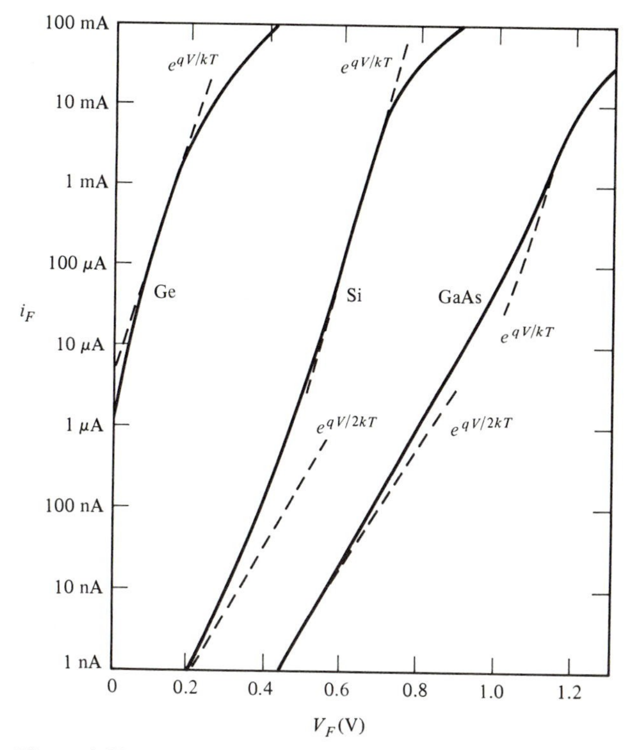

Figure 4-39 The room temperature forward bias current characteristics for a Ge, Si, and GaAs diode.

The voltage across the junction cannot exceed the contact potential. This is accounted for by the series resistance. The neutral region does have a resistance, but it is much smaller than the depletion layer resistance under normal operating conditions. However, as the forward bias increases, the effective resistance of the junction decreases until there is no junction resistance when the junction voltage, $V_j = V_0 - V_a = 0$ where V_a is the applied voltage at the junction. V_a is given by

$$V_a = V - iR_s \tag{4-110}$$

but this simple looking equation is deceptive, since R_s is a function of the current.

READING LIST

1. R. M. Rose, L. A. Shepard, and J. Wulff, *The Structure and Properties of Materials.* Vol. IV. New York: John Wiley, 1964, Chapters 2 and 4 to 7.
2. C. A. Wert and R. M. Thomson, *Physics of Solids.* New York: McGraw-Hill, 1970, Chapters 11 to 13.
3. B. G. Streetman, *Solid State Electronic Devices.* Englewood Cliffs, NJ, Prentice-Hall, 1980, Chapters 3 to 6.

4. D. H. Navon, *Electronic Materials and Devices.* Boston: Houghton Mifflin, 1975, Chapters 5 and 6.
5. A. S. Grove, *Physics and Technology of Semiconductor Devices.* New York, John Wiley, 1967, Chapters 4 to 6.
6. S. R. Forest, "Optical Detectors: Three Contenders," *IEEE Spectrum,* Vol. 23, No. 5, pp. 76–84, May 1986.
7. E. A. Lacy, *Fiber Optics.* Englewood Cliffs, NJ, Prentice-Hall, 1982, Chapter 6.
8. H. Kressel (Ed.), *Topics in Applied Physics, Semiconductor Optical Devices for Optical Communications.* Vol. 39. New York: Springer-Verlag, 1980. Chapters 3 and 4.
9. A. Yariv, *Introduction to Optical Electronics.* New York, Holt, Rinehart and Winston, 1976, Chapter 11.
10. R. J. Keys (Ed.), *Topics in Applied Physics,* Vol. 19, *Optical and Infrared Detectors.* Berlin: Springer-Verlag, 1980.
11. S. M. Sze, *Physics of Semiconductor Devices,* 2d Ed. New York: John Wiley, 1981, Chapter 13.

PROBLEMS

4-1. Calculate the amount of current flowing in a $0.1N$ solution of NaCl when 10 V is applied between the electrodes. The cross-sectional area is 1.0 cm^2, the distance between the electrodes is 5.0 cm, $\mu_p = 52.0 \times 10^{-5}$ cm^2/V · sec, and $\mu_n = 79.0 \times 10^{-5}$ cm^2/V · sec. Assume that the NaCl is completely dissociated and that the volume of the solution containing 0.1 mole of solute is 1 L.

4-2. If the mobility of the H^+ ion is 3.63×10^{-3} cm^2/V · sec and the mobility of OH^- ions is 2.05×10^{-3} cm^2/V · sec, calculate the conductivity of pure water. The fraction of H_2O molecules that are dissociated is 1.81×10^{-9}.

4-3. A solution of $CaCl_2$ has a current of 0.10 A flowing through it when 6.0 V are applied across the electrode. Determine T^+ and the concentration of $CaCl_2$ if $A = 2.0$ cm^2, $L = 3.0$ cm, $\mu_n = 6.16 \times 10^{-4}$ cm^2/V · sec, and $\mu_p = 7.90 \times 10^{-4}$ cm^2/V · sec. Assume that the volume of the aqueous solution does not depend on the amount of solute in solution.

4-4. A copper wire of cross-sectional area 5×10^{-6} m^2 carries a steady current of 50 A. Assuming one free electron per atom, calculate
- **(a)** The density of free electrons.
- **(b)** The average drift velocity.
- **(c)** The relaxation time, τ.

The density of Cu = 8.9×10^{-3} kg/cm^3; atomic weight of Cu = 64.

4-5. The electrical resistivity of pure tantalum wire is 35 $\mu\Omega \cdot$cm at 100 K and 65 $\mu\Omega \cdot$cm at 160 K. Calculate the electrical resistivity at room temperature (20° C) if the resistivity increases linearly with the temperature.

4-6. Aluminum has a room temperature resistivity of 2.65×10^{-8} $\Omega \cdot$m, and copper has a room temperature resistivity of 1.67×10^{-8} $\Omega \cdot$m.
- **(a)** Calculate the resistance of an Al rod 1 m long and 5 mm in diameter.
- **(b)** How long must a Cu rod of the same cross-sectional area be to have the same resistance?

4-7. The work functions for Na, Mg, and Cu are

Na 2.28 V

Mg 3.70 V

Cu 4.48 V

Make a schematic plot of the rate of photoemission for each metal as a function of the wavelength of the incident radiation. Put the curves for each metal on the same graph so that their relative positions may be compared.

4-8. For a cesium-coated oxide-silver sandwich photoemitter, $\phi = 0.9$ V.
- **(a)** What is the longest wavelength which will eject photoelectrons?
- **(b)** If light of 3200-Å wavelength is incident on the emitter, what is the maximum velocity of the ejected electrons?

4-9. The photoelectric threshold is the maximum wavelength of incident light that will produce photoemission. Photoelectric thresholds for polished or clean metal surfaces are: for Al, 4700 Å; for Cu, 3000 Å; for W, 2300 Å; for Na, 5400 Å. Determine the work function in volts.

4-10. **(a)** Explain why the photoelectric emission of photocathodes should decrease when a monolayer of oxygen atoms forms on its surface.
- **(b)** Why might a monolayer of hydrogen atoms increase photoemission?

4-11. In a special type of photomultiplier tube, electrons are ejected by radiation from a Cs_2Sb photoemitter ($\phi = 1.8$ V); the photoelectrons then strike a cesium surface ($\phi = 1.9$ V). The secondary electrons are collected on another electrode. What is the longest wavelength that will give a secondary electron current?

4-12. The resistivities of an intrinsic semiconductor at various temperatures are given in the following table. Calculate the energy gap of this semiconductor.

$(\Omega \cdot m)$	$T(K)$
6×10^{-1}	500
1×10	400
2.1×10^{2}	335
3.0×10^{3}	300

4-13. Calculate the intrinsic conductivity for germanium at temperatures of 300 K, 400 K, and 500 K. For an intrinsic semiconductor.

4-14. What is the percent change in σ for a 1° change at 300 K if $E_g = 1.1$ eV?

4-15. **(a)** What is the maximum Eg an intrinsic semiconductor can have and still have a resistance $\leq 10^6\ \Omega$ at 100 K? The specimen is 1 cm long and has a cross-sectional area of 4 mm^2. Assume $m_n^* = m_p^* = m$ and $\mu_n = \mu_p = 200$ cm^2/V · sec.
- **(b)** What percent change in the resistance will produce an error of 0.1 K at 100 K?

4-16. In the first approximation $Eg = Eg(0) - \alpha T$. $E_g(0)$ for CdS is 2.55 eV and $\alpha = 4.07 \times 10^{-4}$ eV/K.
- **(a)** What is the longest wavelength that is strongly absorbed by CdS at the boiling point of liquid He? (4.2°K)
- **(b)** At what temperature does the crystal become red? (red, 6100 Å)
- **(c)** At what temperature does the crystal become opaque? The wavelength of the longest visible light is 8000 Å.

4-17. **(a)** Determine the maximum value of the energy gap that a semiconductor used as a photoconductor can have if it is to be sensitive to yellow light ($\lambda = 6000 \times 10^{-10}$ m).

(b) A photodetector whose area is 5.0×10^{-6} m^2 is irradiated with yellow light whose intensity is 20 W/m^2. Assuming each photon generates one electron-hole pair, calculate the number of pairs generated per second.

4-18. **(a)** From the known energy gap of gallium arsenide, calculate the primary wavelength of photons emitted from this crystal as a result of electron-hole recombination.

(b) Is this light visible?

(c) Will a silicon photodetector be sensitive to the radiation from a GaAs laser? Why?

4-19. The mobility of electrons in a solid is 5 m^2/V·sec. What is the relaxation time if the effective mass of the electron is 0.9 m?

4-20. Determine E_i for Ge, Si, and GaAs if their respective electron effect mass ratios are 0.55, 0.40, and 0.08, and their respective hole effective mass ratios are 0.37, 0.50, and 0.50.

4-21. The concentration of conduction electrons in an intrinsic semiconductor at different temperatures is shown in the table.

(a) If $\mu_n = 0.17$ and $\mu_p = 0.35$ M^2/V·sec and are independent of temperature, what is the electrical conductivity at 100, 200, and 650 K?

(b) What is E_g for this semiconductor?

K	Elect./cm^3
75	4.55×10^{-20}
100	8.12×10^{-10}
150	1.74×10
200	9.11×10^{4}
300	5.73×10^{9}
450	1.11×10^{13}
650	1.39×10^{15}
900	3.29×10^{16}

4-22. **(a)** Show that the minimum conductivity of a semiconductor sample occurs when $n = n_i\sqrt{\mu_p/\mu_n}$.

(b) What is the expression for the minimum conductivity, σ_{min}?

(c) Calculate σ_{min} for Ge at 300 K and compare it with the intrinsic conductivity.

4-23. A *p*-type silicon semiconductor must have a resistivity of 0.1 Ω·m. What percentage of aluminum must be added to achieve this?

4-25. **(a)** What donor concentration is required to obtain a conductivity of 10 S/m in silicon at room temperature?

(b) What concentration of acceptors is required for *p*-type germanium for the same conductivity?

4-25. Calculate the ratio of electron conduction current to hole conduction current at 300 K in

(a) Intrinsic Ge.

(b) 0.05 Ω-M *p*-type Ge.

4-26. Calculate the position of the Fermi level at 300 K for:

(a) Silicon containing 10^{17} boron atoms/cm^3.

(b) Germanium containing 10^{17} arsenic atoms/cm^3 plus 5×10^{16} atoms/cm^3 of indium.

(c) Repeat part a at 600 K.

4-27. The lifetime of an electron in a particular piece of doped Ge might be of magnitude 10^{-4} sec. Calculate the mean diffusion distance of an electron in such material at room temperature.

$$D_n = 100 \text{ cm}^2/\text{sec}$$

4-28. What dopants would you use to dope the II–VI semiconductors n-type? p-type? Explain.

4-29. **(a)** Determine the intrinsic temperature, T_i, for germanium doped with 10^{16} cm^{-3} of phosphorus.

(b) Determine T_i for a dopant concentration of 10^{17} cm^{-3}.

(c) For a given dopant concentration is T_i for Si greater or less than that for Ge? Explain.

4-30. A sheet of photoelectrons is created at $x = 0$ by a flash of light from a line source. The sheet has an infinitesimal width with a surface concentration of Δn_s cm^{-2}.

(a) What is the differential equation for the rate of change of photoelectrons if no electrons are lost through recombination? Assume $\Delta n \ll p_0$.

(b) Show that

$$\Delta n(x, t) = \frac{\Delta n_s}{2(\pi D_n t)^{1/2}} e^{-x^2/4D_n t}$$

is a solution to this differential equation.

(c) Show that the number of photoelectrons is conserved by this solution.

(d) For small values of x the concentration of photoelectrons is decreasing with time, and for large values of x the concentration is increasing with time. At $t = t_0$ for what value of x is the concentration neither increasing nor decreasing?

4-31. When recombination is included for the experimental setup described in Problem 4-30, the differential equation for the rate of change of the photoelectron concentration is

$$\frac{\partial \Delta n}{\partial t} = D_n \frac{\partial^2 \Delta n}{\partial x^2} - \frac{\Delta n}{\tau_n}$$

and the solution to this equation is

$$\Delta n = \frac{\Delta n_s}{2(\pi D_n t)^{1/2}} \exp\left(\frac{-x^2}{4 D_n t} - \frac{t}{\tau_n}\right)$$

For $t = \tau_n$ make a plot of Δn versus x/L_n for the solution to this equation and for the equation when there is no recombination. Explain why the two curves differ the way they do.

4-32. Explain the mechanism of compensation that occurs when a donor majority exists originally but is compensated by the addition of acceptors.

4-33. Ge has a diamond cubic structure with 8 atoms in a cubic volume 5.66 Å on a side.

(a) How many Ge atoms are there in a cubic centimeter?

(b) Assuming that the substitutional impurity also forms a diamond cubic cell within Ge, what is the size of the unit cell and the interatomic spacing between the impurity atoms at impurity concentrations of 10^{16}, 10^{18}, and 10^{20} cm^{-3}?

(c) What atomic % impurity does an impurity concentration of 10^{20} cm^{-3} represent?

(d) If the impurity is a donor, the discrete donor levels become a donor band at large concentrations. Explain why this is so.

4-34. **(a)** If the energy gap of $In_xGa_{1-x}As$ varies linearly with the composition, what is E_g when $x = 0.53$?

(b) What is the wavelength of the lowest energy photon that can excite an electron into the conduction band?

4-35. A constant radiation source that creates 10^{11} minority carriers/m^3 in a semiconductor is turned off. If the material had a minority carrier lifetime of 1 μsec, how much time is required for the excess concentration to drop to $10^9/m^3$?

4-36. A CdS photodetector receives radiation of 4000-Å wavelength over an area of 2×10^{-6} m^2 and with an intensity of 40 W/m^2. The energy gap is 2.4 eV.

(a) Calculate the number of electron-hole pairs generated per second if each photon generates a pair.

(b) Calculate the increase in the conductance if the electron lifetime is 10^{-3} sec $\mu_n = 10^{-2}$ $m^2/V \cdot$sec, and $L = 2$ mm.

4-37. How many photons per unit area per unit time must strike the surface of an intrinsic CdS film 10 μm thick for the number of optically excited electrons to equal the number of thermally excited electrons in CdS at 300 K? InAs at 300 K? InAs at 77 K? What is the light intensity if $h\nu = E_g$ of the semiconductors? Assume $m_n^* = m_p^* = m$ in both cases. Assume that the decay time in all instances is 10^{-3} sec and $\Delta\, \partial\, \Delta n/\partial t = -\Delta n/\tau$. It is important to determine this carrier concentration because the thermally excited carriers are background noise. You should now know why infrared detectors are cooled to the temperature of liquid nitrogen, 77 K.

4-38. Assume that a photoconductor in the shape of a bar of length, L, and area, A, has a constant voltage, V, applied, and it is illuminated such that g_{op} EPH/$cm^3 \cdot$sec are generated uniformly throughout. If $\mu_n \gg \mu_p$, we can assume the optically induced change in current, Δi, is dominated by the mobility μ_n and the lifetime τ_n for electrons. Show that $\Delta i = qALg_{op}\tau_n/t_{st}$ for this photoconductor, where t_{st} is the transit time of electrons flowing across the length of the bar.

4-39. **(a)** Show that in a photoconductor under high-photon excitation levels ($n_{\text{thermal}} \ll n_{\text{photo}}$) $\partial\, \Delta n/\partial t = -\beta\, \Delta n^2$.

(b) Show that if the photon source is turned off at $t = 0$, $\Delta n = \Delta n_0/(\Delta n_0 \beta t + 1)$ where Δn_0 is the excess electron concentration at $t = 0$. $\Delta n_0 = \sqrt{g_{op}/\beta}$.

(c) Find an expression for the response time, t_0, in terms of Δn_0 and g_{op} if t_0 is defined to be the time at which Δn has decayed to one-half of its original value, Δn_0.

4-40. If Δn_0 electrons are injected into a p-type material at $x = 0$, and all the unrecombined excess electrons are removed at $x = W$, determine the steady state distribution $\Delta n(x)$ when $\Delta n_0 \ll p_0$. Express your answer in terms of sin $h(W - x/L_n)$ and sin hW/L_n.

4-41. Prove that the electron recombination time, L_n, is the average time an excess electron diffuses in a pn type material before it recombines when $\Delta n \ll p_0$.

4-42. Derive expressions for the electron current at a junction when:
(a) $V = 0$
(b) $V < 0$
(c) $V > 0$

4-43. An abrupt Si *pn* junction has $n_a = 10^{17}$ cm^{-3} on one side and $n_d = 10^{15}$ cm^{-3} on the other.
(a) Calculate the Fermi level positions at 300 K in the *p* and *n* regions.
(b) The junction has a circular cross section with a diameter of 100 μm. Calculate x_n, x_p, Q, and ξ_{max} for this junction at equilibrium (300 K). Sketch $\xi(x)$ and the charge density to scale.

4-44. Calculate:
(a) The width of the depletion region, $W = x_n + x_p$, and the maximum value of the electric field in a silicon diode, when $E_{Fn} - E_{Fp} = 1$ eV and $n_a = n_d = 10^{14}/\text{cm}^3$.
(b) Repeat the calculation for $n_d = n_a = 10^{18}$ cm^{-3}.

4-45. A p^+n^+Si junction is doped with $n_d = 10^{16}$ cm^{-3} on the *n* side, where $D_p = 10$ cm^2/sec and $\tau_p = 0.1$ μsec. The junction area is 10^{-4} cm^2. Calculate the reverse saturation current, and the forward current when $V = 0.5$ V.

4-46. A 0.5-V potential is applied to a *pn* junction by connecting the positive terminal of a battery to the *p* side of the junction and the negative terminal to the *n* side. The measured current density is 5 A/cm^2 at 20°C. Calculate the current density that would result if the polarity were reversed.

4-47. Diodes are constructed out of Ge, Si, and GaAs. The *p* sides have $n_a = 10^{16}$ cm^{-3}, and the *n* sides have $n_d = 10^{16}$ cm^{-3}. The diodes are to be operated at 300 K so they will all be extrinsic. Therefore, you can assume that $n_{n0} = n_d$, and $p_{p0} = n_a$. Calculate E_i, E_{Fn}, E_{Fp}, and V_0 for the three diodes.

4-48. A graded junction is one for which $n_d - n_a = -Gx$. Find expressions for Q, ξ, W, and C_j.

4-49. Calculate $V_0 - V$, x_n, x_p, Q, ξ_{max}, and i_f and i_r for the diodes in Problem 4-47 when $V = 0$, $+0.3V$ and $-0.3V$. The cross-sectional area is 10 mm^2, $\epsilon/\epsilon_0 = 16$, 11.8, and 13.2, and $\tau_n - \tau_p = 10^{-4}$, 10^{-5}, and 10^{-8} sec, respectively for Ge, Si, and GaAs.

4-50. The cross-sectional area of a silicon *pn* junction is 1.0×10^{-6} m^2. The *n*-region is 200 μm wide and is doped with 5.0×10^{20} donor atoms/m. The *p*-region is 100 μm wide and is doped with 5.0×10^{25} acceptors/m^3. The minority hole lifetime in the *n*-region is 0.10 μsec. The minority electron lifetime in the *p*-region is 0.05 μsec. The diode current is 1.0 mA.
(a) Determine the injected hole density at the *n*-side edge of the depletion region at 300 K.
(b) Roughly estimate how far these holes penetrate the *n*-region.
(c) Determine the ratio of the density of minority holes injected to the density of majority electrons that are present in the *n*-region.
(d) Calculate the total amount of excess charge in the *n*-region due to injected holes.

4-51. A germanium *pn* junction has 1.0×10^{24} *p*-type impurities/m^3 uniformly distributed on the *p*-side and 9.4×10^{19} *n*-type impurities/m^3 on the *n*-side.

(a) Calculate the junction contact potential under thermal equilibrium conditions at 300 K.
(b) If the impurity concentration on the n-side is increased to $1.7 \times 10^{24}\ \text{m}^{-3}$, what is the new contact potential?
(c) Determine the width of the depletion layer on the n-side for part a. Repeat for the p-side. The relative dielectric constant for germanium is 16.
(d) Calculate the maximum electric field in the depletion region for part a.
(e) Repeat part a at 450 K.

4-52. Show that the terms on either side of the equals sign in Eq. 4-104 are a maximum when $f = \frac{1}{2}$.

4-53. Photocarriers recombine faster at the surface because the surface dangling bonds act as recombination centers. For holes the surface recombination rate is S_p cm/sec. For the case $\Delta p \ll n_0$ and the uniform absorption of radiation generating g_{op} photoholes per cm^3 per sec, the equation for the rate of change of excess holes is

$$\frac{\partial \Delta p}{\partial t} = D_p \frac{\partial^2 \Delta p}{\partial x^2} - \frac{\Delta p}{\tau_p} + g_{\text{op}}$$

The diffusion term must be included because the higher rate at the surface produces a gradient in the carrier concentration. The boundary conditions now are

$$D_p \left. \frac{\Delta\, \partial p}{\partial x} \right|_{x=0} = S_p\, \Delta p(0)$$

since carriers must diffuse to the surface as fast as they recombine, and $\Delta p(\infty) = g_{\text{op}}\tau_p$ because the surface recombination rate has essentially no effect on the carrier concentration far from the surface.

(a) Show that the solution

$$\Delta p = g_{\text{op}}\tau_p \left[1 - \frac{\tau_p S_p \exp(-x/L_p)}{L_p + \tau_p S_p} \right]$$

satisfies these two boundary conditions.
(b) Discuss what happens to Δp as $S_p \rightarrow 0$ and as $S_p \rightarrow \infty$.

4-54. **(a)** If holes are removed at a velocity S_p at $x = 0$, the steady state distribution of holes is given by

$$\Delta p(x) = g_{\text{op}}\tau_p \left[1 - \frac{\tau_p S_p \exp(-x/L_p)}{L_p + \tau_p S_p} \right]$$

Show that this is a solution to the diffusion (continuity) equation.
(b) At the edge of the depletion layer at $x = 0$ $\tau_p S_p \gg L_p$. What is the hole diffusion current at $x = 0$?
(c) What is the velocity at which the holes are diffusing at $x = 0$? Express your answer in terms of L_p and τ_p.
(d) For Si, $L_p = 10\ \mu\text{m}$ and $\tau_p = 10^{-6}$ sec and for GaAs, $L_p = 2\ \mu\text{m}$ and $\tau_p = 10^{-8}$ sec. Calculate their diffusion velocities.

4-55. **(a)** Assuming that the electric field in the depletion region of an abrupt pn junction is uniform and equal to $-\frac{1}{2}\xi_{\max}$, find an expression for the time, t_{sat}, it takes for an electron to reach its saturation velocity if the velocity is proportional to the acceleration.

(b) If after reaching the saturation velocity, v_{sat}, the electron continues to move at a velocity v_{sat}, find an expression for the transit time.

(c) Using the expression derived in (b), compute t_{st} for the diode described in Example 4-8.

(d) For what value of ξ_{max} is $t_{st} = 2 \dfrac{W}{v_{sat}}$?

4-56. Show that $C_j = \epsilon A/W$ for a *pin* diode.

4-57. For a Si *pin* diode, $n_a = n_d = 10^{17}$ cm^{-3}, $L_n = L_p = 10$ μm, $x_i = 2.5$ μm, $A = 1$ mm^2, and $V = -20$ V. Find V_0, W, x_n, x_p, Q, ξ_{max}, τ_n, τ_p, t_{stn}, t_{stp}, and C_j.

4-58. The speed of a photodetector can be increased by creating an electric field outside of the depletion region by varying the doping concentration.

(a) An electric field must exist to create an electric field current that is equal, but opposite to the diffusion current produced by the nonuniform carrier distribution. Show that the ξ field is constant and find an expression for it when the acceptor doping profile is given by $n_a = n_0(0)e^{ax}$.

(b) Find the magnitude of a necessary to generate a field of 10^5 V/m.

(c) Find the ratio of n_a at $x = 1$ μm and n_a at $x = 0$.

(d) Explain in terms of the Fermi level why an inhomogeneous charge distribution and therefore an electric field must exist.

4-59. **(a)** What is the conductivity of *n*-type silicon with $n_d = 10^{15}$ cm^{-3}, *p*-type silicon with $n_a = 10^{15}$ cm^{-3}, and intrinsic silicon with $n_i = 1.2 \times 10^{10}$ cm^{-3}?

(b) Light with an intensity of 0.05 W/cm^2 and wavelength, 0.8 μm, strikes the surface of the *np* detector at normal incidence. What is the photon flux density striking the surface?

(c) What is the thickness of the SiO_2 that will eliminate the reflection losses if ϵ_r for $SiO_2 = 4$?

(d) If the top *n*-layer is 10 μm thick, $L_n = L_p = 10$ μm, all of the photocarriers within L_n and L_p of the depletion region and all photocarriers generated in the depletion region contribute to the photocurrent, and the absorption coefficient for silicon at 0.8 μm is 0.1 μm^{-1}, what fraction of the photocarriers contribute to the photocurrent when $\eta_i = 100\%$ and $V = 5.0$ V? Use $\epsilon_r = 11.8$ for Si.

(e) What is the optical current density?

(f) If an *i* layer 25 μm thick is inserted between the *n* and the *p* layer, what fraction of the light is absorbed in the region where an electric field exists?

4-60. **(a)** When a photocarrier is created, the probability it will recombine between x and $x + dx$ is $Ae^{-x/L}\,dx$ where L is a diffusion length. Find A in terms of L.

(b) If the photocarrier is created a distance x' from the junction, the probability that it will diffuse a distance greater than x' is the probability it will diffuse to the junction. What is this probability?

(c) The number of minority carriers generated in a volume of unit cross-sectional area and thickness, dx, is

$$g_{op}\,dx = -\frac{1}{E_p}\frac{\partial I}{\partial x}\,dx$$

Find an expression for $g_{op}\,dx$ in terms of d, x', and I_0 where d is the distance between the top surface and the depletion layer, x' is the distance between the volume element and the depletion region, and I_0 is the incident intensity.

(d) Show that the number of minority carriers that diffuse to the depletion layer is

$$\frac{\alpha I_0}{E_p(\alpha - 1/L)}\left(e^{-d/L} - e^{-\alpha d}\right)$$

(e) Find the expression for the minority carrier flux when $\alpha = 1/L$.
(f) For what value of d is the result in Part d a maximum? Explain your answer.
(g) Find expressions for the carrier flux as $L \rightarrow \infty$ and $L \rightarrow 0$. Explain your results physically.
(h) Find expression for the carrier flux as $\alpha \rightarrow \infty$ and $\alpha \rightarrow 0$. Explain your results physically.

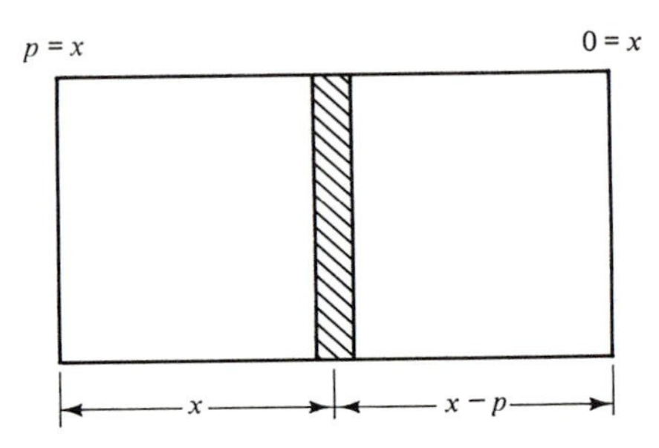

4-61. A reach through Si APD is doped as follows: $n^+ = 10^{18}$, $p = 10^{15}$, $i = 0$, $p^+ = 10^{18}$ cm^{-3}.

(a) Assuming that $V_0 = 1$ V and the p layer is 5 μm thick, what is the minimum reverse bias that must be applied for the ξ field to pass through the p layer to the p^+ layer?
(b) If $V = -50$ V and W = 50 μm, what is the maximum field, ξ_{max}, in the p layer and what is the ξ field, ξ_i, in the intrinsic layer?

You should make some simplifying assumptions for the computations in (a) and (b).

Chapter 5

Other Photodetectors and Noise

5-1 INTRODUCTION

There are many more photodetectors than those mentioned in the previous chapter. In this chapter we examine a few more that contain active semiconductor elements. They include the photovoltaic detector, Schottky barrier diode, phototransistor, photodarlington, and bolometer. Figures of merit and electrical noise are also discussed briefly.

A photovoltaic detector is essentially a photodiode operated with no external bias; the voltage, which is the detector quantity that is measured, is created optically. The charge separation, and therefore the photovoltage, is generated by photoelectrons being swept into the *n*-region and photoholes being swept into the *p*-region by the junction electric field.

A Schottky barrier diode used as a photodetector functions in a manner similar to a photodiode. It is structurally different in that it is a heterojunction device. That is, the junction is composed of two different materials. In this instance the two materials are a metal and a semiconductor. This diode is most frequently used to detect visible and ultraviolet light pulses, and it can operate at frequencies as high as 10 GHz. It can operate at such a high frequency when the photocarriers are created in the depletion region because it is a majority carrier device; there are no minority carrier recombination times to slow the device down.

The advantage of a phototransistor is that it has an internal gain that can exceed 500. Its disadvantage is that it is two to three orders of magnitude slower than a photodiode. Amplification is obtained by forming a second junction—the emitter-base junction. The first junction—the base-collector junction—functions like the photodiode and is reverse biased. The extra carriers are generated at the emitter-base junction by the forward bias produced by the photoelectrons swept into the n-type base from the collector and the photoholes swept out of the base into the collector.

The photocurrent from the phototransistor can be used as the input base current to a second transistor, and this base current can be further amplified. This two transistor combination is called a photodarlington. The disadvantage of the photodarlington is that it is a factor of 10 slower than the phototransistor.

The semiconductor bolometer operates on the principle that the conductivity of a semiconductor is strongly temperature dependent. When light is absorbed by the semiconductor, the temperature, and therefore the conductivity, increases. The conductivity increase is used to determine the temperature change which, in turn, is used to determine the light intensity. Bolometers operating near absolute zero can be extremely sensitive, but their response time is only about 1 msec. They are so sensitive because the thermal noise level is so small.

Thermal noise is due to the random fluctuations of charge carriers that produce random fluctuations in the electrostatic energy. The probability an excess of charge exists at a given position at a given time is the same as that for a deficiency of charge of equal magnitude. Both charge fluctuations result in the same energy fluctuation because the energy is proportional to the square of the charge. The average of all the possible energy fluctuations is the noise equivalent power, NEP. For thermal noise NEP $= kT/\tau$, which illustrates that there is less noise at lower temperatures and when the detector has a narrow bandwidth.

Some other noise sources are generation-recombination, shot, and $1/f$ noise. Generation–recombination noise is due to fluctuations in the rate that carriers are generated or recombine. Shot noise is due to fluctuations in the rate minority carriers are swept out by the electric field at a pn junction, and it is thought that $1/f$ noise, which is inversely proportional to the operating frequency, is due to potential barriers at contacts and surfaces. For an ideal photoconductor at room temperature, $1/f$ noise dominates at the lower frequencies, generation–recombination noise is the largest in the middle frequency range, and thermal noise dominates at the higher frequencies.

The detectivity, D, is the smallest signal that can be detected, and it is equal to the reciprocal of the NEP. To eliminate geometrical and frequency effects, D^* is often used in place of D.

Other figures of merit are the spectral response, response time, gain-bandwidth product, and responsivity. For most photon detectors the responsivity increases with the wavelength up to the absorption edge and then drops precipitously. It is almost wavelength independent for a bolometer.

5-2 PHOTON DETECTORS

5-2.1 Photovoltaic Detector

The photovoltaic detector shown in Fig. 5-1 is a photodiode with no external biasing; the voltage across the junction is produced by the photons themselves.

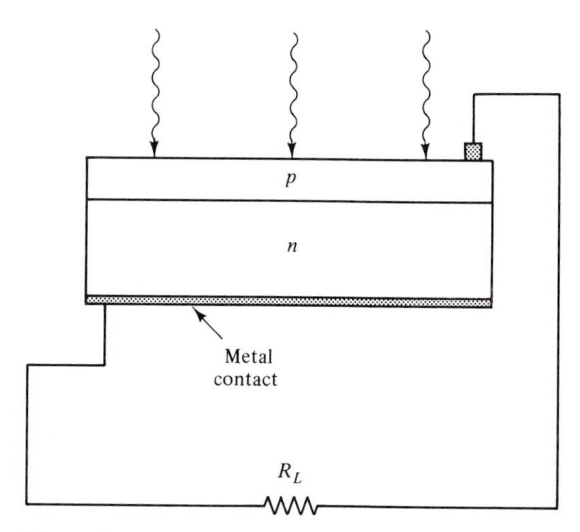

Figure 5-1 Schematic of a photovoltaic detector.

The voltage is generated by the photoelectrons created in the p-layer being swept into the n-layer by the depletion layer electric field, and by the photoholes created in the n-layer being swept into the p-layer. The charge separation of the extra electrons in the n-layer and the extra holes in the p-layer produce a voltage that forward biases the junction.

The forward bias increases the diffusion current, which flows in the direction opposite to that of J_{op}. The magnitude of the diffusion current depends on the value of the junction voltage which, in turn, depends on the resistance of the load resistor in Fig. 5-1. When there is no resistance, there is no increase in the diffusion current so that

$$J_{sc} = J_{op} = \eta_i q \frac{I\lambda}{hc}(1 - R) \tag{4-75}$$

where J_{sc} is the short circuit current density. It is illustrated in Fig. 5-2. When the resistance is infinite, the excess diffusion current and the optical current are equal, but opposite, so that $J = 0$. The voltage across the junction is the open circuit voltage, V_{oc}, and it is also illustrated in Fig. 5-2. From

$$J = J_s(e^{qV/kT} - 1) - J_{op} = 0 \tag{4-74}$$

$$V_{oc} = \frac{kT}{q} \ln\left(1 + \frac{J_{op}}{J_s}\right) \tag{5-1}$$

EXAMPLE 5-1

Monochromatic light with wavelength, $\lambda = 1\ \mu$m, and intensity, $I = 100$ mW/cm^2, strikes a silicon photovoltaic detector. Compute the room temperature values of J_{sc} and V_{oc} if the detector is 90 percent efficient, the p-side is doped to 10^{16} cm^{-3}, and for the n-side, $n_d = 10^{15}$ cm^{-3}

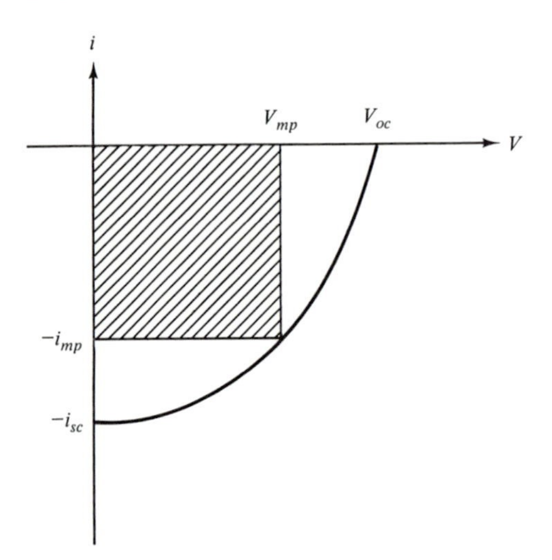

Figure 5-2 The i–V curve for a solar cell with the maximum power rectangle shaded in.

$$J_{op} = \eta q \frac{I\lambda}{hc} = \frac{(0.9)(1.6 \times 10^{-19})(10^3)(10^{-6})}{(6.625 \times 10^{-34})(3 \times 10^8)} = 723 \text{ A/m}^2 = 72.3 \text{ mA/cm}^2$$

From Example 4-10, $J_s = q\left(\frac{D_p}{L_p} P_{n0} + \frac{D_n}{L_\phi} n_{p0}\right) = (2.61 + 0.438) \times 10^{-7} =$

3.048×10^{-7} A/m^2

$$V_{oc} = 0.0259 \ln\left(1 + \frac{7.23 \times 10^2}{3.048 \times 10^{-7}}\right) = 0.559 \text{ V}$$

As seen in Fig. 5-2, the $i - V$ curve lies in the fourth quadrant—a quadrant where the power consumption is negative. Thus, power is generated instead of being consumed; the photodiode operating in this region is a solar cell.

The power output of the cell can be optimized by choosing the correct value for the resistance. The optimum values for i and V are illustrated in Fig. 5-2, and the area of the rectangle determined by them is the maximum power. The ratio of this area and the $J_{sc}V_{oc}$ product is called the fill factor, *FF*, of the cell.

Sunlight, of course, is not monochromatic; the spectrum is shown in Fig. 9-2. To obtain a large J_{sc}, the energy gap of the absorbing semiconductor should be small. This allows a greater percentage of the photons to be absorbed. However, because the photo-excited carriers quickly fall to the bottom of the conduction band, much energy is dissipated in the form of heat. This heat loss is reflected by the small V_{oc} for small bandgap semiconductors. To obtain a large V_{oc}, E_g should be large. Because of the opposing tendencies of J_{sc} and V_{oc}, there is an optimum value of E_g, as is shown in Fig. 5-3. At room temperature the optimum value is 1.35 eV and the maximum theoretical efficiency is 30 percent.

The optimum efficiencies can be approached, but not reached, because there are loss mechanisms. In addition to the reflection and incomplete photo-

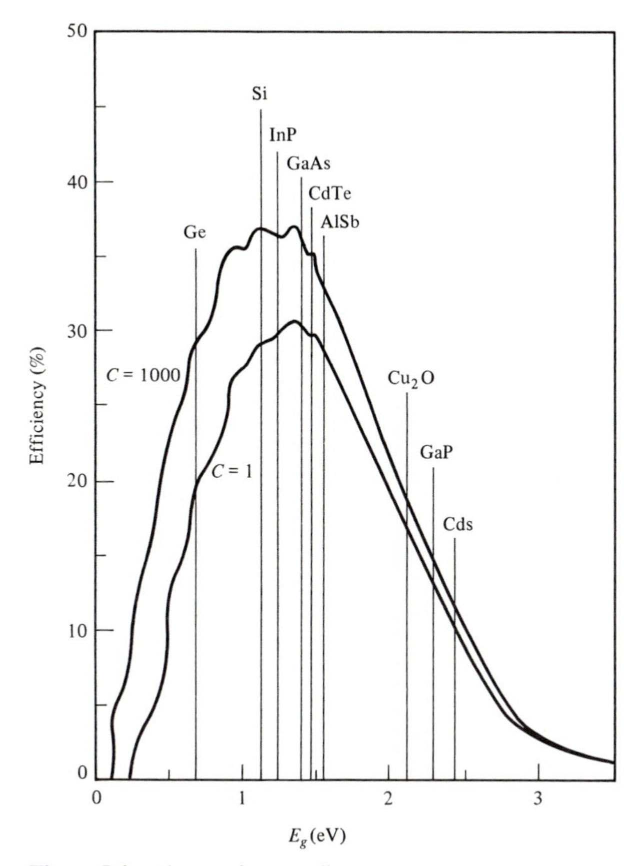

Figure 5-3 The maximum efficiency of a solar cell at 300 K plotted as a function of the energy gap for no concentration, $C = 1$, and for a concentration of 1000 suns, $C = 1000$. (From *Principal Conclusions of the American Physical Society Study Group on Solar Photovoltaic Energy Conversion,* Copyright © 1979, American Physical Society.)

carrier capture discussed in Chapter 4, there is an additional loss due to series resistance in the cell. This results from the need to make the top layer thin so that the photocarriers will be created near the depletion region, and to minimize the contact area to the top layer.

An equivalent circuit for the solar cell in Fig. 5-4 includes this series resistance, R_s. The other elements are the photocurrent source, a shunt resistance, R_{sh}, to account for small leakage currents in the diode, and a capacitor for the junction and diffusion capacitances. For reasonably small values of R_s and large values of R_{sh}, typical values of J_{sc} and V_{oc} for silicon solar cells are 80 mA/cm^2 and 0.6 V. This corresponds to an overall efficiency of 15 percent.

5-2.2 Schottky Barrier Diode

A Schottky barrier diode is formed by a metal and a semiconductor. It is therefore a type of heterojunction; we will discuss semiconductor–semiconductor het-

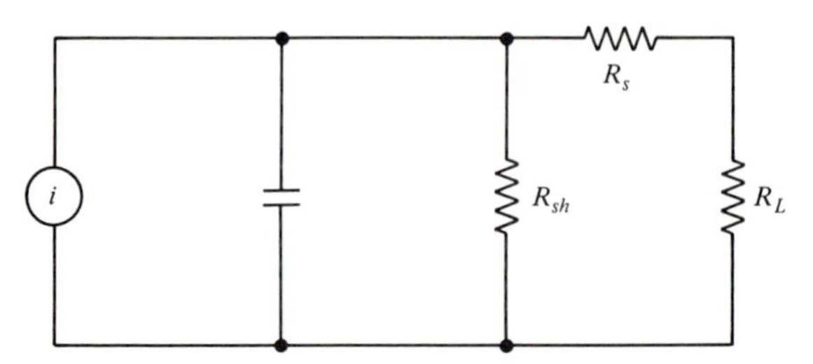

Figure 5-4 Circuit model of a solar cell having a series resistance.

erojunctions in the following chapters. Because a heterojunction involves two different materials, we can no longer use the top of the valence band as the ground point. Instead, the universal ground of a free electron at rest is used. This is illustrated in Fig. 5-5 for a diode formed from a metal and an *n*-type semiconductor.

The Fermi energy rule also applies to heterojunctions. That is, the Fermi energies are everywhere equal at equilibrium. Recall from Section 4-2 that the Fermi energy of a metal, E_{Fm}, is at the highest occupied state in the energy band at 0 K and varies very little with the temperature. Thus, E_{Fm} lies $q\phi_m$ below ground where ϕ_m is again the metal work function. The Fermi level of the *n*-type semiconductor lies $q(\chi + V_n)$ below ground. χ is the electron affinity, and it is the potential difference between the bottom of the conduction band and ground. V_n is the potential difference between the Fermi level and the bottom of the conduction band. Thus

$$V_n = \frac{E_g - E_{Fn}}{q} \tag{5-2}$$

The Fermi levels are again equilibrated by the ionization of the donor electrons in the depletion region—the junction looks much like a p^+n junction. The band bending of the semiconductor due to the potential created by the positively charged donors is

$$qV_0 = q[\phi_m - (\chi + V_n)] \tag{5-3}$$

where V_0 once again is the contact potential. The barrier height, $q\phi_b$, seen by the electrons in the metal moving to the semiconductor, is

$$q\phi_b = q(\phi_m - \chi) \tag{5-4}$$

The barrier height seen by the electrons moving from the semiconductor to the metal depends on how the junction is biased. This is illustrated in Fig. 5-6. When a negative potential is applied to the semiconductor, the energy of the electrons is raised thereby decreasing the barrier height just as it is in a homojunction diode. When a positive potential is applied to the semiconductor, that is, it is reverse biased, the barrier height seen by the electrons in the semiconductor is increased.

The barrier height seen by the electrons moving from the metal to the semiconductor is unaffected by the applied potential: it is fixed at $q(\phi_m - \chi)$. It

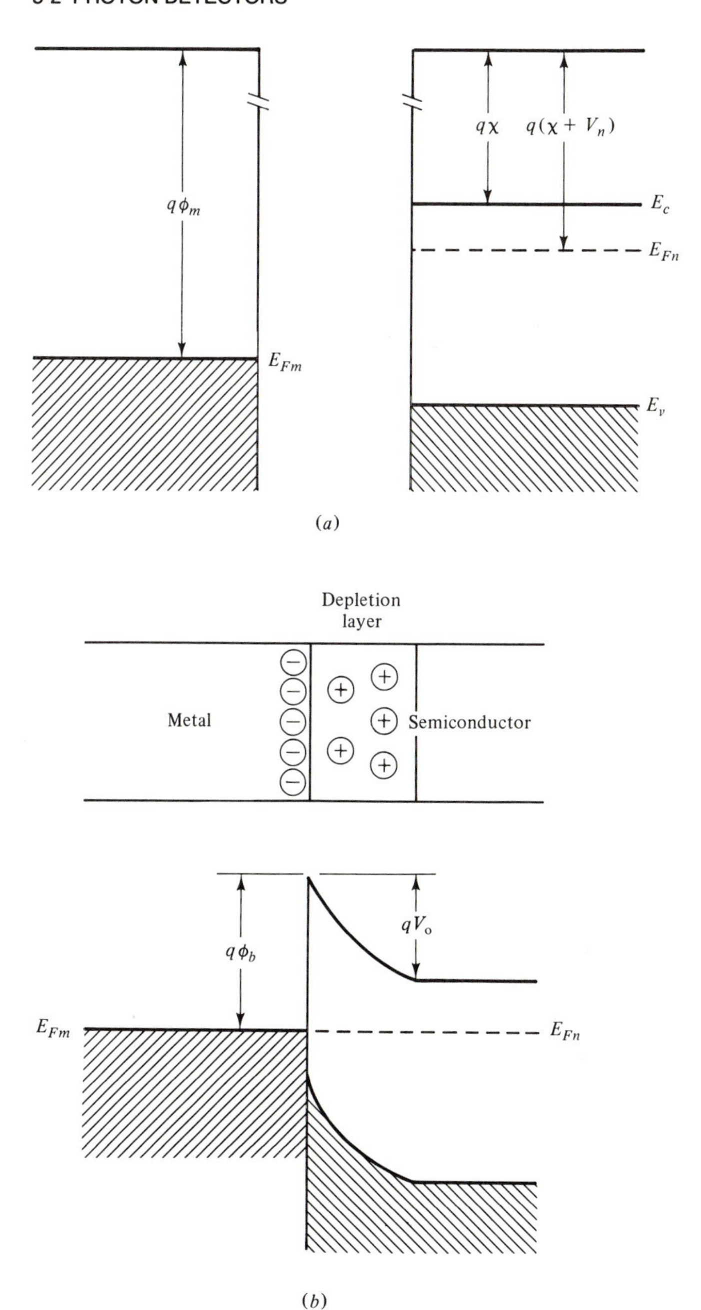

Figure 5-5 Schematic of a metal-n-type semiconductor junction (a) before and (b) after contact that forms a Schottky barrier diode.

does not depend on what the charge in the depletion layer is. Thus, it behaves much like the electric field current in a homojunction diode. It can only be increased by shining light on the diode, as shown in Fig. 5-7. To be counted,

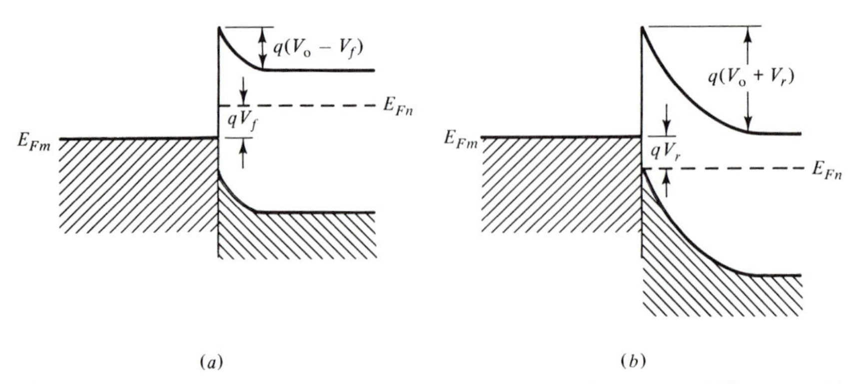

Figure 5-6 The metal-semiconductor junction under (*a*) forward and (*b*) reverse bias.

the photons must have enough energy to excite electrons in the metal over the barrier between the metal and the *n*-type semiconductor, $q\phi_b$. If the photon energy is greater than E_g, EHPs can be created in the depletion region, and the electrons are swept away from the junction while the holes are swept toward it.

The electron current density from the metal to the semiconductor, J_{ms}, is exponentially dependent on the barrier height, $q\phi_b$, because it is proportional to the probability that an electron can jump over the barrier. It can be shown that the equation for J_{ms} is

$$J_{ms} = BT^2 e^{-q\phi_b/kT} \tag{5-5a}$$

where

$$B = 4\pi q m_n^* \frac{k}{h^2} \tag{5-5b}$$

which is 120 $A/cm^2 \cdot K^2$ for a free electron.

At equilibrium the electron current density from the semiconductor to the metal must be the same as the current density flow in the opposite direction. However, it is exponentially dependent on the bias, since the barrier height is $q(V_0 - V)$. Thus

$$J = J_{ms}(e^{qV/kT} - 1) \tag{5-6}$$

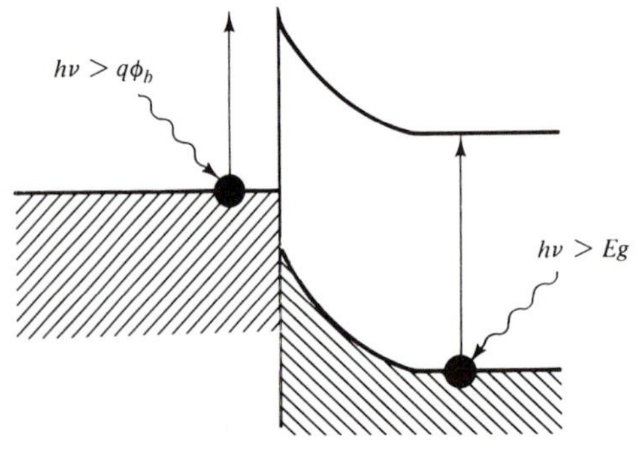

Figure 5-7 The electric field current being increased by exciting electrons over the barrier, $q\phi_b$, or by creating holes in the *n*-type semiconductor.

This equation is similar to that for a homojunction (Eq. 4-73), but it differs in one important way. The current is all majority carrier current. For an n-type semiconductor the current flow is entirely due to electrons; for a p-type it is all hole current. There are no minority currents; hence, there are no delays caused by minority carrier recombination. As a result, Schottky barrier diodes can have rise times of <0.1 nsec. This time is primarily determined by the diode capacitance, which can be reduced by decreasing the cross-sectional area and lightly doping the semiconductor.

EXAMPLE 5-2

Find ϕ_b, V_0, J_{ms}, W, $\xi_{\max}$, and C_j for a Au-Si Schottky barrier diode if the silicon is doped with 10^{15} cm^{-3} with phosphorus. The work function of gold is 4.9 V, the electron affinity of silicon is 4.05 V, and the diode area is 100 μm^2.

$$\phi_b = \phi_m - \chi = 4.9 - 4.05 = 0.85 \text{ V}$$

From Example 4-6, $E_{Fn} = 0.848$ eV

$$V_0 = \frac{\phi_b - (E_g - E_{Fn})}{q} = 0.85 - (1.11 - 0.848) = 0.588 \text{ V}$$

$$J_{ms} = 120\left(\frac{m_n^*}{m_n}\right)T^2 e^{-q\phi_b/kT} = (120)(1.1)(3 \times 10^2)^2 \exp\left(\frac{0.85}{0.0259}\right)$$

$$= 6.61 \times 10^{-8} \text{ A/cm}^2$$

$$W = \left(\frac{2\epsilon V_0}{qn_d}\right)^{1/2} = \left[\frac{(2)(1.045 \times 10^{-12})(0.588)}{(1.6 \times 10^{-19})10^{15}}\right]^{1/2} = 0.877\ \mu\text{m}$$

$$\xi_{\max} = \frac{qn_d x_n}{\epsilon} = \frac{(1.6 \times 10^{19})(10^{15})0.877 \times 10^{-4}}{1.045 \times 10^{-12}} = 1.34 \times 10^4 \text{ V/cm}$$

$$C_j = \frac{\epsilon A}{W} = \frac{(1.045 \times 10^{-12})(10^{-6})}{8.77 \times 10^{-5}} = 1.19 \times 10^{-14} f = 0.0119 \text{ pF}$$

A typical Au-Si Schottky barrier diode is illustrated in Fig. 5-8. The metal layer is thin, about 100 Å, so that very little light is absorbed in it. Only 5 percent

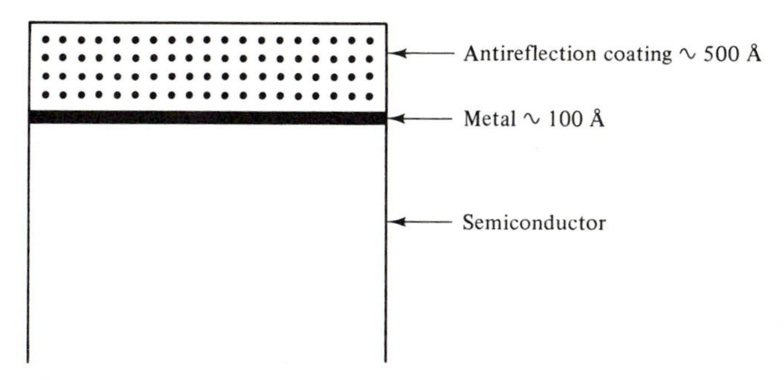

Figure 5-8 A Schottky barrier photodiode with an antireflection coating.

is absorbed when the wavelength is 6328 Å. The ZnSe film is an antireflection coating for visible light; it is 500 Å thick. Losses due to reflection are greatly reduced when the thickness of the antireflection coating is such that waves reflected from the air-coating interface destructively interfere with waves reflected from the coating-semiconductor interface. This will be discussed in more detail in Chapter 8.

This photodiode is particularly efficient in the visible and the ultraviolet spectrum because the light is absorbed in the semiconductor very close to the junction, since the photon energy is larger than that of the direct energy gap. Since the photons are so readily absorbed, most of them are absorbed in the depletion region. Thus, there are essentially no time delays as a result of diffusion. Couple this with the fact that Schottky barrier diodes are majority carrier devices, and you have a very fast diode. It can routinely be made with rise times of 0.1 nsec, and if small areas are used to reduce the junction capacitance, rise times an order of magnitude less can be achieved.

The actual behavior of Schottky barrier diodes deviates more from ideality than it does for homojunctions because the junction contains a large number of defects. If the metal is not deposited on a clean surface under a very high vacuum, the metal will be separated from the semiconductor by a 10 to 20 Å thick oxide. In addition, the metal is polycrystalline, and its bonds do not match up with those of the semiconductor. A number of surface states are thus formed at the interface.

In concluding this section we must note that Schottky barrier diodes can have *pin* type structures, and they can be operated in the avalanche mode.

The metal-*n*-type semiconductor junction in Fig. 5-9*a* will not behave as a diode. Because E_{Fn} lies below E_{Fm}, electrons will flow from the metal to the semiconductor to equilibrate the Fermi energies. These electrons will form a charged plane rather than the distributed charge formed by ionized donors. This

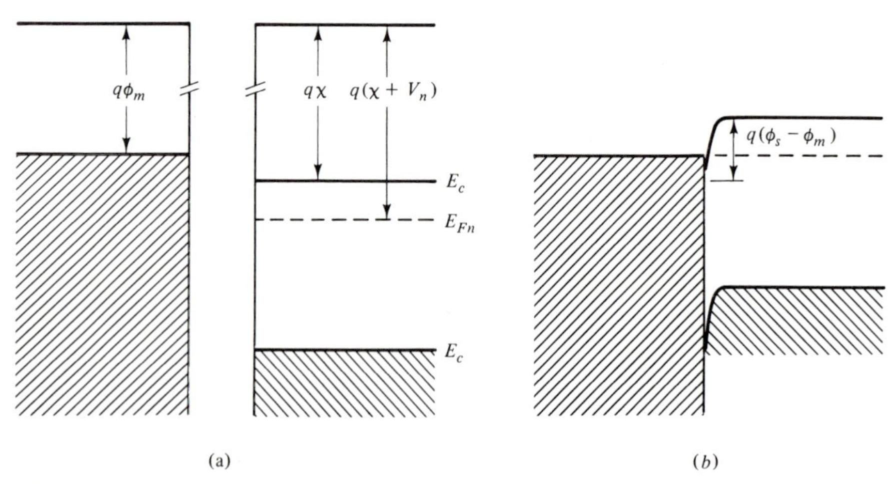

Figure 5-9 A metal-*n*-type semiconductor junction (*a*) before and (*b*) after contact that forms an ohmic contact.

parallel plate capacitor type structure is so thin that electrons can easily tunnel through it. Thus, the current flowing across the junction is ohmic.

Ohmic contacts can also be formed when E_{Fn} lies above E_{Fm} if the semiconductor is doped heavily enough. The depletion region in the heavily doped semiconductor is also thin enough for electrons to tunnel through it.

5-2.3 Phototransistor

The *pnp* phototransistor in Fig. 5-10 is similar to a regular transistor in the common emitter mode in that it is composed of a reverse biased collector-base junction and a base-emitter junction that can easily be forward biased. The phototransistor differs in that there is no electrical base current; the base is said to be floating. The base current is generated optically.

In the dark there is no photocurrent so that there is no bias across the base-emitter junction. The current across the base-collector junction is small because it is reverse biased. However, the reverse current is larger than it would be in an isolated diode because some of the holes diffusing from the emitter to the base reach the base-collector junction before they recombine. Those that do not recombine are swept out by the electric field in the base-collector junction.

The base-collector junction behaves much like a photovoltaic detector when light shines on it. Photoholes created in the *n*-type base are swept into the *p*-type collector while the photoelectrons created in the collector are swept into the base. This negative charge accumulation in the base forward biases the emitter-base junction which, in turn, increases the number of holes diffusing from the emitter into the base. However, as shown in Fig. 5-11, only a small fraction of these holes recombine in the base. The fraction of holes, B, which is often called the base transport factor, diffuse to the collector-base junction where they are swept out by the electric field in this reverse biased junction.

The net current flowing into the collector, i_c, is

$$i_c = i_{co} + i_{op} + Bi_{ep} \tag{5-7a}$$

where i_{co} is the dark collector current. The hole emitter current, i_{ep}, can also be written

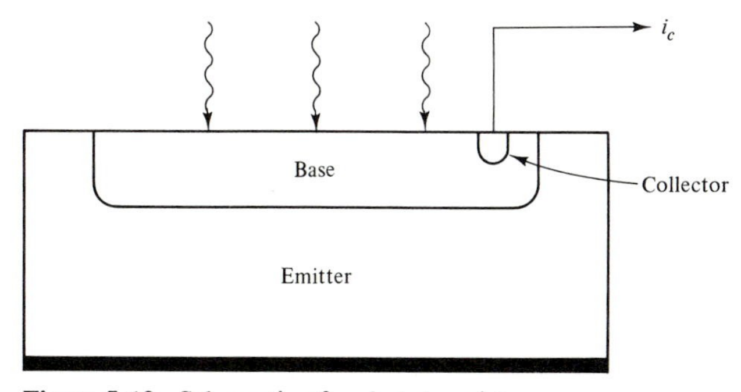

Figure 5-10 Schematic of a phototransistor.

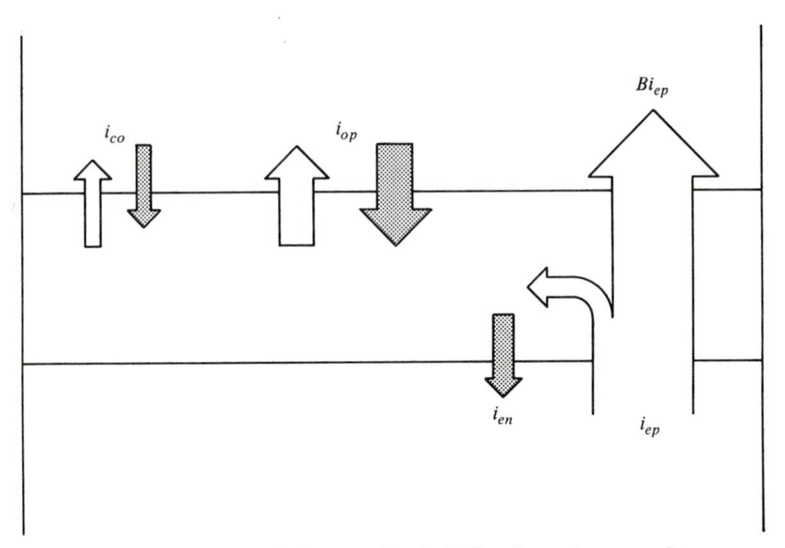

Figure 5-11 The hole and electron (shaded) flow in a phototransistor.

$$i_{ep} = \gamma i_e \tag{5-7b}$$

where γ is the ratio of the emitter hole current and the total emitter current, and it is called the emitter injection efficiency. Thus Eq. 5-7a can be written

$$i_c = i_{co} + i_{op} + \alpha i_e \tag{5-7c}$$

where $\alpha = B\gamma$ is the current transfer ratio. The net current into the base is equal to the net current out under steady state conditions. Thus

$$i_{en} + (1 - B)i_{ep} = (1 - \alpha)i_e = i_{co} + i_{op} \tag{5-8}$$

i_{en} is the electron current diffusing from the base into the emitter. Substituting Eq. 5-8 into Eq. 5-7c yields

$$i_c = (i_{co} + i_{op})\left(1 + \frac{\alpha}{1 - \alpha}\right) = (i_{co} + i_{op})(1 + h_{FE}) \tag{5-9}$$

where h_{FE} is the common emitter gain coefficient. It can have values as large as 500.

The magnitude of B can be computed with the assistance of Fig. 5-12, which illustrates what happens to the Δp versus x curve for low-level injection when the boundary condition, $\Delta p = 0$ at $x = W_b$, is employed. W_b is the base width, which is the distance between the end of the emitter-base depletion region at $x = 0$, and the beginning of the base-collector depletion region. The steady state solution to the differential equation

$$\frac{\partial\, \Delta p}{\partial t} = D_p \frac{\partial^2\, \Delta p}{\partial x^2} - \frac{\Delta p}{\tau_p} = 0 \tag{4-35}$$

must now be found for the new set of boundary conditions. The solution is

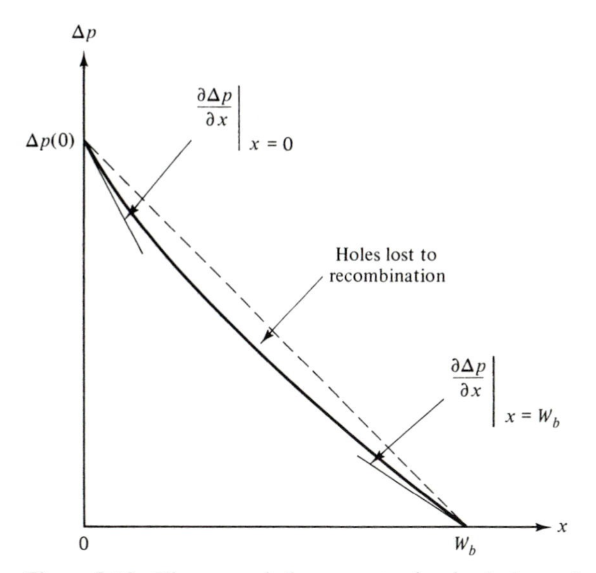

Figure 5-12 The excess hole concentration in the base, the slope of the Δp versus x curve at $x = 0$ and W_b, and the deviation from linearity of the Δp versus x curve due to hole recombination in the base.

$$\Delta p = \Delta p_0 \frac{\sin h\left(\dfrac{W_b - x}{L_p}\right)}{\sin h\left(\dfrac{W_b}{L_p}\right)} \tag{5-10}$$

For $W_b < L_p$ the slope of the Δp versus x curve is practically linear. It is from the small deviation from linearity that B is found.

The hole current diffusing across the emitter-base junction is

$$i_{ep} = -qA\, D_p \frac{\partial\, \Delta p}{\partial x}\bigg|_{x=0} = \frac{qA\, D_p\, \Delta p_0}{L_p} \operatorname{ctn} h \frac{W_b}{L_p} \tag{5-11}$$

and it is nearly all of the total current for a p^+n junction. The hole current diffusing to the base-collector junction is

$$Bi_{ep} = -qA\, D_p \frac{\partial\, \Delta p}{\partial x}\bigg|_{x=W_b} = \frac{qA\, D_p\, \Delta p_0}{L_p} \csc h \frac{W_b}{L_p} \tag{5-12}$$

The fraction of the holes that recombine in the base then is

$$(1 - B)i_{ep} = \frac{qA\, D_p\, \Delta p_0}{L_p}\left(\operatorname{ctn} h \frac{W_b}{L_p} - \csc h \frac{W_b}{L_p}\right)$$

$$= \frac{qA\, D_p\, \Delta p_0}{L_p} \tan h \frac{W_b}{2L_p} \tag{5-13}$$

$$\csc h(x) = \frac{1}{x} - \frac{x}{6} + \frac{7x^3}{360} - \cdots$$

and

$$\tan h(x) = \frac{x - x^3}{3} + \cdots$$

Thus for $W_b \ll L_p$

$$\frac{B}{1 - B} = 2\,\frac{L_p^2}{W_b^2} \tag{5-14a}$$

or

$$B = \frac{1}{1 + (W_b^2/2L_p^2)} \tag{5-14b}$$

This equation confirms that the shorter the base width or, conversely, the larger the diffusion length, the greater is the fraction of emitter holes that diffuse across the n-type base into the collector.

EXAMPLE 5-3

Monochromatic light with wavelength $\lambda = 1.0\ \mu$m, and intensity, $I = 100$ mW/cm^2, strikes a silicon phototransistor. Compute the collector current and the current transfer ratio if the base area is 5 mm^2, 90 percent of the photons create electron-hole pairs, the emitter is doped to 10^{18} cm^{-3}, the base is doped to 10^{15} cm^{-3}, the hole recombination time in the base is 10 μsec, the base width is 1.25 μm, $\gamma = B$, and i_{co} is small enough to be ignored.

$$i_{op} = \eta q \frac{I\lambda}{hc} A = \frac{(0.9)(1.6 \times 10^{-19})(10^3)(10^{-6})(5 \times 10^{-6})}{(6.625 \times 10^{-34})(3 \times 10^8)} = 3.62 \text{ mA}$$

$$L_p = (D_p\tau)^{1/2} = \left(\frac{kT}{q}\,\mu_p\tau_p\right)^{1/2}$$

From Fig. 4-17, $\mu_p = 480$ cm^2/V·sec

$$L_p = [(0.0259)(0.480)10^{-5}] = 1.25 \times 10^{-5} \text{ M} = 12.5\ \mu\text{m}$$

$$B = \frac{1}{1 + 0.5(1.25/12.5)^2} = 0.995$$

$$\alpha = \text{B}\gamma = (0.995)^2 = 0.98025$$

$$h_{FE} = \frac{\alpha}{1 - \alpha} = \frac{0.989025}{0.010975} = 90.1$$

$$i_c = i_{op}(h_{FE} + 1) = 3.62(91.1) = 330 \text{ mA}$$

The major disadvantage of the phototransistor is that it is slower than the photodiode. This is because the speed of the transistor is limited by the time it takes the photocarriers to diffuse to the base-collector junction. Typical minimum response times are 1 μsec.

Further amplification can be achieved by connecting two transistors in series as shown in Fig. 5-13. The emitter current from the phototransistor is the input base current to the second transistor. This arrangement is called a photodarlington. Amplifications as large as 25,000 can be obtained, but unfortunately photodarlingtons are an order of magnitude slower than phototransistors.

5-3 PHOTOTHERMAL DETECTORS

5-3.1 Thermistor

A thermistor is used to measure temperature, particularly low temperatures. This is done by measuring the current through a semiconductor, and then looking up the temperature that corresponds to it in a calibration chart. The reason that a semiconductor is used is that its conductivity can be very temperature sensitive. Therefore, small changes in temperature can correspond to large changes in conductivity.

The parameter that is used to describe the temperature sensitivity is the thermal coefficient of resistivity, α, which is

$$\alpha = \frac{1}{\rho}\frac{\mathrm{d}\rho}{\mathrm{d}T} \tag{5-15a}$$

where ρ is the resistivity. If the conductivity is used instead of the resistivity, then

$$\alpha = -\frac{1}{\sigma}\frac{\mathrm{d}\sigma}{\mathrm{d}T} \tag{5-15b}$$

In Section 4-3.3 it was shown that the conductivity of an intrinsic semiconductor, to a good approximation, is given by

$$\sigma = \sigma_0 c e^{-E_g/2kT} \tag{4-18}$$

The thermal coefficient of resistance for the intrinsic semiconductor, α_{in}, is then

$$\alpha_{\text{in}} = \frac{-E_g}{2kT^2} \tag{5-16}$$

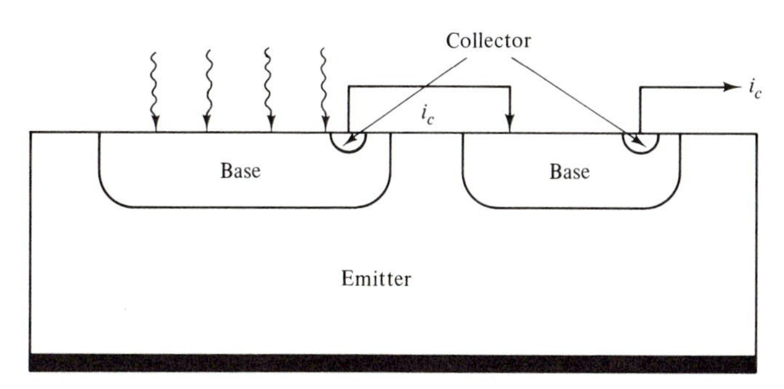

Figure 5-13 Schematic of a photodarlington.

From this equation it can be seen that the thermistor is more sensitive at lower temperatures and wider bandgaps. However, the conductivity is also smaller for these conditions and the intrinsic semiconductor can be too resistive to be useful.

The problem of the large resistance at low temperatures can be circumvented by using extrinsic semiconductors. The electron concentration of an n-type semiconductor at very low temperatures is given by (see Problem 5-9)

$$n = n_c e^{-E^*/kT} \tag{5-17a}$$

where

$$E^* \cong \tfrac{1}{2}E_d + \tfrac{1}{2}kT \ln \frac{n_c}{n_d} \tag{5-17b}$$

Assuming that the temperature dependence of the mobility is cancelled by the temperature dependence of n_c, and ignoring the temperature dependence of E^*, the coefficient of resistance for the extrinsic semiconductor, α_{ex}, is

$$\alpha_{ex} \simeq \frac{-E^*}{kT^2} \tag{5-18}$$

EXAMPLE 5-4

In a circuit in which the resistance of the load resistor is much greater than the resistance of the thermistor, find the temperature change if the change in the voltage across the load resistor is 1.0 mV. The sensor circuit consists of a voltage source with the thermistor and load resistor in series. The initial temperature is 300 K, the energy gap of the intrinsic thermistor is 1.0 eV, the applied voltage is 2.0 V, and the ratio of the thermistor resistance to the load resistance is $R_T/R_L = 10^{-3}$.

$$i = \frac{V}{R_T + R_L}$$

$$\Delta i = \frac{-V\,\Delta R_T}{(R_T + R_L)^2} \simeq \frac{-V\,\Delta R_T}{R_L^2}$$

$$\Delta V_L = R_L\,\Delta i = \frac{-V\,\Delta R_T}{R_L}$$

$$\Delta R_T = \alpha R_T\,\Delta T$$

$$\Delta T = \frac{R_L\,\Delta V_L}{R_T V \alpha} = \frac{2R_L\,\Delta V_L k T^2}{R_T V E_g} = \frac{(2)(10^3)(10^{-3})(0.0259)(3 \times 10^2)}{(2)(1)}$$

$$= 7.77°\ \mathrm{C}$$

5-3.2 Bolometer

A bolometer detects radiation when it is heated by the absorption of radiation. The absorbed heat decreases the resistance of the semiconductor element by raising its temperature, and the resulting increase in the voltage across the load resistor is measured in exactly the same way it is for the photodetector (see Fig. 4-20). The absorber is often carbon black, but any material that is a black body absorber can be used. An obvious advantage that the bolometer has over the photodetector is that a single bolometer element can be used for all wavelengths of light.

For the small signal case the voltage increase across the load resistor is

$$\Delta V = -i\,\Delta R_s \tag{5-19}$$

For the bolometer

$$\Delta R_s = \alpha R_s\,\Delta T \tag{5-20}$$

for small values of ΔT.

When the bolometer is exposed to a constant intensity light source, its temperature will rise exponentially toward a steady state value. The steady state is reached when the energy input equals the energy output. The input energy is the heat absorbed per unit time, ΔH, and the steady state change in the joule heating, ΔP_0, that is produced by the change in the operating temperature. The output at steady state is the thermal energy conducted away per unit time, $\mathcal{G}\,\Delta T_0$, where $\mathcal{G}$ is the thermal conductance of a rod that connects the semiconductor to a temperature bath maintained at T_0, and ΔT_0 is the steady state temperature rise (see Fig. 5-14). Clearly, the semiconductor element is at a temperature $T_1 > T_0$ even when no light is shining on the element because there is joule heating. The ΔT_0 in the following expressions is $T_2 - T_1$ where T_2 is the

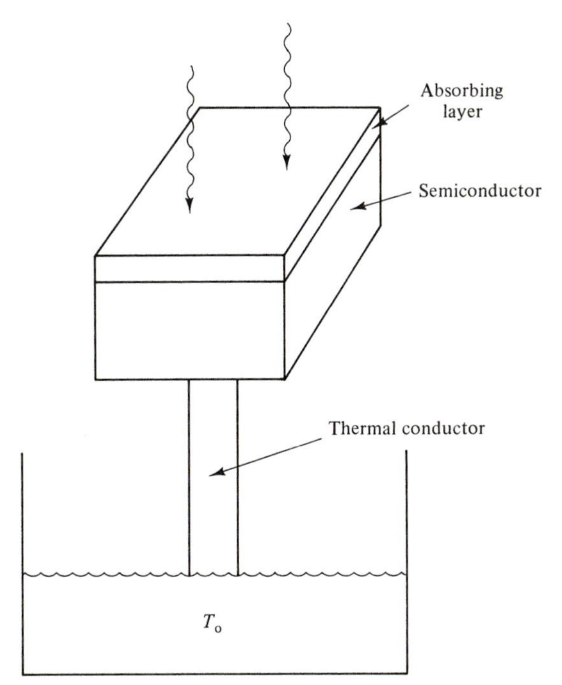

Figure 5-14 The bolometer element and the rod that thermally connects the element to a temperature bath at T_0.

steady state temperature to which the semiconductor is raised when it is exposed to radiation. Equating the input and output power yields

$$\Delta H + \Delta P_0 = \mathcal{G}\,\Delta T_0 \tag{5-21}$$

The change in the joule heating at steady state is

$$\Delta P_0 = i^2\,\Delta R_s = \alpha P\,\Delta T_0 \tag{5-22}$$

Therefore

$$\Delta T_0 = \frac{\Delta H}{\mathcal{G} - \alpha P} \tag{5-23}$$

When the temperature is rising toward its steady state value, there is an additional output term, thermal absorption, and it is equal to $C_p(d\,\Delta T/dt)$ where C_p is the heat capacity, the energy needed to raise the bolometer element one degree. Thus, Eq. 5-21 becomes

$$\Delta H + \Delta P = \mathcal{G}\,\Delta T + C_p\frac{d\,\Delta T}{dt} \tag{5-24}$$

Using Eq. 5-21 to find ΔH and Eq. 5-22 to find an expression for ΔP, Eq. 5-24 becomes

$$(\mathcal{G} - \alpha P)(\Delta T_0 - \Delta T) = C_p\frac{d\,\Delta T}{dt} \tag{5-24}$$

Rearranging and integrating yields

$$\Delta T = \Delta T_0\left[1 - \exp\left(\frac{\mathcal{G} - \alpha P}{C_p}\right)t\right] \tag{5-25}$$

$C_p/(\mathcal{G} - \alpha P)$ is the time constant for the bolometer. For normal operating conditions its value is about 7 msec.

The time constant is longer when the heat capacity is larger because more heat has to be absorbed for a given temperature rise. It is also longer when α is positive. This is because there is more joule heating. In the event that $\alpha P > \mathcal{G}$, heat is created faster than it can be removed. Hence, there is no upper temperature limit. The time constant is shorter when $\mathcal{G}$ is larger because heat is conducted away faster.

EXAMPLE 5-5

A low temperature germanium bolometer operating at 2.15 K with a resistance, $R_s = 12$ kΩ, a thermal coefficient of resistance, $\alpha = -2\ \mathrm{K}^{-1}$, and a heat capacity, $C_p = 6.8 \times 10^{-6}T^3$ mJ/K is connected to a thermal conductor for which the thermal conductance, $\mathcal{G} = 183\ \mu$W/K. If the load resistance is $R_L = 500$ kΩ, and the operating voltage is 30 V, compute the bolometer time constant.

$$P = i^2R_s = \left(\frac{V}{R_L + R_s}\right)^2 R_s = \left[\frac{30}{(5 + 0.12)10^5}\right]^2 1.2 \times 10^4 = 41.2\ \mu\text{W}$$

$$\tau = \frac{C_p}{\mathcal{G} - \alpha P} = \frac{(6.8 \times 10^{-9})(2.15)^3}{[183 + (2)41.2]10^{-6}} = 0.254\ \text{msec}$$

5-4 FIGURES OF MERIT AND NOISE

5-4.1 Figures of Merit

The gain-bandwidth product, GB, is given by

$$\text{GB} = \frac{\text{gain}}{\tau} \tag{5-26}$$

where τ is the response time. For a photodetector this product is given by

$$\text{GB} = \frac{\mu V}{L^2} \tag{5-27}$$

(see Eq. 4-30). Thus, the higher the mobility, the better the performance.

For photon detectors the responsivity, $\mathcal{R}$, is usually defined as the ratio of the photocurrent to the input power (see Fig. 4-3). The responsivity for a bolometer is the ratio of the |voltage output| to the power input. Assuming that the time constant of the circuit is much shorter than the bolometer time constant, the voltage output approaches the change in the voltage drop across the load resistor. From

$$\Delta \text{V} = \left(\frac{R_L}{R_L + R_s + \Delta R_s} - \frac{R_L}{R_L + R_s}\right) V_0 \simeq \frac{R_L V_0}{(R_L + R_s)^2}(-\Delta R) \tag{5-28a}$$

and Eqs. 5-19, 5-20, 5-21, and 5-22

$$\mathcal{R} = \frac{V_0 |\alpha| R_s \,\Delta T R_L}{(\mathcal{G} - \alpha P)\Delta T (R_L + R)^2} = \frac{V_0 |\alpha| R_s R_L}{(\mathcal{G} - \alpha P)(R_L + R_s)^2} \tag{5-28b}$$

Note that for a bolometer $\mathcal{R}$ is independent of the wavelength whereas it increases linearly with λ for a photon detector up to the absorption edge where it drops precipitously.

EXAMPLE 5-6

Compute the responsivity of the bolometer element described in Example 5-5.

$$\mathcal{R} = \frac{V_0 |\alpha| R_L R_s}{(\mathcal{G} - \alpha P)(R_L + R_s)^2} = \frac{(30)(2)(5 \times 10^5)(1.2 \times 10^4)}{[183 + 2(41.2)](10^{-6})(5 + 0.12)^2 10^{10}}$$

$$= 5.17 \times 10^3 \text{ V/W}$$

The detectivity, D, is the reciprocal of the smallest signal that can be detected—the signal for which the signal to noise ratio is equal to one. As we will learn in the next section, the noise is proportional to the bandwidth, Δf, and the area. Both the noise current and noise voltage are proportional to the square root of the noise power. This is used to define another version of the detectivity, D^* (D star) where

$$D^* = \frac{(A\,\Delta f)^{1/2}}{\text{NEP}} \tag{5-29}$$

where NEP is the noise equivalent power.

5-4.2 Thermal Noise

The source of noise is random fluctuations. In the preceding chapter we emphasized that τ is the average recombination time and L is the average diffusion length. Some carriers recombine in a time, $t < \tau$ while others recombine when $t > \tau$, but on the average, $t = \tau$. Another example of a fluctuation is the pressure variation in the room. As an extreme example let us determine the probability that all of the gas molecules are located in one-half of the room. A Maxwell demon can create this situation by placing a molecule on the right side of the room if the flip of a coin produces a head, and on the left side of the room if a tail appears. This is done by fixing the time and flipping the coin N times where N is the number of molecules. The probability a head comes up every time, and therefore all of the molecules are on the right side of the room, is $(\frac{1}{2})^N$. The probability a tail appears each time, and therefore all of the molecules are on the left side of the room, is the same.

On the average there are as many molecules on the right side of the room as there are on the left, since the probability a head will appear is the same as the probability a tail will appear. Thus, having all of the molecules in one-half of the room is the most extreme fluctuation, and it is also the least probable. Fluctuations from the average of varying magnitudes occur at different times. However, the average fluctuation is zero because for every fluctuation that has a deficiency of heads, there is an equally probable fluctuation with an excess of heads equal to the deficiency. Stated mathematically, the bell-shaped curve representing the probability distribution of excess heads is a symmetric function. The average fluctuation in the recombination time, $\langle t - \tau \rangle$, is also zero.

To eliminate the problem of the negative and positive fluctuations "smearing" each other out, one finds the more meaningful value of the square or rms value of the fluctuation. Thus, the negative fluctuation and its positive counterpart will have the same numerical value. Also, $\langle (t - \tau)^2 \rangle$ will not equal zero.

EXAMPLE 5-7

Show that the average fluctuation in the decay time is zero when the excess electrons decay according to the equation $\Delta n = \Delta n_0 \exp(-t/\tau)$. The average fluctuation is the sum of the products of the magnitude of a given fluctuation times the probability this fluctuation occurs. It was shown in Section 4-4.3, the probability recombination occurs between t and $t + dt$ is

$$\frac{\Delta n_0 e^{-t/\tau}\, dt}{\int_0^\infty \Delta n_0 e^{-t/\tau}\, dt} = e^{-t/\tau}\frac{dt}{\tau} \quad \langle t - \tau \rangle = \int_0^\infty (t - \tau) e^{-t/\tau}\frac{dt}{\tau}$$

$$= \int_0^\infty t e^{-t/\tau}\frac{dt}{\tau} - \int_0^\infty e^{-t/\tau}\, dt = \tau - \tau = 0$$

Thermal noise is present even when no current is flowing. Charged particles, which are always in motion, can be distributed in many different ways. The initial particle distribution shown in Fig. 5-15*a* can change into many different configurations after the particles have made one jump, of which three are

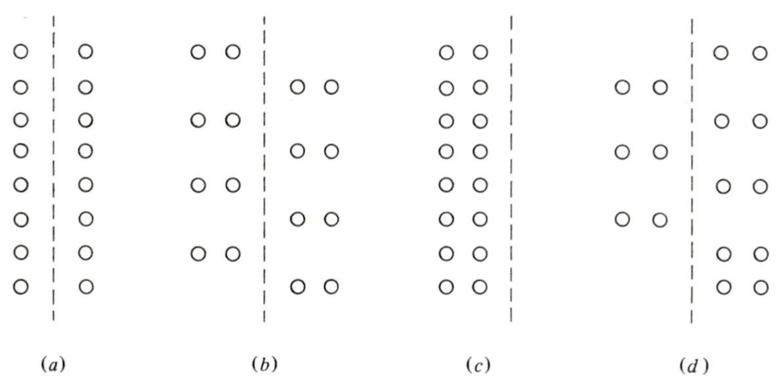

Figure 5-15 The initial particle configuration (*a*) and some particle configurations after the particles have jumped once. The final distributions are designated (*b*) the most probable distribution, (*c*) the least probable distribution, and (*d*) the average fluctuation.

shown in Figs. 5-15*b, c,* and *d.* Although these distributions have different probabilities of occurring, they are all possible. These fluctuations are the source of thermal noise.

The configurations in Fig. 5-15 can be obtained by flipping a coin. The first flip decides the fate of the uppermost pair of particles. If the coin comes up heads, the particle on the left will jump to the right across the imaginary boundary; if the coin comes up tails, the particle on the right will jump to the left. After N flips of the coin have determined the positions of the N pairs of atoms, the atoms will have a new configuration. The distribution shown in Fig. 5-15*b* is one of the many configurations for which there are the same number of particles on either side of the boundary. Because there are more configurations for which the number of particles is the same on each side of the boundary than there are configurations for any other specific distribution of particles, the most probable distribution is one for which the number of particles on either side of the boundary is the same. Another way of saying the same thing is that the most probable number of heads is equal to one-half the number of flips.

A configuration representing the least probable distribution is shown in Fig. 5-15*c.* This configuration occurs if all of the flips come up tails. The only other configuration for which all of the particles would be on one side of the boundary is the one that would be created by all of the flips coming up heads.

The configuration in Fig. 5-15*d* represents the average fluctuation. The average fluctuation is the distribution for which the excess number of particles on the right side is equal to the average number of excess particles on the right side when only those configurations for which there is an excess number of particles on this side are considered. If the configurations for which there was an excess number of particles on the left side were included, the average fluctuation would be zero. This is because for every 3 left-5 right configuration, there is a 5 left-3 right configuration. The average fluctuation is equivalent to the average distance a particle diffusing from a plane source at the origin diffuses, when only those particles that move to the right are considered.

For noise measurements we are more interested in the mean squared fluctuation. Since the fluctuations are mathematically equivalent to diffusion from a plane source, the noise can be quantified by using the mathematics developed for the mean squared diffusion distance. The mean squared distance the particles move is

$$\langle x^2 \rangle = 2Dt \tag{5-30}$$

(See Problem 5-16.) If the particles are charged, this corresponds to a charge fluctuation in a resistor of

$$\langle Q^2 \rangle = (nAq)^2 \frac{\langle x^2 \rangle}{nAL} = \frac{2nAq^2}{L} Dt \tag{5-31}$$

$\langle x^2 \rangle$ is divided by nAL to obtain the average distance moved per particle. This charge fluctuation leads to a capacitance and the equivalent circuit is shown in Fig. 5-16. The energy stored by the charge, Q, on the capacitor plates is

$$E = \frac{Q^2}{2C} \tag{1-18}$$

Recall that the relationship between the mobility and the diffusion coefficient is given by

$$D = \frac{kT\mu}{q} \tag{4-47}$$

Substituting Eq. 4-47 into Eq. 5-31 and then combining the results with Eq. 1-18 yields

$$\Delta E = \frac{GkTt}{C} = \frac{kT}{\tau} t \tag{5-32a}$$

where G is the conductance of the resistor, and τ is the RC time constant. The noise equivalent power, NEP, is equal to the average fluctuation in the electrostatic energy per unit time, and the incoming signal must be greater than or equal to the NEP to be detected above the background noise. The NEP in this instant is

$$\text{NEP} = \frac{kT}{\tau} \tag{5-32b}$$

The actual equation derived by using a more rigorous treatment is

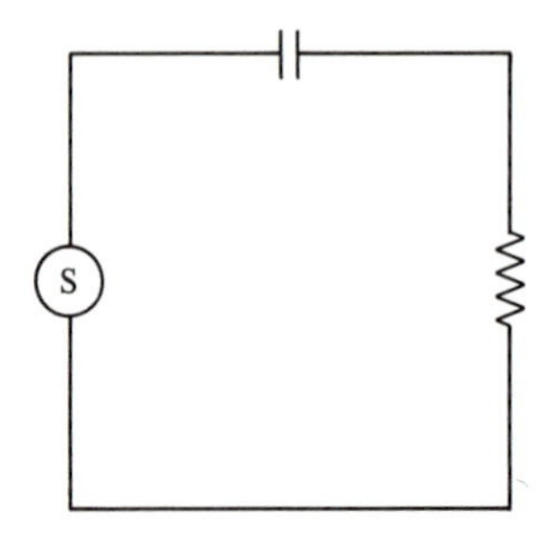

Figure 5-16 The *RC* circuit used to determine the thermal noise signal generated by the fluctuating charge on the capacitor plates.

$$\text{NEP} = \frac{4kT}{\tau} \tag{5-32c}$$

This equation illustrates the general rules that the lower the temperature and the more narrow the bandwidth, the smaller the noise signal.

5-4.3 Other Important Detector Noise Sources

In photomultipliers the dominant noise is usually shot and multiplication noise. Shot noise is due to fluctuations in the current flow, and multiplication noise is due to fluctuations in the secondary electron emission coefficient. A rigorous mathematical derivation of the equations for these noise sources is beyond the scope of this book; we will instead attempt to rationalize them. The interested student should refer to ref. 9 for more detailed explanations.

The square of the fluctuation in the current flow, Δi_{sh}^2, is proportional to the quantized unit of charge, which is the electronic charge, q. It is also proportional to the magnitude of the current flow, i, at a given frequency. The noise is distributed across the range of frequencies so that it is proportional to the bandwidth, $1/\tau$. We complete our "derivation" by pulling the factor of two out of the air so that

$$\Delta i_{\text{sh}}^2 = \frac{2qi}{\tau} \tag{5-33}$$

The fluctuations in the secondary electron photoemission cause the shot noise in a photomultiplier to increase by a factor of $\delta/(\delta - 1)$, where again δ is the secondary electron multiplication factor. Thus, for a photomultiplier

$$\Delta i_{\text{sh}}^2 \text{ (photomultiplier)} = \frac{2qi}{\tau}\frac{\delta}{\delta - 1} \tag{5-34}$$

The minimum detectable signal is the signal with an energy equal to the noise energy. For a photomultiplier the noise energy is proportional to the current; hence, we must find the minimum current. This is the dark current, which is given by

$$i_{\text{D}} = ABT^2 e^{-q\phi_m/kT} \tag{5-35}$$

which is a modification of Eq. 5-5a. The minimum detectable power is then given by

$$P_{\text{min}} = \frac{\Delta i_{\text{sh}}}{\mathcal{R}} = \left(\frac{2qi_D}{\tau}\frac{\delta}{\delta - 1}\right)^{1/2} \frac{1}{\mathcal{R}} \tag{5-36}$$

EXAMPLE 5-8

Compute the minimum detectable power for a photomultiplier tube at 300 K, which has a cathode with an area of 10 cm^2 and a work function of 1.25 eV, and the secondary electron photoemission factor is 5. The incident radiation has a wavelength of 0.5 μm, the internal quantum efficiency is 0.4, and none of the light is reflected.

$$i_D = ABT^2e^{-q\phi m/kT} = (10)(1.20 \times 10^2)(3.00 \times 10^2)\exp\left(\frac{-1.25}{0.025875}\right)$$

$$= 1.14 \times 10^{-15}\text{ A}$$

$$\mathcal{R} = \frac{n_i q\lambda}{hc} = \frac{(0.4)(0.5)}{1.242} = 0.161\text{ A/W}$$

$$P_{\min} = \left[\frac{2qi_D}{\tau}\frac{\delta}{\delta - 1}\right]^{1/2}\frac{1}{\mathcal{R}} = \left[\frac{(2)(1.6 \times 10^{-19})(1.14 \times 10^{-15})}{\tau}\frac{5}{4}\right]^{1/2}\frac{1}{0.161}$$

$$= 1.33 \times \frac{10^{-16}}{\sqrt{\tau}}\frac{\text{W}}{\sqrt{\text{Hz}}}$$

In addition to thermal and shot noise, photodetectors also have generation–recombination and flicker noise sources. Generation–recombination noise is due to fluctuations in the generation and recombination rates. Flicker noise is thought to originate at surface and contact defect states.

The expression for the square of the fluctuation in the generation-recombination current, Δi_{gr}^2, differs from that for the shot noise current in that a factor of 4 replaces the factor of 2 and the gain must be included. Thus

$$\Delta i_{gr}^2 = \frac{4qiG}{\tau} \tag{5-37}$$

As we discuss in Section 4-5, the gain has the frequency dependence

$$G = \frac{G(0)}{1 + \omega^2\tau^2} \tag{5-38}$$

(see Eq. 4-61), so that Δi_{gr} decreases as the frequency increases, but the decrease is minimal until $\omega \sim 1/\tau$.

The minimum detectable signal is again determined by the dark current, i_D, and it, in turn, depends exponentially on the energy gap/temperature ratio. As a rule of thumb, the operating temperature should be no higher than $E_g/25$ K. This is why the narrow bandgap detectors used to detect infrared radiation such as those made from InSb are cooled to liquid nitrogen temperatures (see Problem 4-37).

The square of the flicker, or one over f, noise current, Δi_f^2, is given by the equation

$$\Delta i_f^2 = \frac{B}{\tau f}$$

where $B \sim 10^{-11}\text{ A}^2$. As is illustrated in Fig. 5-17, this noise source is important only at low frequencies. In the mid-frequency range the generation–recombination noise dominates, and as it dies off for $\omega\tau > 1$, the thermal noise dominates.

For a *pin* diode circuit the thermal noise in the load resistor, usually a load resistor for an amplifier, dominates. The signal to noise ratio, S/N, is thus given by

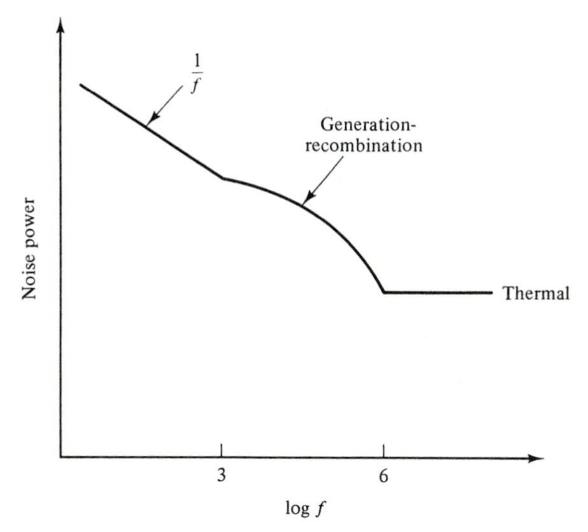

Figure 5-17 The noise power of an ideal infrared photoconductor plotted as a function of the frequency.

$$\frac{S}{N} = \frac{R_L i_{op}^2}{4kT/\tau} \tag{5-39}$$

Using Eq. 4-75 for i_{op} when $F = 1$ and the fact the minimum detectable power occurs for $S/N = 1$,

$$P_{\min} = \frac{hc}{q\lambda\eta_i(1-R)} i_{op(\min)} = \frac{2hc}{q\lambda\eta_i(1-R)} \left(\frac{kT}{R_L\tau}\right)^{1/2} \tag{5-40}$$

From Eq. 5-40 we see that $P_{\min}$ can be decreased by increasing R_L, but this, of course, has the negative effect of increasing the response time.

EXAMPLE 5-9

Compute $P_{\min}$ for the *pin* diode system at 300 K that is described in Example 4-12.

$$P_{\min} = \frac{hc}{q\lambda} \frac{2}{\eta_i(1-R)} \left(\frac{kT}{R_L\tau}\right)^{1/2} = \frac{(1.50)(2)}{0.4} \left[\frac{(1.38 \times 10^{-23})(3 \times 10^2)}{1.5 \times 10^2 \tau}\right]^{1/2}$$

$$= 3.94 \times \frac{10^{-11}}{\sqrt{\tau}} \frac{\text{W}}{\sqrt{\text{Hz}}}$$

For the circuit time constant of 1.38 nsec

$$P_{\min} = 3.94 \times \frac{10^{-11}}{(1.38 \times 10^{-9})^{1/2}} = 1.06 \times 10^{-6} \text{ W}$$

The expression for the fluctuations in the shot noise current for an avalanche photodiode differs because it has an internal multiplication mechanism. It is given by

$$\Delta i_{\text{sh}}^2(\text{APD}) = M^2 \frac{2qiF(M)}{\tau} \tag{5-41}$$

where
$$F(M) = M\left[1 - \left(1 - \frac{1}{r}\right)\left(\frac{M-1}{M}\right)^2\right] \tag{5-42}$$

and r is the larger of the ratio of the electron/hole or hole/electron ionization coefficients. It has a maximum value of M when $r = 1$, and is $\sim M/r$ for large values of M. This is why semiconductors with larger r values have smaller NEPs.

Both the shot noise and the thermal noise are important in the APD. Thus the signal to noise ratio is

$$\frac{S}{N} = \frac{R_L i_{op}^2 \tau}{2R_L q(i_{op} + i_s)M^2 F(M) + 4kT} \tag{5-43}$$

This function has a maximum value for the optimum value of M, M_{opt}, and for $F(M) \approx M/r$ it is

$$M_{\text{opt}} = \left[\frac{4kTr}{q(i_{op} + i_s)R_L}\right]^{1/3} \tag{5-44}$$

EXAMPLE 5-10

Find M_{opt} for an APD with $r = 50$ and parameters similar to those of the photodiode in Example 4-11 and exposed to the same light intensity.

$$M_{\text{opt}} \approx \left[\frac{4kTr}{qi_{op}R_L}\right]^{1/3} = \left[\frac{(4)(2.59 \times 10^{-2})(5 \times 10^1)}{(2.67 \times 10^{-4})(1.5 \times 10^2)}\right]^{1/3} = 5.05$$

Note that for the smaller signals usually encountered in optical fiber systems, M_{opt} is larger. For example, if $i_{op} = 0.267\ \mu\text{A}$, $M_{\text{opt}} = 50.5$.

The noise that is not due to inherent statistical fluctuations in the system is called background noise. Two such noise sources are the unshielded electromagnetic radiation in the air, and the thermal radiation that is produced by all matter that is not at absolute zero. The background noise can be reduced by shielding and by lowering the temperature of the surroundings.

READING LIST

1. B. Chalmers, "The Photovoltaic Generation of Electricity," *Scientific American,* October 1976, pp. 34–43.
2. B. G. Streetman, *Solid State Electronic Devices,* 2d Ed. Englewood Cliffs, NJ: Prentice-Hall, 1980, Chapter 6.
3. J. Wilson and J. F. B. Hawkes, *Optoelectronics, An Introduction.* Englewood Cliffs, NJ: Prentice-Hall, 1983, Chapter 7.
4. S. M. Sze, *Physics of Semiconductor Devices,* 2d Ed. New York: John Wiley, 1981, Chapter 13.

5. R. H. Kingston, *Detection of Infrared Radiation.* Berlin: Springer-Verlag, 1978, Chapters 5 and 6.
6. R. J. Keyes (Ed.), *Topics in Applied Physics,* Vol. 19, *Optical and Infrared Detectors.* Berlin: Springer-Verlag, 1980.
7. R. A. Smith, F. E. Jones, and R. P. Chasmar, *The Detection and Measurement of Infrared Radiation.* London: Oxford University Press, 1968.
8. A. Yariv, *Introduction to Optical Electronics,* 2d Ed. New York: Holt, Rinehart and Winston, 1976, Chapters 10 and 11.
9. W. B. Davenport and W. L. Root, *An Introduction to the Theory of Random Signals and Noise.* New York: McGraw-Hill, 1958.
10. H. Kressel (Ed.), *Topics in Applied Physics,* Vol. 39, *Semiconductor Devices for Optical Communication.* New York: Springer-Verlag, 1982, Chapters 3 and 4.

PROBLEMS

5-1. The maximum power delivered by a solar cell can be found by maximizing the i-V product.

(a) Show that maximizing the power leads to the expression

$$\left(1 + \frac{q}{kT} V_{mp}\right) e^{qV_{mp}/kT} = 1 + \frac{i_{sc}}{i_s}$$

where V_{mp} is the voltage for maximum power, i_{sc} is the magnitude of the short-circuit current, and i_s is the reverse saturation current.

(b) Write this equation in the form $\ln x = C - x$ for the case $i_{sc} \gg i_s$ and $V_{mp} \gg kT/q$.

(c) Assume a Si solar cell with a dark saturation current i_s of 1.0 nA is illuminated such that the short-circuit current is $i_{sc} = 100$ mA. Use a graphical solution to obtain the voltage V_{mp} at the maximum delivered power.

(d) What is the maximum power output of the cell at this illumination?

5-2. For a photovoltaic detector the voltage is

$$V = \frac{kT}{q} \ln\left(1 + \frac{i_{sc} + i}{i_s}\right)$$

Given the cell parameters of Problem 5-1, plot the i-V curve for $-i_{sc} \leq i \leq 0$ and draw the maximum power rectangle.

5-3. A major problem with solar cells is internal resistance, generally in the thin region at the surface that must be only partially contacted. Assume that the cell of Problem 5-2 has a series resistance of 1 Ω, so that the cell voltage is reduced by the iR drop. Replot the i-V curve for this case and compare it with the cell of Problem 5-2.

5-4. An ideal photodiode of unit quantum efficiency is illuminated with 10 mW of radiation of 0.8-μm wavelength; calculate the room temperature current and voltage output when the detector is used in the photoconductive ($V = 0$) and photovoltaic ($i = 0$) modes, respectively. The reverse bias leakage current is 10 nA.

5-5. Draw the band diagram for a metal-p-type semiconductor heterojunction. List the conditions for which the device behaves as a Schottky barrier and as an ohmic contact. Explain your results.

5-6. **(a)** Assume that an ideal Schottky barrier is formed on n-type Si having $n_d = 10^{16}$ cm^{-3}. The metal work function is 4.5 V, and the Si electron affinity is 4.05 V. Draw an equilibrium diagram to scale.

(b) Draw the forward bias and reverse bias diagrams to scale for

$$V_f = 0.1 \text{ V} \quad \text{and} \quad V_r = 3 \text{ V}$$

5-7. The hole diffusion length in the base of a silicon *pnp* transistor is 12.5 μm and the base width is 1.0 μm. Compute B, γ, α, and h_{FE} if the emitter is doped to 10^{18} cm^{-3} and the base is doped to 10^{15} cm^{-3}. Assume $D_n = D_p$ and $L_n = L_p$.

5-8. If the resistivity of an intrinsic semiconductor at 300 K is 10^6 $\Omega \cdot$cm, find the thermal coefficient of resistance if the electron mobility is 1000 $cm^2/V \cdot sec$, the hole mobility is 250 $cm^2/V \cdot sec$, and $m_n^* = m_p^* = m$.

5-9. Derive Eqs. 5-17a and 5-17b.

5-10. Compute the resistivity and α_{ex} at 4.2 K for a thermistor with $E_d = 0.01$ eV, $n_d = 10^{17}$ cm^{-3}, $\mu_n = 10{,}000$ $cm^2/V \cdot sec$, and $m_n = m$.

5-11. The resistance of a semiconductor at very low temperatures is $R_0(T_0/T)^A$.

(a) If $\alpha = -1.79$, at 2.15 K, find the value of A.

(b) Compute the current through the element for applied voltages of 0.2, 0.4, and 0.6 V if it is connected to a temperature bath at 2.15 K by a wire with a thermal conductance of $\mathcal{G} = 183$ μW/K and $R_0 = 1.2 \times 10^4$ Ω.

(c) What is the maximum value of V for which this system is stable?

5-12. **(a)** Calculate the temperature rise in the bolometer element described in Example 5-5 due to joule heating.

(b) If the total temperature rise of the element is 1.12 K and the area of the element is 0.15 cm^2, calculate the intensity of the light striking the surface.

5-13. The bolometer element described in Problem 5-11 is placed in the system described in Example 5-5. Compute the bolometer resistance and responsivity at 0.01, 0.1, 1, and 4.2 K.

5-14. **(a)** Compute the responsivity in terms of ampere/watts for the photoconductor described in Example 4-7.

(b) If the total resistance of the photodetector circuit shown in Fig. 4-20 is 1 MΩ, what is the responsivity in terms of volts/watts if $R_s = 50$ kΩ?

5-15. Pattern noise is due to fluctuations in the doping distribution. Discuss why this will produce electrical noise.

5-16. The concentration profile for a material that has a surface concentration, C_s, at $x = 0$ when $t = 0$ and is zero everywhere else is

$$C(x, t) = \frac{C_s}{2(\pi Dt)^{1/2}} e^{-x^2/4Dt}$$

(a) Show that the average distance moved by an atom is zero and explain physically why this occurs.

(b) Show that the average fluctuation is zero and explain physically why this occurs.

(c) Show that the average of the square of the fluctuation is $2Dt$.

5-17. Compute the NEP per unit bandwidth frequency at 1, 4.2, 10, and 300 K.

5-18. Find an expression for $\langle (t - \tau)^2 \rangle$ for the situation described in Example 5-6.

5-19. **(a)** Find the average value of the number that turns up when a die is rolled if all numbers are equally probable.
(b) Find the average fluctuation.
(c) Find the average of the square of the fluctuation.

5-20. **(a)** Find the average value of the number that turns up when a die is rolled if the probability a number turns up is proportional to the number.
(b) Find the average fluctuation.
(c) Find the average of the square of the fluctuation.

Chapter 6

Light Emitting Diodes

6-1 INTRODUCTION

Light can be created by a semiconductor when an electron injected into a *p*-type material radiatively recombines with a hole, or when a hole injected into *n*-type material radiatively recombines with an electron. The overall efficiency of this process, which in special cases can be as high as 10 to 15 percent, is the fraction of the electrical input power converted to optical power, and it is almost constant over a wide range of current. The internal quantum efficiency is the percentage of the injected carriers that recombine radiatively, and the external quantum efficiency is the internal quantum efficiency times the percentage of the light created that actually escapes the semiconductor. We learned in Chapter 2 that this percentage is quite small because semiconductors have a large index of refraction.

The optical transitions can be band to band, donor to valence band, conduction band to acceptor, or donor to acceptor transitions. Band-to-band transitions have a much higher probability of occurring in a direct gap semiconductor because only an electron and a hole need to collide; in an indirect bandgap material the electron and hole must also simultaneously collide with a phonon. One problem with light created by band-to-band transitions is that it can more readily be absorbed. This problem is somewhat mitigated by using heavily doped materials. They effectively broaden the energy gap with regard to absorption by creating donor and acceptor bands. Also, at low temperatures the electron and hole attract each other to form an exciton and, in so doing, reduce the energy of the emitted photon by an amount equal to the attraction energy. The energy is reduced even more if the exciton is attracted to a donor or an acceptor. At lower temperatures the fraction of the injected carriers that recombine radiatively is also larger.

Donor to valence band and conduction band to acceptor transitions have the advantage that they can as readily occur in indirect bandgap materials as they can in direct gap materials. Moreover, the emitted photon can have an energy significantly less than that of the energy gap, and thus it is not as readily absorbed. However, this process is much less efficient than the band to band process, and it is also more temperature dependent. For a relatively shallow donor the trapped electron has a small probability of being thermally emitted at low temperatures, but at higher temperatures the emission probability can become significant. This is especially a problem for green emitting GaP LEDs.

Donor to acceptor transitions are not transitions that are used very often in LEDs.

Binary semiconductors, those semiconductors containing a single anion and a single cation, that are used in LEDs include SiC, ZnSe, GaN, GaP, GaAs, InP, InAs, and GaSb. The first three are wide bandgap semiconductors and are used to generate blue light. However, they are presently of little importance because LEDs made from them are very inefficient. Many of the visible LEDs are made with GaP containing different dopants. More lightly nitrogen-doped material is used for green emitters whereas more heavily nitrogen-doped material is used for the yellow emitter. When GaP is doped with both zinc and oxygen, it emits red light, which is familiar to many of us with calculators. GaAs, InP, InAs, and GaSb, unlike GaP, are direct gap materials so that they emit light more efficiently. However, they emit in the infrared.

Ternary semiconductors such as GaAsP, GaAlAs, and InGaAs are used because their energy gaps can be tuned to a desired wavelength by picking the appropriate composition. There are, however, no ternary substrates, and the interface defects generated by growing films on a substrate with a different lattice parameter can be very detrimental to the device characteristics. One answer is to use a quaternary such as InGaAsP so that the energy gap can be varied while keeping the lattice parameter at the same value as that of the substrate.

GaAsP LEDs are prepared by using a graded structure. That is, the phosphorus content is slowly increased during the growth process thereby spreading the mismatch over a finite thickness. GaAlAs has the distinct advantage of having almost the same lattice parameter as GaAs, since the lattice parameters of GaAs and AlAs are almost identical. $In_{1-x}Ga_xAs$ is lattice matched to the InP substrate only for the value of $x = 0.47$, and it is used primarily as a detector for light emitted by $In_{1-x}Ga_xAs_{1-y}P_y$ lattice matched to InP. The InGaAs energy gap is smaller than that of all lattice-matched compositions of the quaternary; hence, it can detect the light emitted by any quaternary. The range over which the lattice-matched quaternary can emit is $0.92 - 1.65\ \mu m$, which is ideal for applications to optical fibers.

The bandwidth for the emission peaks from semiconductors is more than four orders of magnitude broader than it is for gaseous atoms and is one to two orders of magnitude larger than it is for other solid state laser materials. This is because energy bands, and not discrete energy states, are involved in the optical transitions. The bandwidth also increases with the doping level and the temperature.

The angular distribution of light from the surface is Lambertian, but it is much more directional when it is emitted from an edge; this allows a greater fraction of the light to be inserted into an optical fiber. However, the edge emission process is less efficient. The light emitted from an edge can be made even more directional by using a double heterostructure that can act as a wave guide. The GaAlAs on either side of the active GaAs region has a smaller index of refraction than the GaAs, since aluminum is a lighter element than gallium. Thus, light can be transmitted down the GaAs much like it is down the core of a step index fiber.

The power output decreases with an increasing frequency, but the decrease is insignificant until $\omega\tau \sim 1$ where τ is the recombination time. The response time is largely determined by the device capacitance. Since LEDs operate in forward bias, we not only have to be concerned with the junction capacitance, we also have to be concerned with the diffusion capacitance. This latter capacitance is due to the charge of the injected carriers. They continue to recombine and emit light after the bias has been reversed. This reverse recovery delay time is approximately equal to the recombination time.

6-2 LED OPERATION

6-2.1 Optical Transitions

As we discussed in Chapter 4, the current under forward bias is given by

$$i = i_s(e^{qV/nkT} - 1) \qquad (6\text{-}1)$$

When the minority carriers recombine, they can emit light, and the conversion efficiency, η, is

$$\eta = \frac{P_{\text{ex}}}{iV} \qquad (6\text{-}2)$$

where P_{ex} is the output (external) power. If all of the recombinations contributed to P_{ex}, η would be the ideal efficiency, η_{id}. However, all of the injected minority carriers do not decay radiatively. The fraction that do is called the internal quantum efficiency, η_i. Also, as is true with lasers, all of the optical power created inside of the LED, P_{in}, does not contribute to the output power. Only the fraction that is transmitted, as opposed to being absorbed, is. This fraction times η_i is called the external quantum efficiency, η_{ex}. The LED efficiency, therefore, can be rewritten as

$$\eta = \eta_{id}\eta_{\text{ex}} \qquad (6\text{-}3)$$

Although it is more of a detector efficiency than an emitter efficiency, the relative luminosity is an important figure of merit for visible LEDs. It is the sensitivity of the eye to visible light. As shown in Fig. 6-1, the eye is much more sensitive to green light than it is to red. This is why there still is a strong interest in knowing how to make efficient green LEDs. To date, however, the red LEDs are so much more efficient that it is usually more cost effective to use them.

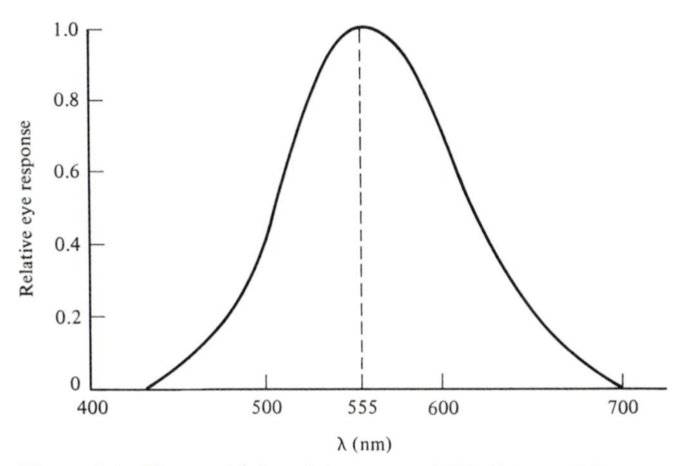

Figure 6-1 The sensitivity of the eye to visible light at different wavelengths.

Light is emitted by band to band, donor to valence band, conduction band to acceptor, and donor to acceptor transitions. The optically most efficient path is a band to band transition if the energy gap is direct. Recall that in Sections 4-4.3 and 4-4.4 when absorption was discussed, we found that an electron could much more easily absorb a photon if both the maximum in the E versus k curve in the valence band and the minimum in the same curve for the conduction band were at the same value of k, $k = 0$. This is because there is no change in k, and therefore no change in momentum, during the transition. Thus, there is no need for the electron and photon to simultaneously collide with each other and the lattice in a three-body collision in order to absorb energy from, or give up energy to, the lattice. What is true for absorption is also true for emission. The probability an electron will collide with a hole is much larger than the probability an electron will collide with a hole and simultaneously with the lattice. This is reflected in the much faster radiative recombination times, τ_r, in direct gap materials.

Recall again from Section 4-4.2 that for a minority carrier electron

$$\tau_{rn} = \frac{1}{Bp_{p0}} \simeq \frac{1}{Bn_a} \tag{4-24}$$

Also, from Table 4-2 we see that B(GaAs) = 7.21×10^{-10} cm^3/sec compared with the indirect materials B(Si) = 1.79×10^{-15} cm^3/sec and B(Ge) = 5.25 $\times 10^{-14}$ cm^3/sec. For InP, which is direct, $B = 1.26 \times 10^{-9}$ cm^3/sec and for GaP, which is indirect, B = 5.37×10^{-14} cm^3/sec.

Band-to-band transitions are more efficient than those that involve donors or acceptors or both because they have a higher probability of occurring. For recombinations involving defect states both the probability a carrier falls into the defect state and the probability that a carrier with the opposite sign will collide with it must be considered. There is also the possibility that a carrier in a defect state will be thermally excited back into a band before it recombines. This is an especially severe problem if the defect depth is of the order of kT.

The reason that a fast recombination time is important is that the longer the carrier stays in an excited state, the greater is the probability that it will find an alternative nonradiative way to lose its energy. The fraction of carriers that do decay radiatively is given simply by

$$\eta_i = \frac{1/\tau_r}{1/\tau_r + 1/\tau_{nr}} = \frac{1}{1 + \tau_r/\tau_{nr}} \tag{6-4a}$$

where τ_{nr} is the nonradiative recombination time. For indirect bandgap materials $\tau_r \gg \tau_{nr}$ so that very little light is created by band-to-band transitions, and for direct gap semiconductors τ_r/τ_{nr} increases with the temperature so that they become less efficient. This is illustrated for the diode in Fig. 6-2.

An example of nonradiative recombination is the Auger process, which is illustrated in Fig. 6-3. The energy given up by an electron falling from the conduction band to the valence band is transferred to another electron in the conduction band which, in turn, is excited to a higher level in the conduction band. It then cascades down to the bottom of the conduction band converting this energy into heat.

For band-to-band transitions one has to be concerned about the absorption of the light before it escapes from the semiconductor. There are a number of reasons why this is not as severe a problem as one would first expect. First, semiconductors used for light emitters are heavily doped to increase the internal quantum efficiency (see Eqs. 4-24 and 6-4). The dopant concentration is so high that the individual dopant atoms are close enough together to interact with each other. As we learned in Section 3-6, when atoms interact with each other, discrete atomic states can be converted into energy bands.

The width of the dopant band can be calculated in a straightforward manner. Equation 4-53b can, for an n-type semiconductor, be rewritten

$$E - E_d = \left(\frac{3n_d}{8\pi}\right)^{2/3} \frac{h^2}{2m_n^*} \tag{6-5}$$

In Example 6-1 we find the donor band thickness as well as the average distance between the atoms for different dopant concentrations.

EXAMPLE 6-1

What is the average distance between the dopants, and what is the width of the donor band for dopant concentrations of 10^{16}, 10^{18}, and 10^{20} cm^{-3}? Assume $m_n^* = m_n$. The average distance, d, between atoms is the cube root of the volume per atom, $V = 1/n_d$

$$d(10^{16}) = \left(\frac{10^{24}}{10^{16}}\right)^{1/3} = 464 \text{ Å}$$

$$d(10^{18}) = \left(\frac{10^{24}}{10^{18}}\right)^{1/3} = 100 \text{ Å}$$

$$d(10^{20}) = \left(\frac{10^{24}}{10^{20}}\right)^{1/3} = 21.5 \text{ Å}$$

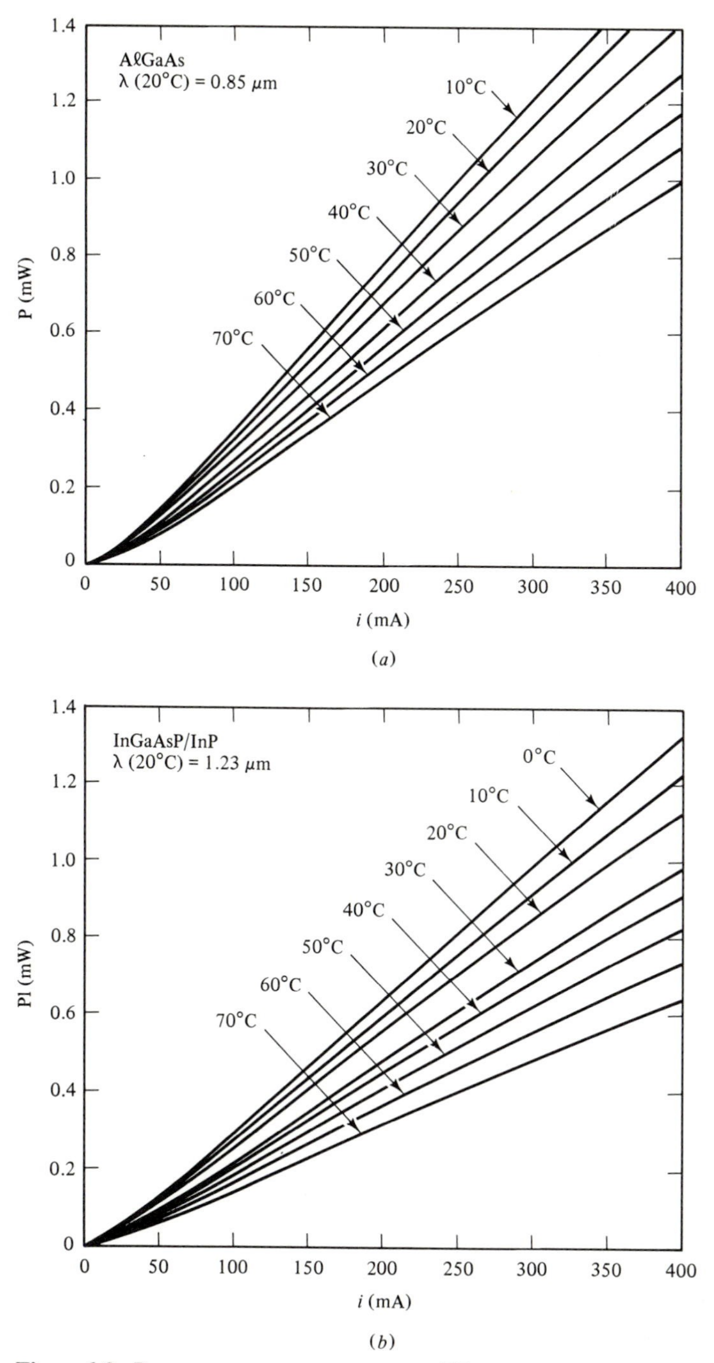

Figure 6-2 Power versus current curves at different temperatures for (*a*) an AlGaAs LED and (*b*) an InGaAsP LED. (From *Topics in Applied Physics,* Vol. 39, *Semiconductor Devices for Optical Communication,* Edited by H. Kressel, Copyright © 1982, Springer-Verlag, New York.)

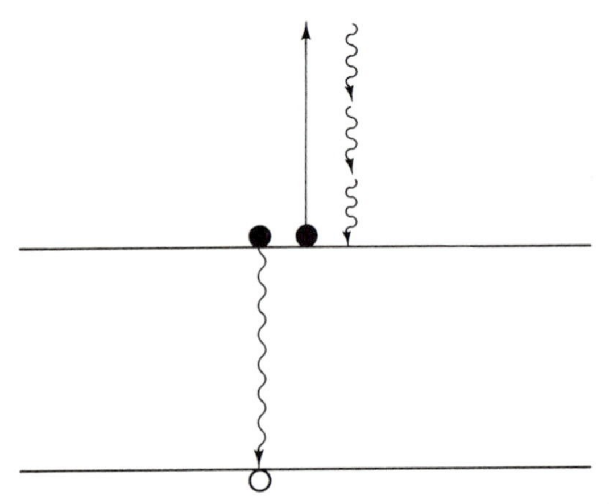

Figure 6-3 The EHP pair recombination energy being converted to heat by the Auger process.

$$E - E_d = \left(\frac{3}{8\pi}\right)^{2/3} \frac{h^2}{2m_n^*} (n_d)^{2/3}$$

$$= \left(\frac{3}{8\pi}\right)^{2/3} \frac{(6.625 \times 10^{-34})^2 (n_d)^{2/3} \times 10^4 \text{ cm}^4/\text{m}^4}{(2)(9.11 \times 10^{-31})(1.6 \times 10^{-19})}$$

$$= 3.6495 \times 10^{-15} n_d^{2/3}$$

$$E - E_d(10^{16}) = 1.69 \times 10^{-4} \text{ eV}$$

$$E - E_d(10^{18}) = 3.65 \times 10^{-3} \text{ eV}$$

$$E - E_d(10^{20}) = 7.86 \times 10^{-2} \text{ eV}$$

As one can see, for doping levels of $>10^{18}$, the donor band spills over into the conduction band.

The emission band is created by electrons dropping from the donor band into the valence band. They can drop from anywhere in the donor band shown in Fig. 6-4*a* so that the emission peak in Fig. 6-4*b* is centered in the middle of the band. However, an optically excited electron must jump from the valence band to the top of the donor band to find an empty state. Thus, the absorption band edge in Fig. 6-4*b* is shifted to shorter wavelengths relative to the emission peak.

The photon energy moves farther away from the absorption band when an exciton is formed. An exciton is an electron bound to a hole by coulombic attraction, and the photon energy is lowered by an amount equal to the attraction energy. The exciton energy is larger if the exciton is also bound to a donor or an acceptor. Usually the attraction energy of both the free and bound excitons is much less than kT at room temperature so that they are broken apart.

In the special case of the nitrogen isoelectronic trap, the binding energy is

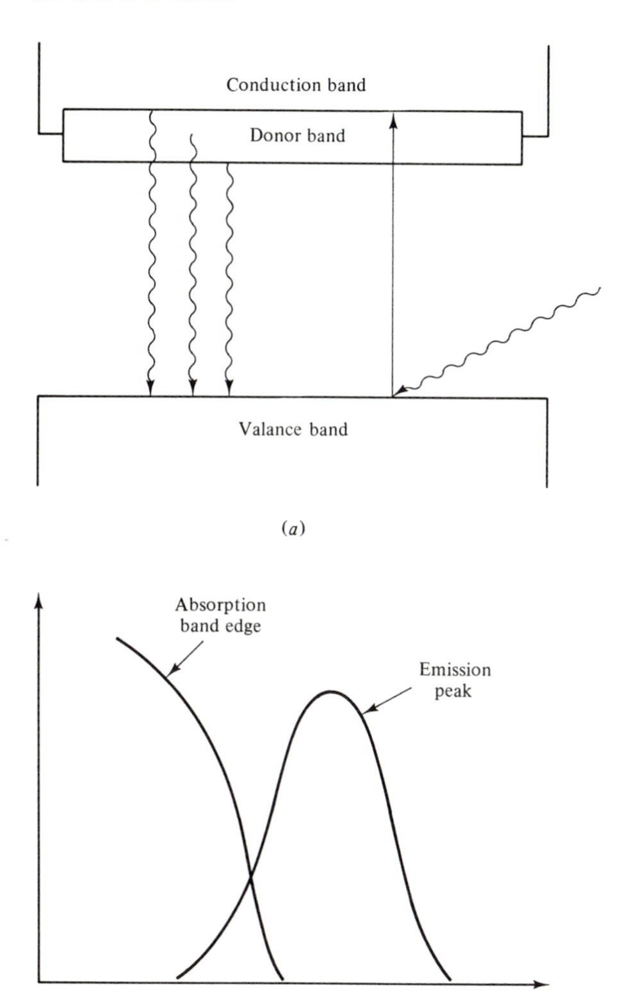

Figure 6-4 (*a*) Photons emitted by electrons dropping from a donor band to the valence band and a valence band electron absorbing a photon. (*b*) The emission peak produced by the donor band to valence band transitions and the absorption band edge plotted as a function of the wavelength.

larger than usual. The trap is formed by substituting nitrogen for phosphorus in GaP. The nitrogen creates a level shown in Fig. 6-5*a* of about 81 meV below the bottom of the conduction band. Electrons injected into *p*-type material also doped with nitrogen can fall into the trap; this charges the trap and attracts a hole that can recombine with the trapped electron.

There is no need for a phonon (lattice vibration) to be involved in the transition because the nitrogen atoms have a relatively precise location. Their position can be known to within a distance "*a*," which is the lattice parameter of the unit cell. Thus from Heisenberg's uncertainty principle

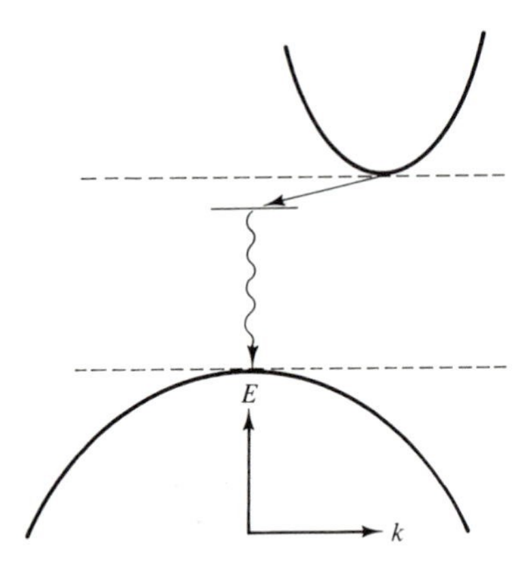

Figure 6-5 The optical transition from a nitrogen isoelectronic trap.

$$\Delta p \, \Delta x > \hbar \tag{6-6a}$$

so

$$\Delta p > \frac{\hbar}{a} \tag{6-6b}$$

where Δp is the uncertainty in the momentum. By using the version of Eq. 3-6,

$$p = \hbar k \tag{6-7}$$

$$\Delta k > \frac{1}{a} \tag{6-6c}$$

This magnitude of Δk is more than enough to make up for any differences in the k values associated with an indirect transition.

The efficiency of the emission process via an isoelectronic trap is strongly temperature dependent, since the electron can be thermally excited out of the trap before it recombines. Of course, this is more likely to happen at higher temperatures.

Recombination through a donor or an acceptor or both is similar to recombination through an isoelectronic trap. There are essentially no restrictions placed on the recombination process by an indirect bandgap, and the efficiency decreases with an increasing temperature. If the donors or acceptors are shallow compared to kT, there is little light emission by this process because the defect states are thermally ionized before they recombine. This is illustrated in Fig. 6-6 where emission from a zinc-tellurium donor acceptor pair, a free exciton, and band to band transitions for $GaAs_{0.41}P_{0.59}$ is shown for different temperatures. (Note that the decrease in E_g with increasing temperature has been subtracted out of the results.)

6-2.2 Current–Voltage–Power Characteristics

As was shown in Fig. 6-2, the output power is approximately proportional to the current, and the proportionality constant decreases as the temperature increases. For LED displays the current levels often range between 10 and 100

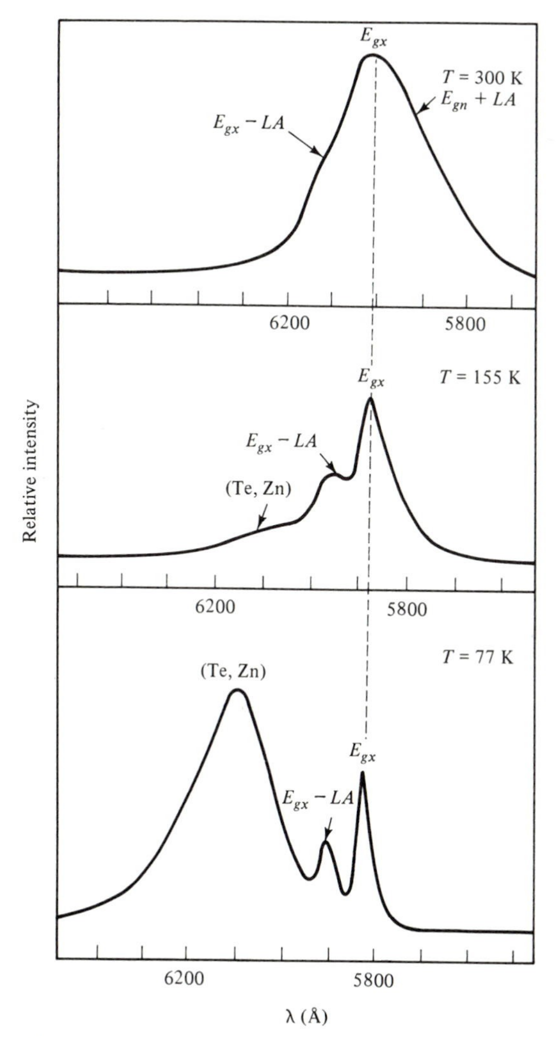

Figure 6-6 The light emission peaks from a (Te, Zn) doped $GaAs_{0.41}P_{0.59}$ LED at 300, 155, and 77 K. The donor acceptor (Te, Zn) transition is dominant at 77 K and the band-to-band transition, E_{gx}, is dominant at room temperature. The third peak, E_{gx}-LA has a longer wavelength than E_{gx} because energy has been given up to the lattice by creating a longitudinal acoustical phonon. [From M. G. Craford, R. W. Shaw, A. H. Herzog, and W. O. Groves, *J. Appl. Phys.*, **43** (4075), 1972.]

mA for continuous operation; for an active area 250 μm on a side the corresponding current densities are 16 and 160 A/cm^2. If the current is raised above 100 mA, heating could become a problem if the diode is not in contact with a heat sink.

For the surface-emitting, Burrus-type diode in Fig. 2-17, which has a fiber optic pigtail, the active area is smaller and the current density is higher. For a 50-μm diameter and a 200-mA current the current density is $\sim 10^4$ A/cm^2. This represents the highest current density that should be used, and the diode must be in close contact with a heat sink.

In the edge emitter in Fig. 6-7*a* the light is emitted parallel, rather than perpendicular, to the junction. As is seen in Fig. 6-7*b*, the edge emitter is less efficient than the surface emitter, but it has the advantage, as we discussed in Section 2-4, that its light is more directional. The stripe contact has a width of

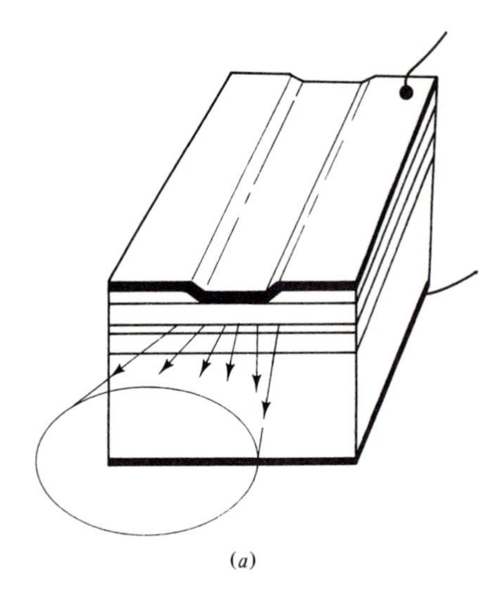

(*a*)

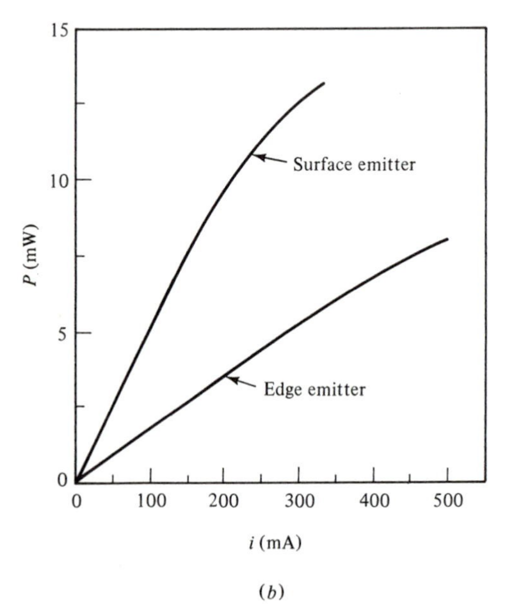

(*b*)

Figure 6-7 (*a*) A stripe contact diode, and (*b*) the power output from a surface and an edge emitter plotted as a function of the current.

$\sim 10\ \mu m$ and a length of $\sim 400\ \mu m$. Thus, for a current of 400 mA, the current density is at its maximum value of $10^4\ A/cm^2$.

Compared with electrical diodes, a lot of power is supplied to the LED. Thus, the junction voltage is almost equal to the contact potential. Since the doping levels are high, $V_0 \sim E_g/q$; the *n*-type Fermi level is near the top of the donor band, and the *p*-type Fermi level is near the bottom of the acceptor band. This is shown in Fig. 6-8.

The *i-V* curve for an LED is not as simple as it is for an electrical diode because heavily doped, wider bandgap materials are used, and the LED often operates at high current levels. When high doping levels are used, we can no longer neglect Pauli's principle as we did when Eq. 4-71 was derived. It can no longer be assumed that the electron concentration in the conduction band is so small that no two electrons will try to occupy the same quantum state. At small forward bias, the current is very small both because the energy gaps are larger and the doping levels are higher. Thus, the generation-recombination current dominates (see Section 4-8), and the diode factor in Eq. 6-1 is $n = 2$. This is shown in the *i-V* curve in Fig. 6-9. Note also at the higher current levels, >10 mA, high injection occurs since $V - V_0$ is small, and there is a significant amount of series resistance.

Also, as predicted by Eqs. 4-71 and 6-1, the voltage needed to keep the current at a fixed value will decrease as the temperature increases because i_s increases faster than $\exp(qV/kT)$ decreases.

LEDs emitting in the visible are typically 0.1 percent efficient. Thus, at 50 mA and 1.5 V the output power is 75 μW. Red emitting GaP LEDs doped with zinc and oxygen and GaAs LEDs doped with silicon can be 10 percent efficient. Their output power would then be 7.5 mW.

For fiber optic applications the directionality as well as the power output are important. The radiance is the output power per unit area per unit solid angle (steradian). Because the light from an edge emitter is more directional, the highest radiance for this structure is $\sim 1000\ W/cm^2 \cdot sr$, whereas for the best surface emitters it is $200\ W/cm^2 \cdot sr$.

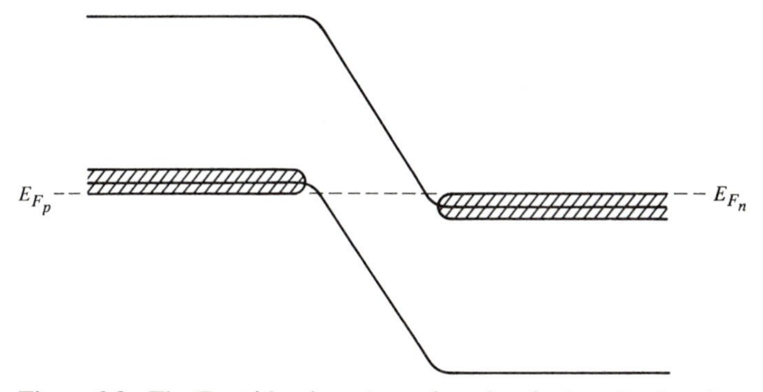

Figure 6-8 The Fermi levels and a *pn* junction for heavily doped materials.

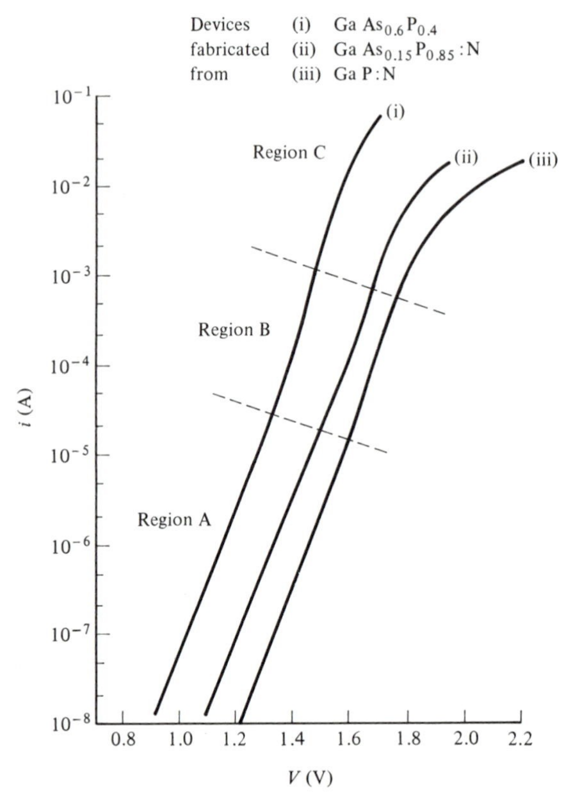

Figure 6-9 Typical *i–V* characteristics for $GaAs_{1-x}P_x$ LEDs. (From E. W. Williams and R. Hall, *Luminescence and the Light Emitting Diode,* Copyright © 1978, Pergamon Press, Oxford.)

As we noted in discussing Fig. 6-2, the diodes become less efficient as the temperature increases. For a given power input the output power varies according to the equation

$$P_{ex} = P_0 e^{E_A/kT} \tag{6-8}$$

where E_A is an activation energy. For a given diode, E_A is often a function of the current. Also, E_A is usually larger for diodes that involve traps in the recombination process, and is larger when the traps are deeper. This is because carriers can be thermally ionized from the traps before recombination occurs, and the probability that they will be thermally ionized is proportional to $\exp(-E_t/kT)$ (see Section 4-8).

EXAMPLE 6-2

20 mA flow through a silicon and a zinc-doped GaAs diode when 1.2 V is applied at 25° C, and 17 mA flow through them at 100° C when 1.1V is applied. **(a)** Find the power emitted at both temperatures from each

diode if the silicon doped LED is 10 percent efficient at 25° C and 5 percent efficient at 100° C, and the zinc-doped LED is one-fourth as efficient as the silicon doped one. **(b)** Find the current densities if the active area is 250 μm on a side. **(c)** Compute the reverse saturation current and current densities assuming the diode factor, $n = 2$. **(d)** Using the results in (c) determine the forward bias necessary to create a current of 10 and 100 mA at 25° C.

(a) For the silicon doped LED at 25° C: $P_{ex} = (20)(1.2)(0.1) = 2.4$ mW

at 100° C: $P_{ex} = (17)(1.1)(0.05) = 0.935$ mW

zinc-doped LED at 25° C: $P_{ex} = (20)(1.2)(0.025) = 0.6$ mW

at 100° C: $P_{ex} = (17)(1.1)(0.0125) = 0.234$ mW

(b) At 25° C: $$J = \frac{2.0 \times 10^{-2}}{(2.5 \times 10^{-2})^2} = 32 \text{ A/cm}^2$$

At 100° C: $$J = \frac{1.7 \times 10^{-2}}{(2.5 \times 10^{-2})^2} = 27.2 \text{ A/cm}^2$$

Thus, the leads can quite easily conduct away the heat created by the device.

(c) At 25° C
$$i_s = ie^{-qV/nkT}$$
$$= (20) \exp\left[\frac{-(1.6 \times 10^{-19})(1.2)}{(2)(1.38 \times 10^{-23})(2.98 \times 10^2)}\right]$$
$$= 1.46 \times 10^{-9} \text{ mA}$$
$$J_s = \frac{1.46 \times 10^{-9}}{(2.5 \times 10^{-2})^2} = 2.34 \times 10^{-6} \text{ mA/cm}^2$$

At 100° C
$$i_s = 17 \exp\left[\frac{-(1.6 \times 10^{-19})(1.1)}{(2)(1.38 \times 10^{-23})(3.73 \times 10^2)}\right]$$
$$= 6.37 \times 10^{-7} \text{ mA}$$
$$J_s = \frac{6.37 \times 10^{-7}}{6.25 \times 10^{-4}} = 1.02 \times 10^{-3} \text{ mA/cm}^2$$

These computed values are much less than the actual reverse current because there is considerable leakage.

(d) $V = \frac{nkT}{q} \ln \frac{i}{i_s}$

For 10 mA

$$V = \frac{(2)(1.38 \times 10^{-23})(2.98 \times 10^{2})}{1.6 \times 10^{-19}} \ln \frac{10^{1}}{1.46 \times 10^{-9}} = 1.164 \text{ V}$$

For 100 mA

$$V = V(10 \text{ mA}) + \frac{nkT}{q} \ln \frac{100}{10} = 1.282 \text{ V}$$

Although these numbers are not accurate because of high carrier injection conditions, they do indicate how sensitive the current is to the voltage.

6-3 SPECIFIC LEDS

6-3.1 Binary Homojunctions

Blue LEDs have been made from SiC, ZnSe, and GaN, which are wide bandgap semiconductors. However, these devices are very inefficient. SiC can have two different crystal structures, and this results in the creation of a number of electronically debilitating crystalline defects. ZnSe and GaN can only be made *n*-type so that a Schottky barrier or a heterojunction structure must be used to create a diode. Many crystalline defects are formed at the interface, and they, too, have debilitating effects.

Green light at 570 nm can be created by nitrogen-doped GaP. The large capture cross-section of the isoelectronic nitrogen trap in the *p*-type material increases the efficiency of this process. However, the trap is only 81 meV below the conduction band, which is only three times the thermal energy at room temperature. Thus, the traps can quite easily be emptied thermally before the electrons recombine. This sensitivity to the temperature is illustrated in Fig. 6-10*a*. The efficiency of these diodes made commercially is only 0.05 to 0.1 percent, and they have been made with an efficiency of 0.7 percent in the laboratory. Their efficiency is also sensitive to the nitrogen-doping levels; it increases with the doping concentration to levels in excess of 10^{18} cm^{-3} by increasing the probability an electron will collide with one of them. At higher concentrations the nitrogen atoms interact with each other in such a way as to reduce the efficiency.

Yellow LEDs emitting at 590 nm are also made from nitrogen-doped GaP; they differ from green LEDs only in that the nitrogen concentration is higher. At these high nitrogen concentrations, the nitrogen atoms interact strongly enough to form complexes. These complexes behave in a more complicated fashion. Their efficiency, as illustrated in Fig. 6-10*b*, can actually increase with the temperature. At room temperature commercial LEDs are <0.05 percent efficient, and they have been made with a 0.1 percent efficiency in the laboratory.

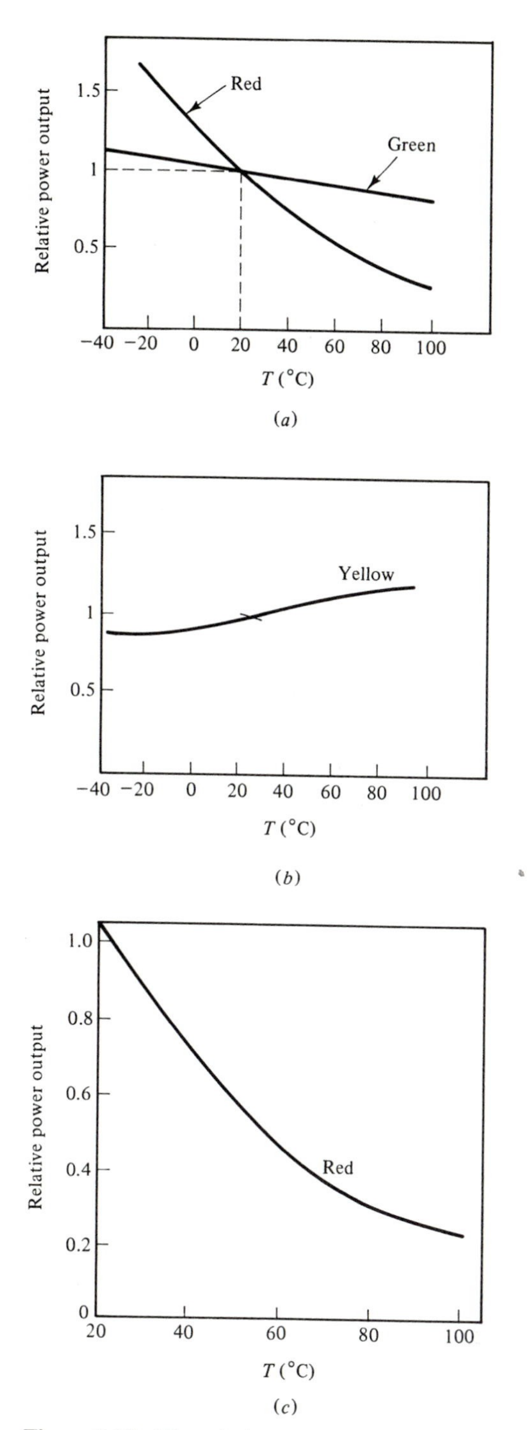

Figure 6-10 The relative power output for a (*a*) green, (*b*) yellow, and (*c*) red GaP LED plotted as a function of the temperature.

GaP red LEDs emitting at 690 nm are made by creating isoelectronic traps with zinc and oxygen complexes. Of course, zinc is an acceptor and oxygen is a donor, so together they compensate each other. The neutral trap is much deeper than the nitrogen trap—it is ~0.3 eV below the conduction band; hence, red LEDs are much more efficient. They can be made as high as 15 percent efficient in the lab, and the efficiency of commercial diodes is 2 to 4 percent. As expected, the output, as shown in Fig. 6-10*c*, decreases with increasing temperature. The output also saturates at relatively low currents. As the current increases, more of the traps become filled thereby slowing down the radiative recombination process, while also the rate of nonradiative recombination increases with increasing Δn.

Since GaAs is a direct bandgap material, it can emit light efficiently by band to band transitions. These diodes emit at 865 nm, and their internal quantum efficiency can be as high as 70 percent. However, their total efficiency is only ~0.1 percent because there is considerable internal absorption. Their efficiency shown in Fig. 6.2 decreases with increasing temperature because (τ_r/τ_{nr}) increases. They are fast LEDs, since a carrier does not have to first fall into a trap; τ_r can be as short as 5 nsec.

A slower, but more efficient, GaAs LED is the silicon-doped diode. At high doping temperatures, silicon substitutes for gallium so it is a donor. At lower temperatures it substitutes for arsenic so it is an acceptor. For the diode an acceptor complex is created about 0.1 eV above the valence band. Light is emitted at 910 to 1020 nm, and the efficiency can be as high as 10 percent because there is little internal absorption. Their recombination time, however, is several hundreds of nanoseconds.

6-3.2 Ternary and Quaternary Homojunctions

Ternary LEDs have the advantage that they can be tuned to emit light at a specific wavelength as the energy gap varies with the composition. To a first approximation, it varies linearly and is given by

$$E_g(A_{1-x}B_xC) = E_g(AC) + [E_g(BC) - E_g(AC)]x \tag{6-8a}$$

whereas a more accurate representation is

$$E_g(A_{1-x}B_xC) = E_g(AC) + [E_g(BC) - E_g(AC)]x - b_{AB}x(1-x) \tag{6-8b}$$

where x is the fraction of component B and b_{AB} is called the bowing parameter. It is an experimentally determined parameter that is always positive; E_g always varies sublinearly. Clearly, either the cations or the anions can be mixed.

Binary semiconductors have both a direct and an indirect bandgap when there are two local minima in the E versus k curve for the conduction band, one of which is at $k = 0$. These values are tabulated in Table 6-1. If the $k = 0$ minimum is the lower, the energy gap is said to be direct. Two contrasting examples of GaAs (direct) and GaP (indirect) are illustrated in Fig. 6-11. When E_g for the ternary is computed, both the direct and the indirect gaps are found and whichever one is smaller is the listed energy gap of the material. For

TABLE 6-1 THE LATTICE PARAMETER, DIRECT AND INDIRECT ENERGY GAPS, AND ELECTRON AFFINITY OF THE III-V SEMICONDUCTOR COMPOUNDS

Material	Lattice parameter a_0 (Å)	Direct energy GaP E_g (eV)	Indirect energy GaP E_g (eV)	Electron affinity χ (V)
AlP	5.451	3.60	2.45	3.5
AlAs	5.661	2.95	2.16	3.65
AlSb	6.135	2.15	1.65	4.3
GaP	5.451	2.75	2.26	4.07
GaAs	5.653	1.43	1.86	4.06
GaSb	6.095	0.73	1.02	4.35
InP	5.869	1.35	2.20	4.9
InAs	6.057	0.36	1.83	4.59
InSb	6.479	0.18	—	—

$GaAs_{1-x}P_x$ the gap is direct for small values of x and indirect for larger values. This is shown in Fig. 6-11*a*. We now find the approximate composition for which the changeover occurs.

EXAMPLE 6-3

Find the composition of $GaAs_{1-x}P_x$ for which the energy gap changes from being direct to being indirect. Also find the approximate value of E_g at this composition by assuming that E_g varies linearly with x.

For the direct energy gap

$$E_g(\text{direct}) = 1.43 + (2.75 - 1.43)x = 1.43 + 1.32x$$

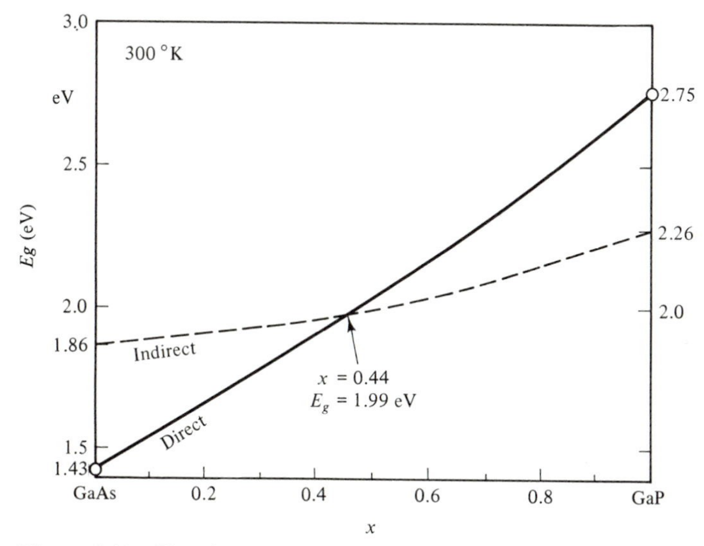

Figure 6-11 The direct and indirect energy gaps of $GaAs_xP_{1-x}$ plotted as a function of the composition.

For the indirect energy gap

$$E_g(\text{indirect}) = 1.86 + (2.26 - 1.86)x = 1.86 + 0.40x$$

The composition, x^*, for which both energy gaps are the same

$$1.43 + 1.32x^* = 1.86 + 0.40x^*$$

$$\therefore x^* = \frac{1.86 - 1.43}{1.32 - 0.40} = 0.467$$

$$E_g(\text{GaAs}_{1-x^*}\text{P}_{x^*}) = 1.43 + 1.32x^* = 2.047 \text{ eV}$$

The correct values are $x = 0.44$ and $E_g = 1.99$ eV.

For band-to-band transitions the efficiency is much greater for the direct gap material. This is illustrated in Fig. 6-12 for GaAsP where the efficiency drops dramatically as the changeover from a direct to an indirect bandgap is approached.

Red LEDs emitting at 649 nm can be made from $GaAs_{0.60}P_{0.40}$. The transitions are band to band, and the process is relatively efficient because the energy gap is direct. The efficiency of commercial diodes is 0.2 percent, and even though this is less efficient than the red emitting GaP LEDs, they are often preferred because they are cheaper to make. In the laboratory their efficiency has been as high as 0.5 percent.

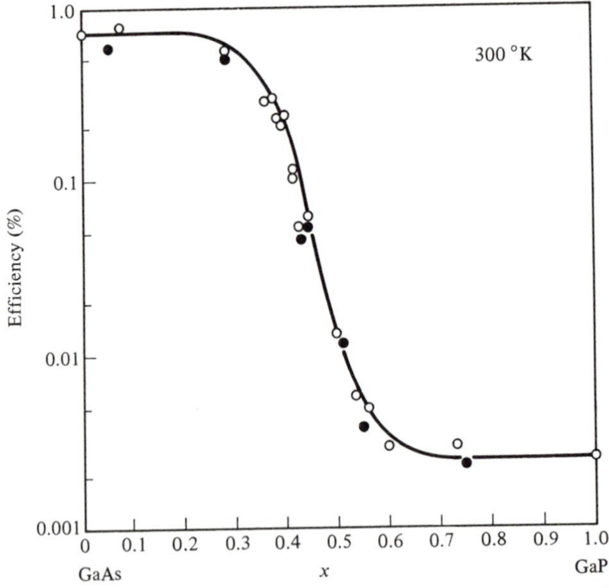

Figure 6-12 The efficiency of a $GaAs_{1-x}P_x$ diode plotted as a function of the composition. [From A. H. Herzog, W. O. Groves, and M. G. Craford, *J. Appl. Phys.*, **40** (1830), 1969.]

Orange LEDs emitting at 632 nm can be made from nitrogen-doped $GaAs_{0.35}P_{0.65}$. The isoelectronic nitrogen trap is needed because the gap is indirect. The best diodes have an efficiency of 0.5 percent, and commercial devices are typically 0.2 percent efficient. When the composition is changed to $GaAs_{0.15}P_{0.85}$, the diode emits yellow light at 589 nm. The efficiency of these devices is 0.2 percent in the laboratory and 0.05 percent commercially.

The primary difficulty in fabricating GaAsP LEDs is in preparing films with as few lattice defects as possible. We saw in Fig. 4-27 that many mismatch dislocations are formed at an abrupt heterojunction when the lattice parameters of the two semiconductors are different. $GaAs_{1-x}P_x$ has a different lattice parameter than either GaAs or GaP as lattice parameters vary linearly with the composition and are given by the equation:

$$a(A_{1-x}B_xC) = a(AC) + [a(BC) - a(AC)]x \tag{6-9}$$

The problem created by the lattice mismatch is to some extent overcome by using a graded junction. When GaAs is used as the substrate, the phosphorus composition is gradually increased over a thickness of 20 to 50 μm. When the composition reaches the desired value, the graded junction acts essentially like a substrate, and a constant composition film is grown on it. In general, if $x < 0.5$ a GaAs substrate is used, and if $x > 0.5$, a GaP substrate is used.

GaAlAs structures have the advantage that GaAs ($a_0 = 5.653$ Å) and AlAs ($a_0 = 5.661$ Å) have almost the same lattice parameter; hence, GaAlAs of any composition can be grown on a GaAs substrate without introducing very many lattice defects. AlAs like GaP has a larger bandgap than GaAs, and it has an indirect gap. Most of the diodes are made with small amounts of aluminum so that E_g is direct. For example, GaAlAs diodes are "tuned" to the 0.82 μm absorption minimum in optical fibers.

Like GaAs diodes the transitions can be either band to band or via deep defect states like those provided by silicon. Again, the band-to-band transitions are faster, but less light is emitted by this process because it is more strongly absorbed.

InGaAs LEDs are fabricated from InGaAs films deposited on InP substrates. The lattice parameter of InAs ($a_0 = 6.057$ Å) is larger than that of InP ($a_0 = 5.869$ Å) whereas the lattice parameter of GaAs is smaller. Thus, $In_{1-x}Ga_xAs$ is lattice matched to InP for a specific value of x, which is $x = 0.47$. Graded junctions with different compositions can be prepared, but rather than do this, the quaternary $In_{1-x}Ga_xAs_{1-y}P_y$ is used. The advantage of using the quaternary is that lattice-matched abrupt junctions can be made. This better allows us to confine the light to a narrow active layer, something that is essential for continuous wave laser diodes.

The quaternary can be lattice matched while at the same time the energy gap can be varied, since now there are two degrees of freedom instead of one. Thus, now both E_g and a_0 can be specified. The lattice parameter can be computed by recognizing that it varies linearly with x or y, but not with both. This is illustrated in Fig. 6-13 and in Example 6-4.

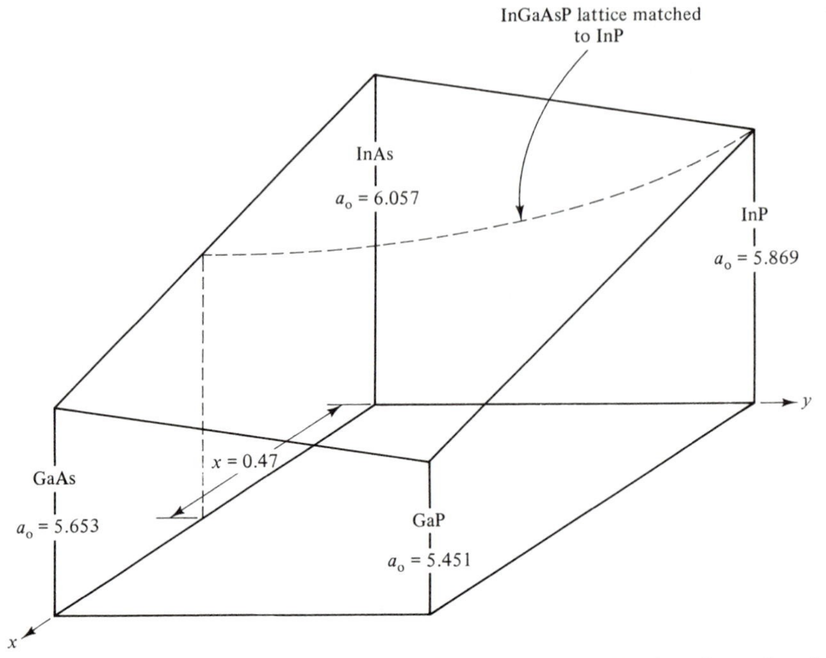

Figure 6-13 The lattice parameter surface for $In_{1-x}Ga_xAs_{1-y}P_y$ plotted as a function of x and y. Note that a_0 for the ternaries varies linearly with the composition.

EXAMPLE 6-4

Find the relationship between x and y for which $In_{1-x}Ga_xAs_{1-y}P_y$ is lattice matched to InP. Find the minimum and maximum values of x and y by using the lattice parameters listed in Table 6-1.

The quaternary is a mixture of $(1-x)(1-y)$ InAs, $(1-x)y$ InP, $x(1-y)$ GaAs, and xy GaP

$$(1-x)(1-y)6.057 + (1-x)y5.869 + x(1-y)5.653 + xy5.451 = 5.869$$

$$6.057 - 6.057x - 6.057y + 6.057xy + 5.869y - 5.869xy$$

$$+ 5.653x - 5.653xy + 5.451xy = 5.869$$

$$0.188 - 0.404x - 0.188y - 0.014xy = 0$$

$$\therefore y = \frac{0.188 - 0.404x}{0.188 + 0.014x}$$

For y = 0

$$0.404x = 0.188 \quad \text{or} \quad x = \frac{0.188}{0.404} = 0.465$$

For y = 1

$$x = 0$$

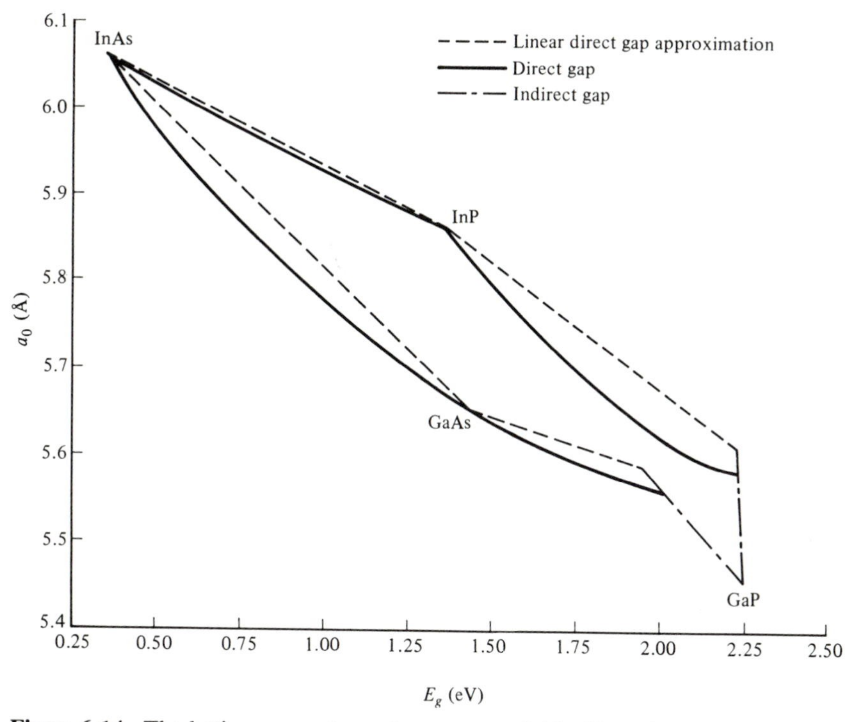

Figure 6-14 The lattice parameter and energy gap field of $In_{1-x}Ga_xAs_{1-y}P_y$.

A diagram similar to Fig. 6-13 could also be constructed if the energy gap varied linearly with the composition. However, as we learned earlier, E_g is sub-linear, and the magnitude of the nonlinearity is characterized by the bowing parameter. Rather than constructing an $E_g(x, y)$ diagram, it is more useful to construct an a_0 versus E_g diagram. This is done in Fig. 6-14 for both the linear approximation (dotted line) and when the bowing parameter is included. Note that of the four binary semiconductors, only GaP has an indirect bandgap. Thus, the boundary marked by the InGaP or GaAsP ternaries is described by two separate equations: one for the direct and one for the indirect bandgap.

Light emitted by the quaternary is via band-to-band transitions. The diodes are doped in the usual manner with the preferred *p*-type dopants being zinc and cadmium while the preferred *n*-type dopants are sulfur and selenium.

Although the quaternary InGaAsP has received the most attention, there is also considerable interest in the GaAlAsSb quaternary. GaSb is a direct bandgap semiconductor with E_g = 0.73 eV, but AlSb, like AlAs, has an indirect gap (1.75 eV). As a result, most of the quaternaries currently being investigated are gallium rich.

Of course, there are many other possible combinations. One combination currently receiving attention is InAlAs because it is a high mobility material, and it has a larger energy gap than InGaAs for lattice-matched compositions.

6-3.3 Heterojunctions

A heterojunction is a junction in which the *p*- and the *n*-type material are different semiconductors. We commented on it briefly in Section 4-7.1, and here we again discuss it briefly. It will be examined in much more detail in the next chapter.

One major advantage of a heterojunction LED is that it can be an edge emitter. In GaAlAs heterojunction the LED's light is created in a central region sandwiched between two regions that contain more aluminum. Because there is a heterojunction on either side of the active region, this structure is called a double heterostructure, DH. The upper and lower confining layers are so designated because they have a lower index of refraction and thus confine the light by reflection at the interface; for large angles of incidence total internal reflection occurs much as it does for step index fibers. The index of refraction of the confining layers is less, since aluminum is a lighter element than gallium and, therefore, has fewer electrons to interact with the electromagnetic radiation. The index difference, Δn, is given by

$$\Delta n \cong 0.62 \, \Delta x \tag{6-10}$$

where Δx is the difference in the aluminum concentration. As one might expect, the larger Δn the greater is the power of the emitted light, but the angular distribution in the plane normal to the junction is also larger. It is larger because the critical angle for total internal reflection is smaller.

A single heterostructure, SH, is sometimes used for a window for surface emitters. A heavily doped relatively thick surface layer containing more aluminum is sometimes used so that a good ohmic contact can be made more easily and the amount of surface recombination, which is nonradiative, can be reduced. The wider bandgap of the surface layer is essentially transparent to the emitted light, and there are very few lattice defects to provide nonradiative recombination centers because the lattice match is very good.

For DH structures the recombination time is given by the equation

$$\frac{1}{\tau} = \frac{1}{\tau_r} + \frac{1}{\tau_{nr}} + \frac{2S}{d} \tag{6-11}$$

where S is the surface recombination rate and d is the width of the active region. Since the number of dangling bonds per unit area (see Fig. 4-27) is proportional to the lattice mismatch, $\Delta a_0/a_0$, S, too, is proportional to it. For GaAs-based materials

$$S = 2 \times 10^7 \frac{\Delta a_0}{a_0} \tag{6-12}$$

Clearly, these interface defects will reduce the efficiency of the light output. Substituting Eq. 6-11 into Eq. 6-4a yields

$$\eta_i = \frac{1}{1 + (2S/d)\tau_r + \dfrac{\tau_r}{\tau_{nr}}} \tag{6-4b}$$

These defects create a large concentration of defect states in the energy gap and in so doing greatly increase the generation–recombination leakage current.

6-4 CHARACTERISTICS OF THE EMITTED LIGHT

6-4.1 Angular Distribution

This topic was covered in some detail in Section 2-4.3. The highlights are as follows.

The external quantum efficiency is much less than the internal quantum efficiency because it is difficult for light to escape from a semiconductor, since semiconductors have a large index of refraction. This, to some extent, can be overcome by coating the surface with an index of refraction lying between that of the semiconductor and air. The fraction of light reflected back into the semiconductor can also be reduced by forming a hemispherical diode. This, however, is an expensive alternative.

For displays and especially for optical fiber applications, it is important to have the light emitted in directions almost parallel to the surface normal. The light can be made more parallel by using a cylindrical plastic cap with a hemispherical top. Also, edge emitters generate light that is more parallel.

6-4.2 Peak Position and Bandwidth

Both the peak position and bandwidth are affected by the temperature. The peak moves to longer wavelength and the peak broadens as the temperature increases. One can attribute the longer wavelength to a smaller energy gap which, in turn, can be attributed to less splitting between the bonding and antibonding orbitals as the atoms move further apart. The effect is relatively small as it is $\sim$0.5 meV/°C for GaAs.

The emission peak bandwidth for semiconductors is much broader than it is for gaseous atoms, and it is even broader than it is for other solid state light emitters such as the ruby laser. The width, typically 1.5×10^{13} Hz, is wide because energy bands, as opposed to discrete atomic states, are involved in the transitions. In the next chapter we will learn that it has a strong effect on the laser threshold current.

The thermal peak broadening is due to a broadened thermal distribution of the electrons in the conduction band or the holes in the valence band or both. They are distributed over an energy band $\sim kT$ thick. The peak broadening of a Zn:O doped GaP diode is illustrated in Fig. 6-15; the bandwidth increases at a rate of 0.46 meV/°C.

The peak width also increases with the doping level, and this is illustrated in Fig. 6-16. The primary cause of this is that the donor and/or acceptor bandwidth increases with the doping level; electrons in the conduction band can fall to any level in the acceptor band. Clearly, there are other considerations, since

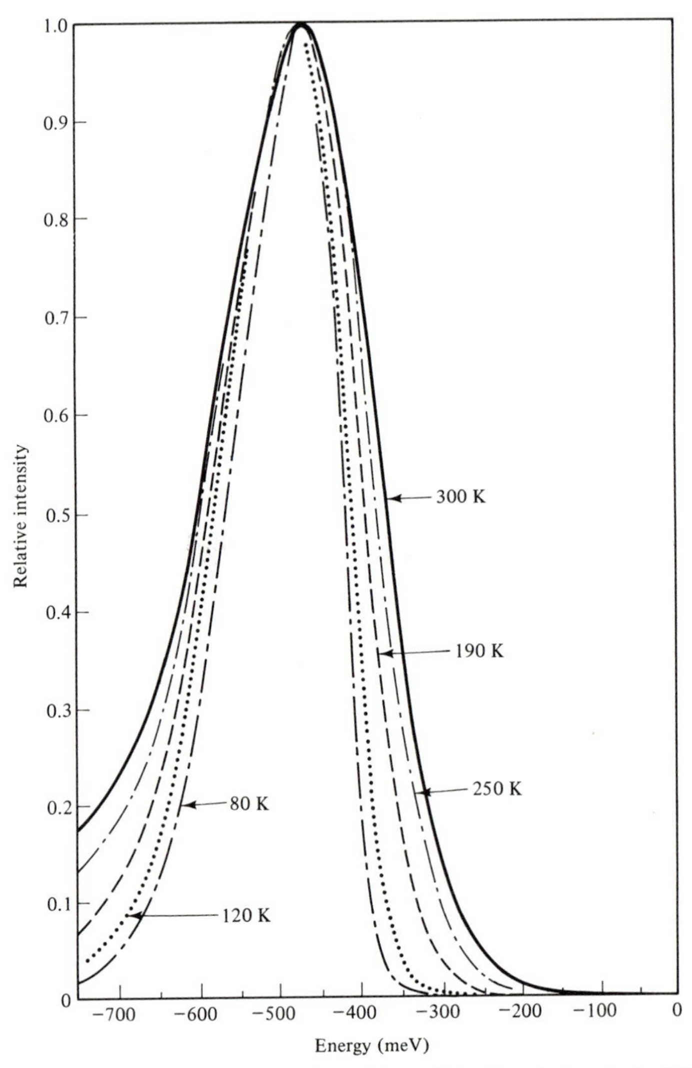

Figure 6-15 Thermal broadening of the red Zn, O emission peak of GaP with an acceptor concentration of 10^{18} cm^{-3}. The horizontal scale is the emission peak energy minus E_g at the given temperature. [From R. Z. Bachrach and J. S. Jayson, *Phys. Rev.*, **B7** (2540), 1973.]

the peak width in Fig. 6-16 does not increase at the same rate with the doping concentration, and the bandwidth depends on which dopant is used. One other important consideration is band tailing, which we mentioned briefly when absorption was discussed. Band tailing is the creation of states just below the conduction band or just above the valence band by dopants and other defects that upset the periodicity of the crystal lattice. The band-tailing effect increases as the doping level increases.

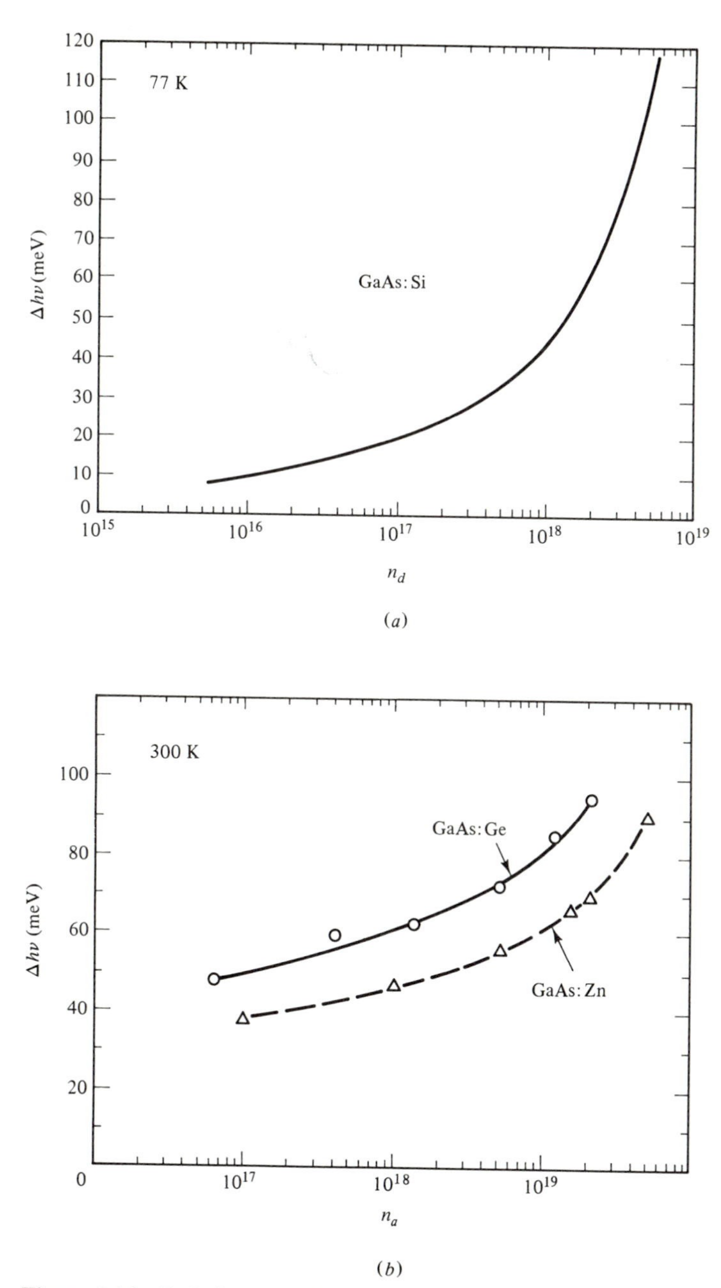

Figure 6-16 Emission peak broadening due to increased doping levels for (*a*) silicon-doped GaAs at 77 K, and (*b*) germanium and zinc doped GaAs at room temperature. [(a) From H. Kressel and J. K. Butler, *Semiconductor Lasers and Heterojunction LEDs,* Copyright © 1977, Academic Press, New York. (*b*) From H. Kressel and M. Ettenberg, *Appl. Phys. Lett.,* **23** (511), 1973.]

6-5 FREQUENCY EFFECTS

6-5.1 Digital Signals

An LED cannot respond instantaneously to a current applied to it because it takes time for the minority carriers to build up to their steady state ON distribution or decay down to their steady state OFF distribution. The primary factor that determines the response time is the decay time—something that you by now should have expected.

Our objective is to find the excess minority charge, $Q(t)$, as a function of time. To reduce the amount of bookkeeping with no loss of generality, we consider a p^+n diode so that essentially all of the current is carried by holes. From the continuity equation,

$$-\frac{\partial J_p}{\partial x} = q\left[\frac{\partial\, \Delta p(x, t)}{\partial t} + \frac{\Delta p(x, t)}{\tau_p}\right] \tag{4-34c}$$

The $\Delta p/\tau$ term is the term that determines the light output, since it is the recombination term. Multiplying by the cross-sectional area and integrating yields

$$-A\int_{J_p(x_n)}^{J_p(\infty)} \partial J_p = [J_p(x_n) - J_p(\infty)]A = i_p = qA\int_{x_n}^{\infty}\left[\frac{\Delta p}{\tau_p}(x, t) + \frac{\partial\, \Delta p}{\partial t}(x, t)\right]dx$$

$$i_p = \frac{Q_p(t)}{\tau_p} + \frac{\partial Q_p(t)}{\partial t} \tag{6-13}$$

$J_p(\infty) = 0$ because all of the current is carried by electrons in n-type material far from the pn junction, and the integrated charge density is the total charge.

When the steady state current, i_0, is turned off at $t = 0$, it will take time for the excess charge to decay. Solving the differential equation, Eq. 6-13, when $i = 0$ yields

$$Q_p(t) = Q_p(0)e^{-t/\tau_p} = i_0\tau_p e^{-t/\tau_p} \tag{6-14}$$

Since the light intensity is proportional to $Q_p(t)/\tau_p$, it will also exponentially decay. It should be noted that the charging of the depletion layer as the forward bias voltage decays has been neglected. In a more sophisticated analysis one would also have to consider the electron transfer from the n- to the p-type region as the depletion layer grew.

A voltage at the junction exists even though the current has been turned off because there is an excess charge. We will now find that voltage as a function of time; it is often called the turn-off transient. To find $V(t)$, we find the relationship between $Q_p(t)$ and $\Delta p(x_n, t)$ and apply it to

$$\Delta p(x_n, t) = p_{n0}[e^{qV(t)/kT} - 1] \simeq p_{n0}e^{qV(t)/kT} \tag{4-69c}$$

In the quasi-steady state approximation it is assumed that the excess charge is removed only by recombination; diffusion and charging and discharging of the junction capacitance are ignored. Thus, since the rate of decay in the volume element, Adx, is proportional to the excess charge in that element, and the initial excess charge density is given by Eq. 4-39, the time-dependent excess charge density is given by

$$\Delta p(x, t) = \Delta p(x_n, t)e^{-(x-x_n)/L_p} \tag{6-15}$$

This is illustrated in Fig. 6-17. Now, from the definition of $Q_p(t)$.

$$Q_p(t) = qA \int_{x_n}^{\infty} \Delta p(x_n, t)e^{(x-x_n)/L_p} = qAL_p \, \Delta p(x_n, t) \tag{6-16}$$

Solving Eq. 6-16 for $\Delta p(x_n, t)$, substituting it into Eq. 4-69c, and using the expression for $Q_p(t)$ given by Eq. 6-14 yields

$$V(t) = \frac{kT}{q} \ln \left(\frac{i_0 \tau_p e^{-t/\tau_p}}{qAL_p p_{n0}} + 1 \right) \tag{6-17}$$

The fact that the lingering excess charge forward biases the junction also is responsible for the reverse recovery transient illustrated in Fig. 6-18. Consider the case where $R_f \ll R \ll R_r$ in Fig. 6-18a. When the forward bias, V_0 (this is not the contact potential), is applied, the current through the circuit is V_0/R, since the forward bias resistance is much smaller than R. So long as there is excess charge, the junction will remain forward biased even though a reverse bias, $-V_0$, has been applied. Thus, the initial reverse current is $-V_0/R$. This current persists until excess charge, and therefore the light intensity, decays to zero. The diode resistance now steadily increases as the junction becomes steadily more reverse biased, and the current through the loop approaches $-V_0/R_r$. The time it takes to drive the light intensity to zero is called the storage delay time, t_{sd}, and its value is $\sim\tau/4$.

6-5.2 Analog Signals

The charging and discharging of the excess charge can be represented by a capacitance, which is called the diffusion capacitance. Again, a p^+n diode will be

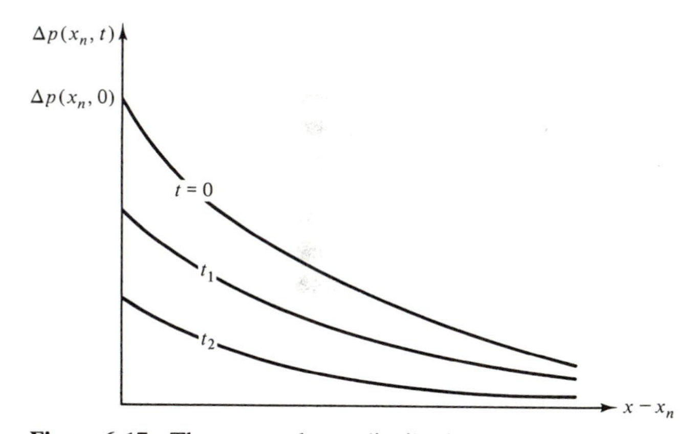

Figure 6-17 The excess charge distribution predicted by the quasi-steady state model plotted for different times.

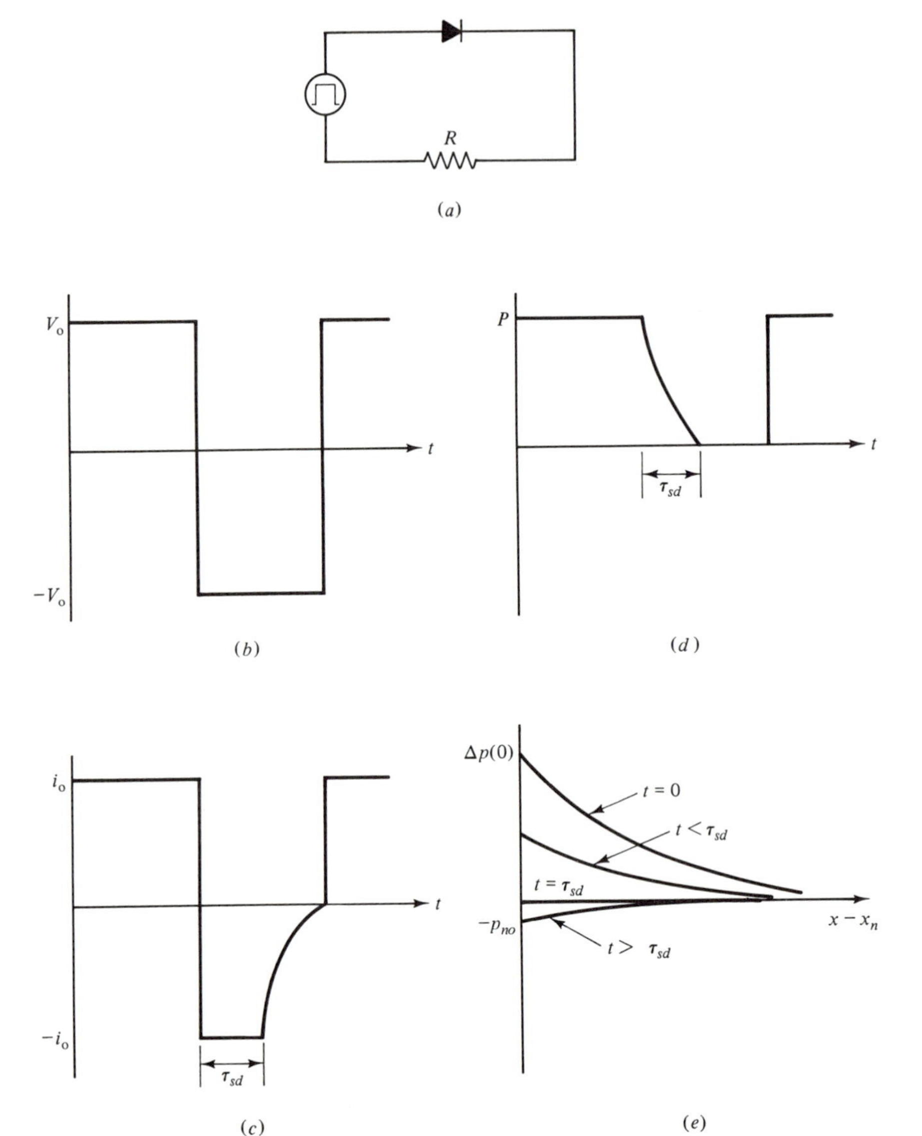

Figure 6-18 (*a*) The circuit used to illustrate the storage delay time. The (*b*) applied voltage, (*c*) current through the circuit, and (*d*) power output of an LED subjected to rectangular voltage pulses plotted as a function of time. (*e*) The quasi-steady state excess charge distribution plotted as a function of distance for different times.

considered. We will only examine small ac signals superimposed on a dc forward bias, and the quasi-steady state approximation will be used.

First, Eq. 4-69c is broken down into its dc and ac components, and we assume that $qV_{ac} < kT$. Thus

$$\Delta p(x_n, t) \simeq p_{n0}e^{q(V_{dc}+V_{ac})/kT} \simeq p_{n0}e^{qV_{dc}/kT}\,\frac{1 + qV_{ac}}{kT} \tag{6-18}$$

Substituting Eq. 6-18 into Eq. 6-13 and assuming the steady state approximation given by Eq. 6-16 yields

$$i_p = \frac{qAL_p}{\tau_p} p_{n0} e^{qV_{dc}/kT}\left(1 + \frac{qV_{ac}}{kT} + \frac{q\tau_p}{kT} + \frac{\partial V_{ac}}{\partial t}\right) \tag{6-19a}$$

which can also be written

$$i_p = i_{dc} + \left(\frac{i_{dc}q}{kT}\right)V_{ac} + \left(\frac{\tau_p i_{dc} q}{kT}\right)\frac{\partial V_{ac}}{\partial t} \tag{6-19b}$$

or

$$i_p = i_{dc} + GV_{ac} + C_d\frac{\partial V_{ac}}{\partial t} \tag{6-19c}$$

Again note that the capacitance is proportional to the response time of the diode.

The amplitude of the ac response is a function of the frequency, and the bandwidth is $\sim 1/\tau_p$. To show this we again neglect the effects of the junction capacitance; thus, Eq. 6-13 can be used. If we assume that the steady state excess charge is given by

$$Q_{p0}(t) = Q_{p0}e^{j(\omega t - \phi)} \tag{6-20}$$

when the current is $i = i_0 \exp(j\omega t)$, then Eq. 6-13 becomes

$$i_0 = \left(\frac{Q_{p0}}{\tau_p} + j\omega Q_{p0}\right)e^{-j\phi}$$

Thus

$$|Q_{p0}| = \frac{i_0\tau_p}{[1 + (\omega\tau_p)^2]^{1/2}} = \frac{Q_{pst}}{[1 + (\omega\tau_p)^2]^{1/2}} \tag{6-21a}$$

where Q_{pst} is the static excess charge when $\omega = 0$, and

$$\tan\phi = \omega\tau_p \tag{6-22}$$

Since the emitted power is proportional to the excess charge, Eq. 6-21a can also be written as

$$P = \frac{P_{st}}{[1 + (\omega\tau)^2]^{1/2}} \tag{6-21b}$$

This is illustrated in Fig. 6-19.

6-5.3 Decay Time

From

$$\tau_p = \frac{1}{Bn_{n0}} \tag{4-24}$$

for a band-to-band transition, it is obvious that the recombination time can be decreased by increasing the doping. For GaAs, $B = 7.2 \times 10^{-10}$ cm^3/sec; hence, for a doping level of 10^{18} cm^{-3}, $\tau_p = 1.4 \times 10^{-9}$ sec. The decay time for donor to valence band or conduction band to acceptor transitions is typically an order of magnitude slower.

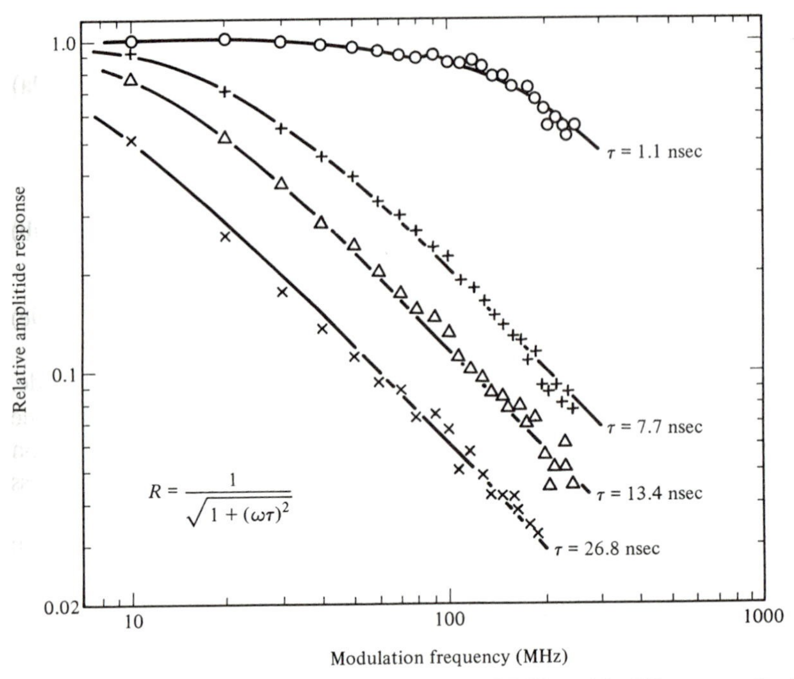

Figure 6-19 The relative frequency response of LEDs with different carrier lifetimes. [From J. P. Wittke, M. Ettenberg, and H. Kressel, *RCA Review,* **37** (159), 1976.]

Decreasing the radiative recombination time by increasing the doping level has the additional advantage of increasing the device efficiency (see Eq. 6-4a). This is illustrated in Fig. 6-20. However, for doping levels greater than $\sim 10^{18}$ cm^{-3} the efficiency decreases with increasing doping. This results from the dopants forming complexes or precipitates or both, which substantially reduces the non-radative recombination rate, τ_{nr}.

The radiative recombination rate can also be increased by using high current levels. Recall from Section 4-4.2 that Eq. 4-24 was developed for low-carrier concentrations. When $\Delta p \gg n_{n0}$, the recombination equation is

$$\frac{\partial\, \Delta p}{\partial t} \simeq -B\, \Delta p^2 \tag{6-23}$$

Thus

$$\tau_p = \frac{1}{B\, \Delta p} \tag{6-24}$$

For steady state conditions to prevail, the carriers must be injected as fast as they recombine. Making the approximation that the carriers are confined to a length, L_p, and they all recombine every τ_p sec, the injected carriers per unit area, J_p/q, must be

$$\frac{J_p}{q} = \Delta p\, \frac{L_p}{\tau_p} \tag{6-25}$$

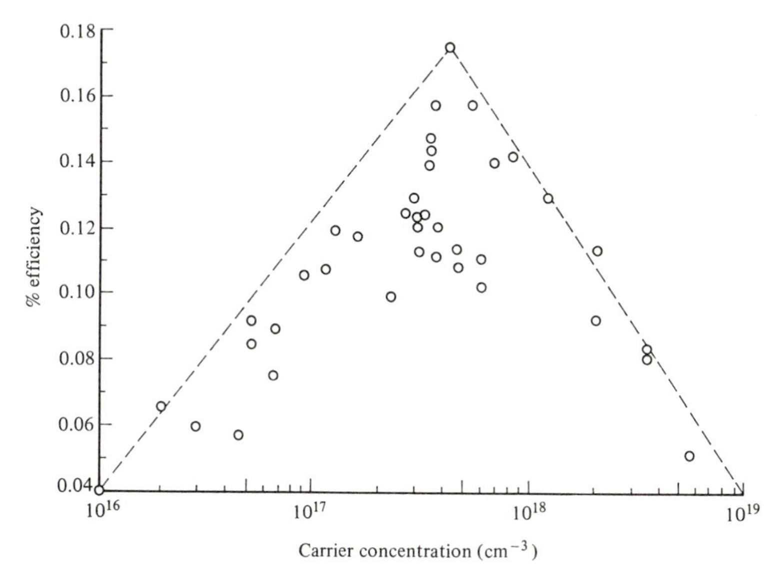

Figure 6-20 The effect of the doping level on the external quantum efficiency of GaAs LEDs. (From E. W. Williams and R. Hall, *Luminescence and the Light Emitting Diode,* Copyright © 1978, Pergamon Press, Oxford, England.)

Solving for Δp and substituting it into Eq. 6-24 yields

$$\tau_p = \left(\frac{qL_p}{BJ_p}\right)^{1/2} \tag{6-26}$$

As was true for increased doping levels, increasing the current increases the bandwidth. This is due to the heating caused by the increased current level.

We end this chapter by noting that the recombination time can be reduced further by confining the excess charge to a layer that is narrower than the diffusion length, L_p. This can be done by depositing a semiconductor film with an energy gap larger than that of the thin ($<L_p$) active semiconductor. As we will learn in the next chapter, confining the excess charge by using a heterostructure revolutionized semiconductor laser technology by greatly reducing the threshold current necessary to achieve lasing.

EXAMPLE 6-5

(a) Compute Δn for a pn^+ GaAs diode at room temperature using both the low- and high-injection formulas for τ_n when $n_a = 10^{18}$ cm^{-3}, $\mu_n = 2 \times 10^3$ cm^2/V·sec, the cross-sectional area, $A = 10^{-4}$ cm^2, and $i = 100$ and 400 mA. **(b)** Using the low-injection formula for τ_n, compute the diffusion length, L_n. **(c)** If the charge is confined by a heterostructure to a length, d, that is one-tenth the value of L_n found in part b, find Δn for the two currents. **(d)** For the conditions described in part c, calculate τ_n for the low- and high-injection conditions for the two currents. **(e)** If the emission peak width at 100 mA is $\Delta\lambda = 225$ Å, and it is $\Delta\lambda = 300$ Å at $i =$ 400 mA, find the frequency bandwidth for the two emission peaks centered at 0.865 μm.

(a) For low injection

$$\Delta n = \frac{J_n \tau_n}{qL_n} = \frac{J_n \tau_n}{q\sqrt{D_n \tau_n}} = \frac{J_n}{q}\left(\frac{q\tau_n}{kT\mu_n}\right)^{1/2} = \frac{i/A}{(qkT\mu_n B n_a)^{1/2}}$$

For $i = 100$ mA

$$\Delta n = \frac{10^{-1}/10^{-4}}{[(1.6 \times 10^{-19})(1.38 \times 10^{-23})(3 \times 10^2) \times (2 \times 10^3)(7.21 \times 10^{-10})(10^{18})]^{1/2}}$$

$$= 3.24 \times 10^{16}\ \text{cm}^{-3}$$

For $i = 400$ mA

$$\Delta n = (4)3.24 \times 10^{16} = 1.30 \times 10^{17}\ \text{cm}^{-3}$$

Thus the low-injection conditions prevail although for $i = 400$ mA the conditions are beginning to go into transition from low- to high-level conditions.

For high-level injection

$$\Delta n = J_n\left(\frac{\tau_n}{qkT\mu_n}\right)^{1/2} = \frac{i/A}{(qkT\mu nB\ \Delta n)^{1/2}}$$

$$\Delta n = \left[\frac{(i/A)^2}{qkT\mu_n B}\right]^{1/3}$$

$$\Delta n = \left[\frac{(10^{-1}/10^{-4})^2}{(1.6 \times 10^{-19})(1.38 \times 10^{-23})(3 \times 10^2)(2 \times 10^3)(7.21 \times 10^{-10})}\right]^{1/3}$$

$$= 1.01 \times 10^{17}\ \text{cm}^{-3}$$

For i = 400 mA

$$\Delta n = 1.01 \times 10^{17}(4)^{2/3} = 2.55 \times 10^{17}\ \text{cm}^{-3}$$

(b)

$$L_n = (D_n \tau_n)^{1/2} = \left(\frac{kT}{q}\frac{\mu_n}{Bn_a}\right)^{1/2}$$

$$= [(2.59 \times 10^{-2})(2 \times 10^3)/(7.21 \times 10^{-10})(10^{18})]^{1/2}$$

$$= 2.68 \times 10^{-4}\ \text{cm} = 2.68\ \mu\text{m}$$

(c) For low-level injection
For $i = 100$ mA

$$\Delta n = \frac{J_n \tau_n}{qd} = \frac{i/A}{q\, dBn_a}$$

$$= \frac{10^{-1}/10^{-4}}{(1.6 \times 10^{-19})(0.268 \times 10^{-4})(7.21 \times 10^{-10})(10^{18})}$$

$$= 3.24 \times 10^{17} \text{ cm}^{-3}$$

For i = 400 mA

$$\Delta n = (4)3.24 \times 10^{17} = 1.30 \times 10^{18} \text{ cm}^{-3}$$

For high-level injection

$$\Delta n = \left(\frac{J_n \tau_n}{qd}\right)^{1/2} = \left(\frac{i/A}{qdB\,\Delta n}\right)^{1/2}$$

For i = 100 mA

$$\Delta n = \left(\frac{i/A}{q\,dB}\right)^{1/2} = \left[\frac{10^{-1}/10^{-4}}{(1.6 \times 10^{-19})(0.268 \times 10^{-4})(7.21 \times 10^{-10})}\right]^{1/2}$$

$$= 5.69 \times 10^{17} \text{ cm}^{-3}$$

For i = 400 mA

$$\Delta n = (4)^{1/2} 5.69 \times 10^{17} = 1.14 \times 10^{18} \text{ cm}^{-3}$$

(d) For low-level injection
For i = 100 mA and 400 mA

$$\tau_n = \frac{1}{Bn_a} = \frac{1}{(7.21 \times 10^{-10})(10^{18})} = 1.39 \times 10^{-9} \text{ sec}$$

For high-level injection:
For i = 100 mA

$$\tau_n = \left(\frac{qd}{BJ_n}\right)^{1/2} = \left[\frac{(1.6 \times 10^{-19})(2.68 \times 10^{-5})}{(7.21 \times 10^{-10})(10^3)}\right]^{1/2} = 2.44 \times 10^{-9} \text{ sec}$$

For i = 400 mA

$$\tau_n = 2.44 \times \frac{10^{-9}}{2} = 1.22 \times 10^{-9} \text{ sec}$$

Thus, one can say that for the GaAs diode doped to 10^{18} cm^{-3} high-injection conditions exist when $d/J_n < 6 \times 10^{-9}$ cm^3/A. For lower doping levels this value is, of course, larger.

(e) $\nu = \dfrac{c}{\lambda}$ or $|\Delta\nu| = \dfrac{c\,\Delta\lambda}{\lambda^2}$

$$\Delta\nu(225) = \frac{(3 \times 10^8)(2.25 \times 10^{-8})}{(0.865 \times 10^{-6})^2} = 9.03 \times 10^{12} \text{ sec}^{-1}$$

$$\Delta\nu(300) = (9.03 \times 10^{12})\left(\frac{3}{2.25}\right) = 1.20 \times 10^{13} \text{ sec}^{-1}$$

READING LIST

1. F. F. Morehead, Jr., "Light Emitting Semiconductors," *Scientific American,* May 1967, pp. 108.
2. R. G. Seippel, *Optoelectronics.* Reston, VA: Reston Publishing Co., 1981, Chapters 1, 2, 4, and 6.
3. B. G. Streetman, *Solid State Electronic Devices,* 2d Ed. Englewood Cliffs, NJ: Prentice-Hall, 1980, Chapter 6.
4. J. Wilson and J. F. B. Hawkes, *Optoelectronics: An Introduction.* Englewood Cliffs, NJ: Prentice-Hall, 1983, Chapter 4.
5. E. W. Williams and R. Hall, *Luminescence and the Light Emitting Diode.* Oxford, England: Pergamon Press, 1978.
6. A. A. Bergh and P. J. Dean, *Light Emitting Diodes.* Oxford, England: Clarendon Press, 1976.
7. T. S. Moss, G. J. Burrell, and B. Ellis, *Semiconductor Opto-Electronics.* London, England: Butterworth, 1973, Chapters 3, 7, 8, and 11.
8. H. C. Casey, Jr., and M. B. Panish, *Heterostructure Lasers: A. Fundamental Principles, B. Materials and Operating Characteristics.* New York: Academic Press, 1978, Chapters 4 to 6.
9. H. Kressel (Ed.), *Topics in Applied Physics:* Vol. 39, *Semiconductor Devices for Optical Communication.* New York: Springer-Verlag, 1982. Chapters 2, 5, and 10.
10. H. Kressel and J. K. Butler, *Semiconductor Lasers and Heterojunction LEDs.* New York: Academic Press, 1977, Chapters 1, 2, and 9 to 14.

PROBLEMS

6-1. A GaAlAs LED emits at 0.85 μm with a power of 0.50 mW when a current of 150 mA flows through the diode and a voltage of 1.50 V is applied.

(a) Find the overall efficiency.

(b) What is the external quantum efficiency if the internal quantum efficiency is 0.10?

6-2. A silicon-doped GaAs LED is forward biased to 1.0 V.

(a) Find the output light intensity if the n side is doped with 5×10^{17} donors and the p side has 10^{15} acceptors, $\tau_n = \tau_p = 10^{-8}$ sec, $n_i = 1.1 \times 10^7$ cm^{-3}, the internal quantum efficiency, $n_i = 0.1$, and the peak intensity is at 1 μm. Assume ideal diode behavior.

(b) How far above the valence band is the silicon hole trap located?

(c) For the same doping levels and bias, how does the junction capacitance for a GaAs diode compare with that of a silicon diode?

(d) How do the capacitances change with temperature? Explain.

6-3. **(a)** If the intensity of the sunlight is 0.10 mW/cm^2, how much current must be supplied to a GaAs diode that is 10 percent efficient and has an applied voltage of 1.2 V such that its intensity is equal to that of the sun? The active area is 250 μm on a side.

(b) Repeat (a) for a GaP diode that is 0.2 percent efficient, has an applied voltage of 1.8 V, and the active region has a diameter of 400 μm.

(c) Calculate the number of photons emitted per second from each device if the GaAs diode emits at 960 nm and the GaP diode emits at 570 nm.

6-4. (a) Using the data in Fig. 6-2*a*, make plots of P versus $1/T$ on semilog paper for $i = 250$ and 350 mA. From your graphs find the activation energy and compute what the output power would be at -10 and 100° C.

(b) On the same graph repeat part a using the data in Fig. 6-2*b*.

6-5. (a) A nitrogen-doped GaP LED emits at 570 nm. Find the depth of the isoelectronic nitrogen trap.

(b) A zinc- and oxygen-doped GaP LED emits at 699 nm. Find the trap depth of the Zn-O complex.

6-6. Find the range of the depth of the silicon complex in GaAs if the range of light emitted is 910 to 1020 nm.

6-7. A $GaAs_{1-x}P_x$ LED emits at 649 nm via a band-to-band transition.

(a) Find the composition of this alloy assuming E_g varies linearly with x.

(b) Find the composition if the bowing parameter, $b = 0.21$.

6-8. $GaAs_{0.35}P_{0.65}$ LEDs doped with nitrogen emit, respectively, at 632 and 609 nm.

(a) Find the trap depths assuming E_g varies linearly with the composition.

(b) Find the trap depths if the bowing parameter, $b = 0.21$.

6-9. In Zn-O doped GaP diodes the efficiency decreases when the isoelectronic traps begin to fill, since the number of recombination centers an electron can collide with is $n_r - \Delta n_r$ where Δn_r is the number of filled recombination centers.

(a) If the rate at which electrons fall into the traps is $B_r(n_r - \Delta n_r)\Delta n$ and the rate at which they leave the traps is $\Delta n_r/\tau_{rb}$, find a steady state expression for Δn_r in terms of B_r, n_r, τ_{rb}, and Δn.

(b) If the rate of injection of eletrons into the p-type material is J_n/qL_n, the rate at which the electrons recombine radiatively is $B_r(n_r - \Delta n_r)\Delta n$, and the rate at which they nonradiatively recombine is $B_{nr}p_{p0}\,\Delta n$, find the quadratic, $A\,\Delta n^2 + B\,\Delta n + C$, for the steady state value of Δn. Express your answer in terms of B_r, n_r, B_{nr}, p_{p0}, τ_{rb}, and J_n/qL_n.

6-10. (a) For the quadratic found in part b of the previous problem, find an expression for Δn in terms of n_r and x if $\tau_{rb} = 1/(9B_rn_r)$, $B_{nr}P_{p0} = B_rn_r/9$, and $J_n/qL_n = xB_rn_r^2$.

(b) For these conditions find an expression for Δn_r in terms of Δn and n_r.

(c) For these conditions find an expression for the internal quantum efficiency in terms of Δn_r and n_r.

(d) Compute values of Δn, Δn_r, and η_i for $x = 0$, 5, 10, 12, 15, and 20.

6-11. Find the composition of $Ga_{1-x}Al_xAs$ that emits at 675 nm.

(a) Assume that E_g varies linearly with x.

(b) Use the bowing parameter, $b = 0.326$.

6-12. The direct and indirect energy gaps for GaAs are 1.43 and 1.86 eV, and the direct and indirect energy gaps for AlAs are 2.96 and 2.16 eV.

(a) Find the composition and the energy gap for which the direct and the indirect gaps are equal assuming that E_g varies linearly with the composition.

(b) Repeat part a if the bowing parameter for the direct gap is 0.326 and is 0.219 for the indirect gap.

(c) Graphically verify your results of parts a and b by plotting E_g versus x for both the direct and indirect gaps. Put all four plots on the same graph.

6-13. **(a)** Find the composition for which $In_{1-x}Ga_xAs$ is lattice matched to InP.
(b) What is the energy gap for the lattice-matched ternary if $b = 0$? if $b = 0.41$?
(c) Calculate the wavelengths associated with each E_g computed in part b.

6-14. **(a)** Find the composition and the energy gap of $In_{1-x}Ga_xP$ for which the direct and indirect bandgap are equal assuming the E_g vary linearly.
(b) Repeat part a using the bowing parameters, b(direct) = 0.76 and b(indirect) = 0.

6-15. **(a)** Find the composition for which InAlAs is lattice matched to InP.
(b) Find E_g for this composition assuming E_g varies linearly with the composition.
(c) Find the composition for which E_g(direct) = E_g(indirect) assuming that both energy gaps vary linearly. Also compute E_g.

6-16. Find the lattice-matched composition to InP of $In_{1-x}Ga_xAs_{1-y}P_y$ that will emit at the zero dispersion wavelength of 1.30 μm.
(a) Assume that the energy gap varies linearly with x or y.
(b) Use the bowing parameters b(InGaAs) = 0.41, b(InAsP) = 0.10, b(InGaP) = 0.76, and b(GaAsP) = 0.21. Assume that $E_g = (1 - y)E_g(\text{InGaAs}) + yE_g(\text{InGaP})$.
(c) Use the bowing parameters in part b and assume that $E_g = (1 - x)E_g(\text{InAsP}) + xE_g(\text{GaAsP})$.

6-17. **(a)** Find the composition of $In_{1-x}Ga_xAs_{1-y}P_y$ lattice matched to GaAs that has a direct energy gap of 1.75 eV. Assume that E_g varies linearly with x or y.
(b) Use the bowing parameters given in Problem 6-16 and assume that $E_g = (1 - y)E_g(\text{InGaAs}) + yE_g(\text{InGaP})$.
(c) Use the bowing parameters and assume that $E_g = (1 - x)E_g(\text{InAsP}) + xE_g \times (\text{GaAsP})$.

6-18. **(a)** Derive the equations for the a_0 versus E_g diagram for $In_{1-x}Ga_xAs_{1-y}P_y$ assuming that both a_0 and E_g vary linearly.
(b) Repeat part a using the bowing parameters listed in Problem 6-16. Assume that the indirect gaps vary linearly.
(c) Using these equations, construct the diagram.

6-19. Find the relationship between x and y for which $In_{1-x}Ga_xAs_{1-y}P_y$ is lattice matched to GaAs. Assume E_g varies linearly with x or y.

6-20. Compute the bandwidths, $\Delta\nu$ and $\Delta\lambda$, for the GaAs LEDs shown in Fig. 6-16 for doping levels of 10^{17}, 10^{18}, and 10^{19} cm^{-3}.

6-21. **(a)** At $t = 0$ a constant current, i_0 is applied to the diode. Use Eq. 6-13 to solve for the charge buildup as a function of time.
(b) The response found in part a is the response to the input, $i_0u(t)$ where $u(t)$ is the step function. The response to a rectangular pulse that is turned on for a time, T, turned off for a time, T, turned on for a time, T, and so forth, is the linear response to the input

$$i_0 \sum_{0}^{m} u(t - 2nT) - i_0 \sum_{0}^{m-1 \text{ or } m} u[t - (2n + 1)T]$$

Find the response for $2mT \leq t \leq (2m + 1)T$ when m is a large integer.
(c) Find the maximum value of Q during this turned on cycle. What is the magnitude of this maximum value when $T/\tau \rightarrow \infty$, $T/\tau \rightarrow 0$, and $T = \tau$?
(d) Find the response for $(2m + 1)T \leq t \leq (2m + 2)T$ when n is a large integer.

(e) Find the minimum value of Q during this turned off cycle. What is the magnitude of this minimum value when $T/\tau \rightarrow \infty$, $T/\tau \rightarrow 0$, and $T = \tau$?

(f) Compute $Q_{max} - Q_{min}$ for $T/\tau \rightarrow \infty$, $T/\tau \rightarrow 0$, and $T = \tau$ using the values found in parts c and e.

6-22. Find the storage delay time, t_{sd}, for a p^+n diode using the quasi-steady state approximation. The rectangular wave has amplitudes of $\pm V_0$.

6-23. A GaAs LED fabricated from quite lightly doped materials has an effective recombination region of width 0.1 μm. If it is operated at a current density of 2×10^7 A/m^2, estimate the modulation bandwidth that can be expected.

Chapter 7

Laser Diodes

7-1 INTRODUCTION

One way laser light differs from light emitted from an LED is that it is coherent. That is, the emitted light has the same phase. This is because laser light is generated by stimulated emission. This type of emission is created by the electromagnetic waves driving electrons in the conduction band to recombine with holes in the valence band.

In order for there to be more stimulated emission than the usual spontaneous emission, the stimulating driving force must be greatly increased. This is done by increasing the amplitude of the exciting electromagnetic waves using mirrors. The stimulated light must also be created faster than it is absorbed. This is accomplished by having more carriers in the upper level than in the lower level of the lasing transition, and this phenomenon is called population inversion.

The population of the upper level increases as the current increases until lasing occurs at the threshold current. For larger currents the population inversion remains fixed at its threshold value. The magnitude of the population inversion at threshold is determined by the absorption losses other than absorption at the lasing transition and the fraction of light transmitted by the mirrors. The population inversion does not increase above its threshold value as the current is increased because the recombination time decreases.

Another way that laser light differs from light emitted by an LED is that it is more directional. Most of the light is emitted normal to the mirrors, since only these rays can be continually reflected back and forth without escaping out the sides. The small divergence of the laser beam allows more of its light to be coupled into an optical fiber.

The mirrors not only select out those rays propagating parallel to the laser axis, they also select out those waves with wavelengths that are half integral multiples of the cavity length by setting up standing wave patterns through constructive interference. Other waves are destructively interfered with so that the mirrors act as a filter that allows only those longitudinal modes with narrow bandwidths and wavelengths that are half integral multiples of the cavity length to pass. The narrow bandwidths of these modes can essentially eliminate wavelength dispersion.

As was the case with optical fibers, transverse modes also exist. For optical fiber applications it is desirable to eliminate the higher order modes because they can cause nonlinearities in the $P - i$ curve to form, and their intensity is large away from the beam center. This makes it more difficult to couple light into a fiber.

Homojunction lasers are made from heavily doped p- and n-type direct bandgap materials and emit at $\lambda \simeq hc/E_g$. Heavily doped materials make it easier for the injected carriers to achieve population inversion. The threshold current density, J_{th}, is relatively large even at 77 K, but it is inordinately large at room temperature because it increases exponentially with the temperature. J_{th} is so large at room temperature that homojunction lasers cannot be operated continuously; they can only be operated in the pulsed mode.

Above i_{th} the power inside the cavity increases linearly with i as does the power emitted by the laser. The emitted power also depends on the reflectivity of the mirrors, and for a given current there is a reflectivity for which the power is a maximum.

Laser diodes for optical fiber applications are usually fabricated with the stripe contact geometry. The top contact is formed by opening a strip 10 to 15 μm wide in an oxide deposited on the top surface before the metal contact is evaporated on. This geometry confines the light, which is emitted normal to the current flow, to a width narrower than the diameter of a multimode fiber.

The mirrors are usually formed merely by cleaving the wafer parallel to its cleavage plane. This process is inexpensive, but the reflectivity is relatively small, 0.3 to 0.35. The reflectivity can be enhanced by depositing either gold or dielectric mirrors.

Lasers are operated at relatively high currents so that they generate a substantial amount of heat. Thus, they must be heat-sinked to stablize their temperature. This is especially important in view of the strong temperature dependence of J_{th}.

Semiconductor heterojunctions like metal-semiconductor junctions are mathematically described by measuring the energies from the universal ground of a free electron at rest.

Heterojunctions differ from homojunctions in that there are discontinuities at the valence and conduction bands. A negative discontinuity in the conduction bands of a pn junction reduces the barrier height for electron injection into the p-type material. A positive discontinuity in the conduction bands of a pp heterojunction generates a barrier to the flow of electrons, and is therefore capable of confining the electrons to a thickness that is less than a diffusion length. In this way J_{th} and its temperature dependence can be greatly reduced below that for a homojunction. Thus, continuous wave heterojunction lasers can be fabricated.

A single heterostructure (SH) laser contains a *pp* or an *nn* heterojunction to respectively confine electrons or holes to the intermediate layer. By reducing the thickness of the region containing the excess minority carriers by a factor of 10, one reduces J_{th} by about the same amount.

In a double heterostructure (DH) laser both junctions are heterojunctions. The outside layers, which usually have the same composition, act as wave guides much like a cladding on an optical fiber because they have a smaller index of refraction than the intermediate layer. Both the difference in the index of refraction and the thickness of the intermediate layer determine what fraction, Γ, of the stimulated light is confined to this layer.

Materials systems that are often used are the GaAlAs/GaAs/GaAlAs and InP/InGaAsP/InP systems. The advantage of the GaAlAs/GaAs system is that GaAs and AlAs have almost the same lattice parameter so that GaAlAs is essentially lattice matched to GaAs over the entire composition range. The advantage of the InP/InGaAsP system is that it can be tuned to zero dispersion wavelength for the fiber at 1.30/μm and the fiber absorption minimum at 1.55 μm. However, the composition must be carefully controlled to ensure that the quaternary is lattice matched to InP.

At the present time the GaAlAs/GaAs lasers have lower threshold currents and are more efficient. There are no fundamental reasons why this should be true, and as the technology of the more infantile InP/InGaAsP system develops, it seems likely that these lasers will become at least as efficient.

The modulation of laser light is a complex dynamic process, since there are a number of nonlinearities. The most obvious one is the knee in the $P - i$ curve at threshold, but kinks in the $P - i$ curve due to mode saturation can also be present. For properly designed lasers, however, small analog signals can be transmitted with good fidelity when the laser is dc biased to above threshold. It is essential to heat-sink the diode since the $P - i$ curves are so temperature sensitive.

Digital signals can be sent without much concern for the nonlinearities. However, if an ON-OFF signal is used, considerable delay is introduced as it takes time for the laser to achieve population inversion. This problem can be eliminated by dc biasing the system above threshold, but in so doing, considerable noise is introduced into the photodetector.

7-2 LASER FUNDAMENTALS

7-2.1 Stimulated Emission

One way light emitted from a laser differs from light emitted from an LED is that the laser light is coherent. That is, the photons are in phase. They are in phase because they are created by stimulated emission. This fact is incorporated into the word, laser, which stands for light amplification by stimulated emission of radiation.

The electromagnetic waves themselves are the driving force for stimulated emission. They drive the atomic oscillators so the light emitted by the oscillators has the same phase as the driver.

The larger the magnitude of the driver (the greater the intensity of the stimulating electromagnetic radiation) is, the larger will be the intensity of the stimulated emission; that is, the probability the excited electron will recombine by stimulated emission is greater when the stimulating photon intensity is larger. The intensity is increased by using mirrors that reflect the emitted light back into the semiconductor cavity.

However, light reflected back into the cavity can be lost by absorption. In order for lasing to occur, the stimulating process must dominate the absorbing process. This occurs for transitions between two levels only when the weighted number of electrons in the upper state is larger than it is in the lower state. This phenomenon is called population inversion, and it is accomplished in semiconductors by injecting a large number of electrons into a *p*-type material containing many holes in the valence band (the lower state).

For illustrative purposes consider the two-level system shown in Fig. 7-1. We learned in Section 3-2 that the number of photons absorbed per unit volume per unit time that excite the electron from state 1 to state 2 is proportional to the density of electrons in state 1 and the number of photons with energy $h\nu_{12}$. It is more convenient to consider the photon energy density per unit frequency, $u(\nu_{12})$, than the number of photons, and they can be interchanged because they are proportional. The number of photons absorbed per unit volume per unit time is therefore proportional to the density of electrons in state 1, n_1, and the photon energy density per unit frequency, and the proportionality constant is B_{12}. Thus

$$\text{number of photons absorbed per unit volume per unit time} = B_{12}u(\nu_{12})n_1 \tag{7-1}$$

In Section 3-2 we also determined that the number of photons emitted *spontaneously* per unit volume per unit time is proportional only to the density of electrons in state 2, and the proportionality constant is A_{21}. Thus

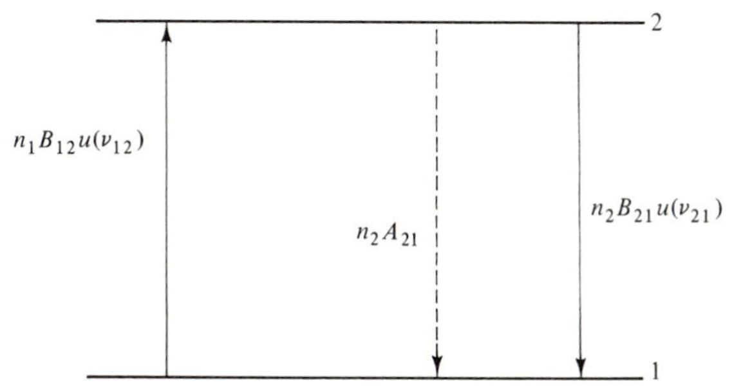

Figure 7-1 A two-level system illustrating absorption and spontaneous (----) and stimulated emission.

number of photons emitted spontaneously per unit volume per unit time

$$= A_{21} n_2 \tag{7-2}$$

The letter, A, is chosen as the proportionality constant to emphasize that spontaneous emission is not a function of the photon energy density.

However, as mentioned above, the stimulated emission does depend on the photon energy density because the electromagnetic waves drive the atomic oscillators that generate stimulated emission. The number of stimulated photons emitted per unit volume per unit time is proportional to the photon energy density and the density of electrons in state 2, and the proportionality constant is B_{21}. Thus

number of stimulated photons emitted per unit volume per unit time

$$= B_{21} u(\nu_{21}) n_2 \tag{7-3}$$

In a more detailed description of lasers in Chapter 9 it will be shown that

$$A_{21} = \frac{8\pi h}{\lambda^3} B_{21} \tag{9-12a}$$

and

$$B_{12} = \frac{\rho_2}{\rho_1} B_{21} \tag{9-11a}$$

where ρ_i is the density of states in level i. Of course, when there are energy bands instead of discrete states, ρ_i is replaced by the density of states function, $\rho_i(\nu)$. Note that Eq. 9-12a shows that the stimulated/spontaneous emission ratio is smaller at shorter wavelengths. Thus, everything else being equal, it is easier to achieve lasing at longer wavelengths. This is why historically masers were built before lasers. The "m" in maser stands for microwave.

For lasing to occur the ratio of stimulated to spontaneous emission must be large. From their ratio

$$\frac{\text{stimulated emission}}{\text{spontaneous emission}} = \frac{B_{21} u(\nu_{21})}{A_{21}} = \frac{1/\tau_{\text{stim}}}{1/\tau_{\text{spon}}} = \frac{\lambda^3}{8\pi h} u(\nu_{21}) \tag{7-4}$$

we see that this happens when the photon energy density is high. The energy density is increased by placing mirrors on opposite ends of the laser that reflect light back into the cavity. Also note in Eq. 7-4 that the ratio is inversely proportional to the recombination times (see Section 3-2.3). Thus the recombination time for a laser is faster than it is for an LED, and it decreases as the photon energy density increases.

The weighted stimulated emission/absorption ratio must also be greater than one for lasing to occur. Using Eq. 9-11a and recognizing that $u(\nu_{12})$ and $u(\nu_{21})$ are the same quantity, we see that their ratio is

$$\frac{\text{stimulated emission}}{\text{absorption}} = \frac{\rho_1 n_2}{\rho_2 n_1} \tag{7-5}$$

Thus, the weighted number of electrons in the upper state must be larger than it is in the lower state for lasing to occur. This is called population inversion.

An approximate relationship between the energy density per unit frequency and the energy density, u, can be found using the rectangular approximation for the peak illustrated in Fig. 3-37. The peak width is $\Delta\nu$ so that

$$u = u(\nu_{12})\Delta\nu \tag{7-6}$$

The normalized peak height is $g(\nu)$, and the area under the curve is one so that

$$g(\nu) = \frac{1}{\Delta\nu} \tag{7-7}$$

To see why lasing will occur only if an optical cavity is used, we turn to Example 7-1.

EXAMPLE 7-1

An LED emits light with a bandwidth of $\Delta\lambda = 40$ nm at 1.0 μm with an intensity of 50 W/cm^2. **(a)** If this is also the light intensity inside of the semiconductor, and the index of refraction of the semiconductor is $n = 3.6$, what is the ratio of stimulated to spontaneous emission? **(b)** To what magnitude must the photon energy density be raised so that the stimulated/spontaneous emission ratio equals one?

$$u(\nu_{12}) = \frac{u}{\Delta\nu} = \frac{u\lambda^2}{c\,\Delta\lambda} = \frac{u\lambda_0^2}{c_0\,\Delta\lambda_0}$$

(a) From Eq. 1-25, $u = I/c = In/c_0$

$$\therefore \text{ratio} = \frac{\lambda^3 u(\nu_{21})}{8\pi h} = \frac{\lambda_0^5 I}{8\pi h n^2 c_0^2\,\Delta\lambda_0}$$

$$= \frac{(10^{-4})^5(5 \times 10^1)}{8\pi(6.625 \times 10^{-34})(3.5)^2(3.0 \times 10^{10})^2(4 \times 10^{-6})}$$

$$= 6.81 \times 10^{-4}$$

(b) For the ratio to equal 1

$$u = \frac{50}{6.81 \times 10^{-4}} = 7.34 \times 10^4 \text{ W/cm}^3$$

This is a large number and can be accomplished only by using mirrors to reflect light back into the cavity.

7-2.2 Optical Gain

In the previous section we established that population inversion has to be achieved for there to be more stimulated emission than there is absorption. This is simply because it is as easy to stimulate the electrons from state 2 to state 1 (emission) as it is to "stimulate" them from state 1 to state 2 (absorption). If there are more electrons in state 2 than there are in state 1, more $2 \rightarrow 1$ than $1 \rightarrow 2$ transitions occur so there is a net gain in the light emitted. In this section we determine what the magnitude of the gain is.

If the optical power flowing into the volume element in Fig. 7-2 is $I(z)\Delta x\ \Delta y$, and the power flowing out is $I(z + \Delta z)\Delta x\ \Delta y$, then the net power absorbed in the element is

$$\text{power absorbed} = I(z)\Delta x\ \Delta y - \left[I(z) + \frac{\partial I(z)}{\partial z}\Delta z\right]\Delta x\ \Delta y$$

$$= \frac{-\partial I(z)}{\partial z}\Delta x\ \Delta y\ \Delta z \qquad (7\text{-}8a)$$

or taking the derivative of

$$I = I_0 e^{-\alpha z} \qquad (3\text{-}18b)$$

yields

$$\text{power absorbed} = \alpha I\ \Delta x\ \Delta y\ \Delta z \qquad (7\text{-}8b)$$

The net power absorbed is also

$$\text{net power absorbed} = [n_1 B_{12} u(\nu_{12}) - n_2 B_{21} u(\nu_{21})]h\nu_{12}\ \Delta x\ \Delta y\ \Delta z \qquad (7\text{-}9)$$

In Example 7-1 we found that

$$u(\nu_{12}) = u(\nu_{21}) = \frac{I}{c\ \Delta\nu} \qquad (7\text{-}10)$$

where $\Delta\nu$ is the emission peak bandwidth. Equating Eqs. 7-8b and 7-9, using Eqs. 9-11a and 9-12a, and remembering that A_{21} is the reciprocal of the spontaneous decay time, τ_{sp}

$$\alpha = \frac{\lambda^2}{8\pi\tau_{sp}\ \Delta\nu}\left(\frac{\rho_2}{\rho_1}n_1 - n_2\right) \qquad (7\text{-}11a)$$

When population inversion occurs and the absorption coefficient is negative, the equation is rewritten

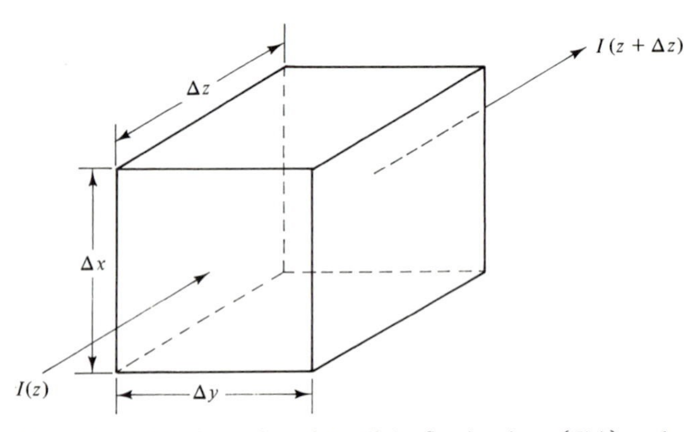

Figure 7-2 The intensity of the light flowing into $\{I(z)\}$ and out of $\{I(z + \Delta z)\}$, the volume element $\Delta x\ \Delta y\ \Delta z$.

$$\alpha_g = \frac{\lambda^2}{8\pi\tau_{sp}\,\Delta\nu}\left(n_2 - \frac{\rho_2}{\rho_1}n_1\right) \tag{7-11b}$$

and α_g is called the gain coefficient.

When a laser is operating in the steady state, the gain must equal the loss. In addition to the "stimulated absorption," which is incorporated into the gain coefficient, there are other losses that include absorption at the mirrors on the ends, absorption in the optical cavity for other than the lasing transition, scattering losses at inhomogeneities in the cavity, diffraction at the mirrors, and light transmitted out of the cavity. The latter loss is the emitted laser light. All but the last loss mechanism are lumped together and represented by an absorption coefficient, α. In semiconductors the dominant absorption mechanism is free carrier absorption. This is caused by the electrons in the conduction band of n-type material or holes in the valence band of p-type material absorbing the light. Since their concentration is proportional to the doping level, α is also proportional to it. The transmission losses are accounted for by the reflection coefficient at each end.

For one round trip in the cavity under steady state conditions

$$\text{gain} = 1 = e^{(\alpha_g-\alpha)L}R_1e^{(\alpha_g-\alpha)L}R_2 \tag{7-12}$$

where L is the cavity length and R_1 and R_2 are the reflection coefficients at normal incidence (see Section 1-4). Only rays normal to the mirrors are continuously reflected back and forth; the other rays eventually escape out the sides of the cavity. Taking the logarithm of both sides and rearranging yields

$$\alpha_g = \alpha - \frac{1}{2L}\ln R_1R_2 \tag{7-13}$$

The magnitude of the population inversion needed to obtain the steady state gain coefficient is called the threshold population inversion concentration, n_t, and is found by equating Eqs. 7-13 and 7-11b and rearranging. Thus

$$n_t = \left(n_2 - \frac{\rho_2}{\rho_1}n_1\right) = \frac{8\pi\tau_{sp}\,\Delta\nu}{\lambda^2}\left(\alpha - \frac{1}{2L}\ln R_1R_2\right) \tag{7-14}$$

Below threshold $[n_2 - (\rho_2/\rho_1)n_1]$ increases approximately linearly with the current as more and more electrons are injected into the conduction band in the p-type material. Also, the light intensity increases slowly with the current as the device is operating as an LED. However, when $[n_2 - (\rho_2/\rho_1)n_1]$ reaches its threshold value, it remains fixed at n_t as the current is increased beyond the threshold current, i_{th}, because it has reached its steady state value. Rather than the population inversion being increased by the additional electrons, the light emission intensity is increased. This is made possible by the fact that the stimulated emission recombination time decreases as the light intensity increases (see Eq. 7-4). For a first approximation one can say that below threshold the injected electrons create light by spontaneous emission and build up the population inversion. Above threshold they generate laser light. This is illustrated in Fig. 7-3. Thus, a laser diode is a more efficient light emitter than an LED.

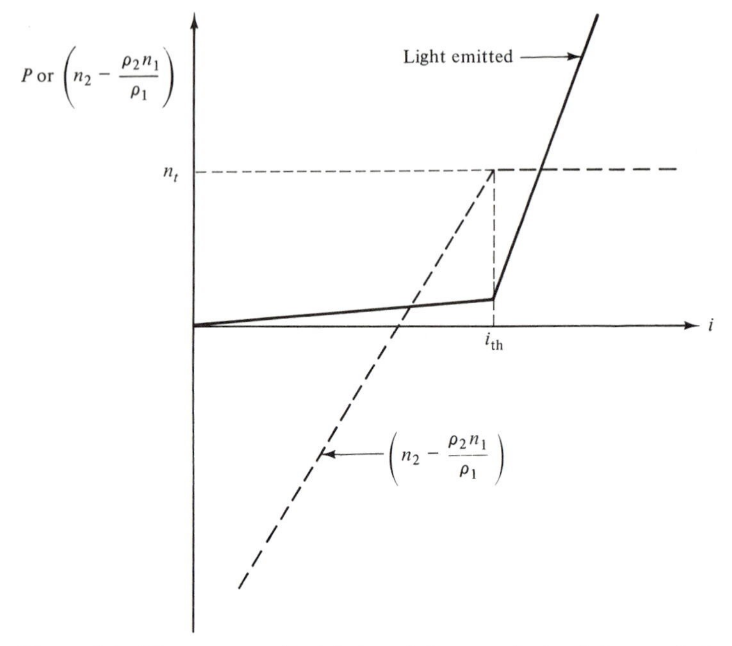

Figure 7-3 The power emitted (—) and the magnitude of the population inversion (----) plotted as a function of the diode current.

EXAMPLE 7-2

Compute the gain constant and the threshold concentration for a GaAs laser if the absorption coefficient is 10 cm^{-1} and the length of the cavity is 500 μm. Light is emitted at 0.84 μm, the bandwidth of the emission peak is 1.5×10^{13} Hz, the spontaneous decay time is 10 nsec, and the index of refraction is $n = 3.6$. The mirrors on either end of the laser are formed by a cleaved GaAs surface.

$$R_1 = R_2 = \left(\frac{n-1}{n+1}\right)^2 = \left(\frac{2.6}{4.6}\right)^2 = 0.32$$

$$\alpha_g = \alpha - \frac{1}{2L} \ln R_1 R_2 = 10 + \frac{1}{0.05} \ln \frac{1}{0.32} = 32.79$$

$$n_t = \frac{8\pi\tau_{sp}\,\Delta\nu\alpha_g}{\lambda^2} = \frac{8\pi(10^{-8})(1.5 \times 10^{13})(32.79)}{(8.4 \times 10^{-5}/3.5)^2} = 1.49 \times 10^{17}\ \text{cm}^{-3}$$

7-2.3 Characteristics of the Emitted Light

Light emitted by a laser has more directionality and a different frequency distribution than an LED because the optical cavity modulates the light. The laser emission spectrum is the product of the portion of the emission peak above threshold times the cavity modulation function.

The light is more directional, since only those rays that are normal to the mirrors are able to build up their energy density to above threshold by continuously being reflected back and forth. Those rays not normal to the mirrors pass out of the cavity before their intensity gets built up.

The directionality, or brightness, is measured in units of power/(area)(steradian). A high radiance edge emitting LED with an output power of 10 mW can have a brightness of $\sim 10^3$ mW/cm$^2 \cdot$ sr, whereas a laser with the same output power typically has a brightness of 10^5 mW/cm$^2 \cdot$ sr. Recall from Section 2-4.3 that only light that is almost parallel to the optical fiber axis can be propagated. Thus, almost two orders of magnitude more light can be injected into an optical fiber using a laser than can be by the best LED with the same output power. It should be noted that although light from a laser diode is much more directional than it is for an LED, it is much less directional than it is for other lasers.

Being parallel to the axis of the optical cavity is a necessary, but not sufficient, condition for the waves to be emitted. The cavity must also be a half integral number of wavelengths long. When this condition is achieved, the multiply reflected waves constructively interfere with each other. Each acceptable wave is called a longitudinal mode, and the order of the mode, m, is given by

$$m = \frac{2L}{\lambda} \tag{7-15}$$

The $m = 1$, 2, and 3 modes are illustrated in Fig. 7-4.

A wave that has a phase change of $m\pi + d\,\delta$ after one pass will have a phase change of $M(m\pi + d\,\delta)$ after M passes. For

$$M = \frac{\pi}{d\,\delta} \tag{7-16}$$

the Mth pass will destructively interfere with the first pass. If the amplitude of the Mth pass is the same as it is for the first pass—and it will be if the mirrors

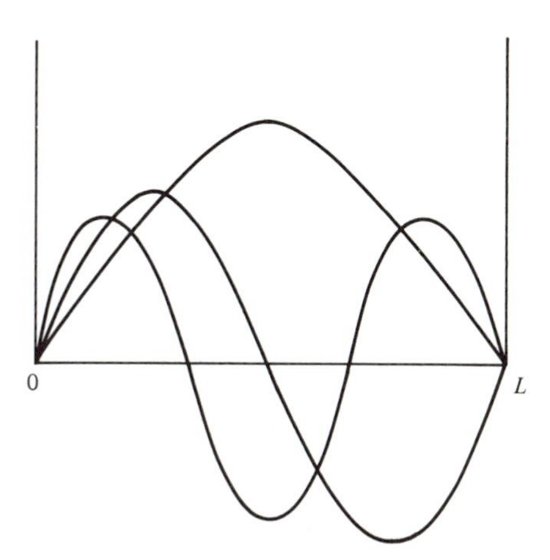

Figure 7-4 The first three modes (standing waves) in a cavity length, L.

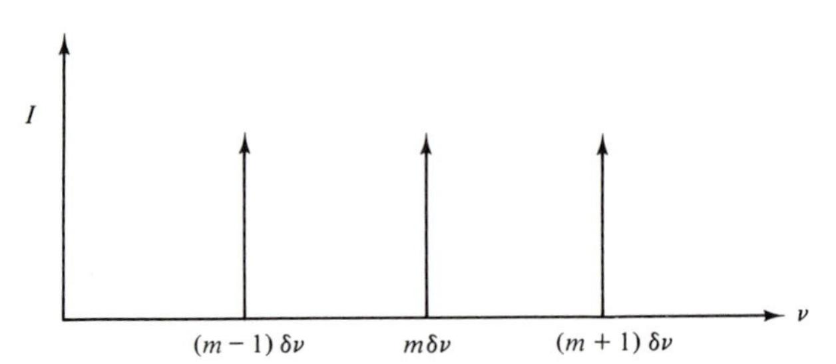

Figure 7-5 The frequency distribution of the light passed by the cavity when the mirrors are perfectly reflecting and there are no absorption losses.

are perfectly reflecting ($R = 1$) and there are no absorption losses ($\alpha = 0$)—the destructive interference will be complete. Thus, for any wave for which $d\,\delta \neq 0$, there will be complete destructive interference so that the optical cavity acts like a filter that will pass impulses of light for which

$$\nu = \frac{c}{\lambda} = \frac{mc}{2L} \tag{7-17}$$

The frequency separation between the impulses, $\delta\nu$, is

$$\delta\nu = \frac{c}{2L} \tag{7-18}$$

and this is illustrated in Fig. 7-5.

When there are transmission and absorption losses, the amplitude of the wave after the Mth pass will not be as large as it is for the first pass. Thus, the destructive interference will not be complete. This causes the impulses for the ideal cavity modulation function to become peaks with finite bandwidths of $d\nu^*$, as is illustrated in Fig. 7-6. $d\nu^*$ is smaller when the losses are smaller since, for a given M, the amplitude of the wave on the Mth pass will be larger and therefore will more strongly destructively interfere with the first pass wave.

A more detailed mathematical description of the effects of the optical cavity is given in Section 8-4.

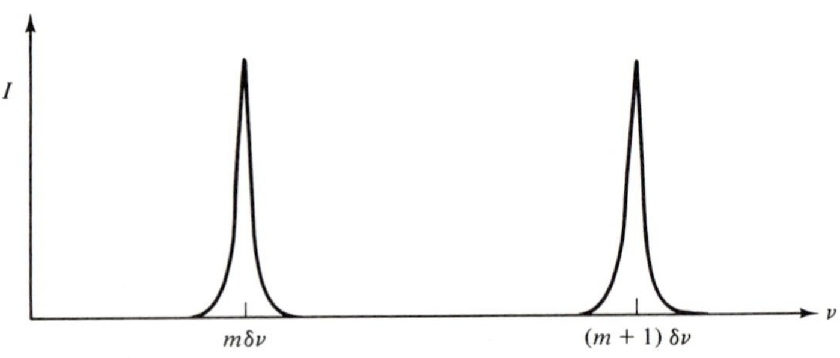

Figure 7-6 The frequency distribution of the light passed by the cavity when the reflectivity of the mirrors is large but less than 1 and/or there is a small amount of absorption losses.

EXAMPLE 7-3

(a) Compute the number of lasing longitudinal modes for a GaAs laser. The peak occurs at $\lambda = 0.84\ \mu$m, the bandwidth for which the intensity is above threshold is 10^{13} Hz, the index of refraction, $n = 3.6$, and the length of the cavity is 500 μm. **(b)** What is the bandwidth in angstroms of a single laser mode if its bandwidth, $d\nu^*$, is equal to $\frac{1}{5}$ of the separation between modes, $\delta\nu$?

(a)
$$\delta\nu = \frac{c}{2L} = \frac{c_0}{2nL} = \frac{3.00 \times 10^{10}}{(2)(3.6)(5 \times 10^{-2})} = 8.33 \times 10^{10}\ \text{Hz}$$

$$\text{number of modes} = \frac{\Delta\nu}{\delta\nu} = \frac{10^{13}}{8.33 \times 10^{10}} = 120$$

(b)
$$\lambda_0 = \frac{c_0}{\nu}$$

$$|d\lambda_0^*| = c_0 \frac{d\nu^*}{\nu^2} = \frac{c_0(\delta\nu/5)}{c_0^2/\lambda_0^2} = \frac{\lambda_0^2 \delta\nu}{5c_0}$$

$$= \frac{(8.4 \times 10^{-5})^2(8.33 \times 10^{10})}{(5)(3 \times 10^{10})} = 0.392\ \text{Å}$$

Laser light like light in optical fibers has an intensity variation in the plane normal to the laser axis. The different allowed intensity variations are called transverse modes. To solve for them exactly, one must solve Maxwell's equations with the appropriate boundary conditions. Approximate solutions can be obtained for a cavity with confocal, as opposed to flat and parallel, mirrors (see Problem 1-57). In this configuration rays that are not quite parallel to the laser axis can be continually reflected back and forth. An analysis of the components of propagation normal to the laser axis will yield the allowed intensity distribution.

7-3 HOMOJUNCTION LASERS

7-3.1 Material and Operating Requirements

To achieve population inversion for electrons injected into a p-type material, there should be a large number of empty states to drop into. This is accomplished at low temperatures in a heavily doped p-type material in which an acceptor band is formed. Likewise, population inversion for holes injected into an n-type material occurs if there are very few holes at the bottom of the conduction band. There are very few holes in the donor band created by heavily doping the n-type material. The Fermi energy of the p-type material is at the bottom of the acceptor band, and the Fermi level of the n-type material is at the top of the donor band; hence, the contact potential is $V_0 \simeq E_g/q$. This is illustrated in Fig. 6-8.

One can arrive at the same answer in a more mathematical way. If f_1 is the probability a state in the p-type material is occupied by an electron, then the probability the state is occupied by a hole is $1 - f_1$. The same statement applies for f_2 and $1 - f_2$ for the n-type material. The probability a state is occupied by an electron is given by

$$f = [e^{(E-E_F)/kT} + 1]^{-1} \tag{7-19}$$

This equation is seen to be consistent with the electron distribution in a metal at 0 K as for $E < E_F$, $f = 1$ and for $E > E_F$, $f = 0$ as it must.

Thus, for the pn junction

$$1 - f_1 = 1 - \frac{1}{e^{(E-E_{Fp})/kT} + 1} = \frac{1}{1 + e^{(E_{Fp}-E)/kT}} \tag{7-20}$$

is large when $E_{Fp} < E$. This can be accomplished for $E = E_v$ if the acceptor band dips into the valence band. Likewise

$$f_2 = \frac{1}{e^{(E-E_{Fn})/kT} + 1} \tag{7-21}$$

is large when $E_{Fn} > E$. This can be accomplished for $E = E_c$ if the donor band extends into the conduction band.

The population of the upper electron and hole levels increases as the current density increases until the threshold current density, J_{th}, is reached, but for $J > J_{\text{th}}$ the occupation of the levels remains fixed at η_t. J_{th} can be computed for very low temperatures at which $n_1 \simeq 0$. From Eq. 6-25

$$n_2 = \frac{J\tau}{qd} \tag{7-22}$$

where d for a homojunction laser is the diffusion length. Inserting Eq. 7-22 into Eq. 7-14, recalling that τ/τ_{sp} is the internal quantum efficiency, η_i, and rearranging yields

$$J_{\text{th}} = \frac{8\pi\ \Delta\nu qd}{\lambda^2 \eta_i}\left(\alpha - \frac{1}{2L}\ln R_1 R_2\right) \tag{7-23}$$

For a number of reasons J_{th} increases quite dramatically with the temperature. Electrons are thermally excited into the acceptor band from the valence band so that n_1 can no longer be ignored. Also, absorption losses increase substantially because the thermally excited electrons can more easily absorb the laser light. η_i also decreases since τ_{sp} increases with the temperature, and the bandwidth, $\Delta\nu$, increases with the temperature. These effects produce a temperature dependence that can be represented by the equation

$$J_{\text{th}} = J_{\text{th}}(0)e^{T/T_0} \tag{7-24}$$

where the temperature, T_0, is experimentally determined.

EXAMPLE 7-4

(a) Compute $J_{\text{th}}(0)$ for a GaAs laser if $\Delta\nu = 5 \times 10^{12}$ Hz, $d = 2\ \mu$m, $\alpha = 8\ \text{cm}^{-1}$, $L = 500\ \mu$m, $\lambda_0 = 0.84\ \mu$m, $\eta_i = 1$, and the cleaved GaAs surfaces act as the mirrors. The index of refraction is 3.6. **(b)** Find T_0, $J_{\text{th}}(77)$ and $J_{\text{th}}(300)$ if $J_{\text{th}}(300)/J_{\text{th}}(77) = 65$.

(a) $R_1 = R_2 = \left(\frac{3.6 - 1}{3.6 + 1}\right)^2 = 0.32$

$$J_{th}(0) = \frac{8\pi\ \Delta\nu q d}{\lambda^2 \eta_i}\left(\alpha - \frac{1}{2L}\ln R_1 R_2\right)$$

$$= \frac{8\pi(5 \times 10^{12})(1.6 \times 10^{-19})(2 \times 10^{-4})}{[(8.4/3.6) \times 10^{-5}]^2(1)}$$

$$\times \left(8 - \frac{1}{5 \times 10^{-2}}\ln 0.32\right) = 242\ \text{A/cm}^2$$

(b)

$$\frac{J_{th}(300)}{J_{th}(77)} = \frac{e^{300/T_0}}{e^{77/T_0}} = 65$$

$$\therefore T_0 = \frac{300 - 77}{\ln 65} = 53.4\ \text{K}$$

$$J_{th}(77) = J_{th}(0)e^{T/T_0} = (242)e^{77/53.4} = 1024\ \text{A/cm}^2$$

$$J_{th}(300) = 65(J_{th}(77)) = 65(1024) = 6.64 \times 10^4\ \text{A/cm}^2$$

The threshold current density is so high in a homojunction laser at room temperature that it is virtually impossible to obtain continuous wave (cw) operation because there is so much joule heating.

7-3.2 Power Output

As we mentioned in Section 7-2.2, the rate of recombination above threshold increases as the current increases in such a way as to keep the population inversion constant. The optical power will increase linearly with the current if the internal quantum efficiency is constant. Thus

$$P_{in} = \frac{i - i_{th}}{q}\eta_i h\nu \tag{7-25}$$

where P_{in} is the stimulated power generated inside of the cavity. The amount of power emitted by the laser, the external power, P_{ex}, is the fraction of the internal power lost through the windows. Thus

$$P_{ex} = \frac{\frac{1}{2L}\ln(1/R_1R_2)}{\alpha + \frac{1}{2L}\ln(1/R_1R_2)}\ \frac{i - i_{th}}{q}\eta_i h\nu \tag{7-26}$$

For a given current there is a value of R_1R_2 for which P_{ex} is a maximum. Everything else being equal, the smaller i_{th} is, the larger P_{ex} will be. This is accomplished by making R_1R_2 closer to unity, since this reduces the loss. However, increasing R_1R_2 decreases the P_{ex}/P_{in} ratio, and in the limit of $R_1R_2 = 1$, $P_{ex} = 0$. On the other hand as R_1R_2 decreases, P_{ex}/P_{in} increases, but i_{th} increases.

If i_{th} exceeds i then, of course, there is no lasing. Between these $R_1 R_2$ values for which $P_{ex} = 0$, there is a value of $R_1 R_2$ for which P_{ex} is a maximum. The optimum value of $R_1 R_2$ depends on i; it increases as i increases (see Problem 7-5).

The external differential quantum efficiency, $\bar{\eta}_{ex}$, is the rate of change above threshold of the number of photons emitted per charge carrier injected. Thus

$$\bar{\eta}_{ex} = \frac{d(P_{ex}/h\nu)}{d(i - i_{th})/q} = \frac{\ln(1/R_1 R_2)}{2\alpha L + \ln(1/R_1 R_2)} \eta_i \tag{7-27}$$

EXAMPLE 7-5

Calculate $\bar{\eta}_{ex}$, the current necessary to produce 5 mW of power from a homojunction GaAs laser emitting at 0.84 μm, and η if $i_{th} = 200$ mA and $\eta_i = 0.9$. The mirrors are cleaved GaAs surfaces, $\alpha = 10\ \text{cm}^{-1}$, $L = 500$ μm, and $V = 2$ V. From Example 7-4, $R_1 = R_2 = 0.32$

$$\bar{\eta}_{ex} = \frac{\ln(1/R_1 R_2)}{2\alpha L + \ln(1/R_1 R_2)} \eta_i = \frac{2 \ln(1/0.32)(0.9)}{(2)(10)(5 \times 10^{-2}) + 2\ln(1/0.32)} = 0.6255$$

$$i = i_{th} + \frac{qP_{ex}}{\bar{\eta}_{ex} h\nu} = 0.2 + \frac{(1.6 \times 10^{-19})(5 \times 10^{-3})(8.4 \times 10^{-7})}{(0.6255)(6.635 \times 10^{-34})(3 \times 10^{8})}$$

$$= 0.2054 \text{ A}$$

$$\eta = \frac{P_{ex}}{iV} = \frac{5 \times 10^{-3}}{(0.2054)2} = 1.22 \times 10^{-2}$$

7-3.3 Device Fabrication and Operation

A broad area and a stripe contact laser are illustrated in Fig. 7-7. They both have metal contacts on the top and the bottom with the current flowing normal to

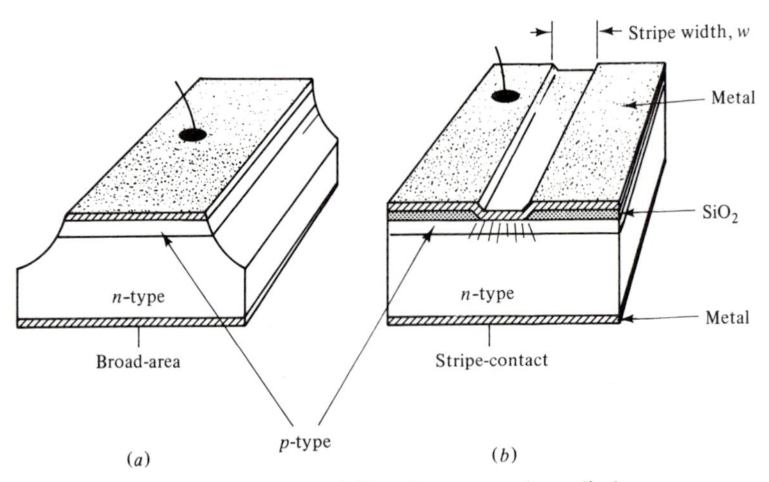

Figure 7-7 A (*a*) broad area, and (*b*) stripe contact laser diode.

the junction and the laser light being emitted parallel to it. The mirrors are usually formed simply by cleaving the crystal parallel to its cleavage plane. The surfaces are very smooth and parallel, but their reflectivity is quite low, typically 0.3 to 0.35. The reflectivity can be greatly improved by evaporating a thin layer of gold on the surface, but gold is easily scratched. As we discuss in Chapter 8, dielectric mirrors can be formed by depositing films of different indices of refraction with precise thicknesses. Dielectric mirrors are formed on lasers by depositing alternating silicon and Al_2O_3 layers. However, this is an expensive process. The two types of lasers are also heat sinked by contacting the top of the laser to a copper block.

The primary difference is that the stripe contact diode emits over a much smaller area. This is accomplished by depositing an oxide on the upper surface and then opening a stripe 10 to 15 μm wide in it. Current flowing from the metal contact is then confined to the narrow stripe. The primary advantages of the stripe contact diode are that the threshold current and the lasing area are much smaller. The lower threshold current reduces the joule heating, and the lasing area is smaller than the cross-sectional area of an optical fiber. This allows much of the emitted light to be injected into the fiber.

Just above threshold most of the longitudinal modes are active, but at higher currents a few of them dominate. This is illustrated in Fig. 7-8. Although the number of longitudinal modes is reduced, more than one is usually present. There currently are research efforts to develop a single mode laser because this would essentially eliminate wavelength dispersion in long length optical fiber systems. This effort is described in more detail in Chapter 10.

The number of transverse modes present depends on the width of the active area and the magnitude of the current. Only the low-order modes are present in stripe contact diodes because the emission width is small. This is desirable because the intensity of the higher order modes is larger away from the center. Thus the injection efficiency into a fiber is less. The optimum condition is for only the fundamental to exist. This can sometimes be accomplished at lower currents, but as the current increases, the mode saturates and the higher order modes become active. As in optical fibers, if the light is confined to a small enough region, only a single transverse mode is present. This is accomplished in a buried heterostructure, which is described in Chapter 10.

In a more precise analysis the region near the junction is divided up into the inside lasing region and the outside nonlasing region. As is shown in Fig. 7-9, the fraction of the light in the lasing region is Γ. The absorption coefficient can be different inside than it is outside so that Eq. 7-13 becomes

$$\Gamma\alpha_g = \Gamma\alpha_{\text{in}} + (1 - \Gamma)\alpha_{\text{out}} + \frac{1}{2L}\ln\left(\frac{1}{R_1R_2}\right)$$

$$= \frac{\Gamma\lambda^2\eta_i}{8\pi\ \Delta\nu qd}[J_{\text{th}} - J(T)] \qquad (7\text{-}28)$$

The $J(T)$ term has been included because the $\rho_2 n_1/\rho_1$ term in Eq. 7-14 can be ignored only at very low temperatures. Rearranging Eq. 7-28, we can write

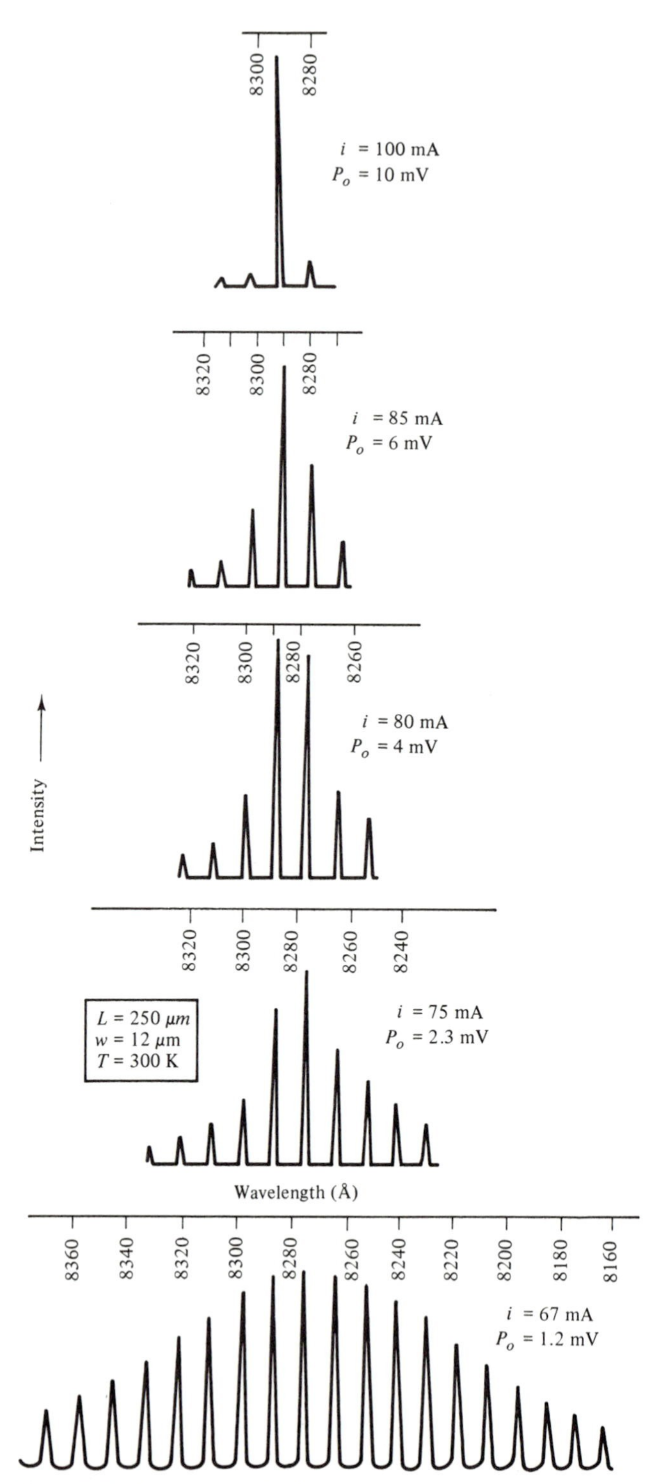

Figure 7-8 The wavelength distribution of the light emitted by a laser at different current input levels. (From H. Kressel, *Topics in Applied Physics,* Vol. 39, *Semiconductor Devices for Optical Communication,* Copyright © 1982, Springer Verlag, Heidelberg, Germany.)

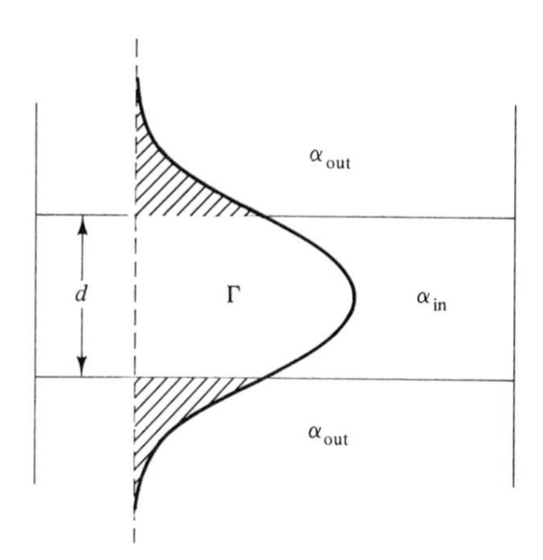

Figure 7-9 The fraction of the light, Γ, confined to the lasing region of width, *d*.

$$J_{\text{th}} = \frac{8\pi\ \Delta\nu qd}{\lambda^2\eta_i}\left[\alpha_{\text{in}} + \left(\frac{1-\Gamma}{\Gamma}\right)\alpha_{\text{out}} + \frac{1}{2L\Gamma}\ln\left(1/R_1R_2\right)\right] + J(T) \quad (7\text{-}29)$$

We see that one way J_{th} can be reduced is by reducing the thickness, *d*, of the lasing section. As will be seen in the next section, *d* can routinely be reduced by a factor of 10 using heterojunctions. The development of double heterostructure (DH) lasers made it possible to build lasers that emit continuously at room temperature.

7-4 HETEROJUNCTION LASERS

7-4.1 Semiconductor Heterojunctions

As is true for the Schottky barrier diode, the energy levels of heterojunctions are defined relative to the universal ground of the free electron at rest. Again the distance to the bottom of the conduction band is $q\chi$; $q\phi$ is the distance to the Fermi energy; qV_n is the separation between E_{Fn} and the bottom of the conduction band; and qV_p is the separation between E_{Fp} and the top of the valence band. They are illustrated for the *pn* junction in Fig. 7-10.

As always the contact potential is the difference between the Fermi levels. Thus, for the *pn* junction

$$\begin{aligned} V_{0pn} = \phi_p - \phi_n &= \left(\chi_p + \frac{E_{gp}}{q} - V_p\right) - (\chi_n + V_n) \\ &= (\chi_p - \chi_n) + \frac{E_{gp}}{q} - (V_p + V_n) \end{aligned} \quad (7\text{-}30a)$$

A very important difference between a heterojunction and a homojunction is that there are discontinuities between the valence and conduction bands at the interface. We will see later that they can be used to confine the injected carriers into a width much narrower than the diffusion length. The discontinuity in the conduction band is

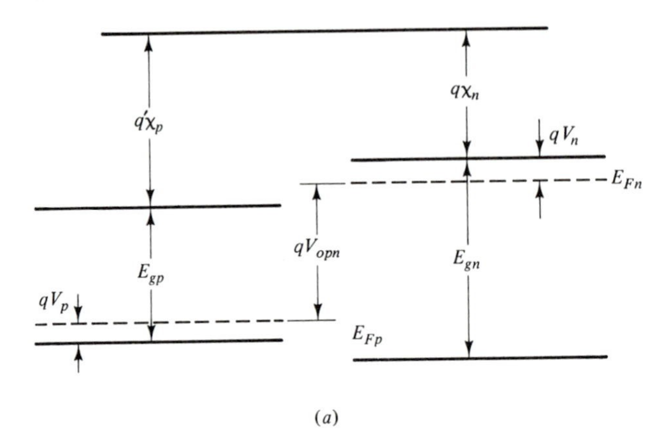

(a)

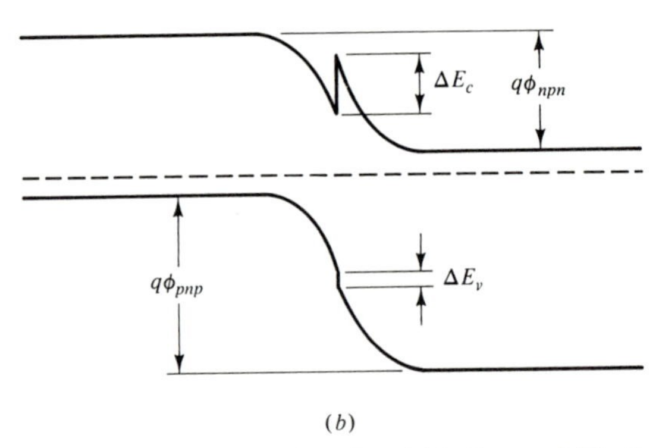

(b)

Figure 7-10 A *pn* heterojunction (*a*) before and (*b*) after contact.

$$\Delta E_{cpn} = q(\chi_n - \chi_p) \tag{7-31a}$$

and the discontinuity in the valence band is

$$\Delta E_{vpn} = q(\chi_n - \chi_p) + (E_{gn} - E_{gp}) \tag{7-32a}$$

In the *pn* junction illustrated in Fig. 7-10, ΔE_{cpn} is negative and it creates an energy spike; ΔE_{vpn} is positive, and it creates an abrupt energy drop.

Also, the barrier height at equilibrium seen by an electron, $q\phi_{npn}$, is not the contact potential. Rather, it is the combination of the contact potential and the energy discontinuity. Thus from Eqs. 7-30a and 7-31a

$$q\phi_{npn} = qV_{0pn} + \Delta E_{cpn} = E_{gp} - q(V_p + V_n) \tag{7-33a}$$

An electron can technically see a larger energy than $q\phi_{npn}$ if ΔE_c is negative and the top of the energy spike at the discontinuity rises above the bottom of the conduction band of the *p*-type material. However, this energy band spike is usually so narrow that the electrons will tunnel through it rather than rise up over it.

Because ΔE_c in Fig. 7-10 has a different sign than qV_0, $\phi_{npn} < V_0$. Thus, electrons can be injected into the p-type material more easily than they can be for a homojunction with the same contact potential. The barrier height for holes, $q\phi_{ppn}$, is

$$q\phi_{ppn} = qV_0 + \Delta E_{vnp} = E_{gn} - q(V_p + V_n) \tag{7-34a}$$

For the case being considered, ΔE_{vnp} has the same sign as qV_0 so that the barrier height seen by the holes is larger than it would be for a homojunction with the same V_0.

The *pp* heterojunction in Fig. 7-11 illustrates that this type of junction also has a contact potential, energy band discontinuities, and carrier barrier heights. The contact potential is achieved by holes in the material with the larger work function flowing over to the other material. This creates a distributed negative charge in the material with the larger work function due to the ionized holes. Planar charge at the interface is also created by a sheet of holes in the material with the smaller work function. For the heterojunction being considered, holes are transferred from the left to form a sheet of positive charge on the right.

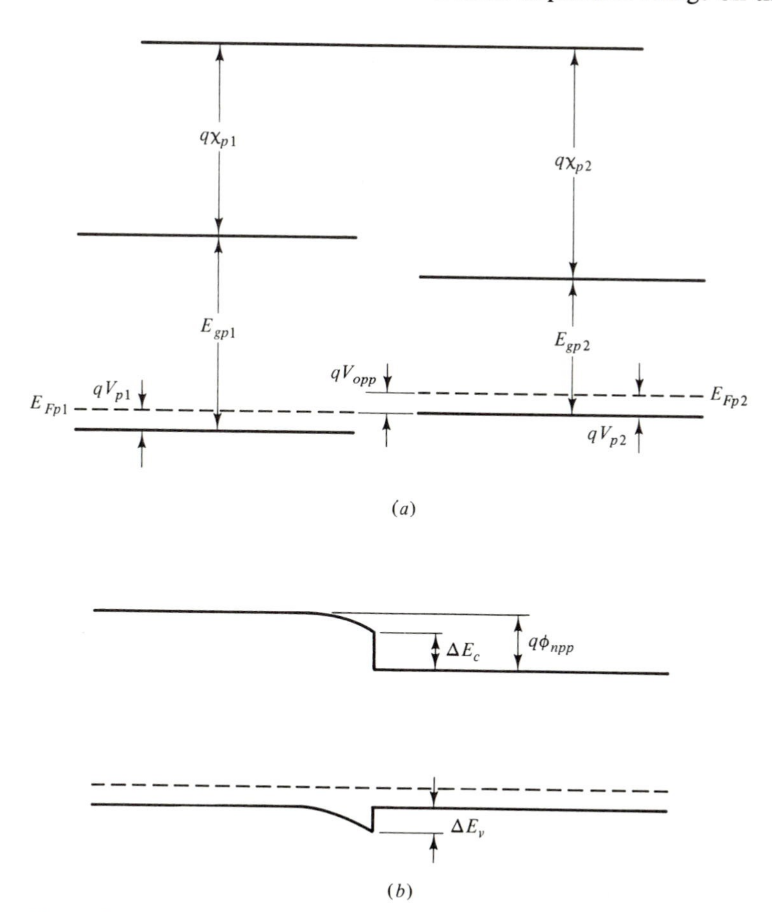

Figure 7-11 A *pp* heterojunction (*a*) before and (*b*) after contact.

The contact potential is given by

$$V_{0pp} = \left(\chi_{p1} + \frac{E_{gp1}}{q} - V_{p1}\right) - \left(\chi_{p2} + \frac{E_{gp2}}{q} - V_{p2}\right)$$

$$= (\chi_{p1} - \chi_{p2}) + \frac{E_{gp1} - E_{gp2}}{q - (V_{p1} - V_{p2})} \tag{7-30b}$$

Note that all of the potential drop occurs on the side with the distributed charge. The discontinuity for the conduction bands is

$$\Delta E_{cpp} = q(\chi_{p2} - \chi_{p1}) \tag{7-31b}$$

whereas

$$\Delta E_{vpp} = q(\chi_{p2} - \chi_{p1}) + (E_{gp2} - E_{gp1}) \tag{7-32b}$$

For the junction being considered, ΔE_c is positive and ΔE_v is negative. The barrier height seen by an electron flowing from right to left is

$$q\phi_{npp} = qV_{0pp} + \Delta E_{cpp} = (E_{gp1} - E_{gp2}) - q(V_{p1} - V_{p2}) \tag{7-33b}$$

This barrier can have a considerable height when ΔE_{cpp} is positive, as it is in Fig. 7-11. Thus, electrons injected into the narrow bandgap material would tend to accumulate there.

The barrier height seen by a hole flowing from left to right when ΔE_{vpp} is negative is qV_0 because there is an energy spike as is illustrated in Fig. 7-11. When ΔE_{vpp} is positive, the barrier height is

$$q\phi_{ppp} = qV_{0pp} + \Delta E_{vpp} = q(V_{p2} - V_{p1}) \tag{7-34b}$$

This equation illustrates the fact that the valence bands are aligned at equilibrium when both materials in the heterojunction have the same Fermi energy relative to the top of the valence band.

The most studied laser system is the GaAs/$Ga_{1-x}Al_xAs$ system. To compute ΔE_c and ΔE_v for the ternary, we will assume that χ and E_g vary linearly with the composition as was done in Section 6-3 for E_g. Thus

$$\chi_{GaAlAs} = \chi_{GaAs} + (\chi_{AlAs} - \chi_{GaAs})x \tag{7-35}$$

As was true for E_g, χ is different for direct and indirect bandgaps. The value listed in the tables is the larger of the two. To find the other, simply subtract $|E_g(\text{dir}) - E_g(\text{ind})|$.

EXAMPLE 7-6

Compute the contact potentials, energy band discontinuities, and carrier barrier heights of a *pp* $Ga_{1-x}Al_xAs$/GaAs and a *pn* GaAs/$Ga_{1-x}Al_xAs$ heterojunction. The materials are doped so that the Fermi levels correspond to the energy band edge, and $x = 0.35$. For smaller amounts of aluminum the ternary is a direct gap semiconductor. Using Table 6-1:

$$\chi_{AlAs}(\text{dir}) = \chi_{AlAs}(\text{ind}) - [E_g(\text{dir}) - E_g(\text{ind})] = 3.5 - (2.95 - 2.16) = 2.71$$

$$\chi\text{GaAlAs} = \chi_{GaAs} + (\chi_{AlAs} - \chi_{GaAs})x$$

$$= 4.07 + (2.71 - 4.07)0.35 = 3.59 \text{ V}$$

Using Table 6.1:

$$E_g = 1.43 + (2.95 - 1.43)0.35 = 1.96 \text{ eV}$$

For the *pp* junction

$$V_{0pp} = (\chi_{p1} - \chi_{p2}) + \frac{E_{gp1} - E_{gp2}}{q} - (V_{p1} - V_{p2})$$

$$= (3.59 - 4.07) + (1.96 - 1.43) - (0 - 0) = 0.05 \text{ V}$$

$$\Delta E_{cpp} = q(\chi_{p2} - \chi_{p1}) = 4.07 - 3.59 = 0.48 \text{ eV}$$

$$\Delta E_{vpp} = q(\chi_{p2} - \chi_{p1}) + (E_{gp2} - E_{gp^1})$$

$$= (4.07 - 3.59) + (1.43 - 1.96) = -0.05 \text{ eV}$$

$$q\phi_{npp} = qV_{0pp} + \Delta E_{cpp} = 0.05 + 0.48 = 0.53 \text{ eV}$$

$$q\phi_{ppp} = qV_{0pp} + \Delta E_{vpp} = 0.05 - 0.05 = 0$$

The holes flowing from left to right will, however, see an energy spike of $qV_0 = 0.05$ eV. For the *pn* junction

$$V_{0pn} = (\chi_p - \chi_n) + \frac{E_{gp}}{q} - (V_p + V_n)$$

$$= 4.07 - 3.59 + 1.43 = 1.91 \text{ V}$$

$$\Delta E_{cpn} = q(\chi_n - \chi_p) = 3.59 - 4.07 = -0.48 \text{ V}$$

$$\Delta E_{vpn} = q(\chi_n - \chi_p) + (E_{gn} - E_{gp})$$

$$= 3.59 - 4.07 + 1.96 - 1.43 = 0.05 \text{ eV}$$

$$q\phi_{npn} = qV_{0pn} + \Delta E_{cpn} = 1.91 - 0.48 = 1.43 \text{ eV}$$

$$q\phi_{ppn} = qV_{0pn} + \Delta E_{vpn} = 1.91 + 0.05 = 1.96 \text{ eV}$$

As is illustrated in Example 7-6 and Fig. 7-12, the sum of the electron-barrier heights and the sum of the hole-barrier heights are both equal to the contact potential between the p_1 and n layers; the intermediate layer acts as a sort of stepping-stone. How the jump from the n to the p_1 layer is broken up by the intermediate layer depends on how the intermediate layer is doped because this determines the contact potentials, V_{0pp} and V_{0pn}, when the width of the intermediate layer is greater than the sum of the portions of the two depletion layers in it.

This is seldom true, however, because the intermediate layer is usually thin and lightly doped. For this more usual condition the $p_2 - n$ depletion layer extends into the p_1 region so that all of the holes in the intermediate p_2 layer are ionized. Thus the bands in the intermediate layer are bent over the entire length, and they are bent in such a way that they impede the motion of electrons moving from the n region to the p_2 region and holes moving from the p_2 region to the n region. The charge distribution is

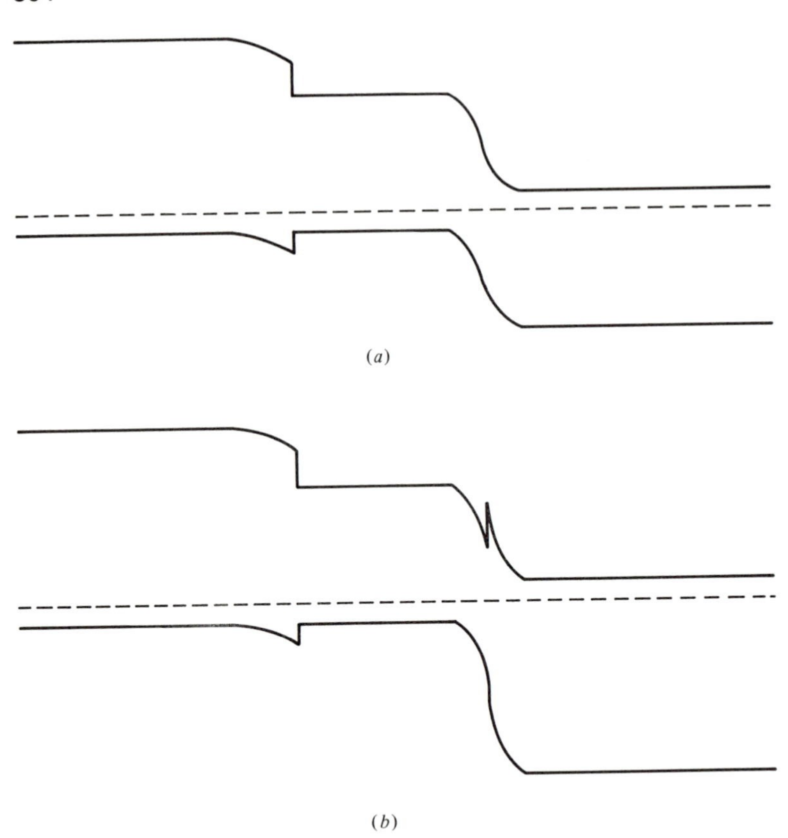

Figure 7-12 The energy band structure for (*a*) a single heterostructure (SH) and (*b*) a double heterostructure (DH) laser when the sum of the junction depletion layer widths is less than the width of the intermediate layer.

$$qn_{a1}x_p + qn_{a2}\,d = qn_d x_n \tag{7-36a}$$

Since $n_{a1}x_p$ is often $\gg qn_{a2}\,d$, Eq. 7-36a can be approximated by

$$qn_{a1}x_p \simeq qn_d x_n \tag{7-36b}$$

If we ignore the fact that ϵ for GaAlAs is different from ϵ for GaAs (see Problem 7-7), the equations developed for the *pin* diode can be used by substituting d for x_i.

If we redefine V_{0pp} to be the potential drop in the p_1 layer rather than the entire potential drop at the p_2n junction and mark this change by a quotation mark

$$V'_{0pp} = \frac{qn_{a1}x_p^2}{2\epsilon} \tag{7-37}$$

Likewise

$$V'_{0pn} = \frac{qn_d x_n^2}{2\epsilon} \tag{7-38}$$

and the potential drop across the intermediate layer is

$$V_{0in} = \frac{qn_d x_n d}{\epsilon} \tag{7-39}$$

The total potential drop between the p_1 and n layers then is

$$V_0 = \frac{E_{F_n} - E_{F_{p1}}}{q} = V'_{0pp} + V_{0in} + V'_{0pn} \tag{7-40}$$

To find x_p and x_n, we use the equations

$$W = x_p + d + x_n = \left(\frac{2\epsilon}{q} V_0 \frac{n_a + n_d}{n_a n_d} + d^2\right)^{1/2} \tag{4-93}$$

$$x_p = \frac{n_d}{n_a + n_d}(W - d) \tag{4-92}$$

and

$$x_n = \frac{n_a}{n_d} x_p \tag{4-64}$$

The electron and hole barrier heights $q\phi_{npp}$, $q\phi_{ppp}$, $q\phi_{npn}$, and $q\phi_{ppn}$ are found by replacing V_{0pp} and V_{0pn} with V'_{0pp} and V'_{0pn} in Eqs. 7-33a, 7-34a, 7-33b, and 7-34b.

The p_1 and n layers are often heavily doped ($\sim 10^{18}$) so that most of the potential drop occurs across the intermediate layer even though it is only $\sim$0.2 to 0.4 μm thick. As a result, most of the carrier confinement is due to V_{0in}, ΔE_{cpp} and ΔE_{vpn} (see Problems 7-19 and 7-20).

7-4.2 Device Operation

The single heterostructure (SH) *ppn* GaAlAs/GaAs laser in Fig. 7-12*a* is a combination of a *pp* heterojunction and a *pn* homojunction. The laser can be treated as a sum of these two junctions if the sum of their depletion layer widths is less than the width of the intermediate layer. This occurs for the small d values used for lasers only when this layer is heavily doped. It differs from the GaAs laser in that there is a p GaAlAs layer that blocks the flow of the electrons in the p GaAs layer injected from the n-GaAs. As we learned in Example 7-6, the blocking barrier is 0.53 eV when the layers are heavily doped, and for this system the barrier is primarily due to ΔE_c. The electrons are confined to the intermediate p layer by this energy barrier so that J_{th} is reduced by making d less than the diffusion length.

The *ppn* double heterostructure laser in Fig. 7-12*b* differs from the SH laser in that the *pn* junction is a heterojunction. For this system the primary electrical effect of this change is that the barrier height for the holes flowing from the p-GaAs layer to the n-GaAlAs layer is increased.

The DH laser also differs optically from the SH laser. GaAlAs has a smaller index of refraction than GaAs because the gallium atom contains more core electrons than the aluminum atom. As a result, the GaAlAs, which usually has the same composition in both layers, acts as a waveguide much like the cladding on a step index fiber (see Section 2-4.1). The light is thus more confined to the intermediate layer so that Γ is a larger fraction.

As illustrated in Fig. 7-13, the magnitude of Γ depends in a complex way on d and the difference in the index of refraction. As one might expect, the larger Δn or d the larger Γ is. For GaAlAs Δn changes linearly with x according to the equation

$$\Delta n \simeq 0.62\ \Delta x \tag{7-41}$$

The DH structure has the additional advantage of having a smaller α_{out} at room temperature, where band-to-band absorption can be a problem for homojunction lasers, since there is no band-to-band absorption in the wider bandgap material.

As in the case of the SH laser, J_{th} is reduced by confining the electrons to the intermediate layer. For larger d values $J_{th} \propto d$, and for this system at room temperature

$$J_{th} \simeq 4800d$$

where d is in microns and J_{th} is in A/cm^2. For smaller values of d, the decrease in Γ with d becomes important. For values of $d < 0.2\ \mu$m, it can become the dominant effect so that further decreasing d will result in an increase in J_{th}. The value of d for which J_{th} is a minimum is smaller when Δn is larger, and this is

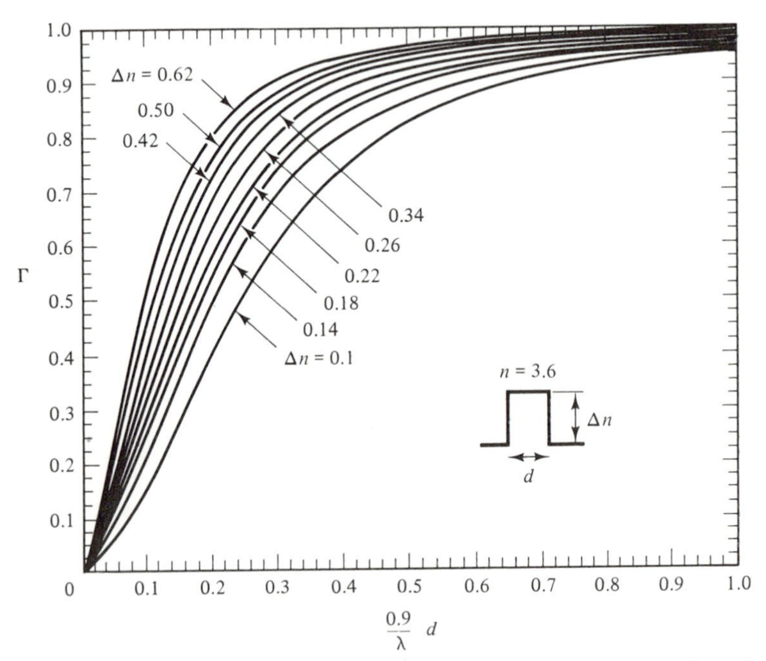

Figure 7-13 The confinement factor, Γ, plotted as a function of the d/λ ratio for various values of the difference in the index of refraction, Δn, between the active and confining layers. (From H. Kressel and J. K. Butler, *Semiconductor Lasers and Heterojunction LEDs*, Copyright © 1977, Academic Press, New York.)

illustrated in Fig. 7-14. The minimum J_{th} is 475 A/cm^2, which occurs for a value of $\Delta n = 0.4$ and $\eta_i = 1$.

The DH lasers not only have a much smaller threshold current at low temperature due to the confinement of carriers, their threshold current has a smaller temperature dependence. This is reflected in the larger T_0 values, which can be as high as 200 K but are more typically 125 to 150 K. The smaller temperature dependence is in part due to the smaller temperature dependence of Γ and α_{out}. The injected carrier distribution does not become more spread out at higher temperatures, since they are confined by the energy barrier. There is a more rapid increase in α_{out} with the temperature in homojunctions because the carriers thermally excited into the donor or acceptor band can absorb the exciting radiation. This does not occur in the GaAlAs confining layers because their energy gap is larger than the energy of the laser radiation.

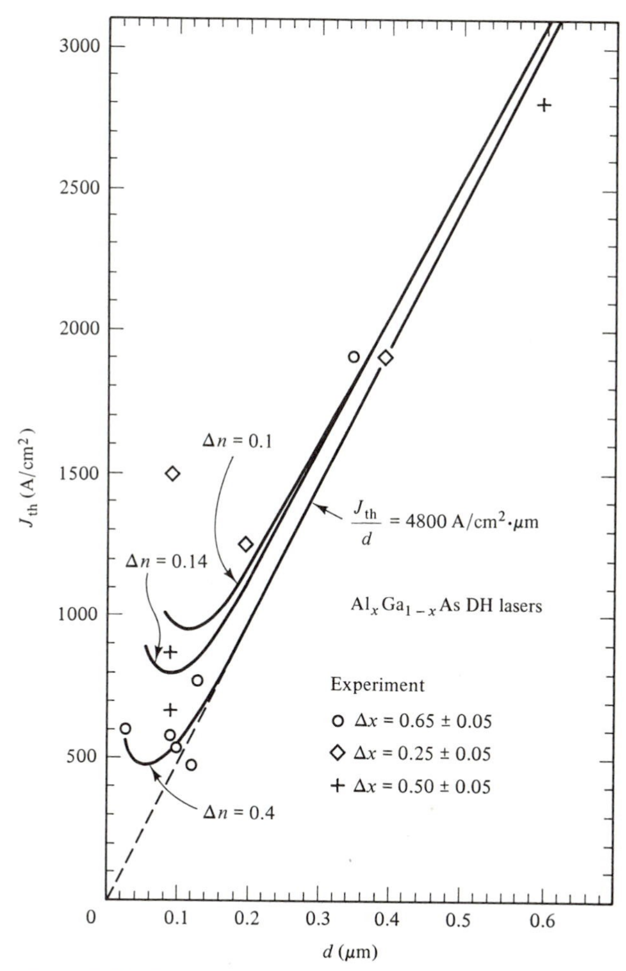

Figure 7-14 The threshold current density, J_{th}, plotted as a function of the intermediate layer thickness, d, for GaAlAs/GaAs/GaAlAs DH lasers. [From H. Kressel and M. Eltenberg, *J. Appl. Phys.*, **47** (3533), 1976.]

EXAMPLE 7-7

Compute $J_{th}(0)$, $J_{th}(77)$, $J_{th}(300)$, and $J(T)$ at 300 K for a DH GaAlAs/GaAs/GaAlAs laser with the following characteristics. $R_1 = R_2 = 0.32$, $\Delta\nu = 5 \times 10^{12}$ Hz, $d = 0.2\ \mu$m, $\alpha_{in} = \alpha_{out} = 8\ \text{cm}^{-1}$, $\Gamma = 0.6$, $L = 500\ \mu$m, $\lambda_0 = 0.84\ \mu$m, $n = 3.6$,

$$\eta_i = 1,\ T_0 = 125\ \text{K and } J(T) = 0 \text{ at } 0\ \text{K}.$$

$$J_{th}(0) = \frac{8\pi\ \Delta\nu q d}{\lambda^2 \eta_i}\left[\alpha_{in} + \left(\frac{1-\Gamma}{\Gamma}\right)\alpha_{out} + \frac{1}{2L\Gamma}\ln\frac{1}{R_1 R_2}\right] + J(0)$$

$$= \frac{8\pi(5 \times 10^{12})(1.6 \times 10^{-19})(0.2 \times 10^{-4})}{(8.4 \times 10^{-5}/3.6)^2 1}$$

$$\times\left[8 + \left(\frac{1-0.6}{0.6}\right)8 + \frac{1}{(5 \times 10^{-2})0.6}\ln\frac{1}{0.32}\right]$$

$$= 37.9\ \text{A/cm}^2$$

$$J_{th}(77) = J_{th}(0)e^{T/T_0} = 37.9e^{77/125} = 70.2\ \text{A/cm}^2$$

$$J_{th}(300) = 37.9e^{300/125} = 417\ \text{A/cm}^2$$

$$J(T = 300) = J_{th}(300) - J_{th}(0) = 417 - 39.7 = 377\ \text{A/cm}^2$$

At current densities $\geq 3J_{th}$ the laser efficiency can be as high as 20 percent. This is because above the threshold almost all of the input energy is converted to light, and η_i increases with J, since τ_r decreases. For a typical stripe contact laser the efficiency at 10-mW output is ~7 percent.

The construction of a GaAlAs/GaAs/GaAlAs laser is illustrated in Fig. 7-15*a*. It is shown upside down to emphasize that they are often used pressed face down on a copper heat sink. Also note that the semiconductor film contacted to the metallic stripe contact is GaAs. This is done because it is easier to make a good ohmic contact to a narrower bandgap material.

The intermediate layer can also be GaAlAs with less aluminum. Thus, the emission wavelength can be tuned by varying the aluminum content. The primary effect of adding more aluminum is to decrease the confining barrier heights which increases J_{th}.

InP/InGaAsP/InP DH lasers are now being built to emit at the zero dispersion wavelength for optical fibers at 1.30 μm and the fiber absorption minima at 1.55 μm, but the technology is not yet as well developed as it is for the GaAs/GaAlAs system. One of the reasons for this is that it is absolutely essential to have the lattices matched. As noted in Chapters 4 and 6, this constraint is no problem for the GaAs/GaAlAs heterostructures because GaAs and AlAs have essentially the same lattice constant. This is not true of the InP/InGaAsP structures so that the composition of the quaternary must be closely controlled.

Everything else being equal, J_{th} for the InP/InGaAsP laser should be lower than it is for of the GaAs/GaAlAs laser because it emits at longer wavelengths. However, this is not true because the technology is not as well developed and

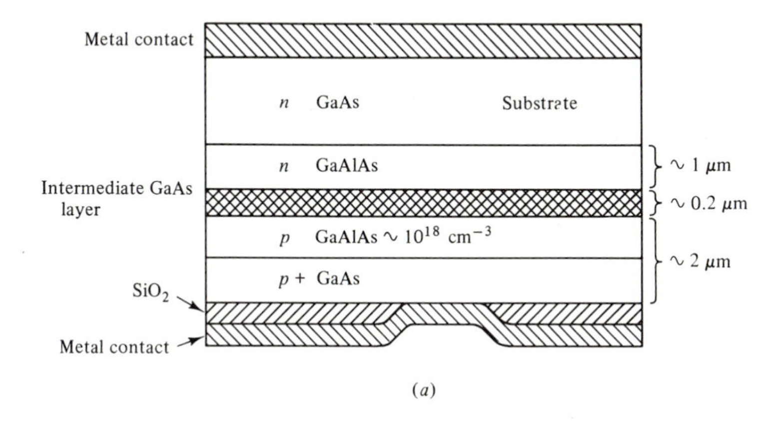

(*a*)

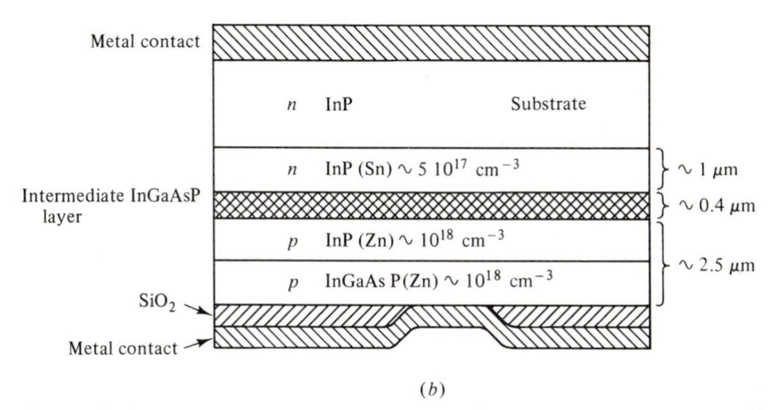

(*b*)

Figure 7-15 Schematics of the (*a*) GaAlAs/GaAs/GaAlAs, and (*b*) InP/InGaAsP/InP stripe contact DH lasers.

the emission peak bandwidth is larger, and because the confining barrier height is less. A typical value of T_0 is 60 K and the smallest room temperature $J_{th} \simeq 730$ A/cm^2. The smaller T_0 is due, in part, to a noticeable temperature dependence of α. The construction of an InP/InGaAsP/InP DH laser is illustrated in Fig. 7-15*b*. Note that again the semiconductor film contacted to the metallic stripe contact is the lower bandgap material.

7-5 FIBER COUPLING AND MODULATION

Stripe contact laser diodes with stripe widths of 10 to 15 μm are used to inject light into optical fibers. The width of the light beam is less than the diameter of multimode fibers, but it is greater than that of the single mode fibers. Thus, the light injection efficiency is higher for multimode fibers: 50 percent compared with ~10 percent for single mode fibers.

The injection efficiency is larger for lasers than for LEDs because the angular spread of the light that is emitted is less. The angular width in the plane parallel to the junction does not significantly depend on the laser geometry and typically is $\theta_{\parallel} = 10°$. On the other hand $\theta_{\perp}$ is a function of both Δn and d. It increases as Δn increases. This can be explained qualitatively by the fact that the angle for total internal reflection is smaller (see Section 2-21). It is also larger when d is larger because Γ is larger. $\theta_{\perp}$ can be as large as 50°, but is more likely to be between 20 and 30°.

Modulation is a very complex dynamic effect; at the level of this book we can examine it only qualitatively. One reason the analysis is complex is that the effects are often nonlinear. An obvious example is the *P-i* curve at threshold.

As was the case with LEDs, analog signals can be generated by intensity modulation. It is more difficult to do using lasers, but lasers have a faster response time. Small signals are sent by dc biasing the signal above threshold and superimposing the ac signal on it. There is little distortion, since the *P-i* curve can be quite linear, but the dc signal does generate considerable noise in the photodetector.

Nonlinearities in the *P-i* curve are produced by temperature changes. It is important to heat-sink the diode well because temperature changes shift the *P-i* curve. This is a particularly severe problem in InP/InGaAsP/InP lasers because they have such a small T_0.

When one mode saturates and another moves through threshold a kink is formed in the *P-i* curve. It is therefore important to have a mode-stable structure. This can more readily be achieved if the carrier mobilities in the intermediate layer are higher, that is, the doping level is low. Also, kinks cannot be formed if only a single mode is stable. The number of lateral modes can be reduced to one by using a narrow stripe. We will discuss how to form single transverse and longitudinal modes in Chapter 10.

When digital signals are sent by using an ON-OFF signal there is considerable delay because of the time it takes the carriers to reach population inversion when a step signal is applied. This delay time is given by

$$t_d = \tau_{sp} \ln \left(\frac{i}{i - i_{th}} \right) \tag{7-42a}$$

but it is reduced by a dc biasing current, i_0, according to the equation:

$$t_d = \tau_{sp} \ln \left(\frac{i}{i + i_0 - i_{th}} \right) \tag{7-42b}$$

The delay time is eliminated for $i_0 > i_{th}$, but this biasing generates more noise in the photodetector.

7-6 LASER DIODE AND LED COMPARISON

In this section we compare laser diodes and LEDs for fiber communications, and the summary is listed in Table 7-1. The items of interest are the power, spectral width, speed, linearity, temperature sensitivity, reliability, and cost.

TABLE 7-1 CHARACTERISTICS OF LEDS AND LASER DIODES USED FOR OPTICAL COMMUNICATION SYSTEMS (Ref. 3, p. 162)

Property	Low radiance LEDs	High radiance LEDs	Laser diodes
Applied voltage (V)	1.5–2.5	1.5–2.5	1.5–2
Forward current (mA)	50–300	50–300	10–300
Threshold current (mA)	NA	NA	5–250
Output power (mW)	1–3	1–10	1–10
Power coupled into fiber (mW)	0.001–0.1	0.05–0.5	0.5–5
Spectral width at 0.8 μm (nm)	35–50	35–50	2–3
Spectral width at 1.3 μm (nm)	70–100	70–100	3–5
$\Delta \ln P_{ex}/\Delta T$ (%/°C)	−1	−1	NA
$\Delta \ln i_{th}/\Delta T$ (%/°C)	NA	NA	1
Brightness (W/cm^2 · sr)	1–10	10–10^3	10^5
Rise time (10% to 90% ns)	5–50	2–20	$\lesssim$1
Frequency response (−3 dB, MHz)	7–70	18–175	350–1000
Nonlinearity (%)	0.03–1	0.3–3	0.3–30
Feedback stabilization required	No	No	Yes
MTTF (*h* at 25° C heat-sink)	10^6–10^9	10^4–10^7	3 × 10^3–10^6
Operating temperature	−55–150	−40–90	−55–70
Chip processing complexity	Low	Medium	High
Packaging complexity	Low	Low	High
Cost	Low	Low	High

LEDs are usually operated with 50 to 300 mA at 1.5 to 2.5 V, and their output power is typically 1 to 10 mW. For surface-emitting LEDs only about 1 percent of the light can be injected into a multimode fiber, but with the proper lenses the amount can be increased to 10 percent. Edge-emitting LEDs by virtue of their narrower spatial emission pattern can couple 10 percent of their light into a multimode fiber. The precise amount of power inserted depends on the diode structure, fiber diameter, and numerical aperture of the fiber.

Laser diodes have similar input and output power levels but, of course, they must be operated above threshold. Although the two types of devices have similar power levels, more light can be coupled into a fiber with a laser because its spacial bandwidth and emission area are smaller. Thirty percent of the laser light can be inserted when no focusing mechanisms are used, but with the proper lens scheme virtually all of the power can be coupled in. The 10- to 20-dB power advantage the laser has makes it a good candidate for long line systems.

The spectral width of GaAlAs LEDs is 35 to 50 nm, which results in an impulse broadening of 2.1 to 3 nsec/km at 0.8 μm due to wavelength dispersion. As a result, the maximum data rates are 150 to 200 Mb/sec · km. Although the spectral bandwidths of InGaAsP LEDs are larger—they can be as large as 100 nm—there is less wavelength dispersion because at 1.30 μm the dispersion is $\sim$0. Thus, the dispersion limited data rate is in excess of 1 Gb/sec · km.

For both types of lasers the spectral width of the lasing modes is 20 times less than that of the LED. When only a single mode is operative, the bandwidth is even less. Thus, the maximum data rate is seldom limited by wavelength dispersion when lasers are used.

The spontaneous recombination time determines the speed of operation of an LED unless special drive circuitry is employed. For GaAs doped to $10^{19} - 10^{-18}\ cm^{-3}$, τ_{sp} = 1–10 nsec. These decay times result in 3-dB bandwidths of 35 to 350 MHz. Lasers have faster recombination times, and if they are dc biased to near threshold, they can operate at data rates that exceed 1 Gb/sec · km.

Provided that the junction is not allowed to heat up, the light output is essentially proportional to the input current. Thus, analog signals can quite faithfully be reproduced. Lasers, on the other hand, can be much more temperature sensitive, and they can have kinks where secondary modes are created. The former can to some extent be compensated for by using negative feedback circuitry, and the latter problem can be reduced by fabricating lasers with more stable mode structures.

The primary effect of increasing the temperature of an LED is to decrease its power output. For a temperature increase from 0 to 70° C the power typically drops by ~2 dB. The temperature dependence of the threshold current in lasers is a more severe problem. In some instances the temperature change is large enough to push i_{th} above the operating current.

Because LEDs are simpler devices and usually operate at lower current densities, they have longer average lifetimes. The mean time to failure (MTTF) for LEDs operating at 70° C can be larger than 10^7 h and is typically in excess of 10^5 h. On the other hand, heat-sinked lasers have lifetimes of 3×10^3–10^5 h. They also are susceptible to mirror damage.

LEDs have a considerable cost advantage over lasers. Thus, for low data rate short lines they are more cost effective. For high data rate long line applications, laser diodes will be the dominant light-emitting device.

READING LIST

1. M. B. Panish and I. Hayashi, "A New Class of Laser Diodes," *Scientific American,* July 1971, p. 32.
2. J. Wilson and J. F. B. Hawkes, *Optoelectronics: An Introduction.* Englewood Cliffs, NJ: Prentice-Hall, 1983, Chapters 6 and 7.
3. H. Kressel (Ed.), *Topics in Applied Physics,* Vol. 39, *Semiconductor Devices for Optical Communication.* Berlin: Springer-Verlag, 1982, Chapters 2, 5, and 7.
4. D. Botez and G. J. Herskowitz, *Proceedings of the IEEE,* **68** (6), 1980, pp. 689–731.
5. A. Yariv, *Introduction to Optical Electronics,* 2d Ed. New York: Holt, Rinehart and Winston, 1976, Chapter 7.
6. H. Kressel and J. K. Butler, *Semiconductor Lasers and Heterojunction LEDs.* New York: Academic Press, 1977.
7. H. C. Casey, Jr., and M. B. Panish, *Heterostructure Lasers: A. Fundamental Principles, B. Materials and Operating Characteristics.* New York: Academic Press, 1978.
8. A. G. Milnes and D. L. Feucht, *Heterojunctions and Metal-Semiconductor Junctions.* New York: Academic Press, 1972.

PROBLEMS

7-1. A GaAlAs/GaAs/GaAlAs DH laser emits 10 mW of power over an area 30 μm wide and 4 μm thick. What is the intensity?

7-2. **(a)** After the first pass in a semiconductor laser 500 μm long the light intensity has increased by a factor of 2. What is α_g if $\alpha = 8\ \text{cm}^{-1}$?

(b) Repeat (a) when the light intensity is increased by a factor of 4 during the first pass.

7-3. Calculate the number of longitudinal modes present in an InGaAsP emission peak at 1.3 μm if $\Delta\lambda = 700$ Å, $n = 3.5$, and $L = 500\ \mu$m.

7-4. Compute the electron diffusion length in GaAs for doping levels of 10^{15} and 10^{18} cm^{-3}. What are the L/d ratios if $d = 0.2\ \mu$m?

7-5. **(a)** Show that P_{ex} for a homojunction laser can be written in the form $P_{ex} = KB\{(J/B)/[\alpha L + \ln(1/R)] - 1\} \ln \frac{1}{R}$ when $R_1 = R_2 = R$. Find an expression in terms of J/B and αL for $\ln(1/R)$ for which P_{ex} is a maximum. (K and B are constants independent of R.)

(b) Find an expression for the optimum value of P_{ex} in terms of KB, J/B, and αL.

(c) For $\alpha = 10^3\ \text{M}^{-1}$, $L = 490\ \mu$m, $d = 2\ \mu$m, $\Delta\nu = 1.5 \times 10^{13}$ Hz, $\lambda_0 = 0.84\ \mu$m, $n = 3.6$, and $\eta_i = 1$, find R_{opt} and J when $J = 4B$.

(d) Repeat part c for the case $J = 9B$.

(e) Find the magnitude of J for which $R_{opt} = 0.32$.

(f) Make a plot of P_{ex}/KB versus R when $J = 4B$ and $J = 9B$ on the same graph.

(g) Find the ratio of $\ln [1/R_{opt}/(\alpha L + \ln 1/R_{opt})]$ when $J = 4B$ and $J = 9B$.

(h) Find the intensity of the emitted light (P_{ex}/A) when $J = 4B$ and $J = 9B$.

(i) Repeat all but part f for the case $R_2 = 1$ (a perfectly reflecting mirror on one end) and $R_1 = R$. Do not grind this one out!

7-6. For which values of R is $P_{ex} = 0$ in the preceding problem? Explain physically what is happening for each zero value.

7-7. For a *pn* heterojunction find expressions for

(a) $q(V_0 - V)$ in terms of χ_p, χ_n, E_{gp}, E_{gn}, n_a, n_d, n_{ip}, n_{in}, kT, and qV. Assume $E_i \simeq \frac{1}{2}E_g$.

(b) The depletion layer charge, $|Q|$, in terms of q, n_a, x_p, and A or q, n_d, x_n, and A.

(c) The electric field on the p side in terms of q, n_a, ϵ_p, x_p, and x, the electric field on the n side in terms of q, n_d, ϵ_n, x_n, and x, and the magnitude of the discontinuity in the ξ field at $x = 0$.

(d) For x_p and x_n in terms of q, n_a, n_d, ϵ_p, ϵ_n, and $V_0 - V$.

(e) The junction capacitance, C_j, in terms of q, n_a, n_d, ϵ_p, ϵ_n, A, and $V_0 - V$.

7-8. Calculate V_{0pn}, ΔE_c, ΔE_v, $q\phi_{npn}$, and $q\phi_{ppn}$ for a *p*-GaAs/*n*-Ge and a *p*-GaAs/n-ZnSe junction. Assume $E_{Fp} = E_v$(GaAs) and $E_{Fn} = E_c$(Ge or ZnSe). These three semiconductors are isoelectronic and are closely lattice matched. χ(ZnSe) = 4.09 V and χ(Ge) = 4.13 V.

7-9. Show that the equations for V_0, ϕ_{npn}, and ϕ_{ppn} for a heterojunction reduce to the equations for these values for a homojunction when $\Delta E_c = \Delta E_v = 0$.

7-10. Find α_g, n_t, $J_{th}(0)$, $J_{th}(77)$, and $J_{th}(300)$ for the homojunction laser described in Example 7-4 if $R_1 = 1$.

7-11. Make a plot of $J_{th}/[J_{th}(\Gamma = 1)]$ versus Γ for $0.1 \leq \Gamma \leq 1$ if $L = 500\ \mu m$, $R_1 = R_2 = 0.32$, $\alpha_{in} = \alpha_{out} = 0.10\ cm^{-1}$, and $J(T) = 0$.

7-12. Make a plot of $J_{th}/[J_{th}(\alpha_{out} = 10\ cm^{-1})]$ versus α_{out} for $10 \leq \alpha_{out} \leq 100\ cm^{-1}$ if $L = 500\ \mu m$, $R_1 = R_2 = 0.32$, $\alpha_{in} = 10\ cm^{-1}$, $\Gamma = 0.5$, and $J(T) = 0$.

7-13. Make a plot of $i_{th}/[i_{th}(L = 1000\ \mu m)]$ versus L for $100 \leq L \leq 1000\ \mu m$ if $\alpha_{in} = \alpha_{out} = 10\ cm^{-1}$, $\Gamma = 0.5$, $R_1 = R_2 = 0.32$, and $J(T) = 0$.

7-14. Make a plot of $J_{th}/[J_{th}(R = 1)]$ versus R for $0.1 \leq R \leq 1$ if $R_1 = R_2 = R$, $L = 500\ \mu m$, $\alpha_{in} = \alpha_{out} = 10\ cm^{-1}$, $\Gamma = 0.5$, and $J(T) = 0$.

7-15. Make the same calculations as those in Example 7-6 for an InP/InGaAsP/InP laser emitting at 1.3 μm. Draw the energy band structure.

7-16. Make the same calculations as those in Example 7-6 for a *pnn* $Ga_{0.65}Al_{0.35}As/GaAs/Ga_{0.65}Al_{0.35}As$ DH laser. Draw the energy band structure.

7-17. A GaAlAs/GaAs/GaAlAs DH laser 130 μm long has a threshold current of 50 mA and emits 8 mW at 60 mA. What is the external differential quantum efficiency if $\alpha = 10\ cm^{-1}$? Compute the internal quantum efficiency if light is emitted at 8400 Å. Find the device efficiency if 2 V are used to generate the 60-mA current.

7-18. The intermediate GaAs layer in Example 7-6 is replaced by a GaAlAs layer that emits at 8000 Å.

(a) Find the composition of this layer assuming that the laser emits bandgap radiation, and E_g varies linearly with the composition.

(b) Compute the parameters calculated in Example 7-6.

7-19. Assuming that the intermediate layer is intrinsic and ignoring the fact that the intermediate layer has a different ϵ than the confining layers, compute W, x_n, x_p, V'_{0pn}, V_{0in}, and V'_{0pp} for the DH laser diode described in Example 7-6. $n_a = n_d = 10^{18}\ cm^{-3}$ and $d = 0.4\ \mu m$.

(b) How good is the assumption that $n_{a1} x_p \gg n_{a2}\, d$ if $n_{a2} = 10^{15}\ cm^{-3}$?

7-20. (a) Make the same calculations as those in Example 7-6 using the data in Problem 7-19.

(b) Construct the charge distribution and the electric field diagrams for the *pnn* DH laser described in the previous problem.

(c) Draw and label the energy band diagram.

7-21. If the intermediate layer of the DH laser described in Example 7-6 is $10^{15}\ cm^{-3}$ zinc, how thick must it be so that all of the acceptors in it are not ionized at zero bias? Assume that $m_n^* = m_p^* = m$.

7-22. A p^+pn^+ InP/InGaAsP/InP DH laser is doped so that the Fermi energies of the p^+ and n^+ InP layers are, respectively, at E_v and E_c, and the intermediate layer is 0.5 μm thick and emits at 1.3 μm. How lightly can the intermediate layer be doped so that all of the acceptors are ionized?

7-23. (a) The index of refraction of the active region of a DH laser is 3.5. Calculate the gain constant, α_g, if the absorption coefficient is $10\ cm^{-1}$ and the length is 500 μm.

(b) Also find the magnitude of $[n_2 - (\rho_2/\rho_1)n_1]$ if $\tau_r = 10$ nsec, the emission peak width is 300 Å, and for the emitted wave $\lambda = 0.8\ \mu m$.

(c) Find the low temperature threshold current if $\eta_i = 1$ and the width of the confinement layer is $d = 0.2\ \mu m$. If the value of J_{th} you have just calculated is J_{th} at 77 K, find J_{th} at 300 K (room temperature) if $T_0 = 100$ K.

(d) What is the power emitted from a stripe contact diode with a width of 12 μm at $J = 2J_{th}$ if the external quantum efficiency is 5 percent and the applied voltage is 1.4 V?

7-24. Using the results of Example 7-7, compute α_g, n_t, n_2, and $(\rho_2/\rho_1)n_1$ if $\tau_{sp} = 10^{-8}$ sec. Be certain to incorporate the effects of Γ.

7-25. If $J_{th} = 730$ A/cm^2 at room temperature, $\Delta\lambda = 60$ nm, $d = 0.1$ μm, $\Gamma = 0.5$, $n = 3.5$, $L = 500$ μm, and $T_0 = 60$ K for an InP/InGaAsP/InP DH laser emitting at 1.3 μm, find $J_{th}(77)$, $J_{th}(0)$, $J_{th}(T = 300)$, α, α_g, and n_t if $\alpha_{in} = \alpha_{out}$. Include the effects of Γ in your computation of α_g and n_t.

7-26. **(a)** Draw to scale an InP/$In_{0.53}Ga_{0.47}As$ *np* photodetector. Both materials are doped to 10^{16} cm^{-1}. Assume $m_n^* = m_p^* = m$ in your calculations. Compute E_{Fn}, E_{Fp}, V_0, ΔE_c, ΔE_v, ϕ_{nnp}, and ϕ_{pnp}.

(b) Explain why this photodetector should have a smaller dark current than the $In_{0.53}Ga_{0.47}As$ homojunction photodetector.

Chapter 8

Optical Cavities

8-1 INTRODUCTION

Large electric fields are necessary to generate enough stimulated emission to amplify the optical signal in a laser. They are obtained by using highly reflective mirrors that reflect most of the stimulated emission back into the cavity. Not only do the mirrors forming the optical cavity and the absorption in the cavity determine the steady state magnitude of the electric field in the cavity, they have a profound effect on the shape of the emission peaks. In general, the smaller the cavity losses, the narrower is the bandwidth of the emission peaks.

To put these concepts on a sound mathematical basis, we first review resonance in a series *LRC* circuit—a system that is mathematically identical to the mass, dash-pot, spring mechanical system discussed in Chapter 2. The normalized amplitude of the charge on the capacitor plates, q_0/q_{st}, at resonance is $1/(\omega_0 RC) = Q_1$, where Q_1 is the quality factor. Thus, the smaller the resistance, and therefore the loss due to i^2R heating, the larger is the charge amplitude.

For high Q_1 systems q_0/q_{st} is very sensitive to small changes in ω in the vicinity of ω_0. The bandwidth, $d\omega^*$, over which the magnitude of $(q_0/q_{st})^2$ drops from Q_1^2 to $\frac{1}{2}Q_1^2$ is ω_0/Q_1. Thus, not only is q_0/q_{st} large in high Q_1 systems, it also has a narrow bandwidth.

We examine the transient solution for q_0 to understand how the amplitude of the charge on the capacitor plates can be larger than the charge injected into the circuit during a cycle. In the beginning when there is not much charge on the plates, more energy is put into the system than is lost through the resistor; the excess energy is stored in the capacitor and inductor. As more charge is put into the system, the loss increases, since more charge is forced through the resistor.

The loss, therefore, increases each cycle, and it asymptotically approaches the energy gain per cycle. At the steady state, of course, these two values must be equal. The steady state value is larger when the resistance is smaller, and it takes a longer time to reach the steady state.

In considering the steady state variation of the electric energy in the capacitor and the magnetic energy in the inductor, we find that their sum is always zero at the resonant frequency. This means that the electromagnetic energy is fed back and forth between the capacitor and inductor and at no time does either element charge the voltage source in the circuit. However, for $\omega < \omega_0$ the capacitor charges the battery during a part of a cycle, and for $\omega > \omega_0$ the inductor charges the battery during a part of the cycle. This, of course, will reduce the energy buildup. At resonance, however, the only barrier to the energy buildup is the loss across the resistor. Thus, at steady state and $\omega = \omega_0$, the energy put into the circuit from the voltage source must equal the energy lost across the resistor.

The optical equivalent to resonance is complete constructive interference between the incident and reflected waves; an incomplete constructive interference is equivalent to charging the voltage source. Complete constructive interference occurs in one dimension when the cavity is a half integral multiple of the wavelength. The optical losses are due to absorption in the cavity and transmission through the mirrors, and the stored energy is the electromagnetic energy in the standing waves, which can easily be calculated. Finally, the quality factor for the optical cavity can be found from the definition, Q_1 = stored energy/energy lost per radian. You will find it is, in general, much larger than it is in electrical systems. Conversely, if Q_1 and the power loss of the cavity are known, the stored energy and, therefore, the magnitude of the electric field can be computed.

The resonant frequencies, or modes, need not be confined to waves in one dimension; they can also be found for two or three dimensions. Although knowing how to mathematically describe these three-dimensional modes is not important in this chapter, it is essential for the understanding of stimulated and spontaneous emission discussed in Chapters 7 and 9.

To understand how the cavity affects the shape of the emission peaks, we must add up all of the waves reflected from, and transmitted through, each of the two surfaces of the cavity. First, one finds that the transmissivity in one dimension is greatest for those waves for which the cavity length is an integral multiple of one-half the wavelength. The bandwidth of the transmission peaks is smaller when the loss is smaller—it is smaller when the mirror reflectivity is larger.

In much the same way that the properties of the laser cavity determine the frequency and bandwidth of the emission peaks, another optical cavity can be used to analyze these emission peaks. This is done by an optical form of convolution. To accurately analyze the laser emission peaks, the reflectivity of the mirrors in the spectrum analyzer must be high so that its transmission peaks approach zero bandwidth. The finesse, F, as well as the quality factor, of these peaks are large.

We end this chapter by examining the absorption coefficient of the cavity and conclude that for lasing to occur, the absorption coefficient must be negative; that is, the beam must gain amplitude as it proceeds rather than be attenuated.

8-2 ELECTRONIC RESONANCE

8-2.1 Steady State Solution

We begin this section by discussing the series *LRC* circuit shown in Fig. 8-1. The student should be familiar with the solutions for the current through the circuit and the voltage across each element. We focus on the voltage across the capacitor because we want to find the charge, q, on the capacitor plates.

To find the steady state value of q when the input voltage is an ac source, we must find the solution to the differential equation:

$$L\ddot{q} + R\dot{q} + \frac{q}{C} = V_0 \sin \omega t \tag{8-1}$$

The steady state solution is

$$q = q_0 \sin (\omega t - \phi) \tag{8-2}$$

where ϕ is the lag angle of q behind V. Both q_0 and ϕ can be found using the phasor diagram in Fig. 8-2, which is created mathematically by letting $q = q_0 \exp j(\omega t - \phi)$. Note that the capacitive and inductive impedances lie on the real axis, and the resistive impedance lies on the imaginary axis. This indicates that the capacitor and inductor are associated with stored energy, and the resistance is associated with loss.

From the phasor diagram

$$-\omega^2 L q_0 + j\omega R q_0 + q_{0/C} = V_0 \tag{8-3a}$$

or

$$q_0 = \frac{CV}{1 - \omega^2 LC + j\omega RC} \tag{8-3b}$$

or

$$\frac{q_0}{q_{st}} = \frac{1}{1 - (\omega/\omega_0)^2 + j\omega\tau} \tag{8-3c}$$

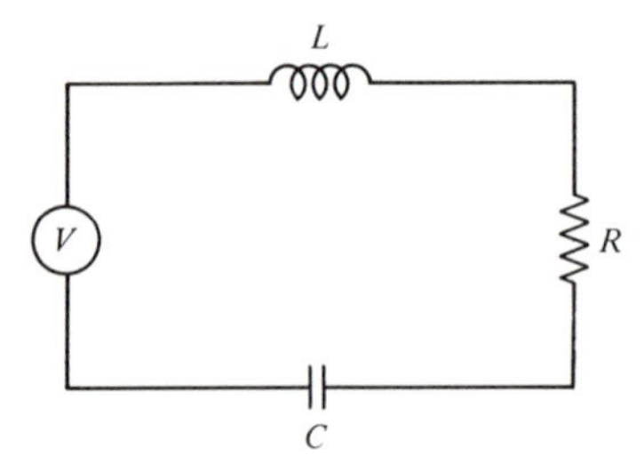

Figure 8-1 A simple *LRC* series circuit.

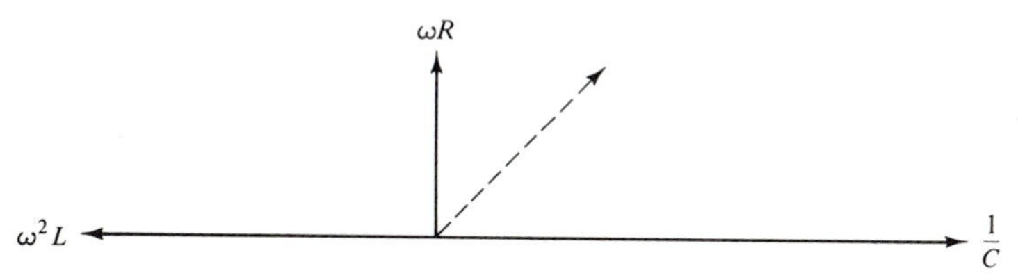

Figure 8-2 The phasor diagram used to find the steady state value of q_o.

where q_{st} is the charge amplitude on the capacitor plates when $\omega = 0$, ω_0 is the resonant frequency, and τ is the decay time constant. As is true for the formation of an induced dipole moment, q_0 can be separated into its real and imaginary parts (see Eqs. 2-26 to 2-29). Of interest to us in this section are the phase angle, ϕ, which is given by

$$\tan \phi = \frac{\omega\tau}{1 - (\omega/\omega_0)^2} \tag{8-4}$$

and the magnitude of q_0/q_{st}, which is given by

$$\frac{q_0}{q_{st}} = \frac{1}{\{[1 - (\omega/\omega_0)^2]^2 + (\omega\tau)^2\}^{1/2}} \tag{8-5}$$

At the resonant frequency, $\omega = \omega_0$, $\phi = 90°$, since the capacitive and inductive impedances are equal but opposite. Also q_0/q_{st} reaches a maximum

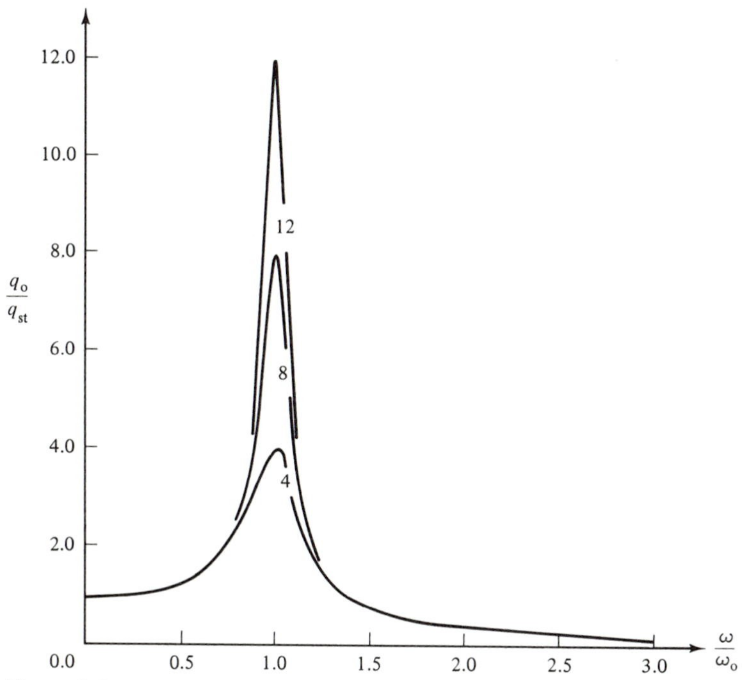

Figure 8-3 q_o/q_{st} plotted as a function of ω/ω_o for $Q_1 = 4.0$, 8.0, and 12.0.

very near resonance if the resistance is relatively small. This is illustrated in Fig. 8-3. Mathematically, at resonance

$$\frac{q_0}{q_{st}} = \frac{1}{\omega_0 RC} = \frac{\omega_0 L}{R} = Q_1 \tag{8-6}$$

where Q_1 is the quality factor. Thus, systems for which q_0/q_{st} at resonance $\gg 1$ are called high Q_1 systems.

Not only is the magnitude of q_0/q_{st} large in high Q_1 systems, the bandwidth, $d\omega^*$, is small. At the half-power points

$$\frac{q_0}{q_{st}} = Q_1/\sqrt{2} = \left\{\left[1 - \left(\frac{\omega_0 \pm d\omega^*/2}{\omega_0}\right)^2\right]^2 + \left[\frac{1}{Q_1}\left(1 \pm \frac{d\omega^*}{2\omega_0}\right)\right]^2\right\}^{-1/2}$$

$$\simeq \left[\left(\frac{d\omega^*}{\omega_0}\right)^2 + \left(\frac{1}{Q_1}\right)^2\right]^{-1/2} \tag{8-7a}$$

Thus
$$d\omega^* = \frac{\omega_0}{Q_1} \tag{8-7b}$$

The reason q_0/q_{st} is larger when Q_1 is larger is that more energy can be stored in the system. As we will show later, Q_1 is defined physically as

$$Q_1 = \frac{\text{energy stored}}{\text{energy lost per radian}} \tag{8-8}$$

There can be more charge on the capacitor plates than is put on during a cycle because the amount of stored energy can be built up over a number of cycles. To understand this, we must turn to the transient solution for our LRC circuit.

EXAMPLE 8-1

(a) Find the resonant frequency that corresponds to $\lambda = 1\ \mu$m. **(b)** If C = 1.0 pf, what is L? **(c)** If $Q_1 = 10^6$, what is R? **(d)** What is $d\omega^*$? **(e)** Find the wavelength bandwidth associated with $d\omega^*$.

(a) $$\omega_0 = kc = 2\pi c/\lambda = \frac{2\pi(3.00 \times 10^{10})}{10^{-4}} = 1.88 \times 10^{15} \text{ rad/sec}$$

(b) $$L = 1/\omega_0^2 C = \frac{1}{(1.884 \times 10^{15})^2 10^{-12}} = 2.82 \times 10^{-19} \text{ H}$$

It is, of course, impossible to generate an inductance this small. This is one indication that optical systems are more refined than electrical systems.

(c) $$R = \frac{\omega_0 L}{Q_1} = \frac{(1.884 \times 10^{15})(2.82 \times 10^{-19})}{10^6} = 5.313 \times 10^{-10}\ \Omega$$

(d) $$d\omega^* = \frac{\omega_0}{Q_1} = \frac{1.884 \times 10^{15}}{10^6} = 1.884 \times 10^9 \text{ rad/sec}$$

(e)
$$\omega = \frac{2\pi c}{\lambda}$$

$$d\omega = \frac{-2\pi c d\lambda}{\lambda^2} \quad \text{or} \quad |d\lambda^*| = \left(\frac{\lambda^2}{2\pi c}\right) d\omega^*$$

$$d\lambda^* = \frac{(1.0^2)(1.884 \times 10^9)}{(2\pi)(3 \times 10^{14})} = 10^{-6}\ \mu\text{m} = 0.01\ \text{A}$$

8-2.2 Transient Solution

The charge amplitude builds up during the transient because more charge is put into the system during a cycle than is lost across the resistor. In the beginning the difference between the energy gained and the energy lost per cycle is large, but as the charge builds up, the energy loss per cycle increases and it asymptotically approaches the energy gain per cycle. Of course, the two are equal when the steady state is reached.

The transient solution is found by solving the homogeneous linear differential equation:

$$L\ddot{q} + R\dot{q} + \frac{q}{C} = 0 \tag{8-9a}$$

The solution is

$$q = K_1 e^{s_1 t} + K_2 e^{s_2 t} \tag{8-10}$$

where the two s values are found by solving the quadratic equation:

$$Ls^2 + Rs + \frac{1}{C} = 0 \tag{8-9b}$$

Thus
$$s_1 = \frac{-R}{2L} - \left[\left(\frac{R}{2L}\right)^2 - \frac{1}{LC}\right]^{1/2} = -\alpha - (\alpha^2 - \omega_0^2)^{1/2}$$

$$= -(\alpha + \beta) \tag{8-11a}$$

and
$$s_2 = -(\alpha - \beta) \tag{8-11b}$$

where α can also be written

$$\alpha = \frac{\omega_0}{2Q_1} \tag{8-12}$$

For high Q_1 systems β is imaginary and is often written

$$\beta = j\omega_d = j\omega_0\left(1 - \frac{1}{(2Q_1)^2}\right)^{1/2} \tag{8-13}$$

ω_d is called the damped natural frequency, and for high Q_1 systems, $\omega_d \simeq \omega_0$.

The complete solution is

$$q = q_0 \sin(\omega t - \phi) + K_1 e^{-(\alpha+\beta)t} + K_2 e^{-(\alpha-\beta)t} \tag{8-14}$$

and the K_1 and K_2 values can be found from the boundary conditions

$$q(0) = 0 \tag{8-15a}$$

and

$$\dot{q}(0) = 0 \tag{8-15b}$$

The first boundary condition is due to no charge being on the plates until the voltage source is turned on, and the second boundary condition results from the fact that $L\ddot{q}$ is finite at $t = 0$. From these boundary conditions we obtain

$$K_2 = \frac{q_0}{2\beta}\,[(\alpha + \beta)\sin\phi - \omega\cos\phi] \tag{8-16a}$$

and

$$K_1 = \frac{q_0}{2\beta}\,[\omega\cos\phi - (\alpha - \beta)\sin\phi] \tag{8-16b}$$

We are interested only in the solution at resonance ($\phi = \pi/2$) in a high Q_1 system (β is imaginary). Thus

$$q = q_0\left[-\cos\omega_0 t + e^{-\alpha t}\left(\cos\omega_d t + \frac{\alpha}{\omega_d}\sin\omega_d t\right)\right] \tag{8-17}$$

The solution for $Q_1 = 8$ is plotted in Fig. 8-4*a* where it is seen that the amplitude of the oscillating solution builds up over time to its steady state value of $q_0/q_{st} = 8$. The envelopes for the local maxima are plotted in Fig. 8-4*b* for $Q_1 =$ 4, 8, and 12, and there one sees that the steady state amplitude is larger in the high Q_1 systems, but it takes a longer time to reach this value.

8-2.3 Energy Considerations

First we consider the electric energy stored between the capacitor plates as a function of time for the steady state. The change in the energy is

$$\Delta E_C = \int_0^{\omega t} V_C\,dq = \int_0^{\omega t} \frac{q_0}{C}\sin(\omega t - \phi) q_0\cos(\omega t - \phi)\,d(\omega t)$$
$$= \frac{q_0^2}{2C}\sin^2 u\,\Bigg|_{-\phi}^{\omega t - \phi} \tag{8-18}$$

Clearly, for $\omega t = 2m\pi$, $\Delta E_C = 0$, as it must for the steady state.

The change in the magnetic energy stored in the inductor is

$$\Delta E_C = \int_0^{\omega t} V_L\,dq = \int_0^{\omega t} -\omega^2 L q_0 \sin(\omega t - \phi) q_0\cos(\omega t - \phi)\,d(\omega t)$$
$$= -\tfrac{1}{2}\omega^2 L q_0^2 \sin^2 u\,\Bigg|_{-\phi}^{\omega t - \phi} \tag{8-19}$$

At the resonant frequency

$$\Delta E_C + \Delta E_L = \tfrac{1}{2} q_0^2 \sin^2 u\,\Bigg|_{-\phi}^{\omega t - \phi}\left(\frac{1}{C - \omega_0^2 L}\right) = 0 \tag{8-20}$$

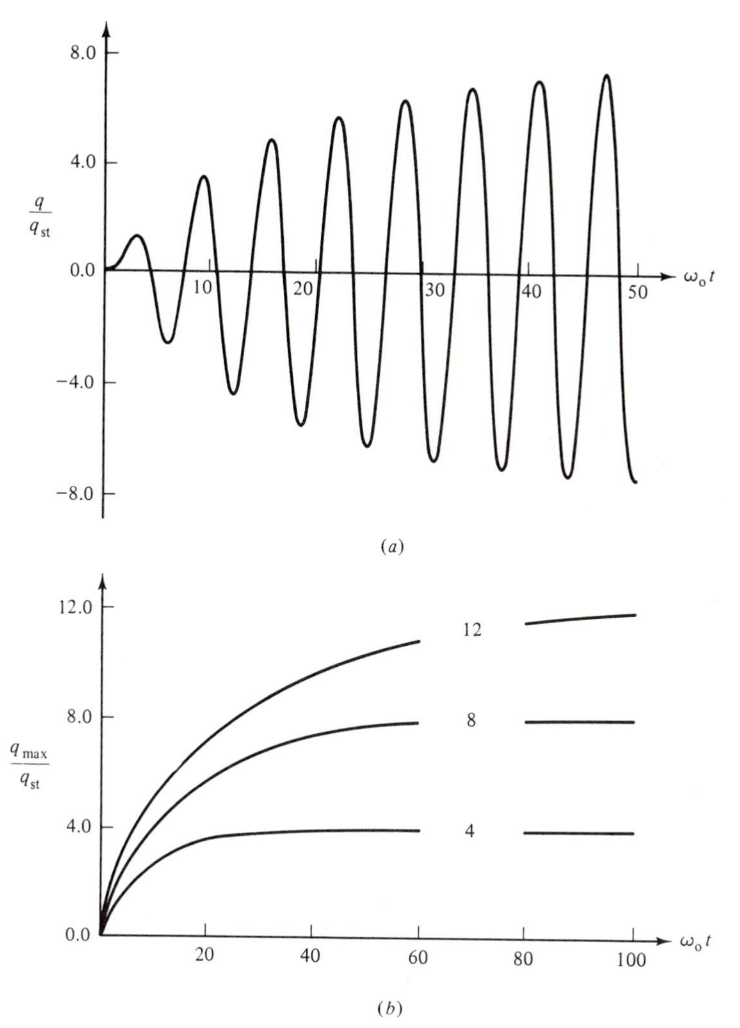

Figure 8-4 (*a*) The reduced charge, q/q_{st}, plotted as a function of $\omega_o t$ for small values of $\omega_o t$ when $Q_1 = 8$. (*b*) The growth in the amplitude of oscillation, q_{max}/q_{st}, at small values of $\omega_o t$ for $Q_1 = 4.0$, 8.0, and 12.0.

Since the sum of these two energies is zero at every point in the cycle, at no time is the voltage source charged by the energy in the circuit: the capacitor and the inductor feed it back and forth between each other. For $\omega < \omega_0$, $1/C > \omega_0^2 L$ so that the capacitor charges the voltage source at some points during the cycle, and for $\omega > \omega_0$, $1/C < \omega_0^2 L$ so that the inductor charges the voltage source at some points during the cycle. The effect of the charging is to reduce the stored energy in the circuit and, therefore, the amplitude of the charge.

The energy loss per cycle due to i^2R heating is

$$\Delta E_R = \int_0^{2\pi} V_R \, dq = \int_0^{2\pi} \omega R q_0 \cos(\omega t - \phi) q_0 \cos(\omega t - \phi) \, d(\omega t)$$

$$= \pi \omega q_0^2 R \tag{8-21}$$

For the steady state the net energy in per cycle must equal the net energy out. The energy in is

$$\Delta E = \int_0^{2\pi} V\,dq = \int_0^{2\pi} V_0 \sin \omega t q_0 \cos(\omega t - \phi)\, d(\omega t)$$

$$= \pi q_0 V_0 \sin \phi \tag{8-22a}$$

$\sin \phi$ can be found from the phasor diagram in Fig. 8-2 and can be substituted into Eq. 8-22a. When this is done,

$$\Delta E = \pi q_0 V_0 \frac{\omega \tau}{\{[1 - (\omega/\omega_0)^2]^2 + (\omega\tau)^2\}^{1/2}} \tag{8-22b}$$

Equating Eqs. 8-21 and 8-22b, one finds that q_0/q_{st} is given by Eq. 8-5.

Equations 8-21 and 8-22a are plotted in Fig. 8-5 for the case $\omega = \omega_0$ where $\sin \phi = 1$. The equilibrium value of q_0 is, of course, the point where the two curves cross. Note that for q_0 less than its equilibrium value, $\Delta E_R < \Delta E$. This is what one should expect, since during the transient when the energy stored in the system is being built up, the energy per cycle put into the system is greater than the energy flowing out.

In ending this section we will verify the energy definition of Q_1 given by Eq. 8-8. The stored energy at resonance is equal to the maximum energy stored in the capacitor or the inductor. Thus

$$Q_1 = \frac{q_0^2/2C}{\pi\omega_0 q_0^2 R/2\pi} = \frac{1}{\omega_0 RC} \tag{8-8}$$

The energy loss per radian is also equal to P/ω where P is the power output. Thus

$$Q_1 = \frac{\text{stored energy}}{P/\omega_0} \tag{8-23}$$

EXAMPLE 8-2

Compute **(a)** the stored energy, and **(b)** the energy dissipated at resonance by the circuit with the parameters computed in Example 8-1 if $V_0 = 1.0$ V.

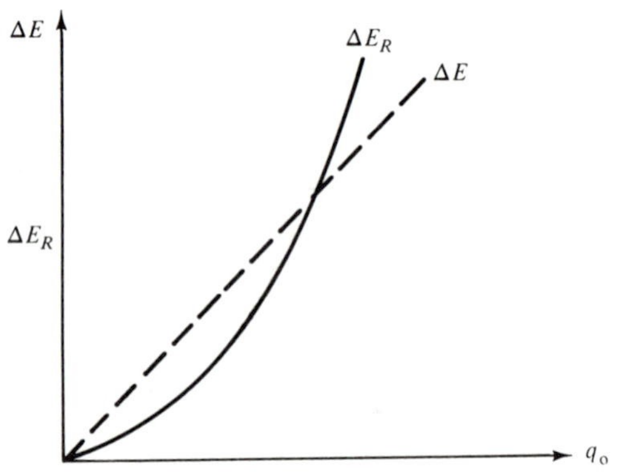

Figure 8-5 The energy supplied by the battery per cycle, ΔE, and the energy dissipated by the resistor per cycle, ΔE_R, plotted as a function of the charge amplitude, q_o.

(a)
$$q_0 = Q_1 q_{st} = Q_1 C V_0 = (10^6)(10^{-12})1 = 10^{-6}\text{ C}$$

$$\text{stored energy} = q_0^2/2C = \frac{(10^{-6})^2}{(2)(10^{-12})} = 0.5\text{ J}$$

(b)
$$P = \omega$$

$$\frac{\text{stored energy}}{Q_1} = \frac{(1.884 \times 10^{15})(0.5)}{10^6} = 9.42 \times 10^8\text{ W}$$

This absurdly large number is a result of the very large operating frequencies, which are never even approached.

8-3 OPTICAL RESONANCE

8-3.1 Energy and Losses in an Optical Cavity

The optical cavity in Fig. 8-6 is composed of a light source emitting light with a constant amplitude that is located between two mirrors of reflectivity R_1 and R_2 that are placed perpendicular to the direction of propagation. The light source is equivalent to the voltage source. The waves reflected back and forth build up the electric field inside of the cavity when they constructively interfere, and the buildup is largest when the mirrors are separated by a distance equal to half integral multiples of the wavelength. This is the resonance condition. The amplitude of the waves builds up until the energy injected by the light source equals the energy dissipated by the cavity. The energy is dissipated within the cavity by absorption mechanisms; this is equivalent to resistive losses. The useful light transmitted through the mirrors is another source of loss, and the electrical equivalent is drawing power out of the circuit by clamping an output device across one or all of the individual elements.

When the cavity length is a half integral multiple of the wavelength, the reflected waves form standing waves described by the equation

$$\xi = \xi_0 \sin \omega t \sin kz \tag{8-24}$$

and

$$k = \frac{m\pi}{L} \tag{8-25}$$

Figure 8-6 Schematic of light being reflected back and forth inside of an optical cavity.

where L is the cavity length. The average electric energy per unit volume is

$$\frac{\langle E \rangle_\xi}{V} = \frac{1}{2\pi}\int_0^{2\pi}\frac{1}{2\pi}\int_0^{2\pi}\frac{1}{2}\,\epsilon\xi_0^2 \sin^2 \omega t \sin^2 kz \; d(\omega t)\; d(kz)$$

$$= \tfrac{1}{8}\epsilon\xi_0^2 \tag{8-26a}$$

The average magnetic energy per unit volume is the same so that the total average electromagnetic energy in the cavity is

$$E = \tfrac{1}{4}\epsilon\xi_0^2 V \tag{8-26b}$$

(This is one-half the energy found in discussing the Poynting vector in Section 1-3 because in this situation the energy is averaged over both time and position.)

Using the definition for Q_1 given by Eq. 8-23, we find that

$$Q_1 = \frac{\omega\epsilon V\xi_0^2}{4P} \tag{8-27a}$$

or, conversely,

$$\xi_0 = \left(\frac{4PQ_1}{\omega\epsilon V}\right)^{1/2} \tag{8-27b}$$

Equations 8-27a and 8-27b implicitly illustrate an important problem that was discussed in Chapter 7 and will be explored in some detail in Chapter 9. That is, the larger Q_1 the larger ξ_0 will be. This will increase the rate of stimulated emission. However, to have a large Q_1 the power loss, and, therefore, the power emitted by the laser is small. Thus, there is a value for the mirror reflectivity for which the emitted power is a maximum. The mirrors must be reflective enough for ξ_0 to be large enough to generate a sufficient amount of stimulated emission, but also transmissive enough so that all of the energy is not locked up in the cavity.

EXAMPLE 8-3

Light is emitted from an optical cavity 10 cm long with a bandwidth of 5×10^7 Hz and a peak wavelength of $\lambda_0 = 6328$ Å. The emitted intensity is 50 mW/cm². Find **(a)** Q_1 of the cavity if all of the losses are emission losses, **(b)** the amplitude of the electric field in the cavity if $\epsilon = \epsilon_0$, and **(c)** the energy density in the cavity.

(a) $$Q_1 = \frac{\nu_0}{d\nu^*} = \frac{c_0}{\lambda_0\, d\nu^*} = \frac{3 \times 10^{10}}{(0.6328 \times 10^{-4})(5 \times 10^7)} = 9.48 \times 10^6$$

(b) $$\xi_0 = \left(\frac{4Q_1 P/A}{\omega\epsilon L}\right)^{1/2} = \left[\frac{(4)(9.48 \times 10^6)(5 \times 10^2)}{\dfrac{2\pi 3 \times 10^8}{0.6328 \times 10^{-6}}(8.854 \times 10^{-12})(0.10)}\right]^{1/2}$$

$$= 2.68 \times 10^4 \; V/M$$

(c) $$\frac{E}{V} = \frac{1}{4}\,\epsilon\xi_0^2 = (0.25)(8.854 \times 10^{-14})(26.8)^2 = 1.59 \times 10^{-11} \text{ J/cm}^3$$

As should be expected, the absorption losses in the cavity are represented by the absorption coefficient. If the mirrors are perfectly reflecting, the attenuation after one round-trip of a beam is $\exp(-2L\alpha)$. For imperfectly reflecting mirrors, however, the attenuation is $R_1 R_2 \exp(-2L\alpha)$. The effective absorption coefficient, α_e, is, therefore,

$$e^{-2L\alpha_e} = R_1 R_2 e^{-2L\alpha} \tag{8-28a}$$

or

$$\alpha_e = \alpha - \frac{1}{2L} \ln R_1 R_2 \tag{8-28b}$$

We now find the relationship between α_e and Q_1. You should expect that increasing α_e decreases Q_1. If the power were shut off, the energy in the cavity would decay. The rate of decay, $-\partial E/\partial t$, would be equal to the power loss. Thus, from Eq. 8-8,

$$Q_1 = \frac{\omega E}{-\partial E/\partial t} \tag{8-29a}$$

From Eq. 8-28a we see that the energy decays exponentially. Thus

$$\frac{-\partial E}{\partial t} = \frac{E}{\tau} \tag{8-30}$$

Since the wave travels a distance $c\tau$ in time τ, the time in which the energy decays to l/e of its original value

$$\alpha_e = \frac{1}{c\tau} \tag{8-31}$$

Substituting Eqs. 8-30 and 8-31 into Eq. 8-29a, we find that

$$Q_1 = \frac{\omega}{c\alpha_e} \tag{8-29b}$$

EXAMPLE 8-4

If all of the losses are transmission losses, what is the mirror reflectivity if $R_1 = R_2$, $\lambda_0 = 1\ \mu\text{m}$, $c = c_0$, $L = 10$ cm, and $Q_1 = 10^6$?

$$\alpha_e = -\frac{1}{2L} \ln R^2 = \frac{\omega}{Q_1 c} = \frac{2\pi}{\lambda_0 Q_1}$$

$$R = \exp\left(\frac{-2\pi L}{\lambda_0 Q_1}\right) = \exp\left[-(2\pi)\frac{10}{10^{-4}}(10^6)\right]$$

$$= \exp(-0.2\pi) = 0.533$$

8-3.2 Longitudinal Modes

The mathematical derivation of the number of modes per unit volume between ν and $\nu + d\nu$ in an optical cavity is similar to the derivation of the number of electron states per unit volume between E and $E + dE$ in the conduction band since, in both instances, the number of standing waves per unit volume is found (see Section 4-4.4). The resonant frequencies, or modes, in one dimension are those associated with wavelengths that are half integral multiples of the cavity length. Thus

$$\nu = \frac{mc}{2L} \tag{8-32}$$

The lowest frequency modes are illustrated in Fig. 8-7.

In three dimensions the component of the k vectors must be an integral multiple of π divided by the cavity length in the component direction (recall Eq. 8-25). For a cubic box

$$k_x = \frac{m_1 \pi}{L} \tag{8-33a}$$

$$k_y = \frac{m_2 \pi}{L} \tag{8-33b}$$

$$k_z = \frac{m_3 \pi}{L} \tag{8-33c}$$

Thus

$$\bar{k} = \frac{\pi}{L}(m_1 i + m_2 j + m_3 k) \tag{8-34}$$

The modes can therefore be represented by a cubic lattice in k space π/L on a side.

Converting the allowed wave vectors to the resonance frequencies, we find that

$$\nu_x = \frac{c}{2L} m_1 \tag{8-35a}$$

$$\nu_y = \frac{c}{2L} m_2 \tag{8-35b}$$

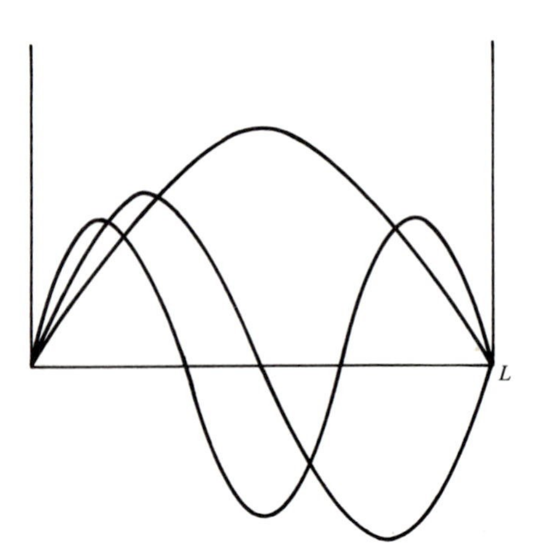

Figure 8-7 The first three standing waves that are parallel to a cube edge.

$$\nu_z = \frac{c}{2L} m_3 \tag{8-35c}$$

They, too, form a cubic lattice in frequency space, and this is illustrated in two dimensions in Fig. 8-8. One can see that the points representing resonant frequencies form cubes $c/2L$ on a side. Thus, the volume associated with each resonant frequency is $(c/2L)^3$. This volume is actually associated with two allowed frequencies: one for each of the two components of the direction of vibration of the ξ field.

For large values of m, there are a number of modes with the same frequency—only the mix of the components is different. Modes with the same frequency can be represented by a vector of constant magnitude, ν^*. The locus of points of this vector is one-eighth of a spherical surface of radius, ν^*, and this is also shown in Fig. 8-8. The total number of resonant frequencies for which $\nu \leq \nu^*$ is then given by the volume of the sphere of radius ν^* in the first octant divided by the volume per state. Thus the number of modes, N, is

$$N = \frac{1}{8} \frac{4/3\pi\nu^{*3}}{1/2(c/2L)^3} = \frac{8\pi L^3 \nu^{*3}}{3c^3} \tag{8-36}$$

EXAMPLE 8-5

What is the number of modes of electromagnetic waves between the wavelengths of 4000 and 4100 Å in a black box that is 5 cm on a side?

The number of modes between frequency ν_2 and ν_1 is

$$N_{\nu 2} - N_{\nu 1} = \frac{8\pi L^3}{3c^3}(\nu_2^3 - \nu_1^3) = \frac{8\pi L^3}{3}\left(\frac{1}{\lambda_2^3} - \frac{1}{\lambda_1^3}\right)$$

$$= \frac{8\pi 5^3}{(3)10^{-12}}\left(\frac{1}{0.4^3} - \frac{1}{0.41^3}\right) = 1.168 \times 10^{15}$$

This illustrates how very many resonant frequencies there are when the cavity length is much longer than the wavelength.

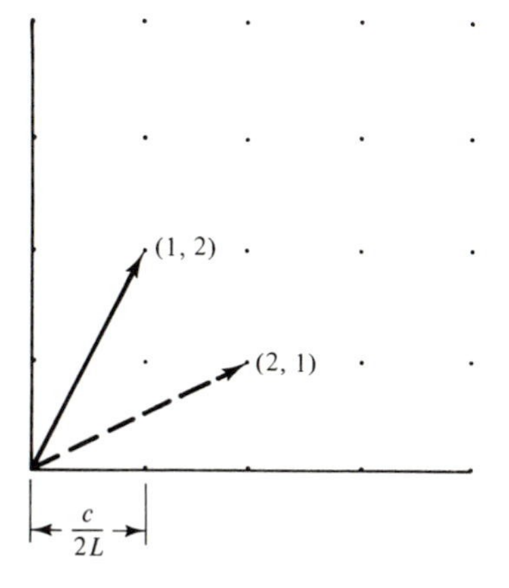

Figure 8-8 The two-dimensional representation of the allowable frequencies for standing waves in a cubical box of length, L, and two frequencies having the same magnitude, but different indices.

8-4 FABRY–PEROT CAVITY

8-4.1 Reflectivity and Transmittance

The amount of light that is transmitted across two interfaces depends on the transmittance at each interface and the interference between waves reflected from the first surface and waves reflected from the second. As is indicated in Fig. 8-9*a*, one must consider an infinite number of reflections; this is mathematically possible because the wave amplitude decreases after each reflection, since some light is also transmitted.

The cavity we will consider in Fig. 8-9*a* has an index of refraction n_2 and is surrounded on either side with a medium of n_1 with $n_1 < n_2$. There also is no absorption in the cavity.

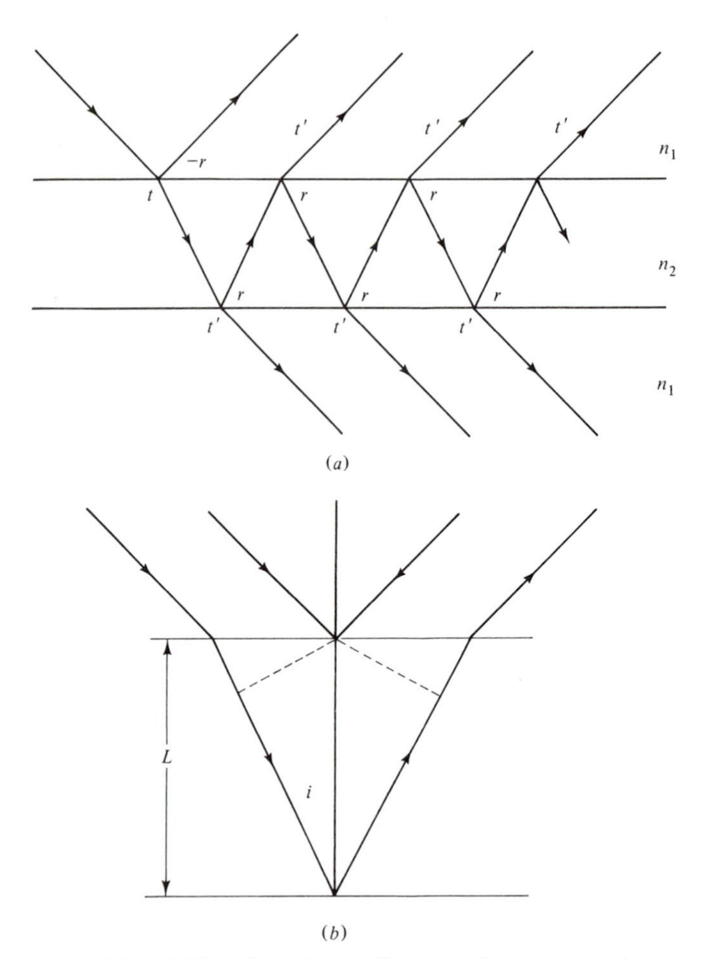

Figure 8-9 (*a*) The reflected waves from and the transmitted waves across two interfaces. (*b*) The diagram used to determine the path-length difference between a wave reflected from the top surface and one reflected from the bottom surface.

The reflection coefficient from the first surface is

$$\frac{\xi_r}{\xi_i} = -r + trt'e^{-2j\phi} + trrrt'e^{-4j\phi} + \cdots$$

$$= -r + tt' \sum_{1}^{\infty} r^{2m-1}e^{-2jm\phi} \qquad \text{(8-37a)}$$

r is the single surface reflection coefficient at the 2-1 interface, t is the single surface transmission coefficient at the 1-2 interface, t' is the single surface transmission coefficient at the 2-1 interface, and 2ϕ is the phase lag between the wave reflected from the first surface and the wave reflected once from the second surface. The minus sign for the first r is a result of the 180° phase change for ξ_r when $n_2 > n_1$. The sum converges since $r < 1$, and it can be found by multiplying it by $r^2e^{-2j\phi}$ and subtracting terms. Thus

$$\sum = re^{-2j\phi} + r^3e^{-4j\phi} + \cdots + r^{2m-1}e^{-2jm\phi} \qquad \text{(8-38a)}$$

and and $r^2e^{-2j\phi} \sum = r^3e^{-4j\phi} + r^5e^{-6j\phi}$

$$+ \cdots + r^{2m-1}e^{-2jm\phi} + r^{2m+1}e^{-2(m+1)j\phi} \qquad \text{(8-38b)}$$

so their difference is

$$(1 - r^2e^{-2j\phi}) \sum = re^{-2j\phi} - r^{2m+1}e^{-2(m+1)j\phi} = re^{-2j\phi} \qquad \text{(8-38c)}$$

for $m \to \infty$

Therefore

$$\sum = \frac{re^{-2j\phi}}{1 - r^2e^{-2j\phi}} \qquad \text{(8-39)}$$

and

$$\frac{\xi_r}{\xi_i} = \frac{-r + re^{-2j\phi}(r^2 + tt')}{1 - r^2e^{-2j\phi}} \qquad \text{(8-37b)}$$

This equation can be reduced further by noting that

$$r^2 + tt' = 1 \qquad \text{(8-40)}$$

Equation 8-40 can be easily verified for normal incidence using Eqs. 1-59 and 1-60. This reduces Eq. 8-37b to

$$\frac{\xi_r}{\xi_i} = \frac{r(e^{-2j\phi} - 1)}{1 - r^2e^{-2j\phi}} \qquad \text{(8-37c)}$$

The transmission coefficient is

$$\frac{\xi_t}{\xi_i} = tt' + trrt'e^{-2j\phi} + \cdots + tt'r^{2m}e^{-2jm\phi} = tt' \sum_{0}^{\infty} r^{2m}e^{-2jm\phi} \qquad \text{(8-41a)}$$

which reduces to

$$\frac{\xi_t}{\xi_i} = \frac{tt'}{1 - r^2e^{-2j\phi}} \qquad \text{(8-41b)}$$

The phase angle, 2ϕ, is 2π times the path length difference between the waves reflected from the first and second surfaces divided by the wavelength. From Fig. 8-9*b* we see that the

$$\text{path-length difference} = 2L \cos i \tag{8-42}$$

Therefore

$$2\phi = \frac{4\pi n_2 L \cos i}{\lambda_0} \tag{8-43a}$$

Because the reflected and incident beams lie in the same medium so their velocities are the same

$$\frac{I_r}{I_i} = \frac{\xi_r}{\xi_i} \cdot \frac{\xi_r^*}{\xi_i^*} = \frac{4R \sin^2 \phi}{(1 - R)^2 + 4R \sin^2 \phi} \tag{8-44}$$

where again $R = r \cdot r^*$.

The energy in the cavity is at its steady state value so the energy in per unit time per unit area must equal the energy out per unit time per unit area. The index of refraction for the reflected and transmitted rays are the same; thus

$$I_t = I_i - I_r \tag{8-45a}$$

or

$$\frac{I_t}{I_i} = 1 - \frac{I_r}{I_t} = \frac{(1 - R)^2}{(1 - R)^2 + 4R \sin^2 \phi} \tag{8-45b}$$

$I_t/I_i = 1.0$ at normal incidence when $2L$ is an integral multiple of λ; $\phi = m\pi$. The intensity of the reflected wave must, therefore, be zero. This is because there is complete destructive interference between a wave reflected from the first surface and a wave reflected from the second surface. Recall that there is a 180° phase change at the first surface because $n_2 > n_1$, but there is no phase change on reflection from the second surface. If there are absorption losses in the cavity and the fractional loss per pass is $(1 - A)$, then $(I_t/I_i)_{\max}$ is reduced to

$$\left(\frac{I_t}{I_i}\right)_{\max} = \frac{(1 - R)^2 A}{(1 - RA)^2} \tag{8-46}$$

The minimum value of I_t/I_i is

$$\left(\frac{I_t}{I_i}\right)_{\min} = \left(\frac{1 - R}{1 + R}\right)^2 \tag{8-47}$$

and this occurs when the path length difference is an odd half integral multiple of the wavelength; $\phi = (2m - 1)\pi/2$. Now the waves reflected from the first and second surfaces completely constructively interfere with each other. Note that the larger R is, the smaller is the minimum value. This is illustrated in Fig. 8-10.

Also shown in Fig. 8-10 is the fact that the bandwidth of the transmission peak is smaller when the reflectivity is larger. We now find the relationship between the bandwidth, $d\phi^*$, and the reflectivity. The phase angle ϕ' for which $(I_t/I_i) = \frac{1}{2}$ is found from Eq. 8-45b to be

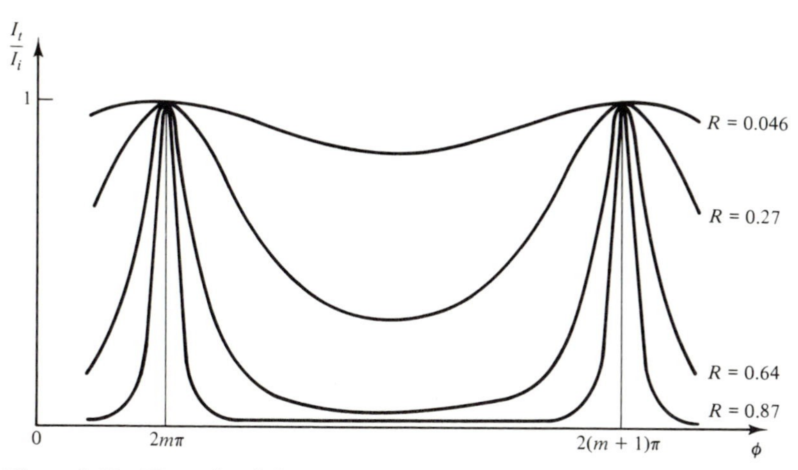

Figure 8-10 The ratio of the transmitted and incident intensities plotted as a function of the phase angle, ϕ, for different reflectivities, R, when there are no absorption losses in the cavity.

$$\sin^2 \phi' = \frac{(1-R)^2}{4R} \tag{8-48}$$

Another expression for $\sin^2 \phi'$ is

$$\sin^2 \phi' = \sin^2\left(\frac{m\pi \pm d\phi^*}{2}\right)$$

$$= \left[\sin m\pi \cos\left(\frac{d\phi^*}{2}\right) \pm \sin\left(\frac{d\phi^*}{2}\right) \cos m\pi\right]^2 = \sin^2\left(\frac{d\phi^*}{2}\right) \tag{8-49}$$

For a narrow bandwidth

$$\sin^2\left(\frac{d\phi^*}{2}\right) \simeq \frac{d\phi^{*2}}{4} \tag{8-50a}$$

so that

$$d\phi^* \simeq \frac{1-R}{\sqrt{R}} \tag{8-50b}$$

8-4.2 Optical Spectrum Analyzer

An optical cavity can be used to analyze an optical signal in much the same way that the impulse function is used to analyze an electrical signal using convolution. An ideal band pass filter with an infinitesimal bandwidth $d\nu^*$ passes all of the light at frequencies between ν and $\nu + d\nu^*$ and blocks all of the other light. The intensity of the light is measured; the band pass frequencies are increased to between $\nu + d\nu^*$ and $\nu + 2\, d\nu^*$; the light intensity is recorded; and so forth. In this way the frequency distribution of the intensity can be determined. The Fabry-Perot cavity can be used as the band pass filter, and it approaches the ideal filter as $R \rightarrow 1$.

To express the phase angle, ϕ, in terms of the frequency, we recall that I_t/I_i is a maximum when the path length difference is an integral number of wavelengths. Thus, the frequencies for which I_t/I_i is a maximum are

$$\nu_{\max} = \frac{mc_0}{2Ln_2 \cos i} \tag{8-51}$$

and the separation between the maximum frequencies, $\delta\nu$, is

$$\delta\nu = \frac{c_0}{2Ln_2 \cos i} \tag{8-52}$$

Eq. 8-43a can also be written

$$\phi = \frac{2\pi n_2 L \cos i}{c_0} \nu \tag{8-43b}$$

so that from the derivative of Eq. 8-43b and Eqs. 8-50b and 8-52 the bandwidth of the transmission peak is

$$d\phi^* = \frac{2\pi n_2 L \cos i}{c_0} d\nu^* = \pi \frac{d\nu^*}{\delta\nu} = \frac{1 - R}{\sqrt{R}} \tag{8-53}$$

The ratio $\delta\nu/d\nu^*$ is called the finesse, F. Thus

$$F = \frac{\delta\nu}{d\nu^*} = \frac{\pi\sqrt{R}}{1 - R} \tag{8-54}$$

The finesse, like the quality factor, is large when the bandwidth is small, but they are not the same quantities because $\delta\nu$ is not the same thing as ν; it is only $1/m$ the value of ν. The integer m is the number of wavelengths in the length $2L$. Thus

$$F = \frac{\lambda_0}{2n_2 L} Q_1 \tag{8-55}$$

The Fabry-Perot cavity is not an ideal filter because $R \neq 1$ and there is more than one transmission peak. Thus, if we are to be able to measure accurately the frequency distribution in an emission peak, the bandwidth of the emission peak, $\Delta\nu$, and the bandwidth of the transmission peaks must be significantly less than the bandwidth of the modes within an emission peak.

The modes within an emission peak from a He-Ne laser centered at 6328 Å are shown in Fig. 8-11. The envelope of the modes would be the emission peak from an electronic transition in neon if the gas were not confined to a laser cavity. The modes correspond to resonant frequencies, and their bandwidths are determined by the losses in the laser cavity. The separation between the modes is given by Eq. 8-52 where now L is the length of the laser cavity and n_2 is the index of refraction of the laser material. These points are illustrated in Example 8-6.

EXAMPLE 8-6

(a) Find the maximum length of the Fabry–Perot cavity that can be used to analyze the output from a He-Ne laser centered at 6328 Å if the emission

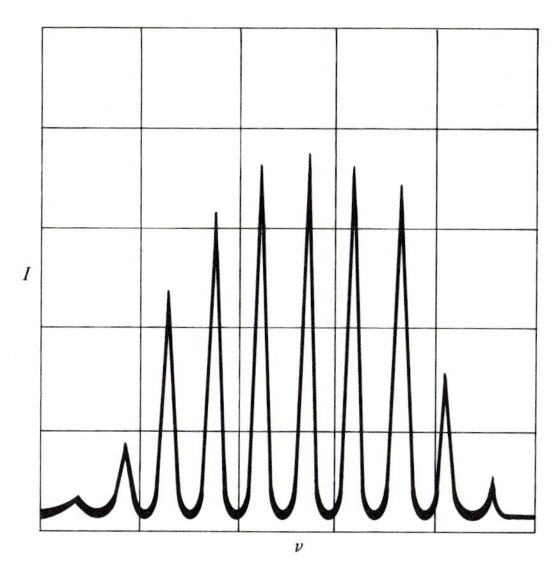

Figure 8-11 Modes from a He-Ne laser. Each square corresponds to a frequency change of 250 MHz.

peak bandwidth is $\Delta\nu = 1.5 \times 10^9$ Hz, $n_2 = 1.0$, and the beam is normally incident on the cavity. **(b)** What is the frequency separation between modes if the laser is 100 cm long? **(c)** What is the minimum acceptable reflectivity of the mirrors, R, if the bandwidth of the transmission peaks for the cavity, $d\nu^*$, is no more than $\frac{1}{10}$ of the separation between the laser modes. **(d)** Find the finesse, F, and the quality factor, Q_1, for this Fabry–Perot cavity.

(a)

$$\delta\nu(\text{cavity}) = \frac{c_0}{2n_2 L} = \Delta\nu(\text{laser})$$

$$\therefore L(\text{cavity}) \le \frac{c_0}{2n_2\,\Delta\nu} = \frac{3.00 \times 10^{10}}{(2)(1)(1.5 \times 10^9)} = 10 \text{ cm}$$

(b)

$$\delta\nu(\text{laser}) = \frac{c_0}{2n_2 L} = \frac{3.00 \times 10^{10}}{(2)(1)(10^2)} = 1.5 \times 10^8 \text{ Hz}$$

(c)

$$d\nu^*(\text{cavity}) = \frac{1-R}{\pi\sqrt{R}}\,\delta\nu(\text{cavity}) = 0.1\delta\nu(\text{laser}) = 1.5 \times 10^7 \text{ Hz}$$

$$R^2 - \left[2 + \left(\pi\frac{d\nu^*}{\delta\nu}\right)^2\right]R + 1 = R^2 - 2.00096R + 1$$

$$R = \tfrac{1}{2}\{2.00096 - [(2.00096)^2 - 4]^{1/2}\} = 0.969$$

(d)

$$F = \frac{\delta\nu}{d\nu^*} = \frac{1.5 \times 10^9}{1.5 \times 10^7} = 100$$

$$Q_1 = \frac{2n_2 L}{\lambda_0}F = \frac{(2)(1)(10)(10^2)}{0.6328 \times 10^{-4}} = 3.16 \times 10^7$$

8-5 FABRY–PEROT LASER

When there are absorption losses in the cavity and the reflection coefficients at the two surfaces are not equal, the transmission coefficient is

$$\frac{\xi_t}{\xi_i} = tt'e^{-j\beta L} + tr_1r_2t'e^{-3j\beta L} + \cdots = \frac{tt'e^{-j\beta L}}{1 - r_1r_2e^{-2j\beta L}} \tag{8-56a}$$

Recalling from Eq. 1-40 that $\beta = k - j\alpha/2$

$$\frac{\xi_t}{\xi_i} = \frac{tt'e^{-(\alpha L/2)-jkL}}{1 - r_1r_2e^{-\alpha L-2jkL}} \tag{8-56b}$$

For $kL = 2m\pi$, ξ_t/ξ_i is given by

$$\frac{\xi_t}{\xi_i} = \frac{tt'e^{-\alpha L/2}}{1 - r_1r_2e^{-\alpha L}} \tag{8-56c}$$

Under normal conditions, that is, when α is positive, $(\xi_t/\xi_i) < 1$. However, if α were negative, ξ_t/ξ_i could be infinity. This physically would correspond to having a transmitted beam when there is no incident beam. It does occur in lasers when the proper conditions are met. Now α is a gain coefficient instead of an absorption coefficient. The value of α for which $(\xi_t/\xi_i) = \infty$ is found from the equation

$$1 - r_1r_2e^{-\alpha^* L} = 0 \tag{8-57a}$$

or

$$\alpha^* = \frac{1}{L} \ln r_1r_2 \tag{8-57b}$$

From Eq. 7-13 we see that (8-58)

$$\alpha^* = \alpha_g - \alpha$$

READING LIST

1. H. H. Skilling, *Electrical Engineering Circuits.* New York: John Wiley, 1960, Chapter 7.
2. J. P. den Hartog, *Mechanical Vibrations,* 4th Ed. New York: McGraw-Hill, 1956, Chapters 1 and 2.
3. R. T. Weidner and R. L. Sells, *Elementary Modern Physics,* 2d Ed. Boston: Allyn and Bacon, 1973, Chapter 11.
4. F. A. Jenkins and H. E. White, *Fundamentals of Optics,* 4th Ed. New York: McGraw-Hill, 1976, Chapters 14, 22, and 23.
5. A. Yariv, *Introduction to Optical Electronics,* 2d Ed. New York: Holt, Rinehart and Winston, 1976, Chapter 4.
6. J. T. Verdeyen, *Laser Electronics.* Englewood Cliffs, NJ: Prentice-Hall, 1981, Chapters 5 and 6.

PROBLEMS

8-1. For a LRC series circuit it has been shown that $q_0/q_{st} = [(1 - y^2)^2 + (\alpha y)^2]^{-1/2}$; $q_0'/q_{st} = (1 - y^2)/[(1 - y^2)^2 + (\alpha y)^2]$ and $q_0'/q_{st} = \alpha y/[(1 - y^2)^2 + (\alpha y)^2]$ where $y = \omega/\omega_0$ and $\alpha = \omega_0 RC$.

(a) Find the value of ω for which q_0/q_{st} is a maximum.
(b) Find the value of ω when q_0'/q_{st} is a maximum and when it is a minimum.
(c) What is the maximum value of q_0/q_{st}?
(d) What is the value of q_0/q_{st} at resonance?
(e) What is the maximum value of q_0'/q_{st}? The minimum value?
(f) Find the value of ω for which q_0'/q_{st} is a maximum.
(g) What is the value of q_0'/q_{st} at resonance? (q_0' is the real part of q_0.)

8-2. In the following calculations assume that the quality factor, $Q_1 = 25$.

(a) What is the percent error in ω in assuming that the maximum in q_0 occurs at ω_0 instead of at ω_{max}?
(b) What is the error in q_0 in assuming that the maximum in q_0 occurs at $\omega = \omega_0$?
(c) What is the error in ω in assuming that the maximum in q_0'' occurs at $\omega = \omega_0$?
(d) What is the error in assuming $Q_1 = \omega_0/(\omega_2 - \omega_1)$ where ω_2 is the value of ω when q_0' is a minimum and ω_1 is the value of ω when q_0' is a maximum? (q_0'' is the imaginary part of q_0.)

8-3. What is the minimum value of R for which q_0/q_{st} is a maximum at $\omega = 0$? What is Q_1 for this R?

8-4. What is $d\phi/dy$ in terms of α and y? (see Problem 8-1.) What is $d\phi/dy$ at $y = 1$?

8-5. A spring, mass, and dash pot system, which has been stretched a distance x_0, is released at $t = 0$. What is the minimum value of β_1 for which no oscillation occurs? (See Section 2-3.2 for the definition of terms.)

8-6. (a) Show that the energy lost per cycle, ΔE, in a mass-dash pot-spring system is $\pi F_0 x_0 \sin \phi$.
(b) Determine the steady state value of x_0 at resonance using an energy balance.
(c) Repeat part b for the case when $\phi = 45°$.

8-7. Make a sketch of the q_0/q_{st} versus V/V_0 hysteresis loops for $\phi = 0, 45, 90, 135$, and $180°$ when $Q_1 = 10$. Use the same scale for all five curves.

8-8. (a) Find an expression for the amplitude of the charge on a capacitor plate, q_0, for an LC series circuit for $\omega \neq \omega_0$ when $V = V_0 \sin \omega t$ and $q(0) = \dot{q}(0) = 0$.
(b) Find an expression for q_0 when $\omega = \omega_0$ and explain physically what is occurring.

8-9. Consider a simple RC series circuit that has a periodic rectangular voltage pulse of magnitude V_0 applied beginning at $t = 0$. The pulse is on for a time T then off for a time T.

(a) What is the charge on the capacitor plate when $t = 2T$ and $t = 4T$?
(b) How much energy was added to the circuit during the first cycle? During the second cycle?

8-10. (a) Make a plot of I_r/I_i versus ϕ for a Fabry–Perot cavity when $R_1 = R_2 = 0.9$.
(b) Calculate the finesse, F.
(c) Calculate the transmission peak width, $d\phi^*$.
(d) Calculate the transmission peak width, $d\nu^*$, for normal incidence, $n = 1$, and $L = 10$ cm.
(e) Find the minimum value of I_t/I_i.

8-11. **(a)** If the fractional intensity loss per pass in a Fabry–Perot cavity is $1 - A$, show that the transmission peak intensity is $(1 - R)^2A/(1 - RA)^2$.
(b) For what value of A is the peak height 0.9?

8-12. Consider a Fabry–Perot cavity in which the top third has a width $L - \Delta$, the middle third has a width L, and the bottom third has a width $L + \Delta$.
(a) For what value of Δ will the adjacent transmission peaks overlap at the half power points? Express your answer in terms of $R = R_1 = R_2$.
(b) Compute Δ for $R = 0.9$ and $\lambda_0 = 1.0\ \mu$m.
(c) Comment on the effects of surface roughness on the width of the transmission peaks.

8-13. Consider a diverging monochromatic beam that is incident on a Fabry–Perot cavity.

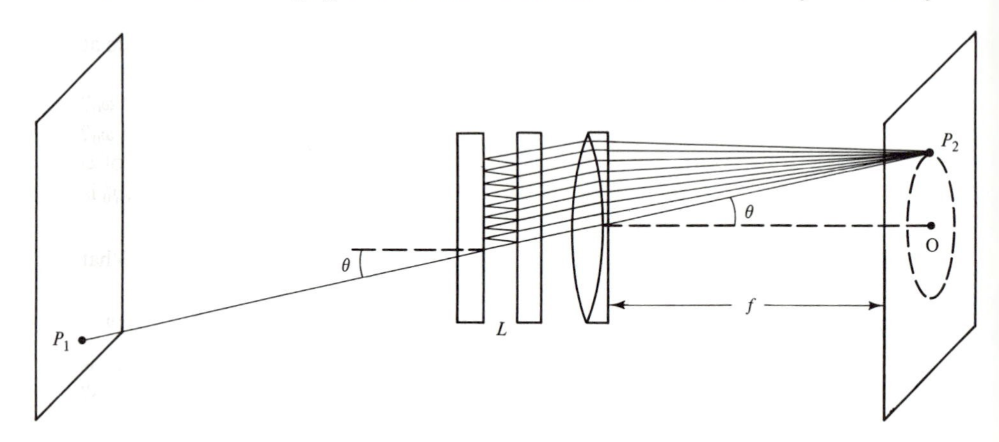

(a) Obtain an expression for the various angles along which the output energy is propagating.
(b) Let the output beam in part a be incident on a lens with a focal length f. Show that the energy distribution in the focal plane consists of a series of circles, each corresponding to a different value of m. Obtain an expression for the radii of the circles.
(c) Consider the effect in part b of having simultaneously two frequencies ν_1 and ν_2 present in the input beam. Derive an expression for the separation of the respective circles in the focal plane. Show that the smallest separation $\nu_1 - \nu_2$ that can be resolved by this technique is given by $(\Delta\nu)_{\min} \sim c/2nLF$.

8-14. **(a)** Using the quantum mechanical value of $\alpha_{st} = 18\ \pi\epsilon_0 a_0^3$ where a_0 = Bohr radius = 0.53 Å for a hydrogen atom, calculate the natural frequency, ω_0, for the electronic polarizability.
(b) If the radiation power of the oscillating hydrogen atom dipole is $P = \sqrt{\mu_0/\epsilon_0}\pi/3(qx_0\omega/\lambda)^2$, determine β_1 at $\omega = \omega_0$ if the only loss is radiative loss.
(c) What is Q_1 for this system?

8-15. A semiconductor laser is typically 500 μm long and the cleaved surfaces act as the mirrors. If $n = 3.4$ and all of the loss is transmission loss, find
(a) F.
(b) $d\nu^*$.
(c) Q_1 for the cavity if light is emitted at 0.84 μm.
(d) Repeat parts a, b, and c for $\alpha = 8\ \text{cm}^{-1}$.

8-16. Consider the semiconductor laser described in Problem 8-15.

(a) How many modes are there in an emission peak with a bandwidth of $\Delta\lambda$ = 300 Å? (This is the LED bandwidth.)

(b) What is the bandwidth in angstroms of an individual mode? (This is the laser bandwidth.) Consider both the cases when $\alpha = 0$ and $\alpha = 8\ \text{cm}^{-1}$.

8-17. The emission peak bandwidth of a neon 0.6328-μm line is $\Delta\nu = 1.5$ GHz.

(a) How many modes are there in this bandwidth if the laser is 30 cm long and $n = 1$?

(b) What is the bandwidth of the modes if the reflectivity of the mirrors is $R = 0.95$?

8-18. If the cavity mirrors have a high reflectivity, is the transient time large or small? Explain.

8-19. A thin film deposited on a bulk material can act as an antireflecting coating if it has the proper thickness and its index of refraction, n_2, is intermediate between that of the surrounding medium, $n_1 = 1$, and the bulk material, n_3.

(a) Find an expression for ξ_r/ξ_i.

(b) Find an expression for $R = (\xi_r/\xi_i) \times (\xi_r^*/\xi_i^*)$.

(c) At normal incidence for what values of the film thickness, L, is R a minimum?

(d) In terms of n_3 for what value of n_2 is $R = 0$ at the minimum?

8-20. An optically bistable device can be constructed from a Fabry–Perot cavity if the index of refraction of the cavity is a function of the light intensity. For normal incidence how much must n_2 change for ξ_t/ξ_i to move from

(a) A maximum to a minimum.

(b) A maximum to a half power point if $n_2 = 2.5$, $L = 1.0$ mm, and $\lambda_0 = 1.0\ \mu$m.

Chapter 9

Laser Fundamentals

9-1 INTRODUCTION

To understand how a laser works, one must first become familiar with stimulated absorption and spontaneous and stimulated emission. These subjects were considered in Chapter 7; in this chapter we will put them on a firmer mathematical footing by examining the Einstein coefficients in some detail. We will show that the absorption and emission rates are proportional to them, and that they are related to each other.

The reason stimulated emission is important in lasing systems and not in many other optical systems is that it becomes significant relative to spontaneous emission only when the light intensity in the cavity is large. This is why a Fabry–Perot cavity is used.

The excitation source for a number of solid state and liquid lasers is an external light source that is often referred to as an optical pump. As the pumping light intensity increases, the ratio of the stimulated emission to the stimulated absorption increases until at the transparency point the ratio is one. At still higher pumping levels more light is emitted than is absorbed, and the excess light is emitted as laser light. In this domain the absorption coefficient is negative so it is now called the optical gain constant.

In order for there to be optical gain, there must be more particles in the upper lasing level than there are in the lower one. This is called population inversion, and the magnitude of the inversion for a given laser is fixed by the losses in the cavity; in the steady state the gain equals the losses.

With the population inversion fixed, the power emitted by the laser increases linearly with the pumping rate above the threshold rate. The actual power output is determined by the reflectivity of the mirrors. Clearly, if the mirrors were perfectly reflecting, there would be no power output. Also, if the transmission through the mirrors were too high, the light intensity in the cavity could not build up to the level needed for population inversion. Thus, there is an optimum value for which the power output is a maximum.

The gain for the lasing mode(s) is fixed in the steady state, but the gain at mode frequencies other than the lasing frequencies increases with the pumping rate until they, too, are pumped to the threshold level. Thus, there can be a number of different lasing frequencies with the separation between these frequencies being determined in the same way that they were determined for a Fabry–Perot cavity.

The most important factor that determines the pumping intensity necessary to reach the lasing threshold is whether the laser is a three- or four-level laser. A three-level laser is one for which the lower laser energy level is the ground state whereas a four-level laser is one for which the lower level is an excited state. The threshold pumping intensity is much larger in a three-level laser because more than one-half of the particles must be excited out of the ground state for population inversion to occur. For a four-level laser not nearly as many particles must be excited into the upper lasing level to achieve population inversion, since there are only a few particles in the lower level.

A ruby laser is a three-level solid state laser. It is operated in the pulsed mode because so much energy is required to pump it. As a result, heat dissipation can be a severe problem. A Nd-YAG laser is a four-level solid state laser which, like the ruby laser, is optically pumped. The He-Ne laser is a four-level gas laser in which the helium atoms are excited by colliding with charged particles that are accelerated by an electric field; the excited helium atoms transfer their energy to neon atoms, and then the neon atoms emit light as the excited electrons decay to the ground state. A CO_2 laser is a four-level gas laser in which light is emitted when CO_2 molecules drop from one excited vibrational level to another. This laser is pumped in much the same way a He-Ne laser is with the N_2 molecule being the originally excited species.

9-2 BLACKBODY RADIATION

The three basic processes that are fundamental to laser operation are absorption, spontaneous emission, and stimulated emission. To develop a mathematical relationship between these three processes, one must know what the equilibrium photon distribution is. This can be done by examining the equilibrium energy distribution among the standing waves in a black box. A black box is simply a material that can emit and absorb radiation at any wavelength; it does not have quantum mechanical restrictions like those produced by a semiconductor energy gap.

A body constantly emits radiation; the higher its temperature the more it emits. Two bodies at different temperatures can come to thermal equilibrium by the radiation process. The warmer body emits more energy than the colder body; hence, the colder body has more energy to absorb. The process will continue until both bodies are at the same temperature so that they emit as much energy as they absorb.

An example of a perfect blackbody is an empty box with perfectly reflecting sides and a small hole through which light is emitted. The only constraint on the wavelengths of light being emitted and absorbed is that their x, y, and z components are half integral multiples of the cavity length in those directions so that they constructively interfere. This is precisely the resonance condition that was discussed in Section 8-2.

The energy density per unit frequency, $u(\nu)$, of the photon flux in the black box with frequencies between ν and $\nu + d\nu$ is

$$u(\nu) = (\text{density of states}) \times (\text{energy of state}) \times (\text{probability that state is occupied}) \tag{9-1}$$

The density of states, the number of states (modes) per unit volume with a frequency between ν and $\nu + d\nu$, $\rho(\nu)$, is simply the derivative of the number of states per unit volume (Eq. 8-36) with respect to the frequency, ν (see Figure 9-1). Thus

$$\rho(\nu) = \frac{dn_\nu}{d\nu} = \frac{8\pi}{c^3}\nu^2 \tag{9-2}$$

The energy of the state is $h\nu$, and the probability that the state is occupied is given by a distribution function. This distribution function, $f_{\mathrm{BE}}(E)$, is called the Bose–Einstein distribution function, and it is given by

$$f_{\mathrm{BE}}(E) = \frac{1}{e^{E/kT} - 1} \tag{9-3}$$

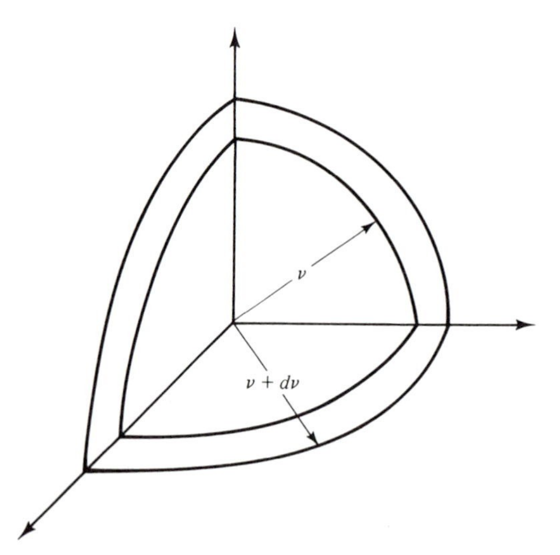

Figure 9-1 The spherical shell in the first octant in frequency space that contains standing waves with frequencies between ν and $\nu + d\nu$.

where k is again the Boltzmann constant. Equation 9-1 now becomes

$$u(\nu) = \frac{8\pi h\nu^3}{c^3} \frac{1}{e^{h\nu/kT} - 1} \tag{9-4}$$

The "particles" that obey Bose–Einstein statistics are those that have an integral quantum mechanical spin, as opposed to the half integral spin that electrons have. Some other "particles" that have integral spin and are often studied by physicists are phonons and some isotopes of helium.

The energy density is sometimes plotted as a function of λ instead of ν. This transformation is simple, since $\lambda = c/\nu$. A plot of $u(\lambda)$ versus λ is made in Fig. 9-2 for the sun that is a blackbody with a surface temperature of about 5200° C. Notice that the maximum intensity is for green light ($\lambda \sim 5550$ Å).

9-3 TWO-LEVEL SYSTEM

9-3.1 Einstein Coefficients

In this section the two-level system discussed briefly in Section 7-2 and shown in Fig. 9-3 is considered. It is an atomic system in that the states are discrete as opposed to being bands. They are separated by an energy ΔE, and it is assumed that $\Delta E \gg kT$. The number of states per atom at level 1 is ρ_1 and the number of states per atom at level 2 is ρ_2.

When light with energy ΔE strikes the two-level system, it can be absorbed by exciting an electron from the lower into the upper level. As we showed in Section 3-2.3 and discussed briefly in Section 7-2.1, the number of photons absorbed per unit volume per unit time is proportional to the photon flux density

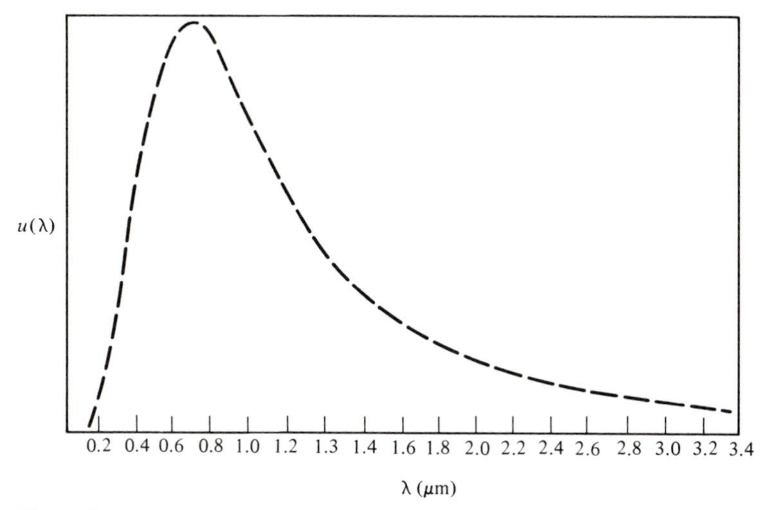

Figure 9-2 The spectral distribution of the radiation from the sun's surface which has an approximate temperature of 5200° C.

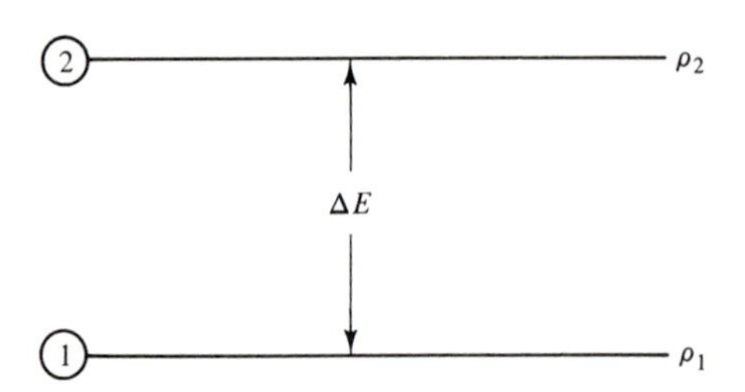

Figure 9-3 The energy level diagram for a two-level system.

with energy ΔE, which in this case is $u(\nu_{12})$. The rate of absorption is also proportional to the density of electrons, n_1, in level 1. The constant of proportionality is the Einstein coefficient, B_{12}, so that the

$$\text{rate of absorption} = B_{12}n_1u(\nu_{12}) \tag{9-5}$$

The rate of spontaneous emission produced by electrons dropping from level 2 to level 1 is proportional to the density of electrons, n_2, in level 2. The proportionality constant is the Einstein coefficient, A_{21}, so that the

$$\text{rate of spontaneous emission} = A_{21}n_2 \tag{9-6}$$

The rate of stimulated emission must be proportional to $u(\nu_{12})$, since it is proportional to the driving force of the oscillator. It must also be proportional to the number of electrons in state 2. The Einstein proportionality constant is B_{21} so that the

$$\text{rate of stimulated emission} = B_{21}n_2u(\nu_{12}) \tag{9-7}$$

When the steady state is reached, photons are created at the same rate that they are destroyed. Therefore

$$B_{12}n_1u(\nu_{12}) = A_{21}n_2 + B_{21}n_2u(\nu_{12}) \tag{9-8}$$

Because it was assumed that $\Delta E \gg kT$, the relative probability that the electron is in state 2 and not in state 1 is proportional to $\exp(-\Delta E/kT)$. The relative probability is also proportional to the ratio of the density of states, ρ_2/ρ_1; the more states there are available, the more probable it is an electron will be in that state. The relationship between n_1 and n_2 is then

$$n_1 = \frac{\rho_1}{\rho_2}n_2e^{h\nu_{12}/kT} \tag{9-9}$$

Substituting Eq. 9-9 into Eq. 9-8 and rearranging yields

$$u(\nu_{12}) = \frac{A_{21}/[(\rho_1/\rho_2)B_{12}]}{e(h\nu_{12}/kT) - B_{21}/[(\rho_1/\rho_2)B_{12}]} \tag{9-10}$$

By correlating coefficients with the expression for $u(\nu_{12})$ given in Eq. 9-4, one obtains

$$\rho_1B_{12} = \rho_2B_{21} \tag{9-11a}$$

and

$$A_{21} = \frac{8\pi h\nu^3}{c^3}B_{21} \tag{9-12a}$$

It can be shown that these results are true for any two levels in a multilevel system so that

$$\rho_i B_{ij} = \rho_j B_{ji} \tag{9-11b}$$

and

$$A_{ji} = \frac{8\pi h\nu^3}{c^3} B_{ji} \tag{9-12b}$$

9-3.2 Light Amplification

The equations developed in Section 7-2 for the semiconductor laser are also applicable to the other types of lasers. However, the magnitudes of the parameters are quite different. The emission peak bandwidth is narrower, the decay time is often slower, the cavity length is longer, and the mirror reflectivities are usually larger because metallic or dielectric mirrors are used. Some of these points are illustrated in Example 9-1.

EXAMPLE 9-1

Estimate the gain constant, α_g, in a ruby laser for which

$$\rho_2 = \rho_1 = 4,\ n_2 - n_1 = 5 \times 10^{17}\ \text{cm}^{-3}$$

$$\Delta\nu = 6 \times 10^{10}\ \text{Hz},\ \tau_{sp} = 3 \times 10^{-3}\ \text{sec}$$

$$\lambda_0 = 0.6934\ \mu\text{m and } n = 1.77$$

Using Eq. 7-11

$$\alpha_g = \frac{\lambda_0^2(n_2 - \rho_2/\rho_1 n_1)}{8\pi\tau_{sp}\,\Delta\nu n^2} = \frac{(0.6934 \times 10^{-4})^2(5 \times 10^{17})}{8(3 \times 10^{-3})(6 \times 10^{10})(1.77)^2} = 0.17\ \text{cm}^{-1}$$

9-4 THREE-LEVEL LASER

9-4.1 Transparency Point

The two-level system discussed in the previous section is not a good lasing system. This is because, to excite electrons into the upper level, B_{12} should be reasonably large, but this also requires that B_{21} be large. Therefore, the electrons will return to the lower level almost as soon as they are excited into the upper level. Thus, it will be difficult to obtain population inversion. This problem, to a large extent, can be avoided by using an appropriate three-level system which is shown schematically in Fig. 9-4.

In an ideal three-level system the electrons are excited from the 1 level into the 3 level. They then fall quickly into the metastable 2 level where, after a much longer period of time, they fall back down into the 1 level. The latter is the lasing transition. The reason that level 2 should be metastable is that this will allow the electron population to build up at this level so that population inversion can be obtained more easily.

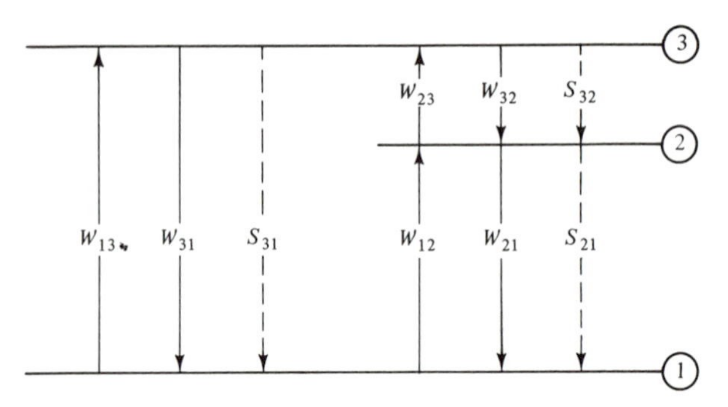

Figure 9-4 The energy level diagram for a three-level system.

In Fig. 9-4 one can see that the transition rates are designated by W_{ij} and S_{ij}. The W_{ij} transition rates are the absorption and stimulated emission rates per electron: rates that are proportional to $u(\nu_{ij})$. The rate of spontaneous emission and nonradiative recombination transitions per electron are combined to form the S_{ij} transitions. Recall that nonradiative transitions are those transitions that produce no light.

In order that the electrons do not fall back from 3 to 1, S_{32} must be very large compared with S_{31} and W_{31}. However, as is shown by the relationship between B_{13} and B_{31} (Eq. 9-11b), a small W_{31} implies that W_{13}, and therefore the absorption of light, will be small. This problem, to some extent, can be alleviated by making $\rho_3 \gg \rho_1$. Also, if S_{32} is very large, then W_{23} and W_{32} will be small because $u(\nu_{23})$ will be small since there will be so few electrons in level three. The difference between S_{32} and W_{32} and W_{23} will be even larger if a substantial component of S_{32} is due to nonradiative recombinations.

Since the 2 level is a metastable level, S_{21} must necessarily be small. This implies that A_{21}, and therefore B_{21} and B_{12}, are also small. However, W_{21} and W_{12} can be large if $u(\nu_{12})$ is large, and it will be large if the radiation is confined to a resonant cavity and there is optical gain.

The rate equations for the second and third levels are

$$\frac{dn_3}{dt} = W_{13}n_1 + W_{23}n_2 - (W_{31} + S_{31} + W_{32} + S_{32})n_3 \tag{9-13a}$$

$$\frac{dn_2}{dt} = W_{12}n_1 - (W_{23} + W_{21} + S_{21})n_2 + (W_{32} + S_{32})n_3 \tag{9-14a}$$

Under steady state conditions and the assumptions that W_{23} and W_{32} are negligible, the above equations become

$$W_{13}n_1 = (W_{31} + S_{31} + S_{32})n_3 \tag{9-13b}$$

$$W_{12}n_1 = (W_{21} + S_{21})n_2 - S_{32}n_3 \tag{9-14b}$$

Eliminating n_3 by combining Eqs. 9-13b and 9-14b yields

$$\frac{n_2}{n_1} = \left(\frac{W_{13}S_{32}}{W_{31} + S_{31} + S_{32}} + W_{12}\right)\frac{1}{S_{21} + W_{21}} \tag{9-15a}$$

But in an ideal three-level system $S_{32} \gg W_{31}$ and S_{31} so that Eq. 9-15a becomes

$$\frac{n_2}{n_1} \simeq \frac{W_{13} + W_{12}}{S_{21} + W_{21}} \tag{9-15b}$$

Multiplying both sides of the equation by ρ_1/ρ_2 yields

$$\frac{\rho_1 n_2}{\rho_2 n_1} = \frac{\rho_1 W_{13} + \rho_1 W_{12}}{\rho_2 S_{21} + \rho_2 W_{21}} \tag{9-15c}$$

At the point where level 2 becomes inverted relative to level 1—sometimes called the transparency point

$$n_1 \rho_2 = n_2 \rho_1 \tag{9-16a}$$

At this point

$$\rho_1 W_{13} + \rho_1 W_{12} = \rho_2 S_{21} + \rho_2 W_{21} \tag{9-16b}$$

Recalling from Eq. 9-11b that $\rho_i W_{ij} = \rho_j W_{ji}$, Eq. 9-16b becomes

$$\rho_1 W_{13} = \rho_2 S_{21} \tag{9-16c}$$

EXAMPLE 9-2

Find the intensity of the light, I_p, necessary to pump a ruby laser to the transparency point at room temperature if $\rho_1 = \rho_2$, $S_{21} = 330 \text{ sec}^{-1}$, and the absorption cross section, $\sigma_{13} = 10^{-19} \text{ cm}^{-2}$. The energy absorbed per unit time per unit area in length dz is, from Eq. 3-18a

$$-dI_p = \alpha_{13} I_p \, dz = n_1 \sigma_{13} I_p \, dz$$

Also
$$-dI_p = n_1 W_{13} h\nu_{13} \, dz$$

Equating and solving for I_p

$$I_p = \frac{W_{13} h\nu_{13}}{\sigma_{13}}$$

The exciting radiation for a ruby laser is green light, which has an energy of 2.25 eV per photon. Also, since $\rho_1 = \rho_2$, $W_{13} = S_{21}$ at the transparency point. Therefore

$$I_p = \frac{(3.30 \times 10^2)(2.25)(1.6 \times 10^{-19} \text{ J/eV})}{10^{-19}} = 1200 \text{ W/cm}^2$$

If only 1 percent of the pumping radiation is absorbed, the intensity necessary to reach the transparency point is 120,000 W/cm^2.

9-4.2 Saturation Gain

Pumping beyond the transparency point inverts the population and increases the gain. However, increasing the gain increases the intensity of the stimulated emission which, in turn, decreases the gain. The saturation gain constant, α_g^0, would be the gain if there were no stimulated emission. It, therefore, represents the maximum value α_g can approach.

From Eq. 9-15b

$$n_2 = \left(\frac{W_{13} + W_{12}}{S_{21} + W_{21}}\right) n_1 \tag{9-15d}$$

subtracting $(\rho_2/\rho_1)n_1$ from each side, recalling from Eq. 9-11 that $\rho_1 W_{12} = \rho_2 W_{21}$, and simplifying yields

$$n_2 - (\rho_2/\rho_1)n_1 = \left[\frac{(W_{13}/S_{21}) - (\rho_2/\rho_1)}{1 + (W_{21}/S_{21})}\right] n_1 \tag{9-17a}$$

W_{21} is proportional to the stimulated emission intensity, I_{in}, and S_{21} is proportional to the spontaneous emission intensity, I_s. Thus Eq. 9-17a can also be written

$$n_2 - \frac{\rho_2}{\rho_1} n_1 = \left[\frac{(W_{13}/S_{21}) - (\rho_2/\rho_1)}{1 + (I_{in}/I_s)}\right] n_1 \tag{9-17b}$$

The saturated population inversion, Δn^0, the inversion when $I_{in} = 0$, is

$$\Delta n^0 = \left(\frac{W_{13}}{S_{21}} - \frac{\rho_2}{\rho_1}\right) n_1 \tag{9-18}$$

From Eq. 7-11b

$$\alpha_g^0 = \frac{\lambda^2 \, \Delta n^0}{8\pi \tau_{sp} \, \Delta\nu} \tag{9-19a}$$

EXAMPLE 9-3

Calculate Δn^0 and α_g^0 for a ruby laser at room temperature if $W_{13} = 2S_{21}$, $n_1 = 10^{19}$ cm^{-3}, $\Delta\nu = 6 \times 10^{10}$ Hz, $n = 1.77$, and $\lambda_0 = 0.6934$ μm. Recall from Example 9-1 that $\rho_1 = \rho_2$ and $\tau_{sp} \approx 1/S_{21} = 1/330$.

$$\Delta n^0 = (2 - 1)n_1 = 10^{19} \text{ cm}^{-3}$$

$$\alpha_g^0 = \frac{(0.6934 \times 10^{-4})^2(10^{19})3.3 \times 10^2}{(8\pi)(6 \times 10^{10})(1.77^2)} = 3.36 \text{ cm}^{-1}$$

9-4.3 Internal Power

When a constant pumping intensity is applied, the population inversion will reach a steady state threshold value, n_t. As was pointed out in Section 7-2.2, at this point the net optical energy gained via stimulated emission is equal to the energy lost by transmission out of, or absorption in, the cavity.

The number of transitions per second per unit volume that contribute to the power generated inside of the cavity is $n_t W_{21}$, the energy per transition is $h\nu_{21}$, and the volume of the mode is V. Thus, the internal power, P_{in}, is

$$P_{in} = n_t W_{21} h\nu_{21} V \tag{9-19b}$$

Equation 9-17a can be rewritten as

$$W_{21} = S_{21}\left(\frac{W_{13}n_1}{S_{21}n_t} - \frac{\rho_2}{\rho_1}\frac{n_1}{n_t} - 1\right) \tag{9-20}$$

Multiplying both sides by $n_t h\nu_{21} V$ yields

$$P_{\text{in}} = P_s\left(\frac{R_p}{R_s} - \frac{\rho_2}{\rho_1}\frac{n_1}{n_t} - 1\right) \tag{9-21}$$

where $P_s = n_t h\nu_{21} V S_{21}$, $R_p = W_{13}n_1$ is the pumping rate, and $R_s = S_{21}n_t$ Clearly

$$R_p \geq R_s\left(\frac{\rho_2}{\rho_1}\frac{n_1}{n_t} + 1\right)$$

for lasing to occur.

9-4.4 Optimum Coupling

Because both beneficial (transmission) and parasitic (absorption and diffraction) losses are present, there is a transmission coefficient, T_{opt}, for which the power output is a maximum. For $T < T_{\text{opt}}$, the energy in the cavity is larger, but because the dominant effect is the smaller transmission coefficient, the output power is less. For $T > T_{\text{opt}}$ the transmission coefficient is larger, but since the dominant effect is the energy in the cavity, the output power will be less.

The first step in finding T_{opt} is demonstrating the approximate equivalency of the loss per pass, Λ, and $\alpha_g L$. From Eq. 7-13

$$\alpha_g = \alpha - \frac{1}{2L}\ln R_1 R_2 \tag{7-13}$$

$$\alpha_g L \simeq \tfrac{1}{2}(1 - R_1 R_2) + \alpha L \tag{9-22}$$

for R values near 1. The loss per round trip is

$$2\Lambda = 1 - R_1 R_2 e^{-2\alpha L} \tag{9-23a}$$

Thus

$$\Lambda \simeq \tfrac{1}{2}[1 - R_1 R_2(\Gamma - 2\alpha L)] \simeq \tfrac{1}{2}(1 - R_1 R_2) + \alpha L \tag{9-23b}$$

Because $\alpha_g \propto n_t$, $P_{\text{in}} \propto I_{\text{in}}$, and $P_s \propto I_x$, Eq. 9-17 can be written

$$\alpha_g L = \frac{\alpha_g^0 L}{1 + \dfrac{P_{\text{in}}}{P_s}} \tag{9-24a}$$

Rearranging and noting the equivalence of Λ and $\alpha_g L$

$$P_{\text{in}} = P_s\left(\frac{\alpha_g^0 L}{\Lambda} - 1\right) \tag{9-24b}$$

The loss per pass can be written as the sum of the transmission loss per pass, T, and the absorption loss per pass, Λ_a. Therefore

$$\Lambda = T + \Lambda_a \tag{9-25}$$

noting that $P_s \propto \Lambda$ because it is proportional to n_t, and that the fraction of the internal power that is the external output power, P_{ex}, is

$$P_{ex} = \frac{T}{\Lambda} P_{in} \tag{9-26a}$$

Eq. 9-26a becomes

$$P_{ex} = \left(\frac{\alpha_g^0 L}{\Lambda_a + T} - 1\right) KT \tag{9-26b}$$

where K is a combination of constant terms. Taking the derivative and setting it equal to zero yields

$$T_{opt} = \sqrt{\alpha_g^0 L \Lambda_a} - \Lambda_a \tag{9-27}$$

Substituting T_{opt} into Eq. 9-26b to find the maximum power output, we find that

$$P_{opt} = K(\sqrt{\alpha_g^0 L} - \sqrt{\Lambda_a})^2 \tag{9-28}$$

9-5 FOUR-LEVEL LASER

The primary difference between a three- and four-level laser is that the terminal lasing level is not the ground state level so that population inversion can be achieved at much lower pumping levels; the lower lasing level is virtually empty under equilibrium conditions because it is $\gg kT$ above the ground state level. This is illustrated in Fig. 9-5. The physical situation is similar to that for a three-

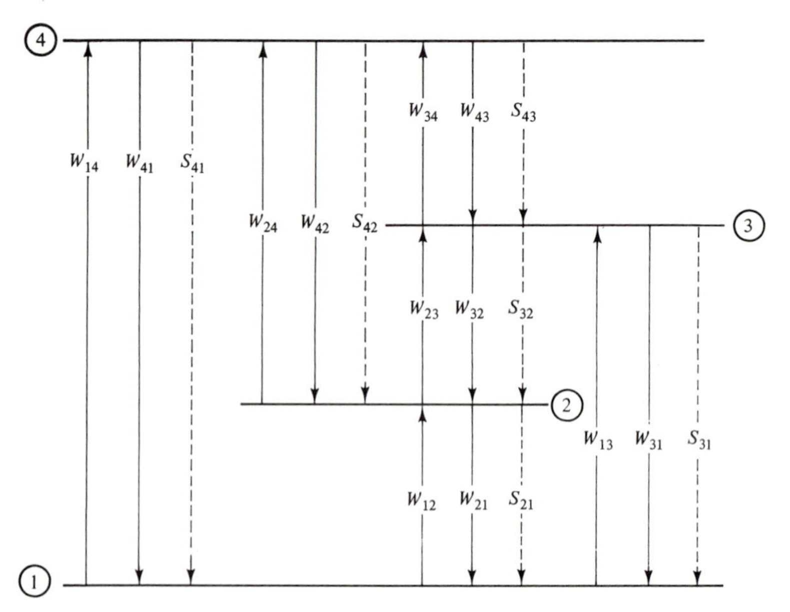

Figure 9-5 The transition rates for a four-level system.

level laser and is as follows. The electrons are pumped from level 1 to level 4 quite easily, and then they very rapidly drop to level 3. They sit in level 3 for a relatively long time, and then drop to level 2. In so doing they emit light at the lasing frequency. The electrons then drop rapidly back to the ground state so that population inversion can be achieved more easily. Because the electrons rapidly drop from level 2 to level 1, it is likely that the Einstein coefficient A_{21} is large which, in turn, means that B_{12} is relatively large. Thus, electrons can be pumped from the ground state to level 2 as well as to level 4. The threshold pumping rate is larger than it would be without this parasitic pumping.

The power needed to pump a four-level laser to its transparency point is computed in a different manner than it is for a three-level laser. If we assume that there are n_t electrons per unit volume in level 3 and no electrons in level 2, then in order to maintain this population level, we must pump electrons into level 3 at least as fast as they spontaneously decay to level 2. The transparency power, P_t, is therefore

$$P_t(\text{4-level}) \simeq n_t S_{32} h\nu_{14} V \tag{9-29}$$

where V is again the volume. In a three-level laser at least one-half of the valence electrons, $n_0/2$, must be in level 2. Thus, the transparency power for a three-level laser is

$$P_t(\text{3-level}) \simeq \frac{n_0}{2} S_{21} h\nu_{13} V \tag{9-30}$$

The pumping power necessary to achieve lasing in a four-level laser is $\sim 10^3$ less than that for a three-level laser, and this is illustrated in Example 9-4.

EXAMPLE 9-4

(a) Find the transparency power per unit volume of a Nd-YAG laser with the following characteristics. $h\nu_{14} \simeq 2.25$ eV, $L = 20$ cm, $\Lambda = 0.04$, $n = 1.82$, $\Delta\nu = 1.8 \times 10^{11}$ Hz, $\tau_{sp} = 5.5 \times 10^{-4}$ sec, and $\lambda_0 = 1.06\ \mu$m. **(b)** Also find the $P_t(\text{ruby})/P_t(Nd)$ ratio if for ruby $n_0 = 2 \times 10^{19}$ cm^{-3} and $S_{21} = 330$ sec^{-1}. Assume $h\nu_{13} = h\nu_{14}$.

(a) From Eq. 7-11b and the equivalency of $\alpha_g L$ and Λ

$$n_t = \frac{8\pi\tau_{sp}\,\Delta\nu\Lambda n^2}{\lambda_0^2 L}$$

$$= \frac{(8\pi)(5.5 \times 10^{-9})(1.8 \times 10^{11})(4 \times 10^{-3})(1.82)^2}{(1.06 \times 10^{-4})^2(20)}$$

$$= 1.75 \times 10^{15} \text{ cm}^{-3}$$

From Eq. 9-29

$$\frac{P_t}{V} = \frac{(1.75 \times 10^{15})(2.25)(1.6 \times 10^{-19})}{5.5 \times 10^{-4}} = 1.15 \text{ W/cm}^3$$

(b)
$$\frac{P_t(\text{ruby})}{P_t(Nd)} = \frac{n_0 S_{21}\tau_{32}}{2n_t} = \frac{(10^{19})(3.3 \times 10^2)(5.5 \times 10^{-4})}{1.75 \times 10^{15}} \cong 1000$$

The equation for the saturated gain, α_g^0, for the four-level laser is only slightly different. The complete set of rate equations is

$$\frac{dn_4}{dt} = W_{14}n_1 + W_{24}n_2 + W_{34}n_3 - (W_{41} + S_{41} + W_{42} + S_{42} + W_{43} + S_{43})n_4 \quad (9\text{-}31a)$$

$$\frac{dn_3}{dt} = W_{13}n_1 + W_{23}n_2 - (W_{31} + S_{31} + W_{32} + S_{32} + W_{34})n_3 + (W_{43} + S_{43})n_4 \quad (9\text{-}32a)$$

$$\frac{dn_2}{dt} = W_{12}n_1 - (W_{21} + S_{21} + W_{23} + W_{24})n_2 + (W_{32} + S_{32})n_3 + (W_{42} + S_{42})n_4 \quad (9\text{-}33a)$$

We note that W_{24} and W_{34} have very small values and that S_{43} is a very large number; thus, Eq. 9-31a under steady state conditions becomes

$$W_{14}n_1 = S_{43}n_4 = R_p \quad (9\text{-}31b)$$

where R_p is the pumping rate. Because the pumping rate in four-level lasers is relatively small, $n_1 \simeq n_0$ for all pumping rates. Thus, R_p increases linearly with the pumping light intensity. In Eq. 9-32a, $W_{13}n_1$, $W_{31}n_3$, $S_{31}n_3$, and $W_{43}n_4$ can be neglected so that the steady state equation becomes

$$0 = W_{23}n_2 - (W_{32} + S_{32})n_3 + R_p \quad (9\text{-}32b)$$

$W_{34}n_2$, $W_{42}n_4$, and $S_{42}n_4$ can be neglected in Eq. 9-33a so that the steady state equation becomes

$$0 = R_2 - (W_{23} + S_{21})n_2 + (W_{32} + S_{32})n_3 \quad (9\text{-}33b)$$

where the parasitic pumping rate, R_2, is

$$R_2 = W_{12}n_1 - W_{21}n_2 \quad (9\text{-}34)$$

Solving Eq. 9-32b for n_2 and Eq. 9-33b for n_3, we find that

$$n_2 = \frac{R_p + R_2}{S_{21}} \quad (9\text{-}35)$$

and

$$n_3 = \frac{(W_{23}/S_{21})R_2 + R_p[(W_{23}/S_{21}) + 1]}{W_{32} + S_{32}} \quad (9\text{-}36)$$

Combining the last two equations yields

$$n_t = n_3 - \frac{\rho_2}{\rho_1} n_2 = \frac{R_p\{1 - [(\rho_3/\rho_2)(S_{32}/S_{21})][1 + (R_2/R_p)]\}}{W_{32} + S_{32}} \quad (9\text{-}37a)$$

Note that $(\rho_3/\rho_2)S_{32} < S_{21}$ if n_t is to be positive. Of course, this means that electrons must fall out of level 2 faster than they do out of level 3; otherwise population inversion could never be achieved.

Equation 9-37a can also be written

$$n_t = \frac{(R_p/S_{32}) - (\rho_3/\rho_2/S_{21})(R_p + R_2)}{1 + P_{in}/P_s} \tag{9-37b}$$

$$= \frac{\Delta n^0}{1 + P_{in}/P_s} \tag{9-37c}$$

Thus, α_g^0 has the same form as Eq. 9-19 for a three-level laser.

The internal power, P_{in}, can be found by rewriting Eq. 9-37a as

$$n_t = \frac{R}{W_{32} + S_{32}} \tag{9-37d}$$

Below threshold W_{32} is zero. Thus, n_t increases as the effective pumping rate, R, increases until it reaches the threshold value at which point it ceases to grow. Now, W_{32} increases in such a way that the ratio, $R/(W_{32} + S_{32})$, remains constant (see Fig. 7-3).

Rearranging Eq. 9-37d, we find that

$$W_{32} = S_{32}\left(\frac{R}{n_t S_{32}} - 1\right) \tag{9-38}$$

Multiplying both sides by $n_t h\nu_{32} V$ and recognizing that $n_t S_{32}$ is the spontaneous pumping rate, R_s, Eq. 9-38 becomes

$$P_{in} = P_s\left(\frac{R}{R_s} - 1\right) \tag{9-39}$$

It is clear that the effective pumping rate must exceed the spontaneous rate for there to be laser emission.

The equations for the output power, P_{ex}, for a four-level laser are the same as those for the three-level laser.

9-6 OSCILLATION FREQUENCIES

We learned in Chapter 7 that the allowed frequencies for the standing waves inside of a one-dimensional cavity are

$$\nu_m = \frac{mc}{2L} \tag{7-17}$$

The frequency at which the optical output is a maximum, ν_0, however, does not necessarily coincide with a standing wave frequency. When this occurs, the lasing frequency is pulled toward ν_0 and it is given by the equation

$$\nu = \nu_m - \frac{\nu_m - \nu_0}{F} \tag{9-40}$$

where F is again the finesse of the cavity. The result is illustrated in Fig. 9-6.

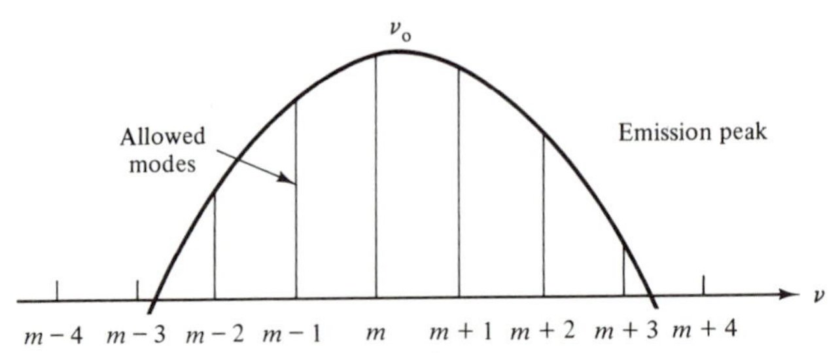

Figure 9-6 The relationship between the maximum in the fluorescence peak and the allowed modes near it.

When the pumping rate is large enough, there can be more than one lasing mode as is shown in Fig. 9-7. Below threshold the gain increases with the pumping intensity. However, once the threshold is reached the gain becomes fixed at the threshold value. The first mode to reach threshold is the mode with frequency ν_m, which is the frequency closest to the frequency with the maximum optical output, ν_0. As the pumping rate increases, the gain for the $m + 1$ and $m - 1$ modes reaches the threshold value so that laser emission also occurs at these frequencies. The number of lasing modes continually increases as the pumping increases and the gain of more and more modes reaches the threshold value. A spectrum of this mode structure for a He-Ne laser is shown in Fig. 9-8.

9-7 EFFICIENCIES

There are several independent efficiency factors, and their product is the overall device efficiency. These include the atomic quantum efficiency, η_a, the pumping quantum efficiency, η_p, the internal quantum efficiency, η_i, and the external

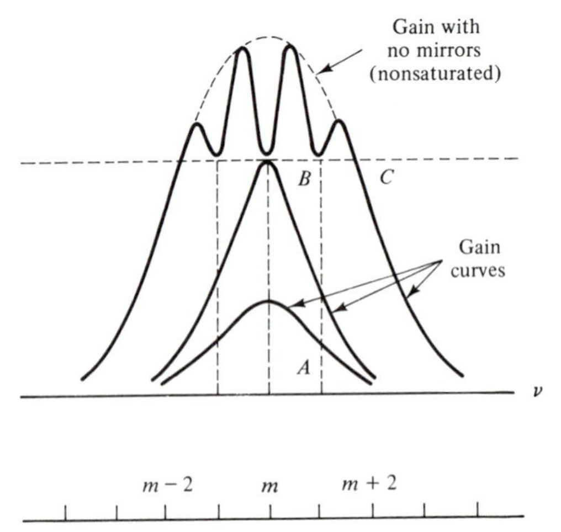

Figure 9-7 The gain curves below threshold (a) at threshold (b) and above threshold (c).

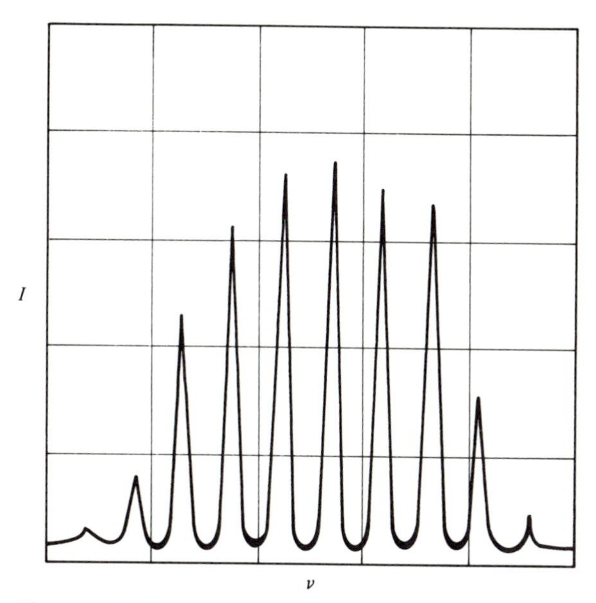

Figure 9-8 The output modes of a He-Ne laser lasing at 6328 Å.

quantum efficiency, η_{ex}. For solid state lasers the overall efficiency is typically 1 percent, for CO_2 lasers it can be as high as 30 percent, and for semiconductor lasers it can approach 50 percent.

The atomic quantum efficiency is the ratio of the energy of the lasing transition and the pumping transition. For a three-level laser

$$\eta_a = \frac{\Delta E_{21}}{\Delta E_{13}} \equiv \frac{\nu_{21}}{\nu_{13}} \tag{9-41a}$$

and for a four-level laser

$$\eta_a = \frac{\Delta E_{32}}{\Delta E_{14}} = \frac{\nu_{32}}{\nu_{14}} \tag{9-41b}$$

Clearly, for this efficiency to be large, the lasing levels should be low-lying levels so that ΔE_{13} or ΔE_{14} will not be too much larger than ΔE for the lasing transition.

The pumping efficiency is the fraction of the particles pumped to the highest level that fall into the upper lasing level. For a three-level laser

$$\eta_p = \frac{S_{32}}{S_{32} + W_{31} + S_{31}} \tag{9-42a}$$

and for a four-level laser

$$\eta_p = \frac{S_{43}}{S_{43} + W_{41} + S_{41}} \tag{9-42b}$$

The internal quantum efficiency is the fraction of the particles in the upper lasing level that decay by stimulated emission to the lower level. For a three-level laser

$$\eta_i = \frac{W_{21}}{W_{21} + S_{21}} \tag{9-43a}$$

and for a four-level laser

$$\eta_i = \frac{W_{32}}{W_{32} + S_{32}} \tag{9-43b}$$

The external quantum efficiency is the ratio of the output power to the stimulated emission power; it is the fraction of the stimulated emission that escapes the cavity. Thus

$$\eta_{ex} = \frac{dP_{ex}}{dP_{in}} \tag{9-44a}$$

For the simple case treated in the previous section η_{ex} is a constant and is given by

$$\eta_{ex} = \frac{T}{\Lambda} \tag{9-44b}$$

9-8 SOLID STATE AND LIQUID LASERS

9-8.1 Ruby Laser

As we mentioned briefly in Chapter 3, ruby is Al_2O_3 doped with chromium. The Cr^{+3} ion substitutes for the Al^{+3} ion, and the Cr^{+3} concentration is usually 0.05 percent (1.58×10^{19} cm^{-3}). It is pink in color and the primary room temperature lasing radiation is in the red at 6943 Å.

The aluminum oxide has a rhombohedral crystal structure that can be derived by a small distortion along the diagonal of a cube. The Cr^{+3} ion has a number of *d* electron energy states in the energy gap of Al_2O_3 because the *d* electrons have a number of states associated with them. These atomic states can be separated by the internal field produced by the atoms surrounding the chromium atom, and these levels can be further separated by a more anisotropic (noncubic) crystal field. Five of these levels are shown in Fig. 9-9.

Level 1 is the ground state. It is a narrow level that is one of the three

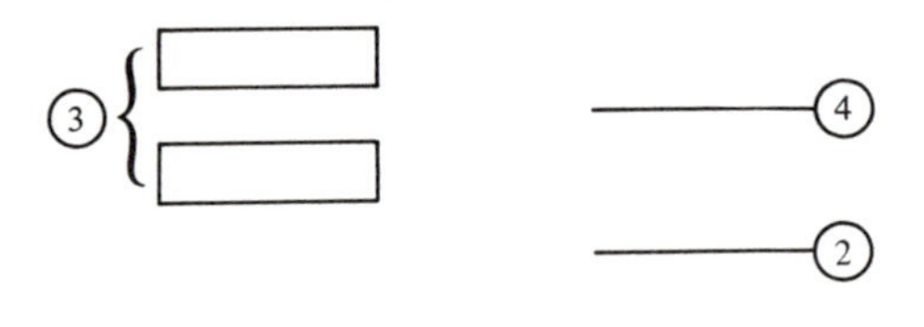

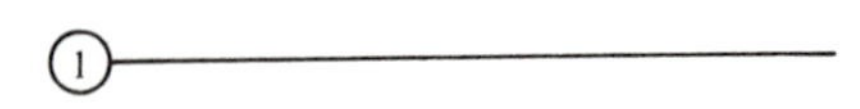

Figure 9-9 The energy level diagram for a ruby laser.

levels into which the ground state atomic level is broken by a cubic crystal field, and it has a multiplicity, $\rho_1 = 4$. The multiplicity of this state can be determined from group theory once the total multiplicity of the original atomic level is known.

The 3 level is composed of two energy bands that are separated by a forbidden region. These two bands are the remaining two levels into which the original atomic ground state is broken by the crystal field, and each band has a multiplicity of 12. The reason that these levels form bands rather than discrete levels is that the internal fields vary rapidly with time. The levels move up and down with the field fluctuations thereby creating the appearance of forming bands. These bands allow electrons in the ground state to absorb light over a range of energies. The absorption due to these transitions is shown in Fig. 9-10*a* where it can be seen that there are two absorption bands: one between 3600 and 4500 Å and the other between 5100 and 6000 Å. It is also interesting to note that the absorption spectrum is different when the incident light is parallel to the [111] direction (sometimes called the *c* axis) than it is when it is perpendicular to this direction. This serves to show that many of the optical properties of noncubic crystals are anisotropic. Although optical anisotropy is not important in this chapter, it is used in many optical applications.

We must point out that the absorption peaks associated with the 1-3 transitions are much larger than those associated with the 1-2 transitions. This is because the Einstein coefficients for the 1-3 transitions are larger and the multiplicity of the states in level 3 is larger. Although the 1-3 and therefore the 3-1 transitions have a reasonable probability of occurring, these transition probabilities are much less than S_{32} so that the ideal three-level laser operation is approached. At room temperature $S_{32} \simeq 2 \times 10^7\ \text{sec}^{-1}$ and $S_{31} \simeq 2 \times 10^5\ \text{sec}^{-1}$.

These values are much greater than S_{21} so that level 2 is a metastable level, as it must be. It is metastable because the transition is a phonon-assisted transition. That it is a metastable state is also supported by Fig. 9-10*a* where we can see that the absorption associated with this transition is small.

The fluorescence spectrum for the 2-1 transition is shown in Fig. 9-10*b*. There it can be seen that the energy levels must be narrow because the fluorescence peaks are narrow. This time there are two levels instead of one because the noncubic crystal field breaks the atomic level into two separate, but closely spaced (29 cm^{-1}) levels each having two states. There is, however, usually only one lasing transition, which is the transition associated with the peak at 6943 Å. This is because the electrons in the upper level quickly replace those in the lower level when the lower level empties.

The 4 level has little influence on the laser operation. It does not steal many electrons from the upper band in the 3 levels because most of the excited electrons in it drop rapidly into the lower band where they are quickly swept into the 2 levels. Also, few electrons are excited directly into it from the ground state because the transition probabilities are low.

The threshold of the ruby laser is higher at higher temperatures. This is primarily due to the increase in S_{21} with temperature. Its value at 77 K is 230 sec^{-1} whereas it is 330 sec^{-1} at room temperature. S_{21} increases with the tem-

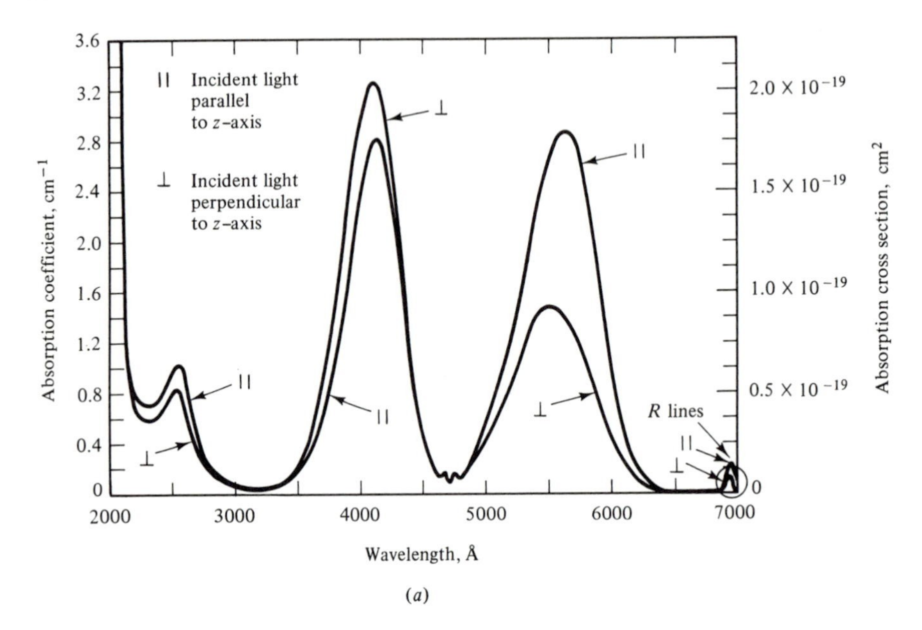

(*a*)

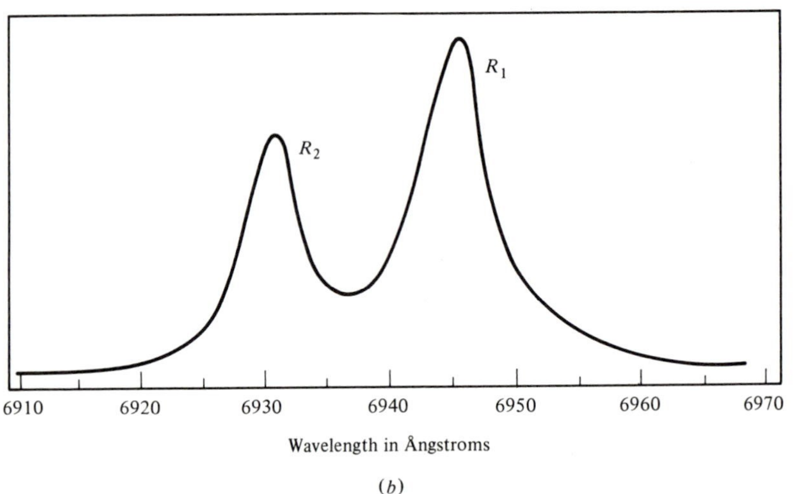

(*b*)

Figure 9-10 (*a*) The absorption spectrum for a pink ruby. (*b*) The fluorescence spectrum for the lasing transitions—the R_1 and R_2 lines—in pink ruby. (B. A. Lengyel, *Lasers,* 2d Ed., Copyright © 1971, Wiley-Interscience, New York.)

perature because there are more phonons available at the higher temperature to assist this transition. The threshold is also higher at higher temperatures because the fluorescence peak broadens.

It should also be pointed out that the fluorescence peaks shift to longer wavelengths as the temperature is increased. The shift in the lasing line is given by

$$\lambda = 6943.25 + 0.068(T - 20) \tag{9-48}$$

where T is in degrees celsius.

A typical ruby laser is shown in Fig. 9-11. The ruby rod has a diameter of 0.5 to 1 cm and a length of 2 to 10 cm. The ends are smooth and parallel to each other, and they have mirrors deposited directly on them. Also, the crystal axis makes an angle of 60° or 90° with the [111] direction so that the light output will be plane polarized.

The helical xenon flash tube is the optical pump. Light is produced in the flash tube by accelerated charged particles, which are created by dielectric breakdown, colliding with, and thereby exciting, electrons in the atoms. The light, which is generated in millisecond pulses, has a range of wavelengths and the maximum intensity is in the green. The flash tube is surrounded by mirrors that reflect the outward moving radiation so that it is reflected back toward the crystal.

The laser must have a good cooling system because most of the input energy, which is considerable, becomes heat. Since so much heat must be removed, ruby lasers are pulsed and not operated continuously. They must also be operated at a constant temperature because changing the temperature changes the lasing wavelength, and it changes the length of the ruby rod which, in turn, changes the wavelengths of the standing wave in the cavity.

9-8.2 Other Solid and Liquid Lasers

As is shown in Table 9-1, there are many different hosts. They are almost all ionic compounds because ionic compounds are almost always transparent to visible and to near infrared radiation. The most popular hosts are garnets, tungstates, and fluorides. There are also many different active ions, but they are either transition or rare earth ions.

One of the more popular lasers listed in Table 9-1 is the Nd:YAG (neodymium-yttrium aluminum garnet-$Y_3Al_5O_{12}$) laser. It contains from 0.5 to 2 percent Nd and it differs from a ruby laser in that it is a four-level laser. The energy-level diagram is shown in Fig. 9-12. The other major difference is that

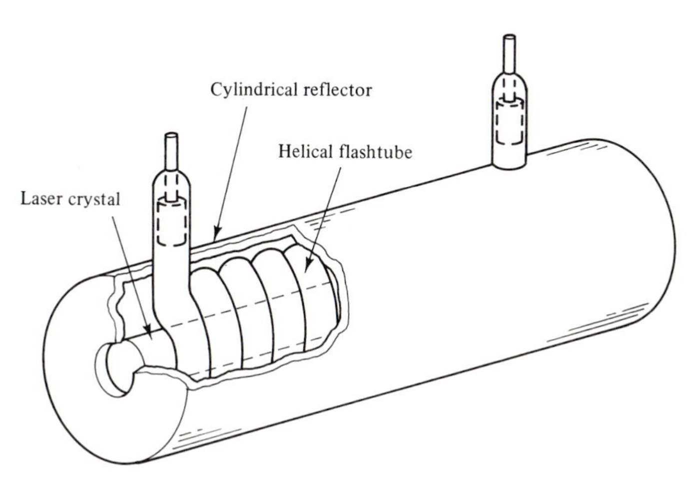

Figure 9-11 The ruby laser.

TABLE 9-1 SOME SOLID ION LASERS AND THEIR TYPICAL HOST MATERIALS

Ion	Typical host	Wavelengths (μm)	Note
Cr^{3+}	Sapphire	0.6943	Pink ruby
Cr^{3+}	Sapphire	0.7009, 0.7041	Red ruby
Co^{2+}	MgF_3, ZnF_3	1.75, 1.80, 1.99, 2.05	Phonon-assisted
Ni^{2+}	MgF_2	1.62	Phonon-assisted
Pr^{3+}	$CaWO_4$	1.0468	
Nd^{3+}	$CaWO_4$, glass	0.9142	
Nd^{3+}	$CaWO_4$, CaF_3, YAG, glass	1.04–1.07	Several nearby lines around 1.06 μm
Nd^{3+}	$CaWO_4$, glass	1.34–1.39	Several lines
Sm^{3+}	SrF_3	0.6969	20 K and below
Sm^{2+}	CaF_2	0.7083	20 K and below
Dy^{3+}	CaF_2	2.36	
Ho^{3+}	CaF_3	0.5512	
Ho^{3+}	$CaWO_4$, $CaMoO_4$, CaF_3, glass	2.05–2.07	Several nearby lines: host dependent
Ho^{3+}	YAG	2.09–2.12	Several nearby lines
Er^{3+}	$CaWO_4$, CaF_2	1.61	
Er^{3+}	YAG	1.654–1.660	
Er^{3+}	CaF_3	2.69	
Tm^{2+}	CaF_3	1.116	Continuous at 20 K
Tm^{3+}	$CaWO_4$, $Ca(NbO_3)_3$	1.91	
Tm^{3+}	SrF_3	1.97	
Yb^{3+}	YAG	1.0296	
U^{3+}	SrF_3	2.41	
U^{2+}	CaF_2	2.24, 2.51, 2.57, 2.61	

there are three widely spaced lasing transitions instead of two closely spaced ones. However, the lasing transition at 1.064 μm is the dominant one. The Nd:YAG laser also differs from the ruby laser in that its threshold decreases as the temperature is increased from 77 K to room temperature. This is because

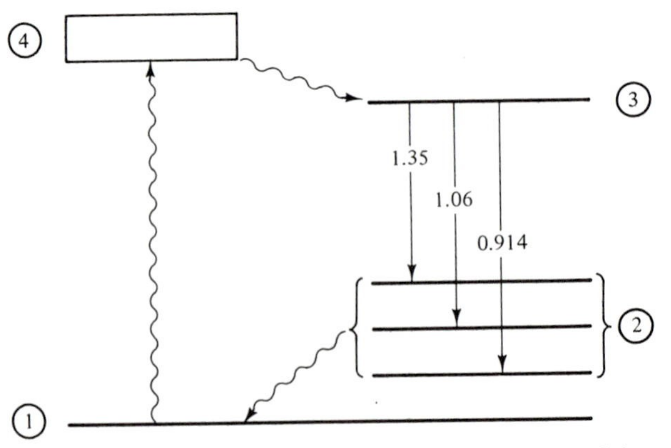

Figure 9-12 The energy level diagram for the Nd:YAG laser.

the ground state level broadens as the temperature increases, thus enabling it to absorb a greater percentage of the pumping radiation.

The pumping power necessary to obtain lasing in a four-level laser is much less than that needed to obtain lasing in a three-level laser, since population inversion can more easily be achieved. This is illustrated in the following example.

EXAMPLE 9-5

(a) Find the threshold pumping power per unit volume for a Nd:YAG laser. **(b)** Compare this power with the light intensity necessary to reach the transparency point in the ruby laser described in Example 9-4. For the Nd:YAG laser $\Delta\nu = 1.8 \times 10^{11}$ Hz, $n = 1.5$, $R_1 R_2 = 0.96$, $\lambda_0 = 1.064\ \mu$m, and the length of the cavity is $L = 20$ cm.

(a) The threshold pumping power for a four-level laser is P_s

$$\frac{P_s}{V} = \frac{n_t h\nu_{32}}{\tau_{sp}} = \frac{n_t h c_0}{\lambda_0 \tau_{sp}}$$

$$n_t = \frac{8\pi\tau_{sp}\alpha_g\ \Delta\nu n^2}{\lambda_0^2}$$

$$\alpha_g = \frac{\Lambda}{L} = \frac{1 - R_1 R_2}{L}$$

when there are no absorption losses.

$$\therefore \frac{P_s}{V} = \frac{8\pi h c_0\ \Delta\nu n^2 \Lambda}{\lambda_0^3 L}$$

$$= \frac{(8\pi)(6.63 \times 10^{-34})(3 \times 10^{10})(1.8 \times 10^{11})(1.5)^2(0.04)}{(1.06 \times 10^{-4})^3(2.0 \times 10^1)} = 0.34\ \mathrm{W/cm^3}$$

(b) The intensity necessary to pump a ruby laser to the transparency point is 1200 W/cm^2. When the fraction of the intensity absorbed is small, $P/V = \alpha I_p = n_1\sigma_{13}I$. For a ruby laser $n_1 = 2 \times 10^{19}$, $\sigma_{13} = 10^{-19}$

$$\therefore \frac{P}{V} = 2I_p = 2400\ \mathrm{W/cm^3}$$

The ratio of the power computed in (b) to that computed in (a) is 7060. This is a typical ratio for three- and four-level lasers.

The Nd^{+3} ion is also the active ion in a glass and a liquid laser. In the glass laser the Nd^{+3} energy levels are broader than they are in the crystal host, but this is compensated for by the fact that glass lasers are easier to fabricate. The liquid laser in which it is the active ion is the liquid composed of Nd_2O_3, $SeOCl_2$, and $SnCl_4$. The selenium oxychloride is used instead of water because there is an OH vibration energy level that has an energy similar to that of level 3. This enables some of the electronic energy in level 3 to be transferred to the vibrating OH radical, thereby making it more difficult to achieve population inversion.

We will say more about vibration energy levels when we discuss the CO_2 laser in the next section.

Another liquid laser of interest is the liquid dye laser. The electronic, vibrational, and rotational levels are illustrated in Fig. 9-13, and the absorption and emission peaks of a specific dye, rhodamine 6G in an ethanol solvent, are found in Fig. 9-14. Both peaks are very broad. This is because many closely spaced but different transitions contribute to them; what we see is the integrated effect.

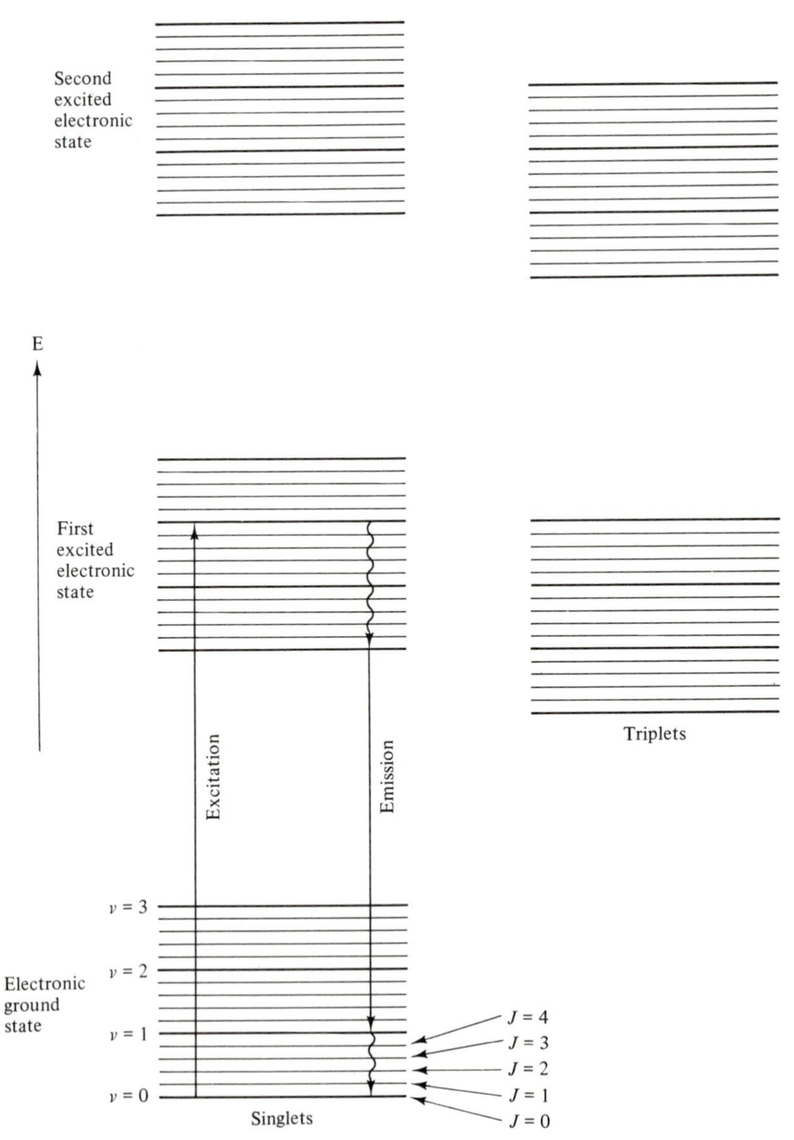

Figure 9-13 The electronic–vibrational–rotational energy level diagram for a liquid dye laser.

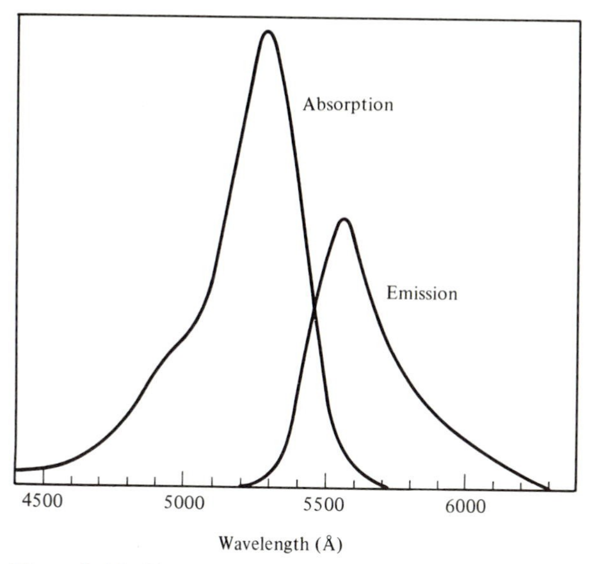

Figure 9-14 Singlet state absorption and emission spectra of rhodamine 6G.

Because the peaks are so broad, dye lasers usually have to be pumped by other lasers to achieve population inversion. The advantage of the broad emission peak is that a large number of modes can be driven above threshold so that dyes can lase over a range of wavelengths. To control which mode lases and also to eliminate the parasitic effects of other lasing modes, a diffraction grating can be used as one of the mirrors as is shown in Fig. 9-15.

The mode that is selected is the one that obeys the equation

$$2d \cos \theta = m\lambda \tag{9-45}$$

where d is the diffraction grating line spacing, θ is the angle between the laser axis and its projection on the diffraction grating, and m is an integer. Only

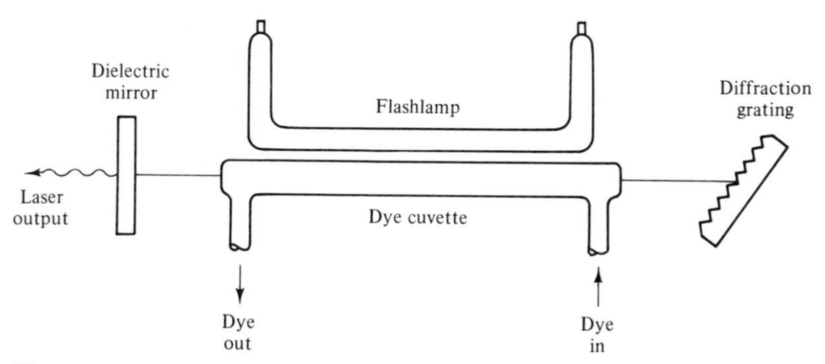

Figure 9-15 A dye laser with the dye flowing through a cuvette, a flashlamp, and the external mirrors, one of which is a diffraction grating.

the modes that obey this equation will be reflected parallel to the laser axis and will therefore be able to resonate (see Section 10-2.2 for a derivation of Eq. 9-45). Using this setup, rhodamine dye lasers have lased over the range 5550 to 5950 Å.

It is important to point out that the absorption peak of the rhodamine 6G lies at a shorter wavelength than does the emission peak. This is because emission is usually more than a one-step process. For example, a molecule could be excited from the ground state into a specific electronic–vibrational–rotational state, and then the molecule could decay to a lower rotational state before it decayed to the ground state.

The fact that the emission and absorption peaks do not coincide is important because, if they did, the emitted light would be readily absorbed. It would therefore be difficult, if not impossible, to achieve population inversion. We have already seen that the shift between the absorption and emission peaks is also important in semiconducting light-emitting devices.

9-9 GAS LASERS

9-9.1 Helium Neon Laser

The He-Ne laser is a four-level laser in which the active ingredient is the neon. The neon energy-level diagram in Fig. 9-16 differs from that of the Nd:YAG laser in that there is no equivalent four level and there are additional levels between the lower lasing level and the ground state level. We will see later that the 2^1S_0 or 2^3S_1 levels in helium will take the place of the four level. Also, to remain consistent with the notation for four-level lasers, the level that the electron decays to from the lower lasing level is designated the one level. The ground level will then be called the zero level to avoid confusion.

There are three dominant laser transitions, 0.6328, 1.1523, and 3.3913 μm, and these transitions are illustrated in Fig. 9-16. The 6328-Å line is generated when an electron in the 5*s* level drops to a 3*p* level. As we learned in Section 3-4.3, the *s* levels in the rare gases other than helium are composed of 4 sublevels, and the *p* levels are composed of 10 sublevels. Thus, the precise transition for the 6328-Å line is from the $^1P_1$ sublevel in the 5*s* level to the $^3P_2$ sublevel in the 3*p* level.

The lifetime of the upper lasing level, 10^{-7} sec, is longer than the lifetime in the lower lasing level, 10^{-8} sec. Thus the condition, $S_{32} < S_{21}$, which is necessary to achieve population inversion, is fulfilled. The electron in the lower lasing level does not drop to the ground state. Rather, it drops to a 3*s* level, which is a relatively long-lived state. Thus, it is possible for this electron to be pumped back up to the lower level, and this will lower the gain because it reduces the magnitude of the population inversion. Electrons are removed from the 3*s* levels primarily by generating heat by collision with the wall. There are more collisions with the wall when the tube diameter is smaller, so that the gain is larger in smaller diameter tubes.

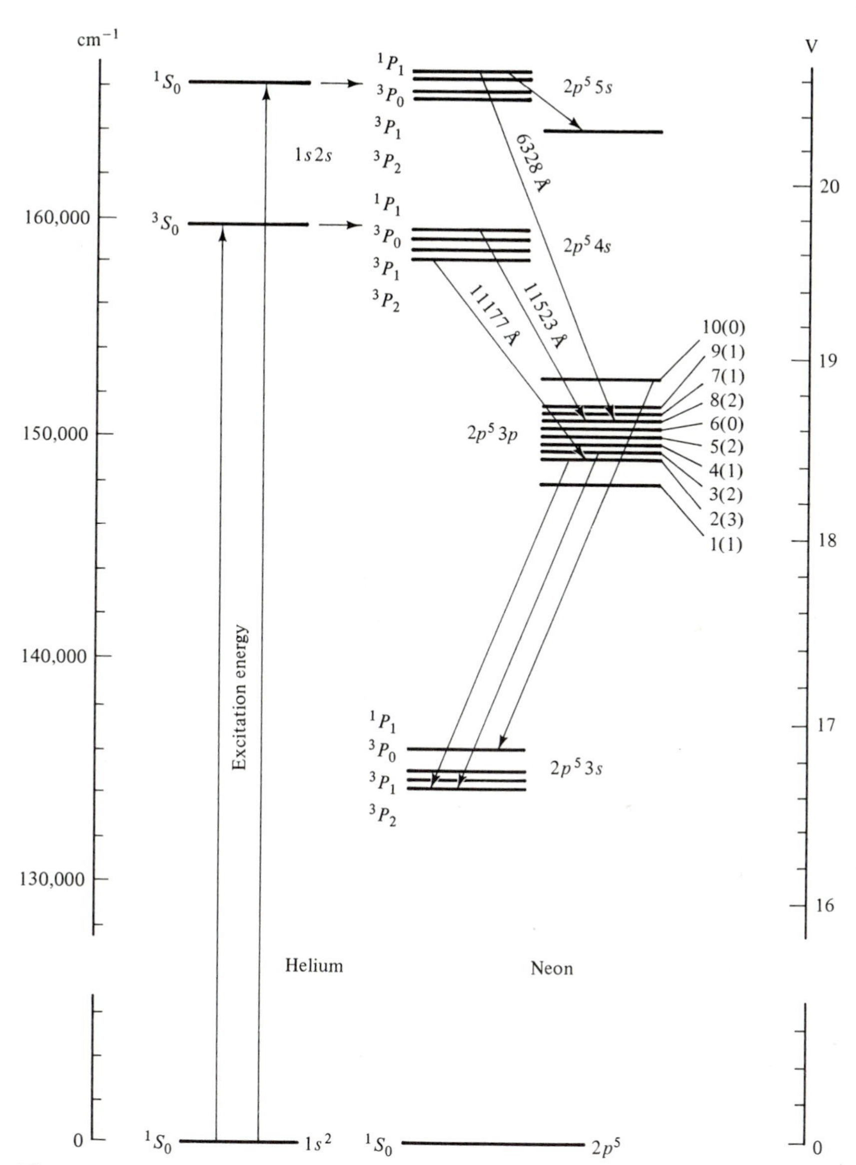

Figure 9-16 Electronic energy levels for helium and neon which are involved in the lasing transitions of the He-Ne laser. (Note that the energies are measured from the ground state.)

The 1.15-μm line is generated by an electron dropping from the $^1P_1$ sublevel of the 4s levels to the $^3P_2$ sublevel of the 3p levels—the same terminal level as the 0.63-μm line. The electron in the lower lasing level also drops to a 3s level and not to the ground state level.

The 3.39-μm line originates at the same level as the 0.63-μm line, but it terminates in the $^3P_2$ sublevel of the 4p levels instead of the 3p. As a result, these two emission peaks are competitors. If no constraints are applied, the 3.39 line will win out. First, the ratio of the stimulated to the spontaneous emission ratio is proportional to ν^{-3} (see Eq. 9-12); this clearly favors the longer wavelength line. Second, $\Delta\nu$ in gas lasers is proportional to ν because the dominant pulse broadening effect is Doppler broadening (see Section 3-8). The threshold population inversion is proportional to $\Delta\nu$ so that the 3.39-μm line begins to lase at a lower pumping level. Furthermore, this depletes the number of electrons in the upper lasing level of the 0.63-μm line making it even more difficult to achieve population inversion for this lasing transition.

The 3.39-μm line can be suppressed in a couple of different ways. In the laboratory methane (CH_4) can be included in the gas mixture. The C-H vibrations will preferentially absorb the 3.39-μm wave. Methane is not used in a commercial device because it tends to leak out. Instead, a pair of Brewster windows is often used. The laser illustrated in Fig. 9-17 has glass windows that make an angle equal to the Brewster angle with the tube axis. Recall that at the Brewster angle, $r_{\|} = 0$. Therefore, all of the components of the electric field vibrating parallel to the plane of the paper will be transmitted parallel to the axis with only a small offset as it passes through the window. The light will then be reflected back into the laser cavity by the external mirrors. There is essentially no absorption of the 0.63-μm line, but the lattice vibrations in the glass absorb the 3.39-μm line thereby suppressing it. Also, the laser light of the emitted 0.63-μm line will be polarized because much of the component of the electric field vibrating perpendicular to the plane of the paper will be reflected off the axis by the Brewster windows.

The method of excitation is illustrated in Fig. 9-17, and the excitation process is shown in Fig. 9-16. The He-Ne mixture is excited by discharging a capacitor via the dielectric breakdown of the gases. The charged particles created by this process are accelerated by the fields of the capacitors and, during their journey, they collide with atoms and excite them electronically.

The primary excitation mechanism is as follows. Helium atoms are excited directly to the 2^1S_0 or 2^3S_1 states, or the electrons cascade down to them. These are relatively long-lived states, 5×10^{-6} sec for the 2^1S_0 state and 10^{-4} sec for the 2^3S_1 state, because electrons in these states cannot easily return to the ground state, since $\Delta L = 0$ instead of ± 1. Also, electrons in the higher-lying 2^1S_0 state will not easily drop into the 2^3S_1 state because this would require a flipping of the electron spin.

Because these states are relatively long-lived, there is sufficient time for the excited helium atoms to transfer their energy to the neon atoms by colliding

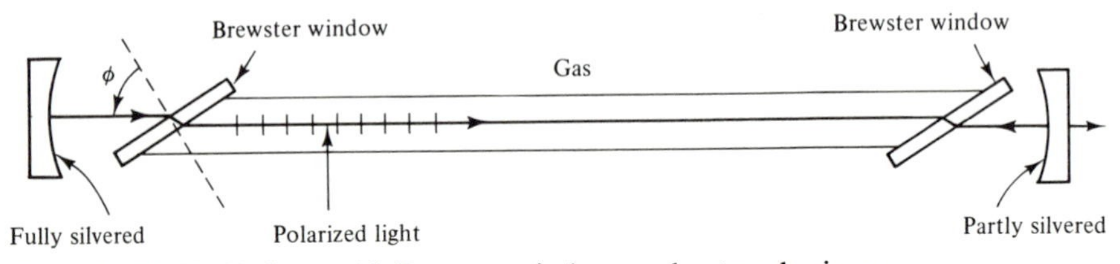

Figure 9-17 He-Ne laser with Brewster windows and external mirrors.

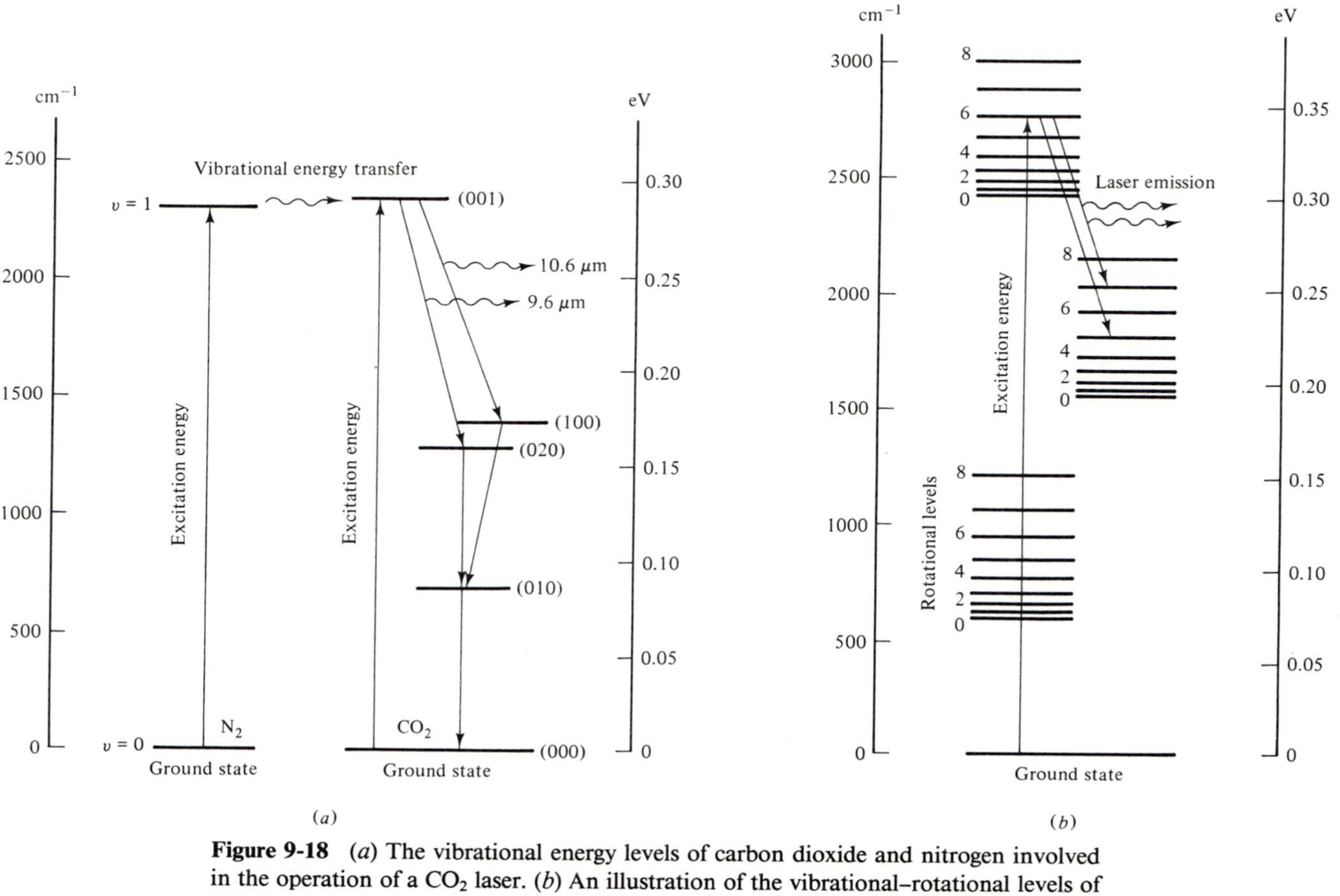

Figure 9-18 (*a*) The vibrational energy levels of carbon dioxide and nitrogen involved in the operation of a CO_2 laser. (*b*) An illustration of the vibrational–rotational levels of CO_2.

with them. The energy transfer from a helium atom in the 2^1S_0 state to a neon atom in the $^1P_1$ sublevel in the 5*s* level is an efficient process because the excitation energies of these two states differ by only $\sim$400 cm^{-1}. For the same reason that the energy transfer from a helium atom in the 2^3S_1 state to a neon atom in the $^1P_1$ sublevel of the 4*s* levels is an efficient process, the energy difference is $\sim$300 cm^{-1}. Both of these energies are only slightly larger than kT at room temperature so that they can be readily absorbed by, or readily absorbed from, the kinetic energy of the gas.

For He-Ne lasers that lase at 3.39 or 0.6328 μm, the optimum He:Ne ratio is 5:1. The total pressure is determined by the product of the tube diameter and the pressure. This product is $\sim$3 Torr mm, and the tube diameter is 1 to 10 mm. For lasers that produce the 1.15-μm line, the optimal gas pressures are He = 1 Torr and Ne = 10 Torr, and the tube diameters range from 5 to 8 mm.

9-9.2 CO_2 Laser

The CO_2 laser, with levels with the same nomenclature as that of the He-Ne laser, differs from the lasers discussed previously in that the transitions are vibrational–rotational and not electronic. The vibrational levels are illustrated in Fig. 9-18 and are discussed in more detail in Section 3-7. A molecule is excited into the (001) vibrational state, which is 2331 cm^{-1} above the ground state level. Rather than quickly falling into the next lower state, it is much more likely that, below threshold, the molecule will radiatively return to the ground state. This vibration mode does form a significant dipole, and this is reflected in the relatively large Einstein coefficient, A_{40} = 210 sec^{-1}. The 4 to 0 transition is not a lasing transition since, to achieve population inversion, more than one-half of the CO_2 molecules will have to be excited out of the ground state.

Laser emission can occur when there is a transition from an (001) level to a (100) level at 1337 cm^{-1} or a (020) level at 1286 cm^{-1}; in the former instance the emission is at 10.6 μm, and in the latter it is at 9.6 μm. The spontaneous transition to the (100) state is not probable in that A_{43} = 0.34. However, the transition probability can be greatly enhanced via stimulated emission.

The molecule also can change its rotational energy during the transition. For a lasing transition $\Delta J = +1$, it is a P transition (see Section 3-7.2); hence, the rotational energy is increased. The lasing transition is usually a P transition, since it is easier to achieve a population inversion as the lower rotational levels are more highly populated. Lasing transitions have been observed at room temperature between $P(12)$ and $P(38)$ with the most probable transition being $P(22)$. The wavelength of the latter transition is 10.6118 μm. Whereas it is possible for more than one rotational mode to be simultaneously lasing, usually there is only one that is.

For transition to the (020) level, the spontaneous transition is improbable as A_{43} = 0.20, and the room temperature lasing transitions lie between $P(22)$, (λ = 9.5691 μm), and $P(34)$, (λ = 9.6762 μm).

For lasing to occur, states must empty from the lower level faster than they build up in the upper level. In both instances the molecules fall to the (010) level at 667 cm^{-1}. The spontaneous radiative transition rate is small, however, in both cases as $A_{21} = 0.2$. This is because the dipole moments for these particular oscillations are small. The transitions are greatly facilitated by collisions with unexcited CO_2 molecules. For the reactions

$$CO_2(100) + CO_2(000) \rightarrow 2CO_2(010) + 3\ cm^{-1} \tag{9-46a}$$

$$CO_2(020) + CO_2(000) \rightarrow 2CO_2(010) - 48\ cm \tag{9-46b}$$

the small amounts of energy necessary for the reactions can be supplied by the surrounding gas.

Removing molecules from the (010) state before they become reexcited into the (020) or (100) states can also be a problem. The most effective way to accomplish this is to insert helium into the gas mixture to collide with the $CO_2(010)$ molecule and to convert vibrational energy into translational energy. Water vapor and hydrogen can also be effectively used for this process.

The CO_2 molecule can be excited into the (001) state either by direct collision with a charge carrier or by collision with a vibrationally excited N_2 molecule. In the former instance, molecules excited into the $(00v_3)$ state cascade down in steps of $\Delta v_3 = 1$ as dictated by the selection rule. They can do this by giving up energy to a CO_2 molecule in the ground state, since the vibrational levels are approximately evenly spaced.

The $(0v_20)$ levels can also be excited by collisions with accelerated charged particles, but these collisions are less likely. The physical explanation for this is that for a particle moving parallel to the molecular axis, the most probable distortion on collision with the molecule is that of a $(00v_3)$ vibration. The preference for the $(00v_3)$ excitation does depend on the accelerating voltage of the charged particle. The voltage is, therefore, picked to enhance this transition.

More of the CO_2 molecules are excited into the (001) state by colliding with an N_2 molecule in its first excited vibrational state. The energy can be readily transferred, since the difference in the two energies is only 18 cm^{-1}. The N_2 molecules are excited by accelerated charged particles, and they are able to remain in the excited state long enough to transfer their energy to the CO_2 molecule because the N_2 molecule has no dipole moment. It, therefore, does not radiatively decay.

Thus, there are three gases in the CO_2 laser. The optimum partial pressure of each is 3 Torr CO_2, 5 Torr N_2, and 14 Torr He.

READING LIST

1. *Lasers and Light—Readings from Scientific American.* San Francisco: W. H. Freeman, 1969.
2. F. A. Jenkins and H. E. White, *Fundamentals of Optics,* 4th Ed. New York: McGraw-Hill, 1976, Chapter 30.
3. R. T. Weidner and R. L. Sells, *Elementary Modern Physics,* 2d Ed. Boston: Allyn and Bacon, 1973, Chapter 11.
4. B. A. Lengyel, *Lasers,* 2d Ed. New York: Wiley-Interscience, 1971.

5. J. Wilson and J. F. B. Hawkes, *Optoelectronics: An Introduction.* Englewood Cliffs, NJ: Prentice-Hall, 1983, Chapters 5 and 6.
6. J. T. Verdeyen, *Laser Electronics.* Englewood Cliffs, NJ: Prentice-Hall, 1981, Chapters 8 to 10.
7. A. Yariv, *Introduction to Optical Electronics,* 2d Ed. New York: Holt, Rinehart and Winston, 1976, Chapters 5 to 7.
8. A. K. Levine and A. J. DeMaria, *Lasers.* New York: Marcel Dekker, Vol. 1—Chapters 1 to 3, Vol. 2—Chapter 1, Vol. 3—Chapters 2 and 3.
9. A. Yariv, *Quantum Electronics,* 2d Ed. New York: John Wiley, 1975.

PROBLEMS

9-1. The Wien displacement law is $\lambda_{max} T$ = constant, where λ_{max} is the wavelength for which the radiated energy is a maximum.

(a) Derive this relationship and find the numerical value of the constant in angstroms minus degrees kelvin.

(b) What is λ_{max} for the sun if the surface of the sun is 5200 K? Assume $h\nu \gg kT$.

9-2. (a) Find an expression for the ratio of spontaneous to stimulated emission in terms of $h\nu/kT$ under conditions of thermal equilibrium.

(b) Compute the ratio of spontaneous to stimulated emission for $\nu = 5 \times 10^{14}$ Hz and $T = 2000$ K.

9-3. When $\Delta E \gg kT$, the equilibrium probability an electron has energy between ΔE and $\Delta E + dE$ is $Ae^{-\Delta E/kT}\, dE$.

(a) Knowing that the probability an electron has an energy E between 0 and ∞ is 1, find an expression for A.

(b) What is the probability an electron has an energy $\geq \Delta E^*$ where ΔE^* is some fixed value of ΔE?

(c) The lower-lasing level of Nd^{+3} in a Nd:YAG laser has an energy 2111 cm^{-1} above the ground state level. What is the ratio of the probability that an electron is in this level compared with the probability it is in the ground state at $T = 300$ K?

9-4. An atom has two energy levels with a transition wavelength of 694.3 nm. Assuming that all of the atoms in an assembly are in one or another of these levels and the density of states for the two levels is the same, calculate the percentage of the atoms in the upper level at room temperature, $T = 300$ K, and at $T = 500$ K.

9-5. At what temperature are the rates of spontaneous and stimulated emission equal under equilibrium conditions when $\lambda = 500$ nm? At what wavelength are they equal at room temperature, $T = 300$ K?

9-6. If the light intensity doubles after passing once through a laser amplifier 0.5 m long, calculate the gain coefficient assuming no losses. If the increase in intensity is only 5 percent what would α_g be?

9-7. Calculate the mirror reflectances required to sustain laser oscillations in a laser that is 0.1 m long given that the gain coefficient is 1 m^{-1} (assume the mirrors have the same reflectance).

9-8. Calculate the Doppler broadened linewidth for the CO_2 laser transition ($\lambda = 10.6$ μm) and the He-Ne laser transition ($\lambda = 632.8$ nm) assuming a gas discharge temperature of about 400 K. Take the relative atomic masses of carbon, oxygen, and neon to be 12, 16, and 20.2, respectively.

9-9. Consider a laser transition whose gain coefficient, at the line center, exceeds the loss (prorated over the length) by some specified factor; $\alpha_g(\nu) = \alpha_g(\nu_0) \exp[-(4 \ln 2)(\nu - \nu_0)^2/\Delta\nu_D^2]$; $\alpha_g(\nu_0)/\alpha$ = specified = K.

(a) Derive an expression for the bandwidth over which oscillation is possible.

(b) If the length of this laser is d units, how many modes can oscillate? Assume that the central mode is at the line center.

(c) Find the equation describing the amplitude of a mode at the frequency ν in terms of $\alpha_g(\nu)I_s$ and α.

(d) Derive an explicit formula for I_ν/I_s in terms of the gain coefficient at the line center $\alpha_g(\nu_0)$, α, and the frequency separation from the line center.

9-10. Consider the three-level system for which level 2 is 1.1 eV and level 3 is 3.4 eV above level 1, and $A_{21} = 10^8$, $A_{31} = 2 \times 10^7$, and $A_{32} = 5 \times 10^7$ sec^{-1}.

(a) What are the wavelengths for the various transitions?

(b) What is the lifetime of state 3?

(c) What fraction of the electrons falling from level 3 fall to level 2?

(d) Suppose that 10^{14} atoms/cm^3 are excited from state 1 to state 3 at $t = 0$ by some external mechanism. Describe the time evolution of the various populations in state 2 and state 3. Assume $n_2(t) = k_1(1 - e^{-k_2t})e^{-k_3t}$ if $A_{21} \neq A_{32} + A_{31}$.

(e) Suppose that this external agent is strong enough to keep a steady state population of 10^{14} atoms/cm^3 in state 3. (1) How much power is required? (2) What is the steady state population of state 2? (3) What power is radiated spontaneously in the 3 to 2 transition? (4) What is the atomic quantum efficiency of the 3 to 2 transition?

9-11. Suppose that we consider a four-level laser system as shown in the accompanying diagram. Assume the density of states at all levels is the same. The total density of active electrons is 10^{16}/cm^3. Although level 2 is distinct from state 1, collisional processes keep the two in equilibrium, so that $n_2/n_1 = \exp(-\Delta E/kT) = 0.05$. The external agent pumps the atoms from state 1 to 4 at a rate proportional to the difference in populations. Once an atom is in state 4, it can radiatively decay back to state 1($A_{41} = 10^8$ sec^{-1}) or transfer over to state 3($S_{43} = 5 \times 10^8$ sec^{-1}). An atom in state 3 can transfer back to state 4 ($W_{34} = 10^8$ sec^{-1}) or decay spontaneously down to state 2 ($A_{32} = 10^7$ sec^{-1}). [If the population difference $(N_2 - N_1)$ is large enough, stimulated emission will also affect the population of state 3.]

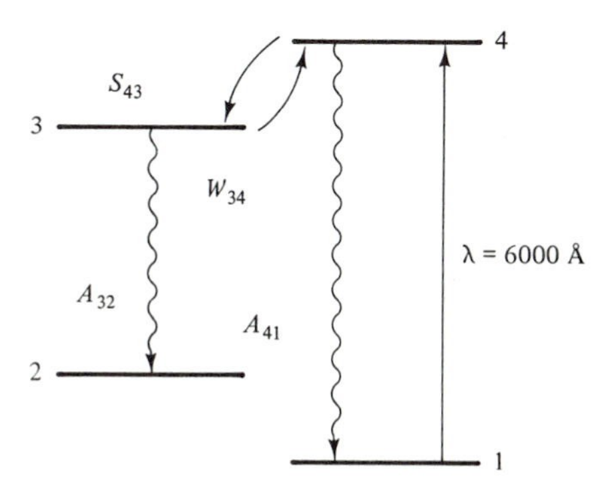

(a) How much power must the external agent deposit in the medium to obtain equal populations in states 2 and 3, that is, zero gain?

(b) If we had an infinitely strong pump ($R \longrightarrow \infty$) and could prevent stimulated emission from inducing transitions, what are the populations in the various states?

(c) If the transition 3 − 2 is 100 Å wide (FWHM), what is the gain for an infinite pump? Express your answer in decibels per centimeter.

(d) What fraction of the absorbed light is reemitted by the 3 to 2 transition?

(e) If the laser medium is 20 cm long and 1 cm^2 in cross section, how much power must the external agent supply to achieve threshold in a cavity with an average loss of 0.01 cm^{-1}?

9-12. Consider the possibility of pumping a ruby laser rod, 1 cm in diameter, placed at the focal point of a cylindrical parabolic reflector aimed at the sun. The spectrum from the sun has the shape of a blackbody distribution with T = 5000 K and a total integrated intensity (over all frequencies) of 1.4 kW/m^2. Estimate the size of the parabola needed to achieve equal populations in the upper and lower laser levels.

9-13. (a) What is the optimum value of the transmissivity if $\alpha_g^0\, L = 0.12$ and $\Lambda = 0.055$?

(b) What is the ratio of the output power at $T = 0.01$ and $T = T_{opt}$? At $T = 0.04$ and $T = T_{opt}$?

(c) For what value of T is $P_{ex} = 0$?

9-14. Determine α_g at room temperature for a ruby laser with the following characteristics: $n_2 - (\rho_2/\rho_1)n_1 = 5 \times 10^{17}$ cm^{-3}, $\Delta\nu = 2 \times 10^{11}$ Hz, $\tau_{sp} = 3 \times 10^{-3}$ sec, and $n = 1.77$.

9-15. Consider the Nd:YAG laser described in Example 9-5. How much power must be supplied to the exciting lamp if the Nd:YAG crystal has a 0.25 cm diameter and a 3 cm length, 5 percent of the light of the lamp has the proper frequencies for absorption, 5 percent of the light with these frequencies is actually absorbed, the ratio of the average lasing frequency to exciting frequency is 0.5, and 50 percent of the power supplied to the lamp is converted into optical power?

9-16. Determine n_t and P_s for the 1.06-μm line in a Nd:glass laser with the following characteristics: $\Delta\nu = 3 \times 10^{12}$ Hz, $L_3 = 10$ cm, $\alpha \cong 0$, $R_1 = R_2 = 0.95$, $\tau_{sp} = 10^{-4}$ sec, $n = 1.5$, and $V = 10$ cm^3. Why are these values larger than they are for the Nd:YAG laser?

9-17. Determine n_t and P_s for the 1.06-μm line in a Nd:glass laser with the following characteristics: $\Delta\nu = 200$ cm^{-1}, $L = 20$ cm, $\Lambda = 0.02$, $\tau_{sp} = 3 \times 10^{-4}$, $n = 1.5$, and $V = 1$ cm^3.

9-18. Assuming that 10 percent of the light from the lamp emits at frequencies that can be absorbed by the laser, 10 percent of this light is actually absorbed, the product of the pumping efficiency and the internal quantum efficiency is 0.4, the atomic quantum efficiency is 50 percent, and the lamp is 50 percent efficient, find the power that must be supplied to the lamp to get the Nd:glass laser described in Problem 9-17 to lase.

9-19. The gain for the transition in the He-Ne laser at 3.39 μm is very large: 30 dB in a 1.0-m-long tube.

(a) Assume that the gas mixture is at 400 K and compute the population inversion necessary to provide this gain if the emission peak bandwidth is determined by Doppler broadening.

(b) Suppose that this same population inversion was obtained on the common "red" line at 6328 Å. What would be the gain in a 1.0-m tube? Why is the gain so different?

(c) What is the saturation intensity for the two laser transitions. A_{32} (3.39 μm) = 2.86×10^6 and A_{32} (6328 Å) = 6.56×10^6 sec^{-1}?

9-20. Calculate the saturation power P_s of the He-Ne laser operating at 6328 Å. Assume $V = 2$ cm^3, $\Lambda = 1$ percent per pass, $L = 30$ cm, and $\Delta\nu = 1.5 \times 10^9$ Hz. Calculate the steady state population inversion, n_t, if $\tau_{sp} = 10^{-7}$ sec.

9-21. Determine n_t for the 6328-Å line for a He-Ne laser with the following properties: $\Delta\nu = 10^9$ Hz, $L = 10$ cm, $\alpha \simeq 0$, $R_1 = R_2 = 0.98$, $\tau_{sp} = 10^{-7}$ sec.

9-22. A typical He-Ne laser operates at 2500 V dc and 10 mA. If the light output is 5 mW, what is the overall efficiency?

9-23. Calculate the degree of population inversion required to give a gain coefficient for a CO_2 laser ($\lambda = 10.6$ μm) of 0.5 m^{-1}. Take the Einstein A coefficient for the upper laser level to be 200 sec^{-1} and assume the emission peak bandwidth is determined by Doppler broadening at 400 K.

9-24. Consider a pulsed ruby laser exposed to uniform pumping radiation of 2.25 eV per photon for a time, $t_f = 5 \times 10^{-4}$ sec.

(a) Show that the number of electrons per unit volume excited by absorbing this radiation is

$$n_2 = t_f \int_0^\infty \alpha_g(\nu) I(\nu) \eta(\nu) \frac{d\nu}{h\nu_{13}}$$

(b) Why can it be assumed that all of the excited electrons reside in level 2?

(c) Assuming that only a small amount of the radiation is absorbed so $I(\nu)$ is constant and that $\alpha_g(\nu)\eta(\nu)\,d\nu$ can be approximated by a rectangular pulse of width $\Delta\nu$, find the pumping energy per unit area necessary to reach the transparency point if the chromium concentration is 2×10^{19} cm^{-3}, the collision cross section is $\sigma = 10^{-19}$ cm^{-2}, and $\eta = 1$. Repeat for $\eta = 0.01$.

Chapter 10

New Developments

10-1 INTRODUCTION

The rate at which optical information is processed and the efficiency of transmitting it is constantly being improved. Manufacturers are improving the devices that have already been developed by improving the quality control in the manufacturing process, and researchers are creating faster and more efficient devices. But the specifications for these new devices are ever more exacting.

One such device is the single transverse mode laser. Lasers with only a single transverse mode can transmit optical signals at a higher rate over longer distances because there is essentially no modal dispersion when only a single mode is present. Single mode lasers are made by reducing the effective lateral dimension of the laser cavity to a size where only one mode is stable. This can be accomplished by etching a mesa structure and regrowing an insulating semiconductor layer around it, and by depositing the active layer on a concave confining layer. The former structure is called a buried heterostructure (BH), and the latter is called a constricted double heterostructure (CDH).

Data can also be transmitted at a higher rate over longer distances when a single frequency laser is used. A single frequency laser is a laser that emits only one longitudinal mode. This essentially eliminates wavelength dispersion. Single frequency lasers can be built by adding a second cavity to the system with a length such that one, but only one, wavelength is a standing wave in both cavities. This can be done by etching a groove in the active layer, cleaving the cavity into two pieces, or etching a periodic structure into the substrate. A single frequency can also be obtained by exciting the cavity with another laser with a narrow emission bandwidth like that emitted by a He-Ne laser.

The laser threshold current density for a quantum well laser can be substantially below that of a regular DH laser. Quantum well lasers are fabricated from heterostructures in which the active region material has a smaller energy gap than the confining layers, and its width is less than a de Broglie wavelength—about 500 Å for electrons in GaAs. The discontinuity in the energy gap between the active and confining layers creates a potential well in the conduction band for electrons and in the valence band for holes. When the thickness of the active region is less than a de Broglie wavelength, the electron states in the conduction band and the hole states in the valence band no longer form a continuum. Rather, the states are discrete, and the density of states curve in three dimensions has a staircase structure in which the density of states for the lowest energy state in the conduction and valence bands is larger than it is in a normal structure. The larger density of states at the lowest energy levels makes it easier to obtain population inversion so that J_{th} is lowered.

Another exciting area being developed is integrated optics. Light emitters, transmitters, and detectors can be built into the same chip. The transmitters are simply waveguides that are created much in the same way DH structures are made. Information can be encoded on the transmitted beams by modulating the signals in the waveguide. This is often done by splitting a beam, changing or not changing the velocity of the beam through one of the branches using the electro-optic effect, and then recombining the light from the two branches.

Bistability, the last topic to be discussed, results from the fact that the index of refraction decreases slightly as the intensity increases. This is caused by a decrease in the absorption of bandgap radiation as the electron states at the bottom of the conduction band begin to saturate. The change in the index of refraction shifts the transmittance peaks of the cavity. There is an abrupt increase in the transmittance when n has a value such that the wave in the cavity has a wavelength that is equal to a half integer multiple of the cavity length, L. By taking advantage of the bistability and properly optically biasing the cavity, it can be used as an AND, OR, or NOT gate and these gates can be switched in 10's of picoseconds.

10-2 SINGLE MODE, SINGLE FREQUENCY LASERS

Single mode lasers have only one transverse mode whereas a single frequency laser has only one longitudinal mode. Only a single transverse mode can exist in an optical cavity when the lateral cavity dimensions are made small enough. Limiting the emission to a single longitudinal mode can be accomplished by making the cavity length small enough, by dividing a single laser cavity into two cavities for which only a single mode obeys the boundary conditions in both cavities, by externally selecting a mode using diffraction mechanisms, and by externally pumping the laser with a single frequency.

A primary application for single mode, single frequency lasers is long-distance, high-data-rate optical communication systems. Having a single transverse mode essentially eliminates modal dispersion and having a single longitudinal mode essentially eliminates wavelength dispersion. As a result, the information carrying capacity of a single mode fiber can theoretically be greater than 200 Gb/sec·km; rates greater than 70 Gb/sec·km have already been realized. Because practical modulation rates are limited to ~2 Gb/sec, this would allow repeater stations to be located ~100 km apart—a very attractive feature to those who lay optical cables under the sea. Signals can be sent this distance when the laser emits at 1.55 μm, since fibers can be fabricated for which the attenuation is only 0.2 dB/km at this wavelength.

EXAMPLE 10-1

(a) If a 5-mW signal with $\lambda_0 = 1.55\ \mu$m is inserted into a single mode optical fiber, what is the power of the signal 100 km away if the attenuation is 0.2 dB/km? **(b)** What photocurrent will it produce in a photodiode detector in which $\alpha = 1/L = 1/(d + x_n)$, $R = 0.1$, and $\eta_i = 0.9$?

(a) $\text{dB} = 10 \log (P_0/P) = (0.2)(100) = 20$

$$P = \frac{P_0}{100} = 50\ \mu\text{W}$$

(b) From Eqs. 4-75 and 4-80

$$J_{op} = \frac{(1-R)\eta_i q P \alpha L e^{-\alpha(d+x_n)}}{E_p(1+\alpha L)} = \frac{(0.9)(0.9)(5\times10^{-5})(1)e^{-1}}{(1.242/1.55)(1+1)} = 9.30\ \mu\text{A}$$

10-2.1 Single Mode Lasers

As in a planar dielectric, if the lateral width, *d*, is small enough, only a single transverse mode will be stable in the laser cavity; the direction normal to the junction of a DH laser (0.15 − 0.4 μm) will always be small enough to limit the cavity to a single mode. This single mode will have a Gaussian distribution, which is described by the equation

$$I = I_0 \exp\left(\frac{-x}{w/2}\right)^2 \tag{10-1}$$

where w is the cavity width, and x is the distance from the cavity axis.

A narrow laser cavity can be fabricated by using an etch and growth technique to form what is called a buried heterostructure (BH). The GaAs/GaAlAs BH laser in Fig. 10-1*a* is formed by etching away the different layers and then regrowing an insulating GaAlAs layer around the remaining mesa structure. The etched surface must be thoroughly cleaned before the regrowth to ensure that the shunt currents along the interface are small. The InP/InGaAsP BH laser in Fig. 10-1*b* differs from the GaAs/GaAlAs laser in that the regrown InP is too conductive because it is always slightly *n*-type. As a result, the shunt currents have to be eliminated by forming *pn* junctions at the regrowth interface.

The active region of a semiconductor laser can also be made small enough to sustain only a single transverse mode by growing the structures on a nonplanar substrate. These structures are called constricted double heterostructures (CDHs),

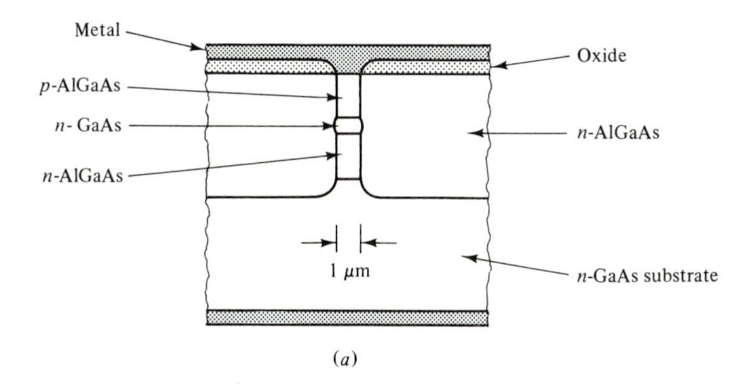

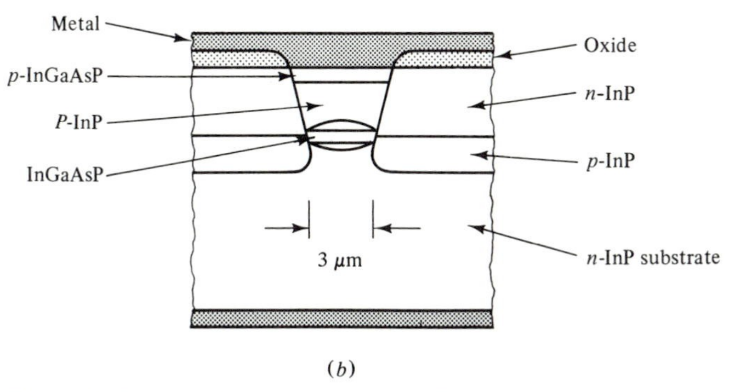

Figure 10-1 Buried heterostructure (*a*) GaAs/GaAlAs, and (*b*) InP/InGaAsP single mode lasers.

and they are formed during a single growth sequence, which reduces the problem of shunt currents encountered in BH lasers. The CDH laser in Fig. 10-2*a* is made by first etching a concave channel in the *n*-type substrate. The *n*-type GaAlAs confining layer is thicker where it grows over the channel, but the surface above the channel is still concave. The active region is only ~0.2 μm thick in the concave region, and it is so thin away from the center that it is effectively cut off from the thicker region. The structure is completed by growing a *p*-GaAlAs confining area on top of the active layer, and a *p*-GaAs contacting layer is grown on top of it.

Another way to form a CDH laser is to deposit films on a mesa with specific exposed crystallographic planes. Growth on the inclined planes in Fig. 10-2*b* is slower than it is on the horizontal planes so that the *n*-confining layer forms a convex surface. The active region is so thin in the inclined plane regions that the portion growing above the horizontal plane of the mesa is effectively cut off. Again the laser structure is completed by growing *p*-confining and capping layers.

One advantage a CDH laser has over a BH laser is that it can be wider and still be a single mode laser. This is because the effective index of refraction change in this structure is smaller. Thus, CDH lasers have widths of 4 to 5 μm, whereas the widths of BH lasers are typically 1 to 3 μm. The power output of the CDH lasers can be larger because they are wider.

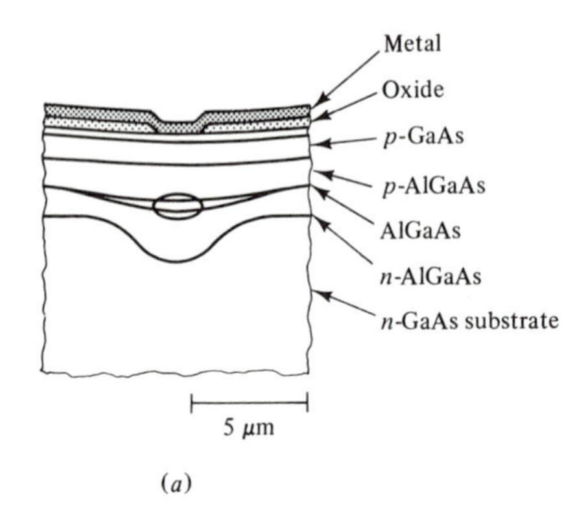

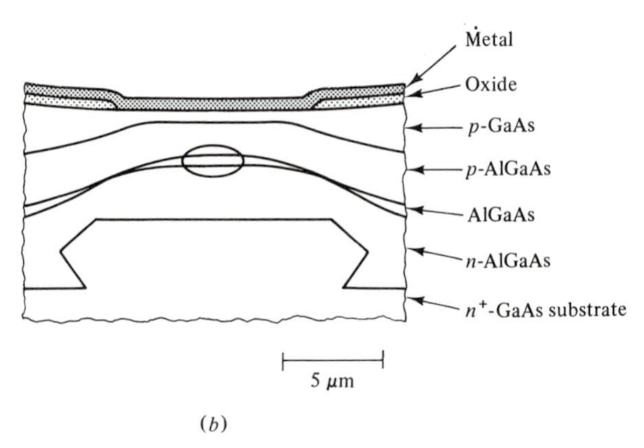

Figure 10-2 Constricted heterostructure GaAs/GaAlAs lasers formed by growing (*a*) over an etched channel, and (*b*) on a mesa with specifically exposed crystallographic planes.

EXAMPLE 10-2

If the equation for the maximum width of a planar dielectric for which only one transverse mode is stable can be applied to a DH laser cavity, what is the maximum width of the cavity for **(a)** a GaAs/$Ga_{0.70}Al_{0.30}As$ single mode laser emitting at 0.84 μm, and **(b)** an InP/$In_{0.59}Ga_{0.41}As_{0.88}P_{0.12}$ single mode laser emitting at 1.55 μm? Assume that the index of refraction of the InGaAsP is the weighted average of the binary components. n(InAs) = 3.82, n(InP) = 3.52, n(GaAs) = 3.63, n(GaP) = 3.33.

(a) From Eq. 7-36

$$n(Ga_{0.70}Al_{0.30}As) = n_{GaAs} - 0.62\ \Delta x = 3.6 - 0.62(0.3) = 3.444$$

From Example 2-2

$$w < \frac{3\lambda_0}{2(n_1^2 - n_2^2)^{1/2}} = \frac{3(0.84)}{2[(3.63)^2 - (3.344)^2]^{1/2}} = 1.10\ \mu\text{m}$$

(b) $$n(\text{InGaAsP}) = (1-x)(1-y)n(\text{InAs}) + (1-x)yn(\text{InP}) + x(1-y)n(\text{GaAs}) + yn(\text{GaP})$$
$$= (0.59)(0.88)3.82 + (0.59)(0.12)3.52 + (0.41)(0.88)3.63 + (0.41)(0.12)3.33 = 3.706$$

$$w < \frac{3\lambda_0}{2(n_1^2 - n_2^2)^{1/2}} = \frac{3(1.55)}{2[(3.706)^2 - (3.52)^2]^{1/2}} = 2.01\ \mu\text{m}$$

Since single mode lasers are narrower, $\theta_{\|}$ is a little larger than it is for a multimode laser. Also, the reduced width leads to a smaller threshold current. A typical value of i_{th} is 20 mA, but it can be as low as 10 mA. The power output is from 1 to 7 mW. The upper limit, particularly for GaAs/GaAlAs lasers, is determined by the linear power density for which catastrophic heating occurs. When it reaches 1.5 mW/μm, the surface states cause the energy gap to narrow so that the laser light at the surface is strongly absorbed.

One way to achieve a larger power output is to use nonabsorbing mirrors. This can be accomplished as shown in Fig. 10-3 with a deep zinc diffusion that does not extend to the ends of the laser cavity. The zinc doping decreases the energy gap enough so that the laser light created by band-to-band transitions in the heavily doped region will not be absorbed at the surface even at high power levels.

Another way higher power levels can be achieved is by widening the lateral direction and selectively absorbing the higher order modes that are now stable. Recall from Sections 2-2.2 and 2-4 that the intensity of the higher order modes is larger away from the cavity axis. They are thus more strongly absorbed by absorbing materials at the edge of the cavity.

This can be accomplished by the buried heterostructure laser in Fig. 10-4*a*. The GaAs absorbing layer near the $Ga_{0.95}Al_{0.05}As$ active region more strongly absorbs the higher order modes. The same thing can be accomplished by growth over a channel in the absorbing GaAs substrate. This is illustrated in Fig. 10-4*b*. These lasers can emit from 10 to 50 mW; their threshold currents range from 60 to 100 mA; and their differential quantum efficiencies are 30 to 40 percent.

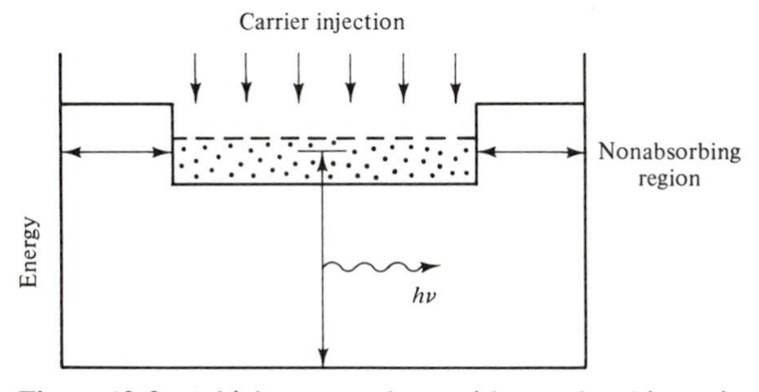

Figure 10-3 A higher power laser with nonabsorbing mirrors.

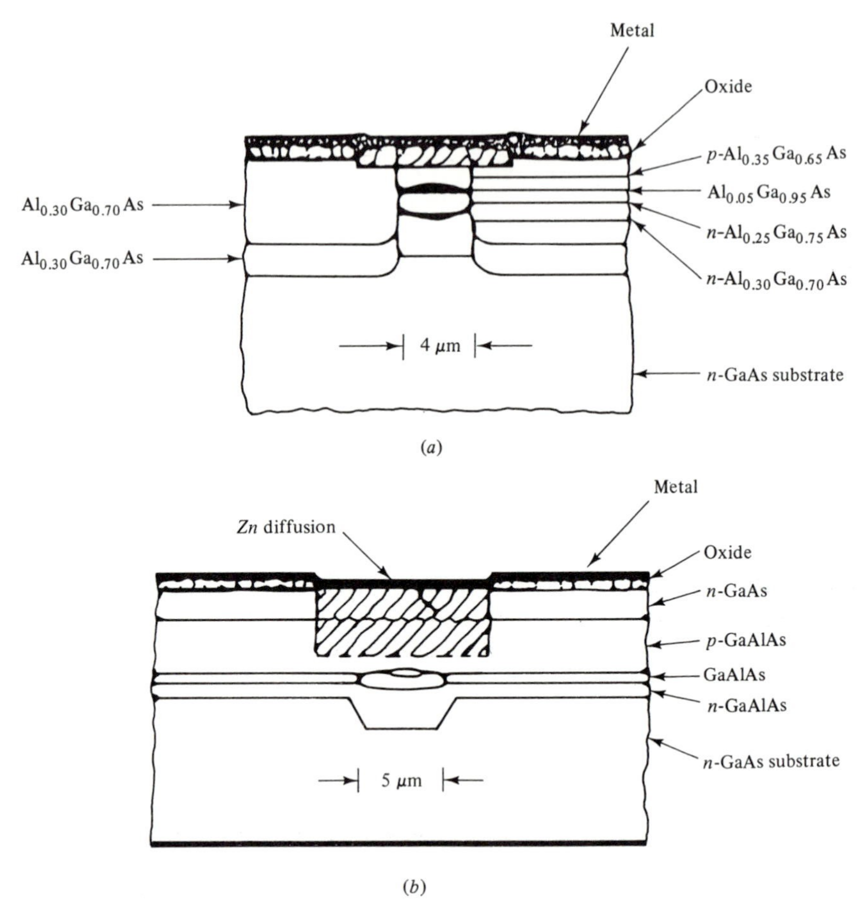

Figure 10-4 Higher power single mode lasers that preferentially absorb the higher order modes in (*a*) a regrown GaAs layer, and (*b*) an absorbing substrate with an etched channel.

This is lower than it is for a multimode DH laser because less of the current is confined to the active region, and there is more absorption in the cavity.

10-2.2 Single Frequency Laser

As we learned in Chapters 7 and 8, the output from a normal laser consists of a number of modes with closely, but evenly spaced frequencies (refer to Figs. 7-8 and 8-11). The spacing, $\delta\nu$, is the frequency difference between adjacent standing waves in the cavity, and the number of modes is determined by the bandwidth of the emission peak with an intensity above threshold. Lasing at a single frequency is accomplished by limiting the cavity to a single mode that is above threshold or by external means favoring one of the modes over the others.

One possible, but not viable, solution is to use a filter. First, the filter would have to be an extremely narrow bandpass filter, and even if one could find such a filter, the laser performance would be mediocre because the partition noise due to mode hopping would be unacceptably large. Mode hopping involves the transfer of energy back and forth between the neighboring modes inside the cavity. Thus, the amplitude of the desired mode would vary significantly as energy was transferred to and from the modes filtered out of the output signal.

Mode hopping, which was also mentioned in Chapters 7 and 9, is sufficiently important to devote a few paragraphs to it here. We will do so by using the mechanical analog illustrated in Fig. 10-5. The system is composed of two masses, m_1 and m_2, connected to each other by a spring with a spring constant of k_3 and to the rigid walls by a spring with a spring constant of k_1 or k_2.

The equations of motion for the masses are

$$\Sigma F = m_1\ddot{x}_1 = -k_1x_1 - k_3(x_1 - x_2) \tag{10-2a}$$

$$m_2x_2 = -k_3(x_2 - x_1) - k_2x_2 \tag{10-2b}$$

Assuming the solutions for x_1 and x_2 are

$$x_1 = A_1 \cos \omega t \tag{10-3a}$$

$$x_2 = A_2 \cos \omega t \tag{10-3b}$$

and substituting them into Eqs. 10-2a and 10-2b yields

$$\frac{A_1}{A_2} = \frac{-k_3}{m_1\omega^2 - k_1 - k_3} = \frac{m_2\omega^2 - k_2 - k_3}{-k_3} \tag{10-4a}$$

or

$$\omega^4 - \left(\frac{k_1 + k_3}{m_1} + \frac{k_2 + k_3}{m_2}\right)\omega^2 + \frac{k_1k_2 + k_2k_3 + k_3k_1}{m_1m_2} = 0 \tag{10-4b}$$

To reduce the algebraic complexity, we let $m_1 = m_2$, $k_1 = k_2 = k$, and $k_3 = K$. The solutions for Eq. 10-4b then are

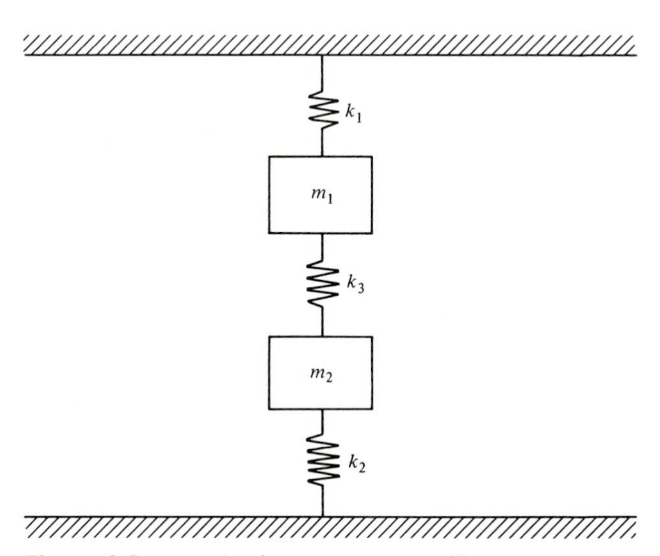

Figure 10-5 A mechanical analog used to illustrate mode hopping.

$$\omega_1^2 = \frac{k}{m} \tag{10-5a}$$

$$\omega_2^2 = \frac{k + 2K}{m} \tag{10-5b}$$

From Eq. 10-4a, $A_1 = A_2 = A$ when $\omega = \omega_1$ and $A_1 = -A_2 = A$ when $\omega = \omega_2$.

The frequencies ω_1 and ω_2 are the natural frequencies of the system and are equivalent to two longitudinal modes in an optical system. If the initial displacements are A_1 and A_2 and they have precisely the same values, both masses will oscillate in unison with $\omega = \omega_1$ forever and a day. They are not coupled because the middle spring, K, is never compressed or extended. Thus, each mass sees an effective spring constant, k. On the other hand, if the initial displacements are equal, but opposite, then the two masses will oscillate in opposing directions at $\omega = \omega_2$ forever. The two masses are not coupled because the midpoint of the middle spring remains fixed. Because the spring constant is inversely proportional to the effective length of the spring, the middle spring contributes $2K$ to the effective force constant seen by each mass.

For any other initial displacement, however, the two masses will be coupled and energy will be transferred back and forth between them. Consider the situation where the initial displacements are $x_1(0) = A$ and $x_2(0) = 0$. The initial displacement and the subsequent motion can be described by the linear combination of the two natural frequency solutions,

$$x_1 = \frac{A}{2}(\cos \omega_1 t + \cos \omega_2 t) \tag{10-6a}$$

$$x_2 = \frac{A}{2}(\cos \omega_1 t - \cos \omega_2 t) \tag{10-6b}$$

If we define the energy of the first system, E_1, to be the potential energy stored in spring 1, one-half the potential energy stored in spring 3, and the kinetic energy of mass 1, and we define the energy of the second system, E_2, to be the potential energy stored in spring 2, one-half the potential energy stored in spring 3, and the kinetic energy of mass 2, then

$$E_1 = \tfrac{1}{2}kx_1^2 + \tfrac{1}{4}K(x_1 - x_2)^2 + m\ddot{x}_1^2 \tag{10-7a}$$

$$E_2 = \tfrac{1}{2}kx_2 + \tfrac{1}{4}K(x_1 - x_2)^2 + m\ddot{x}^2 \tag{10-7b}$$

Substituting Eqs. 10-6a and 10-6b into Eq. 10-7a and 10-7b and using the results of Eqs. 10-5a and 10-5b yields

$$E_1 = \tfrac{1}{4}A^2[k + K + k \cos \omega_1 t \cos \omega_2 t + (k^2 + 2kK)^{1/2} \sin \omega_1 t \sin \omega_2 t] \tag{10.8a}$$

$$E_2 = \tfrac{1}{4}A^2[k + K - k \cos \omega_1 t \cos \omega_2 t - (k^2 + 2kK)^{1/2} \sin \omega_1 t \sin \omega_2 t] \tag{10.8b}$$

Note that the total energy, E_T, is constant and has the value

$$E_T = E_1 + E_2 = \tfrac{1}{2}(k + K)A^2 \tag{10-9}$$

One can easily show that E_1 and E_2 have local maxima or minima or are at an inflection point when

$$t = \frac{m\pi}{\omega_1} \tag{10-10a}$$

$$t = \frac{(2m - 1)\pi}{2\omega_2} \tag{10-10b}$$

and their magnitudes at these points depend on the K/k ratio. When $K = 0$, there will be no coupling and all of the energy will be retained by m_1 and k_1, and when $k = 0$, E_1 and E_2 will be equal to $\frac{1}{2}E$. For insight into other values we turn to Example 10-3.

EXAMPLE 10-3

(a) For what ratio of K/k is $\omega_2/\omega_1 = \frac{3}{2}$? **(b)** Make a plot of E_1/E_T versus $\omega_1 t$. **(c)** For what values of t is E_1 a maximum? A minimum? **(d)** What are the maximum and minimum values of E_1/E_T? **(e)** Repeat parts a to d for $\omega_2/\omega_1 = 2$.

(a)
$$\omega_2^2 = \frac{k + 2K}{m} = \frac{9}{4}\,\omega_1^2 = \frac{9}{4}\frac{k}{m}$$

$$K = \tfrac{5}{8}k$$

(b) See Fig. 10-6.

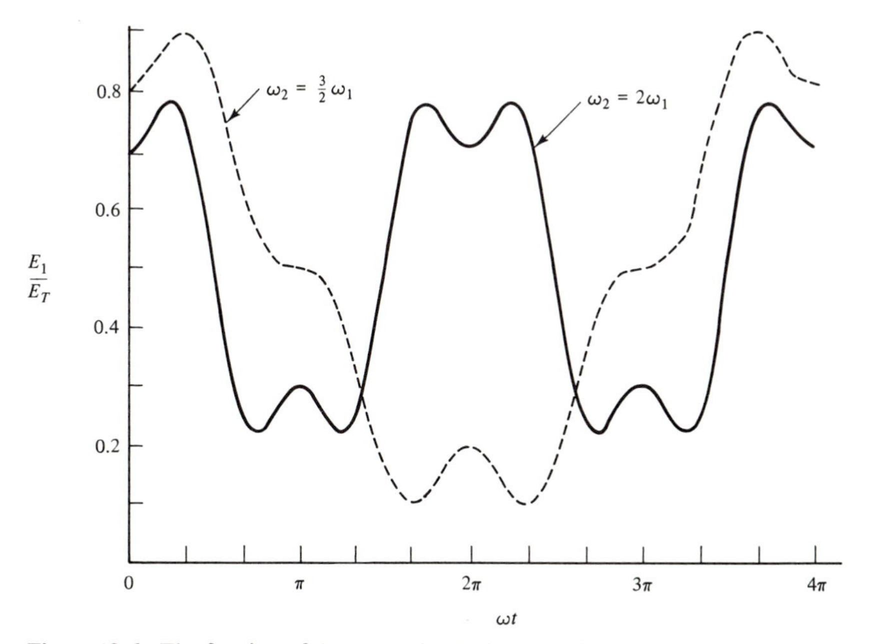

Figure 10-6 The fraction of the energy in the first "mode" plotted as a function of time for $\omega_2 = \frac{3}{2}\omega_1$ (----) and $\omega_2 = 2\omega_1$ (—).

(c) From Fig. 10-6 at $t = 0$, E is a local minimum

$$t = \frac{\pi}{3\omega_1}, \; E_1 \text{ is a maximum}$$

$$t = \frac{\pi}{\omega_1}, \; E_1 \text{ is at an inflection point}$$

$$t = \frac{5\pi}{3\omega_1}, \; E_1 \text{ is a minimum}$$

$$t = \frac{2\pi}{\omega_1}, \; E_1 \text{ is a local maximum}$$

For $2\pi/\omega_1 < t < 4\pi/\omega_1$, E_1 is a mirror reflection of the values for $0 < t < 2\pi/\omega_1$ and the period for this function is $4\pi/\omega_1$.

(d) At $t = \pi/3\omega_1$

$$\frac{E_1}{E_T} = \frac{A^2/4[k + K + k \cos(\pi/3)\cos(\pi/2) + \sin(\pi/3)\sin(\pi/2)]}{A^2/2[k + K]}$$

$$= 0.900$$

From symmetry arguments the minimum value is

$$\frac{E_1}{E_T} = 1 - 0.900 = 0.100$$

(e) (i)

$$\omega_2^2 = \frac{k + 2K}{m} = 4\omega_1^2 = \frac{4k}{m}$$

$$K = \frac{3k}{2}$$

(ii) See Fig. 10-6

(iii) At $t = 0$, E_1 is a local minimum

$$t = \frac{\pi}{4\omega_1}, \; E_1 \text{ is a maximum}$$

$$t = \frac{3\pi}{4\omega_1}, \; E_1 \text{ is a minimum}$$

$$t = \frac{\pi}{\omega_1}, \; E_1 \text{ is a local maximum}$$

For $\pi/\omega_1 < t < 2\pi/\omega_1$, E_1 is a mirror function of the values for $0 < t < \pi/\omega_1$, and the period for this function is $2\pi/\omega_1$.

(iv) At $t = \pi/4\omega_1$

$$\frac{E_1}{E_T} = \frac{A^2/4[k + K + k \cos(\pi/4) \cos(\pi/2) + 2k \sin(\pi/4) \sin(\pi/2)]}{A^2/2[k + K]}$$

$$= 0.783$$

The minimum value is $1 - 0.783 = 0.217$

Now that it has been established that only one longitudinal mode can be emitted, let us return to the methods used to accomplish this goal. Perhaps the most obvious way is to widen $\delta\nu$ by shortening the cavity. If $\delta\nu > \Delta\nu_{\text{th}}$, where $\Delta\nu_{\text{th}}$ is the bandwidth of the emission peak above threshold, then only one mode is stable. The problem with this method is that L is too small for the diode to be conveniently handled, and the output power is too small.

EXAMPLE 10-4

Find the maximum cavity length for a GaAs laser that ensures that only one longitudinal mode will be present if $\Delta\nu_{\text{th}} = 10^{12}$ Hz. The index of refraction for GaAs is $n = 3.6$.

From Eq. 7-18

$$\delta\nu = \frac{c_0}{2nL} = \Delta\nu_{\text{th}} = 10^{12}$$

$$L = \frac{c_0}{2n\Delta\nu_{\text{th}}} = \frac{0.3 \times 10^{14}}{(2)(3.6)(10^{12})} = 41.7\ \mu\text{m}$$

A second cavity can also be used to accomplish our goal. As illustrated in Fig. 10-7, an external mirror, a groove through the active layer, or a cleaved cavity can be used to create the second cavity. The second cavity adds the additional boundary condition that the lasing mode must be a standing wave in it as well as the laser cavity. For this to occur, $\delta\nu_2$ for the second cavity, which is the shorter of the two, should be an integral multiple of $\delta\nu_1$ for the laser cavity. For a mode to be a standing wave in both cavities

$$\nu = \frac{m_1 c_0}{2n_1 L_1} = \frac{m_2 c_0}{2n_1 L_2} \tag{10-11}$$

EXAMPLE 10-5

(a) What is the length of the second cavity for which $\delta\nu_2 = 10^{12}$ Hz if the second cavity is composed of air and a mirror? **(b)** What is the length of the laser cavity if $\delta\nu_1 = \delta\nu_2/10$ and $n_1 = 3.6$?

(a)

$$L_2 = \frac{c_0}{2n_2\delta\nu_2} = \frac{3.00 \times 10^{14}}{(2)(1)(10^{12})} = 150\ \mu\text{m}$$

(b)

$$\frac{10c_0}{2n_1 L_1} = \frac{c_0}{2n_2 L_2} \quad \text{or} \quad L_1 = \frac{10n_2 L_2}{n_1}$$

$$L_1 = \frac{(10)(1)(150)}{3.6} = 417\ \mu\text{m}$$

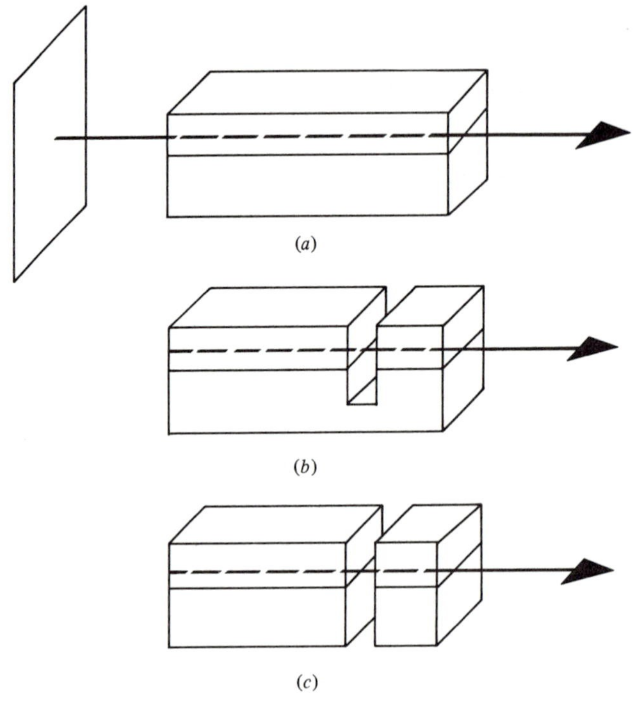

Figure 10-7 Coupled cavity single frequency lasers fabricated using (*a*) an external mirror, (*b*) a groove through the active layer, and (*c*) a cleaved cavity.

The output from the laser is very sensitive to the position of the mirror. It is so sensitive that small changes in the temperature, which produce small changes in the dimensions due to thermal expansion, can cause the laser to cease lasing. Lasing would cease if L_2 were increased by an amount that would decrease $\delta\nu_2$ by an amount $d\nu_1^*$, which is the bandwidth of a longitudinal mode.

EXAMPLE 10-6

For what change in the length of the second cavity, ΔL_2, will $\delta\nu_2$ be changed by $d\nu_1^*$ if $d\nu_1^* = \delta\nu_1/5$, $\delta\nu_1 = \delta\nu_2/10$, $\delta\nu_2 = 10^{12}$ Hz, $L_2 = 150$ μm, and $\lambda_0 = 1.0$ μm?

$$\nu = \frac{m_2 c_0}{2n_2 L_2} \quad \text{or} \quad d\nu^* = \frac{-m_2 c_0}{2n_2 L_2^2}\,\Delta L_2 = \frac{-c_0}{\lambda_0}\frac{\Delta L_2}{L_2} = \frac{\delta\nu_2}{(5)(10)} = 2 \times 10^{10}$$

$$\Delta L_2 = \frac{-2 \times 10^{10}\lambda_0 L_2}{c_0} = \frac{-(2 \times 10^{10})(1)(1.5 \times 10^2)}{3.0 \times 10^{14}} = -0.01\ \mu\text{m}$$

The laser output is equally sensitive to small changes in the index of refraction. This feature can be exploited by the cleaved cavity laser, since n is a slowly varying function of the current passing through the cavity. The current can thus be used to modulate the output of the laser.

Another method used to pick out a single longitudinal mode is to use a diffraction grating. The grating can be separate from the substrate, etched into the portion of the substrate that extends beyond the laser cavity (Bragg reflecting laser), or etched into the substrate prior to the deposition of the semiconductor films (distributed feedback laser). These lasers are illustrated in Fig. 10-8.

The grating illustrated in Fig. 10-9 is used to reflect a single longitudinal mode back along the axis. The mode is determined by the angle, θ, made between the grating and the horizontal axis. When the path length difference up to and back from the grating from adjacent grating lines is an integrable number of wavelengths, the waves reflected by the grating will constructively interfere. The grating line spacing is d so that the condition for constructive interference is

$$2d \cos \theta = m\lambda \tag{10-12}$$

EXAMPLE 10-7

(a) If the grating has 10,000 lines per centimeter and $\lambda_0 = 1.0$ cm, for what value of θ is the wave reflected back along the horizontal axis? **(b)** For what change in θ will an adjacent mode be reflected along the horizontal axis if $\delta\nu_1 = 10^{11}$ Hz?

(a)

$$\cos \theta = \frac{\lambda}{2d} = \frac{10^{-4}}{(2)(10^{-4})} = 0.5$$

$$\theta = 60.0°$$

(b)

$$-2d \sin \theta \, \partial\theta = d\lambda = d\left(\frac{c}{\nu}\right) = \frac{-c\,\partial\nu}{\nu^2} = \frac{\lambda^2\,\partial\nu}{c}$$

$$-\partial\theta = \frac{\lambda^2\,\partial\nu}{2dc \sin \theta} = \frac{(10^{-4})^2(10^{11})}{(2)(10^{-4})(3 \times 10^{10})\sqrt{3}/2}$$

$$= 1.925 \times 10^{-4} \text{ rad} = 0.662'$$

The Bragg reflection single frequency laser selects out the mode with a wavelength for which

$$d = \frac{m\lambda}{2} \tag{10-13}$$

both in the groove and in the semiconductor ridge. When this condition is satisfied, the waves reflected from all the surfaces constructively interfere. This condition also prevails in the distributed feedback laser. The only difference is that ridges and the grooves have the same width because they have the same index of refraction.

The final single frequency laser that will be mentioned is the mode locked laser. In this laser a particular mode is selected by exciting the semiconductor laser with another laser with a narrow bandwidth. One possibility is the He-Ne laser lasing at 1.52 μm, which is very near the desired wavelength of 1.55 μm.

At the present time one type of single frequency laser does not dominate

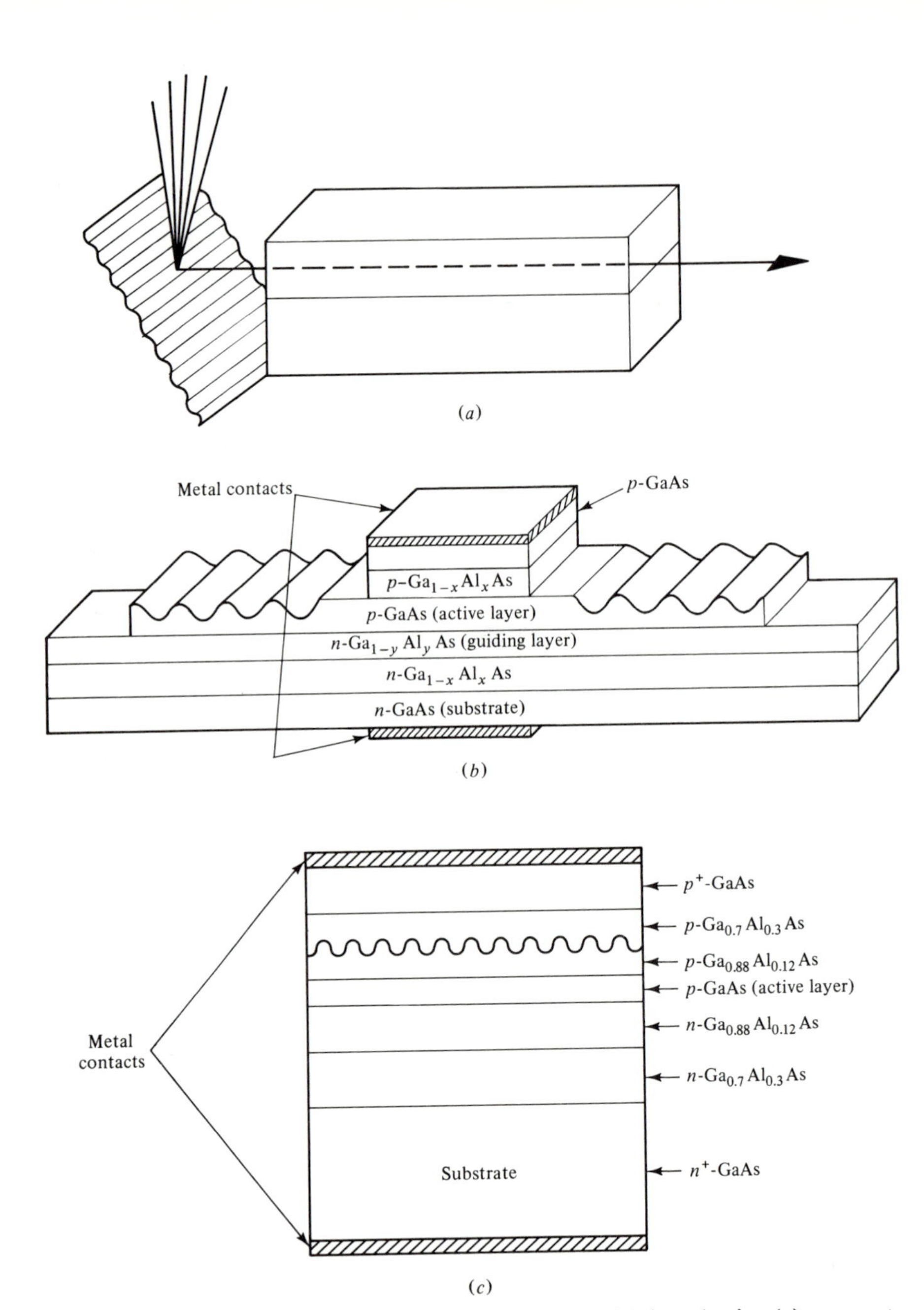

Figure 10-8 Frequency selective single frequency lasers fabricated using (*a*) a separate grating, (*b*) grating etched in the substrate external to the laser cavity, and (*c*) grating etched in the substrate prior to film deposition.

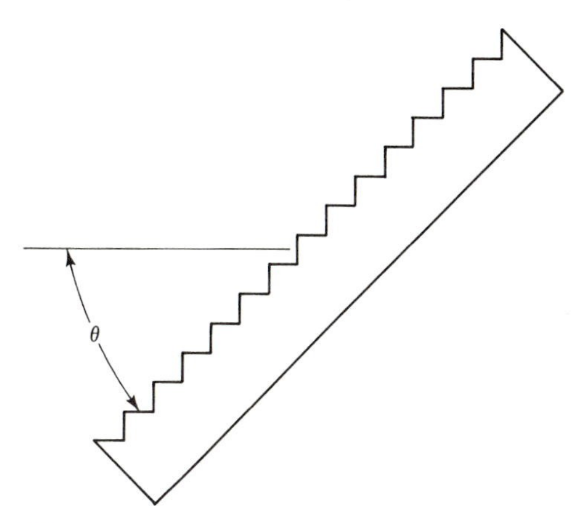

Figure 10-9 A grating used to reflect a single mode parallel to the laser axis.

the others. The one that will ultimately emerge from the pack is the one that can maintain its spectral purity under high-speed modulation.

10-3 QUANTUM WELL LASER

A quantum well laser is made by reducing the thickness of the active region to a point where the quantum mechanical effects become important. An approximate distance for which this occurs is the de Broglie wavelength

$$\lambda = \frac{h}{mv} \tag{3-6}$$

which for an electron in GaAs with its small mass is ~500 Å.

The electrons and holes in the GaAs layer sandwiched between two GaAlAs layers in Fig. 10-10 are confined by the larger GaAlAs energy gap. Recall from Eqs. 4-53 and 4-54 that the energy levels of a confined particle are quantized, and the smaller the confinement distance, the larger is the split between adjacent energy states.

Because the electron states in the conduction band and the hole states in the valence band cannot be represented by a quasi-continuum of states in all three dimensions, the density of states curve is no longer a smooth parabola. Rather, it has the staircase structure illustrated in Fig. 10-11. The most significant difference is that the lowest energy in the staircase density of states has more states than the lowest energy in the quasi-continuum. As a result, population inversion is more readily achieved with the former so that the threshold current density is smaller.

Deriving the equations that illustrate these concepts is quite a straightforward process. Recalling that the electron energy in the conduction band is kinetic energy, defining the energy to be zero at the bottom of the conduction band, and using Eq. 3-6

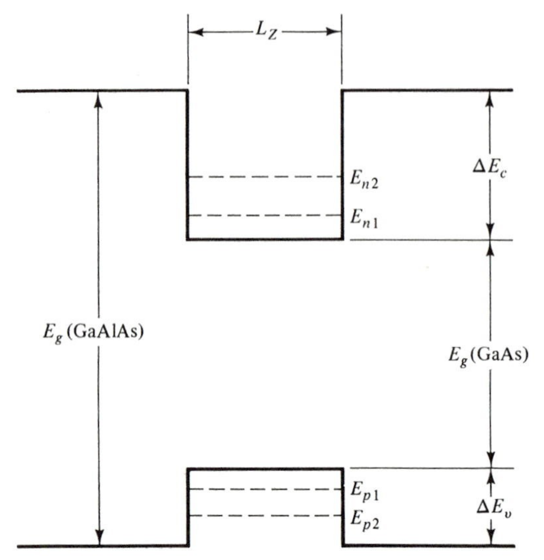

Figure 10-10 Energy level diagram for a quantum well laser.

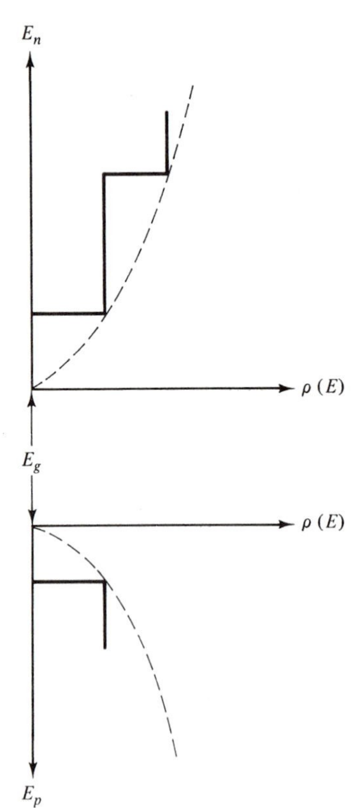

Figure 10-11 The stairwell (—) and quasi-continuum (----) density of states curves.

$$E = \tfrac{1}{2}mv^2 = \frac{\hbar^2 k^2}{2m^*} \tag{10-14}$$

For an infinite potential well and to a very good approximation near the bottom of the well, only standing electron waves are allowed. Thus, k is quantized since only wavelengths that are half integral multiples of the cavity length, L, are allowed. This yields

$$k_n = \frac{2\pi}{\lambda_n} = \frac{\pi}{L} n \tag{10-15}$$

By substituting Eq. 10-14 into Eq. 10-15, we see that the quantized energy levels are given by

$$E_n = \frac{h^2 n^2}{8m^* L^2} \tag{10-16a}$$

for one dimension, while in three dimensions

$$E_n = \frac{h^2}{8m^*}\left[\left(\frac{n_x}{L_x}\right)^2 + \left(\frac{n_y}{L_y}\right)^2 + \left(\frac{n_z}{L_z}\right)^2\right] \tag{10-16b}$$

EXAMPLE 10-8

(a) Find how far the first level is above the bottom of the well for an electron in GaAs in a well 1.0 mm long and 100 Å long, and a hole in GaAs in a well 100 Å long. **(b)** Calculate the wavelength of the light emitted by the transition from E_{n1} to E_{p1} at room temperature.

(a)
$$E_{n1}(1\text{ mm}) = \frac{h^2}{8m_n^* L^2}$$

$$= \frac{(6.625 \times 10^{-34})^2}{(8)(0.068)(9.11 \times 10^{-31})(10^{-3})^2(1.6 \times 10^{-19})}$$

$$= 5.54 \times 10^{-12}\text{ eV}$$

$$E_{n1}(100\text{ Å}) = \frac{(5.54 \times 10^{-12})(10^{-3})}{(10^{-8})^2} = 55.4\text{ meV}$$

$$E_{p1}(100\text{ Å}) = \frac{(55.4)m_n^*}{m_p^*} = \frac{(55.4)(0.068)}{0.5} = 7.53\text{ meV}$$

(b)
$$\lambda = \frac{1.242}{E_g + E_{n1} + E_{p1}} = \frac{1.242}{1.43 + 0.0554 + 0.00753} = 0.832\ \mu\text{m}$$

Equation 10-16 is exact only when the height of the potential well, $V_0 = \infty$. The physical reason for this is that an electron will penetrate into the well wall for any other V_0 so that the electron waves are not standing waves. This is the source of the quantum mechanical effect called tunneling. A more precise analysis shows that the acceptable values of k are found from the transcendental equation

$$\frac{k_1}{k_2} = \tan\left(\frac{k_2 L}{2}\right) \tag{10-17a}$$

for the odd integral values of k, and

$$\frac{k_1}{k_2} = -\cot\left(\frac{k_2 L}{2}\right) \tag{10-17b}$$

for the even integral values. $k_1 = [2m^*(V_0 - E)]^{1/2}/\hbar$ and $k_2 = (2mE)^{1/2}/\hbar$ (see ref. 8).

When the energy levels form a quasi-continuum of states as they do in a normal solid, we found that the density of states, $\rho(E)$, is given by

$$\rho(E) = \frac{4\pi(2m^*)^{3/2}E^{1/2}}{h^3} \tag{4-54}$$

and it is plotted for both electrons and holes in Fig. 10-11. To find out how $\rho(E)$ is affected by confinement in the z direction, we write the equation for the energy as

$$E = E_n + E_k = E_n + \frac{\hbar^2}{2m_n^*}(k_x^2 + k_y^2) \tag{10-18}$$

The number of states per unit area, N/L^2, in Fig. 10-12 for which $k < k^*$ is

$$\frac{N}{L^2} = \frac{\frac{1}{4}\pi k^2}{\frac{1}{2}\left(\frac{\pi}{L}\right)^2 L^2} = \frac{k^2}{2\pi} \tag{10-19a}$$

when one accounts for the fact that there are two electron states—spin up and spin down—per k state. Substituting Eq. 10-16 into Eq. 10-19a yields

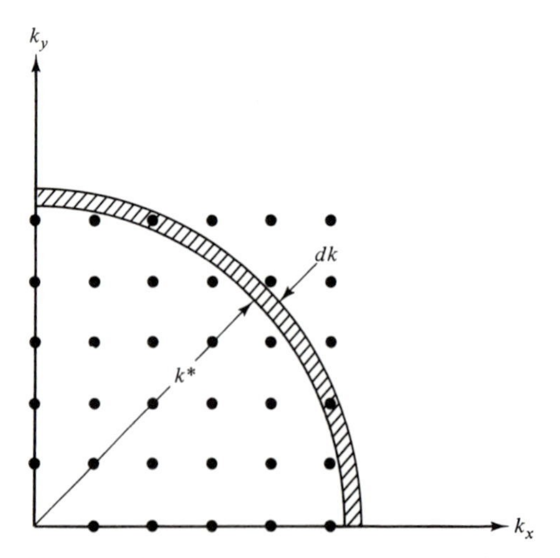

Figure 10-12 Figure used to find the density of states in two dimensions.

$$\frac{N}{L^2} = \frac{4\pi m^*}{h^2} E_k \tag{10-19b}$$

The number of states per unit area with energy between E_k and $E_k + \Delta E_k$ is then

$$\rho(E_k) = \frac{d(N/L^2)}{dE_k} = \frac{4\pi m^*}{h^2} \tag{10-20}$$

which is independent of E_k. Again accounting for the two electrons per k state, the number of states per unit area with energy E_1 is

$$2\rho(E_k) = \frac{8\pi m^*}{h^2} \tag{10-21}$$

This is also the number of states per unit area with energy E_2, E_3, ... Thus, the density of states curve has a staircase shape when the particles are confined in one dimension.

As mentioned above, the larger number of electron and hole states with the smallest energy in the quantum well laser reduces J_{th}. Values as low as 370 Å/cm^2 at room temperature have been obtained. This is in large part due to the large T_0 values, which can be as high as 220° C. The emitted light also often has an energy larger than the energy gap by an amount approximately equal to E_{n1} + E_{p1}. The energy is larger when recombination occurs between the $n = 2$ states, and the energy is smaller when lattice phonons are created during the transition.

E_{n1} and E_{p1} are made larger by decreasing L. However, L cannot be less than about 100 Å in a single quantum well laser because the carriers will drift across the well before they are scattered into the lowest energy state. The problem is more severe at lower temperatures where there are fewer phonons to scatter the carriers.

The well width can be reduced by using multiple quantum wells. Each additional well is created by growing an additional GaAlAs barrier and a GaAs well. Both the confining and well layers can be as thin as 50 Å. They can be this thin because carriers not captured by the first well can be captured in subsequent wells. Also, even when the barriers are thin, the energy levels in the adjacent wells are only loosely coupled except near the top of the well. As a result, the energy levels calculated for a single well are almost the same as those in the multiquantum well structures.

One problem we faced in Chapter 7 but have not yet faced in this chapter is confining the electromagnetic energy to the active region where it stimulates further radiation. In Fig. 7-14 we saw that J_{th} actually increased for $d \leq 1000$ Å because the confinement factor, Γ, decreased rapidly. Γ is small for quantum well lasers, but it can be increased by an order of magnitude using the separate confinement graded index structure illustrated in Fig. 10-13. The 200 Å GaAs well confines the electrons, and the adjacent 0.6 μm $Ga_{1-x}Al_xAs$ layers in which x varies parabolically with distance from $x = 0.2$ to $x = 0.5$ confines the electromagnetic radiation. Adjacent confining layers with an abrupt change in composition, which are easier to grow, can also be used but they are less effective, since electromagnetic radiation "abhors" discontinuities.

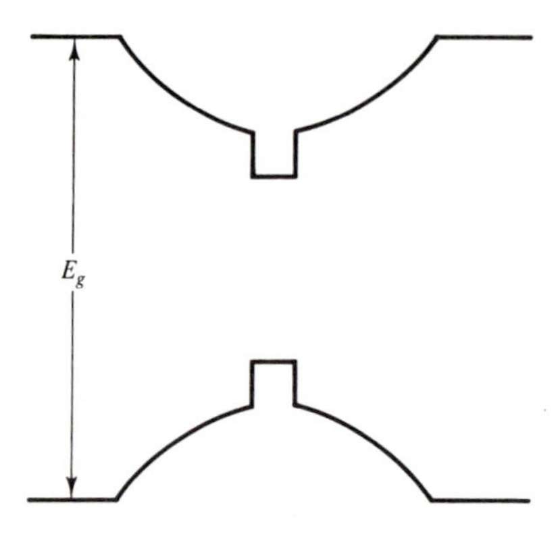

Figure 10-13 The energy band structure for a separate confinement graded index quantum well structure.

10-4 INTEGRATED OPTICS

An integrated optical circuit (IOC) like an integrated circuit is a chip that contains a number of individual devices. Some obvious advantages of integration are smaller weight and size, lower power consumption, batch fabrication economy, improved alignment, and immunity from vibration. Light is generated by a laser, detected by a photodiode, and transmitted through, modulated in, and switched to and from waveguides.

Hybrid IOCs, which are made using devices fabricated on different substrates and then bonded together, can be made by bonding a GaAs laser and a silicon detector onto a lithium niobate ($LiNbO_3$) substrate that can act as a waveguide. The attraction of $LiNbO_3$ is that, being ferroelectric, it has large electro-optic coefficients which, as we will shortly see, makes it easier to switch and/or modulate the light beam.

Monolithic IOCs will be made from light-emitting semiconductors such as GaAs. Instead of using cleaved surfaces to form a laser cavity, distributed feedback lasers will be used to make it easier to inject light into the waveguides. Light will be detected using a photodiode, but there is an added advantage that the light will be injected parallel to the junction. Light can be transmitted in a semiconductor waveguide much like it is in a double heterostructure laser, and it can be modulated and switched using the electro-optic effect.

The distributed feedback laser was already described in the previous section on single frequency lasers so we discuss them here only briefly. They are made by masking and then etching a grating into the active layer or an adjacent confining layer. Because the electromagnetic beam is more spread out than the electrons and holes are, forming the grating in an adjacent layer is almost as effective. This configuration has the advantage of having the defects induced during the fabrication of the grating away from the recombination region. The most frequently used grating separation is $\lambda/3$. Lasers can be made that emit at different wavelengths and then are multiplexed into the same waveguide and then are separated after they leave the guide and before they are detected.

Detection is accomplished using a photodiode, and the light is inserted parallel to the junction. For this geometry the absorption coefficient need not be large, since all of the light will be absorbed in the depletion region no matter how long the diode is. If smaller devices are desired, the absorption coefficient can be increased by making the detector out of a narrower bandgap semiconductor deposited at the end of the waveguide, proton damaging the detector material, or narrowing the energy gap by applying a large electric field. The proton bombardment creates states in the energy gap that allow the semiconductor to more strongly absorb near bandgap radiation. A large electric field will effectively narrow the energy gap. This is known as the Franz–Keldysh effect, and ΔE_g can be found from the equation,

$$\Delta E_g = \frac{\frac{3}{2}(qh\xi)^{2/3}}{m^{*\,1/3}} \tag{10-22}$$

(See ref. 10.) As seen in Fig. 10-14, α at 9000 Å increases from 25 cm^{-1} to 10^4 cm^{-1} when an electric field of 1.3×10^5 V/cm is applied.

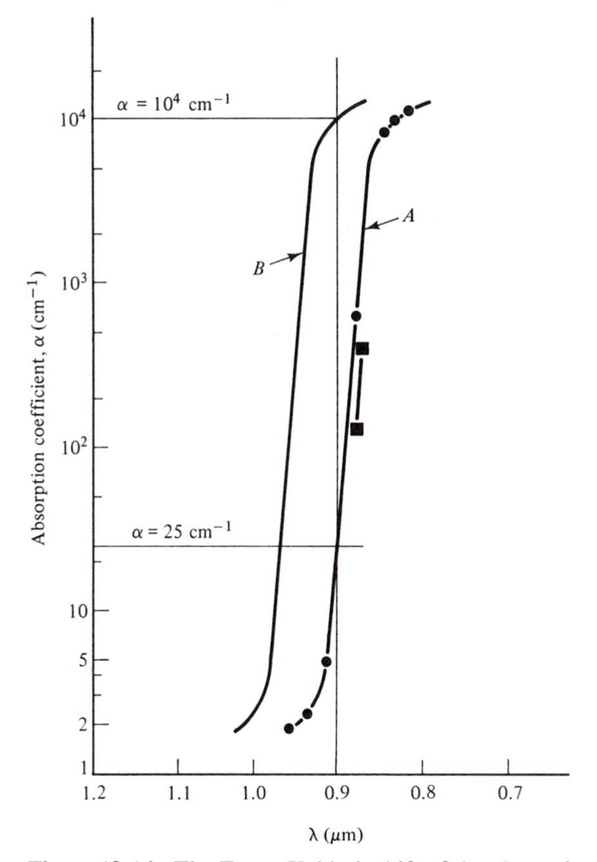

Figure 10-14 The Franz–Keldysh shift of the absorption edge of GaAs. Curve *A* is the absorption edge with no field and curve *B* is the absorption edge with a field of 1.3×10^5 V/cm. (From R. G. Hunsperger, *Integrated Optics: Theory and Technology,* Copyright © 1982, Springer-Verlag, New York.)

The photodetector is again a reverse biased diode with a photocurrent proportional to the input optical power. A large enough reverse bias is applied to create a depletion region that is wider than the waveguide; hence, the photocarriers are always created in the depletion region where they are swept out rapidly by the electric field—there are no delays due to the diffusion of carriers. Thus, the delay time is short, since the carriers can drift across the 3-μm depletion region at 10^7 cm/sec in 33 psec. Also, because the detector need not be any wider than the 3-μm stripe waveguide, the area of the gate and, therefore, the junction capacitance are small. For a 3-μm by 3-mm gate the capacitance is 0.32 pF. Thus, the RC time constant is 48 psec when $R = 150\ \Omega$. With these small time delays the detector can operate effectively at frequencies in excess of 10 GHz.

The waveguides are similar to DH lasers and the planar dielectric discussed in Chapter 2. They differ only in that they are not symmetric. The material above the guide is air, and the material below is a semiconductor with an index of refraction, n_2, slightly less than that of the guide, n_1. One major difference between this type of guide and a symmetric guide is that the guide thickness, d, must exceed a certain thickness for even one mode to propagate in it. For m modes to propagate,

$$d \geqq \frac{\lambda_0(m + 1/2)}{2(n_1^2 - n_2^2)^{1/2}} \tag{10-23}$$

(See ref. 10.)

EXAMPLE 10-9

Find the values of d for which one and only one mode propagates in a waveguide for which $\Delta n = 0.01$ and in a GaAs waveguide when the underlying layer is $Ga_{0.9}Al_{0.1}As$ when $\lambda_0 = 8400$ Å.

For $\Delta n = 0.01$

$$d_{\text{min}} = \frac{0.84(1 + 1/2)}{2[(3.6)^2 - (3.59)^2]^{1/2}} = 2.35\ \mu\text{m}$$

$$d_{\text{max}} = \frac{0.84(2 + 1/2)}{2[(3.6)^2 - (3.59)^2]^{1/2}} = 3.92\ \mu\text{m}$$

Thus, one mode will propagate when $d_{\text{min}} < d < d_{\text{max}}$.

From Eq. 7-36, Δn for the GaAs/GaAlAs interface is $\Delta n = 0.63\Delta x = (0.63)(0.10) = 0.062$

$$d_{\text{min}} = \frac{0.84(1 + 1/2)}{2[(3.6)^2 - (3.538)^2]^{1/2}} = 0.940\ \mu\text{m}$$

$$d_{\text{max}} = \frac{0.84(2 + 1/2)}{2[(3.6)^2 - (3.538)^2]^{1/2}} = 1.57\ \mu\text{m}$$

The waveguides illustrated in Fig. 10-15 are the slab, stripe mesa, and buried stripe waveguides. Slab waveguides are most easily fabricated by depositing a higher index material on one with a smaller index. A good example is a GaAs layer on top of a GaAlAs layer.

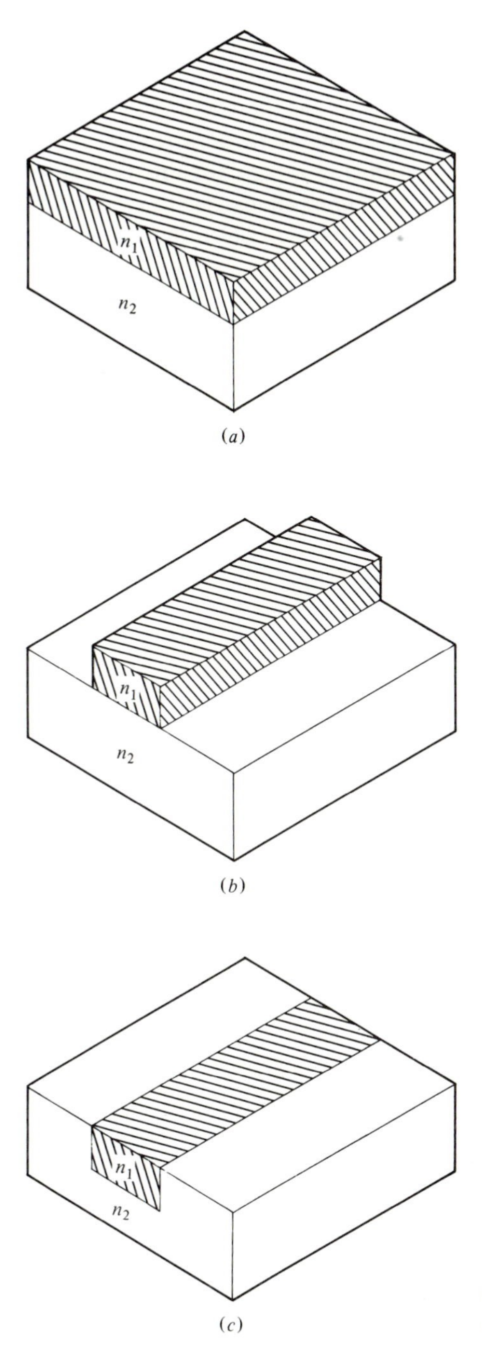

Figure 10-15 The (*a*) slab, (*b*) stripe mesa, and (*c*) buried stripe waveguides.

Mesa stripes, which are often $\sim 3\ \mu m$ wide, are formed by masking and etching the uppermost film. An ion beam etch produces a smooth, but damaged guide, and a chemical etch produces a rougher, but less damaged guide. Also, the guides need not be straight, but the radius of curvature must not be small or there will be a considerable radiation loss.

Buried stripe waveguides are often made out of the same material that surrounds them; the index of the guide is higher because its net carrier concentration is smaller. This is accomplished by diffusing n-type dopants through a mask into p-type material or vice versa, and by creating carrier traps through proton bombardment. The index of refraction of the material with a net carrier concentration, N, is

$$n = n_0 - \frac{Nq^2}{2n_0 E_0 m^* \omega^2} \tag{10-24}$$

with n_0 being the index of the intrinsic material. Δn is not very large as is illustrated in Example 10-10.

EXAMPLE 10-10

Find the change in the index of refraction of n-type GaAs when the carrier concentration is reduced from 10^{18} cm^{-3} to 10^{17} cm^{-3}. Let

$$n_0 = 3.6, \lambda_0 = 8400 \text{ Å}, \quad \text{and} \quad m_n^* = 0.068 \text{ m}$$

$$\Delta n = \frac{(N_1 - N_2)q^2}{2n_0 E_0 m_p^* \omega^2}$$

$$= \frac{(10^{24} - 10^{23})(1.6 \times 10^{-19})^2(0.84 \times 10^{-6})^2}{(2)(3.6)(8.854 \times 10^{-12})(0.5)(9.11 \times 10^{-31})(2\pi)^2(3 \times 10^8)^2}$$

$$= 0.00116$$

The most straightforward method of coupling light into another waveguide is to use a narrow tapered section. One can see in Fig. 10-16 that the incident angle for the bottom interface, ϕ_2', increases by $2\theta_T$ after every reflection, where θ_T is the taper angle. ϕ_2' decreases to a value less than the critical angle so that a fraction of the light is transmitted. The lower waveguide is sufficiently thick so that light is not reflected back into the tapered guide. Up to 70 percent of the light can be transferred in this way. The major deficiency of this method is that the transmitted rays diverge from each other.

The dual channel coupler works on the principle of optical tunneling, which is illustrated in Fig. 10-17*a*, and it can be explained with the mechanical

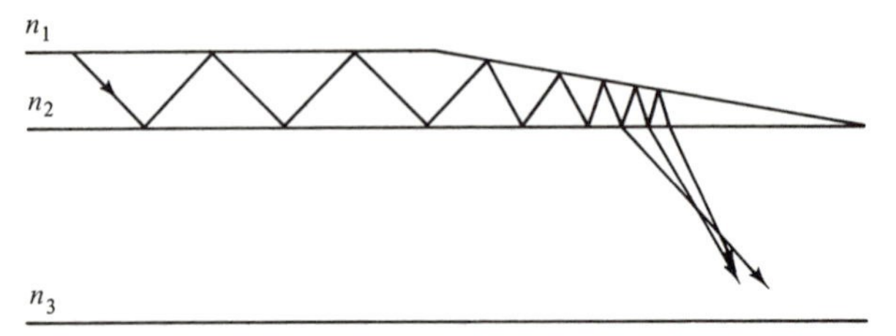

Figure 10-16 A tapered waveguide section used to couple light into an adjacent waveguide.

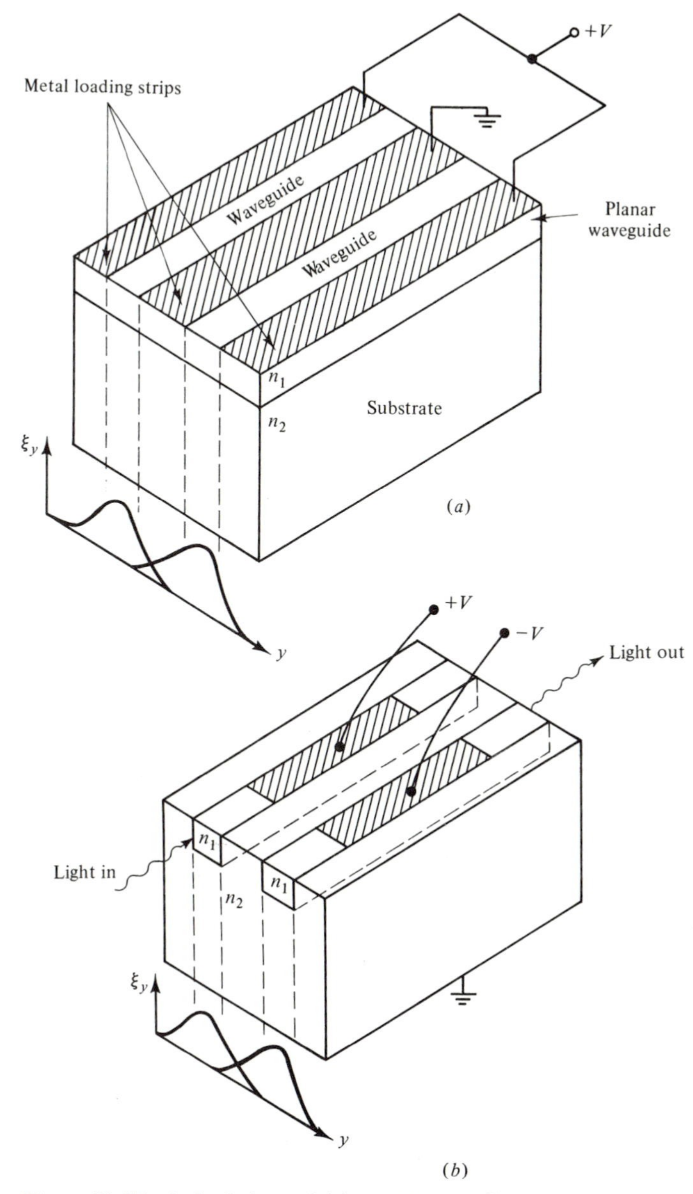

Figure 10-17 A dual channel (*a*) coupler and (*b*) switch.

model used to describe mode hopping in a laser. Equations 10-6a and 10-6b, the equations that describe the motions of the two masses with the original displacements $x_1(0) = A$ and $x_2(0) = 0$, can be rewritten in exponential form

$$x_1 = \frac{A}{2} e^{j\omega_1 t}(1 + e^{j\,\Delta\omega t}) \tag{10-25a}$$

$$x_2 = \frac{A}{2} e^{j\omega_1 t}(1 - e^{j\,\Delta\omega t}) \tag{10-25b}$$

where $\Delta\omega = \omega_2 - \omega_1 = 2C$. The electric fields in each waveguide have similar expressions when they are expressed as a function of position, z, instead of time, t, so that the squares of the fields are

$$\xi_1^2 = A^2 \cos^2 Cz \tag{10-26a}$$

$$\xi_2^2 = A^2 \sin^2 Cz \tag{10-26b}$$

Since the electromagnetic energy is proportional to ξ^2, one can see that the energy is fed back and forth between the coupled guides with a period of π/C. If the length, L, over which the two guides are coupled is $L = \pi/2C$, all of the energy will be transferred from the first guide to the second.

The above discussion applies to the situation when the index of refraction in both waveguides is the same. When it is not, Eq. 10-26b must be modified to

$$\xi_2^2 = A^2 \frac{C^2}{C^2 + \delta^2} \sin^2(C^2 + \delta^2)z \tag{10-27}$$

where

$$\delta = \frac{\pi\,\Delta n}{\lambda_0} \tag{10-28}$$

(see ref. 10).

From Eqs. 10-27 and 10-28 we see that the amount of energy transferred to the second guide is a function of Δn for a fixed value of L. Thus, the amount of energy transferred can, to some extent, be controlled by controlling Δn. When this is done, the coupler is a switch.

The Δn is created by increasing n in one channel by using the electro-optic effect as is illustrated in Fig. 10-17*b*. The crystal structure of GaAs and most other III-V semiconductor compounds is such that when an electric field is applied in certain crystallographic directions, the index of refraction for a wave traveling normal to the direction of the field is increased. For crystals with the GaAs crystal structure

$$\Delta n = \tfrac{1}{2} n^3 r \xi \tag{10-29}$$

where r is the electro-optic coefficient. For GaAs $r = 1.6 \times 10^{-12}$ m/V.

Instead of relying on coupling energy into or out of a channel to perform the switching operation, interference between two waves can be used as it is in the Mach-Zehnder interferometer switch in Fig. 10-18. The beam is split evenly between the two paths, these two beams travel the same distance, and then they rejoin. If no electric field is applied to the upper path, the two beams will constructively interfere with each other. However, if a field is applied, destructive interference will occur, and it will be complete when the phase lag, $\Delta\phi = \pi$.

$$\Delta\phi = \Delta k L = k_0\,\Delta n L = \frac{\pi n^3 r \xi L}{\lambda_0} \tag{10-30}$$

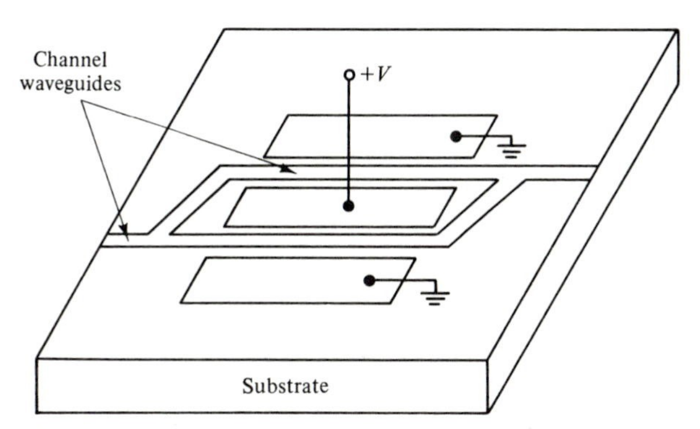

Figure 10-18 A Mach–Zehnder interferometer switch.

EXAMPLE 10-11

(a) Find Δn in GaAs when 10 V is applied to a channel waveguide 3.3 μm thick. **(b)** What is δ? **(c)** Find the value of L for which $\Delta\phi = \pi$ if $\lambda_0 = 0.84$ μm and 1.0 V are applied.

(a) $$\Delta n = \frac{\frac{1}{2}n^3 rV}{d} = \frac{(0.5)(3.6)^3(1.6 \times 10^{-12})(10)}{3.3 \times 10^{-6}} = 1.12 \times 10^{-4}$$

(b) $$\delta = \frac{\pi\,\Delta n}{\lambda_0} = \frac{\pi(1.12 \times 10^{-4})}{0.84 \times 10^{-4}} = 4.19\ \text{cm}^{-1}$$

(c) $$L = \frac{\lambda_0\,\Delta\phi}{\pi n^3 r}\left(\frac{V}{d}\right) = \frac{0.84 \times 10^{-4}}{(2)(1.12 \times 10^{-5})} = 3.75\ \mu\text{m}$$

A different type of switch, the Bragg type, is shown in Fig. 10-19. When a voltage is applied to the equally spaced electrodes, the index of refraction in the semiconductor material under them is different than it is some distance away. Thus a periodic variation in n is created, and the waves that satisfy the condition

$$\sin\theta = \frac{m\lambda_0}{2nd} \tag{10-31}$$

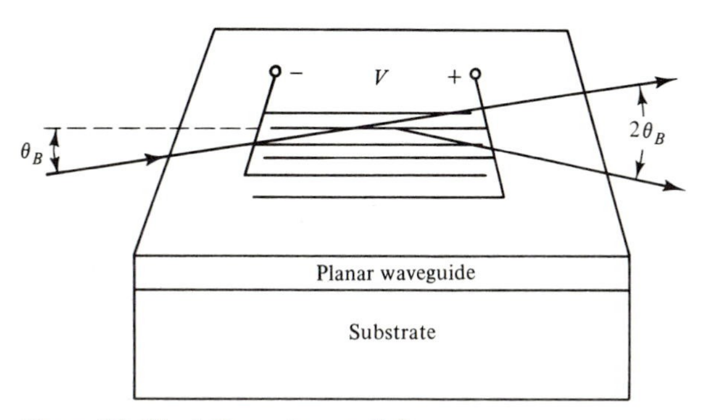

Figure 10-19 A Bragg-type switch.

where d is the electrode spacing, will be reflected in the same way they are in a distributed feedback single frequency laser. The amplitude of the reflected ray is larger when the applied voltage and, therefore, Δn is larger.

As we learned in Chapter 7, semiconductor lasers can be modulated by modulating the input current. The light can also be modulated externally in much the same way it is switched. The beams can by modulated by electroabsorption with the amount of energy absorbed being proportional to the applied field. Likewise, the amount of light switched out of a channel is proportional to the field applied to a branch of the dual channel system, and the attenuation in a Mach-Zehnder interferometer type or a Bragg type of switch is proportional to an applied field.

10-5 OPTICAL BISTABILITY

In an optically biased semiconductor cavity, a small increase in the incident intensity can produce a large change in the transmitted intensity. This is because the index of refraction is a function of the incident intensity. As the index changes, the transmittance curve (see Fig. 8-10) shifts from a position of low transmittance at the given wavelength to a transmittance that is high.

A switching device that can switch in picoseconds can be made using this effect. Moreover, more than one optical signal can be processed by a single semiconductor element, since the rays do not mix together as electrical currents do. Another potential advantage is that there are a number of bistable regions in the transmittance versus wavelength curve. This could be exploited by developing computer logic with a base greater than two. The cavity can also be detuned so that a hysteresis loop is formed in the bistable region. As is true for the magnetic hysteresis loop, the optical hysteresis loop can be used as a memory element.

Optical bistability is an appropriate topic on which to end this book. We show that the index of refraction decreases at higher intensities because the absorption coefficient decreases. Recall from Chapter 1 that n is a function of α. The absorption coefficient decreases at higher intensities because the electronic states at the bottom of the conduction band begin to saturate. This phenomenon can be explained by using material in the middle of the book. The transmittance curve shifts when n changes, and to describe this effect we must draw on material from the last chapters.

The dependence of n on I is written explicitly as

$$n = n_1 + n_2 I \tag{10-32}$$

By using this equation we can determine how large the intensity must be to change n by an amount δn, which is the change in n necessary to shift the m_{th} transmittance peak when $I = 0$. The number of transmittance peaks, m, is

$$m = \frac{2Ln}{\lambda_0} \tag{10-33}$$

thus for $\Delta m = 1$

$$\delta n = \frac{\lambda_0}{2L} \tag{10-34}$$

Also of interest is the change in n, dn^*, necessary to shift the transmittance curve by an amount equal to the bandwidth of an individual transmittance peak. For normal incidence Eq. 8-43a can be written

$$\phi = \frac{2\pi L n}{\lambda_0} \tag{8-43}$$

Thus, using Eq. 10-34

$$dn^* = \frac{\lambda_0}{2\pi L}\, d\phi^* = \frac{d\phi^*}{\pi \delta n} \tag{10-35a}$$

By using Eq. 8-50b, which was developed assuming that the transmittance peak was narrow, we can also write Eq. 10-35a as

$$dn^* = \frac{1 - R}{\pi\ \delta n \sqrt{R}} \tag{10-35b}$$

Another version of Eq. 8-54 is thus

$$\frac{\delta n}{dn^*} = \frac{\pi\sqrt{R}}{1 - R} = F \tag{10-36}$$

where once again F is the finesse.

EXAMPLE 10-12

(a) Find the change in the intensity necessary to shift n by amounts of δn and dn^* if the cleaved cavity is an InSb sheet 125 μm thick at $T = 77$ K where the wave number for the excitation beam is 1852 cm^{-1}, $n = 4.0$, and $n_2 = -3.0 \times 10^{-3}$ cm^2/W. **(b)** Find the corresponding changes in the excitation power if the excitation beam has a radius of 150 μm.

(a) $$\delta n = \frac{\lambda_0}{2L} = \frac{1}{(2)(1.25 \times 10^{-2})(1.852 \times 10^3)} = 2.16 \times 10^{-2}$$

$$\Delta I = \frac{\delta n}{n_2} = \frac{2.16 \times 10^{-2}}{3.0 \times 10^{-3}} = 7.2 \text{ W/cm}^2$$

Ignoring the effects of absorption

$$dn^* = \frac{\delta n}{F}$$

For the cleaved cavity

$$R = \left(\frac{n-1}{n+1}\right)^2 = \left(\frac{3}{5}\right)^2 = 0.36$$

$$F = \frac{\pi\sqrt{R}}{1 - R} = \frac{\pi\sqrt{0.36}}{1 - 0.36} = 2.945$$

$$dn^* = \frac{\delta\nu}{F} = \frac{2.16 \times 10^{-2}}{2.945} = 7.33 \times 10^{-3}$$

$$\Delta I = \frac{dn^*}{n_2} = \frac{7.33 \times 10^{-3}}{3 \times 10^{-3}} = 2.44 \text{ W/cm}^2$$

(b)

$$\Delta P(\delta n) = \pi r^2 \, \Delta I = (\pi)(1.5 \times 10^{-2})^2 7.2 = 5.09 \text{ mW}$$

$$\Delta P(dn^*) = \frac{\Delta P(\delta n)}{F} = \frac{5.09}{2.945} = 1.73 \text{ mW}$$

The transmitted versus the incident intensity curve illustrated in Fig. 10-20*a* can be understood with the assistance of the transmittance versus phase curve in Fig. 8-10. If the cavity length and the excitation wavelength are such that the transmittance is at the minimum halfway between the *m*th and the *m* + 1th peaks, the transmitted intensity increases slowly as the incident intensity is increased because T is small and relatively constant in the region near its minimum. The curve, however, is superlinear because the decreasing index of refraction shifts the *m*th peak toward the operating wavelength. As I continues to increase, the superlinearity increases because T increases as the curve moves away from the minima. The increase is very rapid when the transmittance peak is within dn^* of the peak where the transmittance changes rapidly with a change in n.

The hysteresis results from the fact that it is the light intensity inside of the cavity that affects α and therefore n. As the resonant condition for the transmittance peak lying at the excitation wavelength is approached, more electromagnetic energy is being stored in the cavity. The larger amount of energy being stored in the cavity further decreases n. Now, when the incident intensity is reduced, the large intensity in the cavity keeps n in the cavity relatively high so that the transmittance peak shifts only a little. However, if the incident intensity decreases below a critical value, the light intensity in the cavity can no longer keep n from changing much. Once n does change sufficiently, the cavity shifts from being near the resonance condition, and the light intensity inside the cavity is greatly reduced. This, of course, produces a large change in n, which results in a sharp drop in the transmitted intensity as the transmittance peak moves away from the operating point fixed by the incident wavelength.

To understand why α is a function of the intensity, we refer to Fig. 10-21. If the wavelength of the incident radiation is tuned to the energy gap of the semiconductor, electrons are excited from the top of the valence band to the bottom of the conduction band. Since the probability that the excitation can occur is proportional to the probability the site in the conduction band is vacant, the transition probability, and therefore α, decreases as more and more states at the bottom of the conduction band become populated. An example of a system where this can occur is a CO laser tuned to the bandgap of InSb at 77 K.

For a given geometry the bistable region can be reached at lower intensities in narrower bandgap materials. This is because the transition probability of an electron across the energy gap is proportional $1/\nu^2$ (see ref. 13) where ν is the

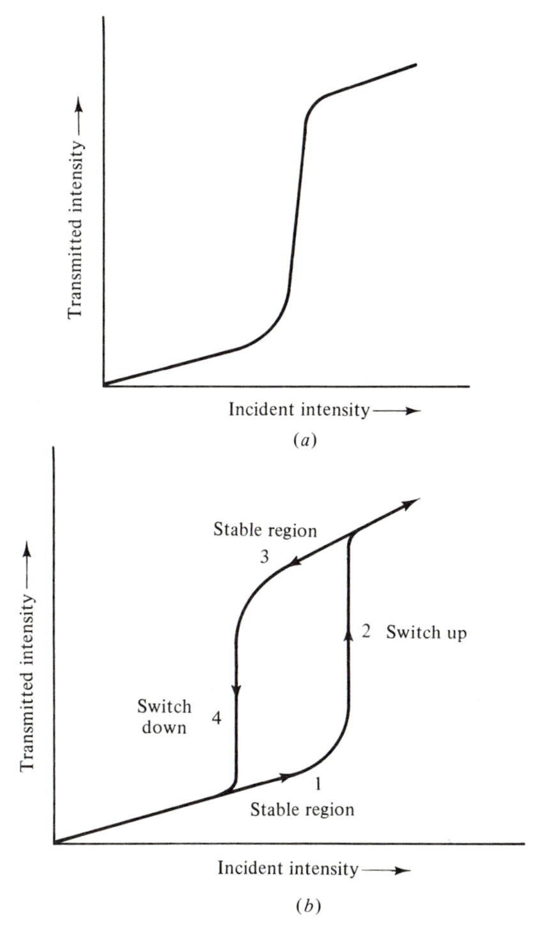

Figure 10-20 The transmitted versus the incident intensity curve of an optically bistable device (*a*) without and (*b*) with an hysteresis loop.

frequency of the exciting radiation. Also, the number of photons in a constant intensity beam is proportional to $1/\nu$. Thus, everything else being equal $n_2 \alpha 1/E_g^3$. As is often true, everything else is not always equal. In GaAs there is some exciton assistance (see ref. 13) so that $|n_2|$ for GaAs is not as much smaller than n_2 for InSb as one might first expect. However, it is substantially smaller as $n_2(\text{GaAs}) = -4 \times 10^{-4}\ \text{cm}^2/\text{W}$ whereas for InSb $n_2 = -3 \times 10^{-3}\ \text{cm}^2/\text{W}$.

Assuming low-level excitation (see Eq. 4-25), the steady state excess electron concentration in the conduction band, Δn_0, is

$$\Delta n_0 = \frac{\eta_i \alpha I \tau}{h\nu} \tag{10-37}$$

since $g_{op} = \alpha I$. If we define the change in the index of refraction per excited electron to be σ, then

$$\sigma\ \Delta n_0 = -n_2 I \tag{10-38}$$

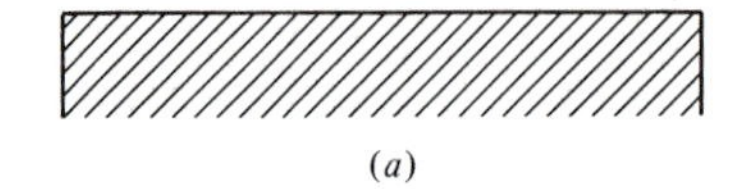

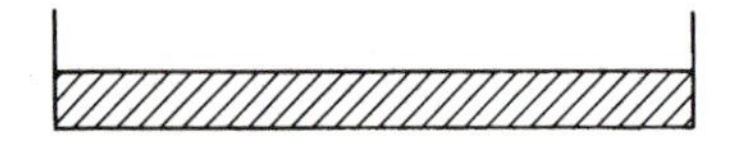

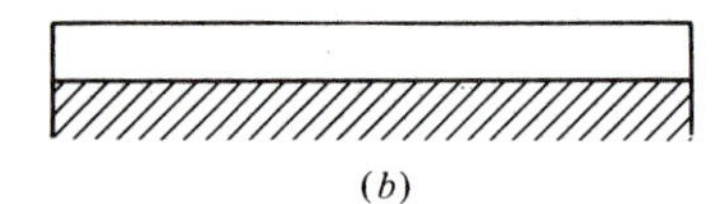

Figure 10-21 The electron distribution in the valence and conduction bands for (*a*) no incident light, and (*b*) incident light with a large intensity.

Combining Eqs. 10-37 and 10-38 yields

$$n_2 = \frac{-\eta_i \alpha \tau \sigma}{h\nu} \tag{10-39}$$

EXAMPLE 10-13

(a) Using the intensity necessary to change n by an amount δn that was computed in the previous example, find the steady state excess electron concentration if $\alpha = 80 \text{ cm}^{-1}$, $\tau = 300$ nsec, $\eta_i = 1$, and $1/\lambda_0 = 1852 \text{ cm}^{-1}$. **(b)** Compute σ if $n_2 = -3 \times 10^{-3}$.

(a)
$$\Delta n_0 = \frac{\eta_i \alpha I \tau \lambda_0}{hc} = \frac{(1)(8 \times 10^1)(7.2)(3 \times 10^{-7})}{(6.635 \times 10^{-34})(3.0 \times 10^{10})(1.852 \times 10^3)}$$
$$= 4.69 \times 10^{15} \text{ cm}^{-3}$$

(b)
$$\sigma = \frac{-n_2 I}{\Delta n_0} = \frac{(3 \times 10^{-3})(7.2)}{4.69 \times 10^{15}} = 4.61 \times 10^{-18} \text{ cm}^{-3}$$

The optically bistable cavity can be operated as a switch by biasing it to near the bistable region with a continuous wave (CW) light source. This, of course, corresponds to dc biasing. A small light input into this system would then greatly increase the transmittance by creating a total intensity that is just to the high side of the bistable region. In this way a small input signal can create a large output signal much as it does in a transistor.

An AND gate can be created by having two input signals whose combined intensity is on the high side of the bistable region, but whose individual intensities are below. An OR gate can be created by using input beams either of which has an intensity that is on the high side of the bistable region. Finally, a NOT gate can be created by using the reflected beam. As is shown in Fig. 10-22, the reflectance drops rapidly in the region where the transmittance increases rapidly.

The switching time, τ_{sw}, is rapid as it is the time it takes the beam to travel a distance $1/\alpha$. Thus

$$\tau_{sw} = \frac{n}{\alpha c_0} \tag{10-40}$$

When $\alpha = 80\ \text{cm}^{-1}$ and $n = 4$, $\tau_{sw} = 1.7$ psec.

The amount of energy consumed by the switching process is an important parameter because it is often difficult to dissipate the heat rapidly. The energy, E_s, needed to change n by an amount δn is

$$E_s = \frac{\delta n \pi r^2 L h\nu}{\sigma} \tag{10-41}$$

where $\delta n/\sigma$ is the number of photons absorbed per unit volume and r is the beam radius. By using a minimum beam radius of 20 μm, the parameters listed in Example 10-13, and assuming that E_s is $\frac{1}{4}$ of the incident energy, the incident energy is only 2.0 FJ. However, even if this number were increased by three orders of magnitude, switching could occur in nanoseconds for milliwatts of power.

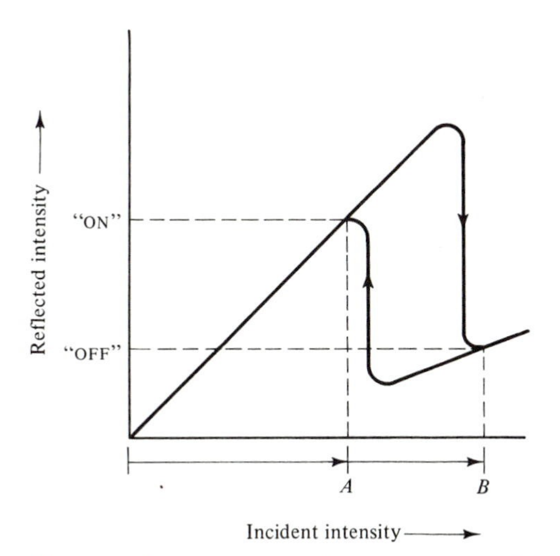

Figure 10-22 The reflected wave intensity of the bistable device plotted as a function of the incident intensity.

REFERENCES

1. D. Botez, "Single Mode Lasers for Optical Communications," *IEEE Proceedings,* **129**(6) 1982, pp. 237–251.
2. T. E. Bell, "Single Frequency Semiconductor Lasers," *IEEE Spectrum,* **20**(12), 1983, pp. 28–45.
3. D. Botez, "Laser Diodes Are Power Packed," *IEEE Spectrum,* **22**(6), 1985, pp. 43–53.
4. D. Botez and G. J. Herskowitz, "Components for Optical Communications Systems: A Review," *IEEE Proceedings,* **68**(6), 1980, 689–731.
5. C. H. Henry (Ed.), "Special Issue on Semiconductor Lasers," *IEEE Journal of Quantum Electronics,* **21**(6), 1985.
6. N. Holonyak, Jr., R. M. Kolbas, R. D. Dupuis, and P. D. Dapkus, "Quantum-Well Heterostructure Lasers," *IEEE Journal of Quantum Electronics,* **16**(2), 1980, pp. 170–185.
7. D. S. Chemla and A. Pinczuk, "Special Issue on Quantum Well Structures," *IEEE Spectrum,* **20**(12), 1983, pp. 38–45.
8. R. M. Eisberg, *Fundamentals of Modern Physics.* New York: John Wiley, 1964, Chapters 7 and 8.
9. T. Tamir (Ed.), *Integrated Optics, Topics in Applied Physics,* Vol. 7. New York: Springer-Verlag, 1975.
10. R. G. Hunsperger, *Integrated Optics: Theory and Technology.* New York: Springer-Verlag, 1982.
11. R. C. Alferness and J. N. Walpole (Eds.), "Special Issue on Integrated Optics," *IEEE Journal of Quantum Electronics,* **21**(6), 1985.
12. E. Abraham, C. T. Seaton, and S. D. Smith, "The Optical Computer," *Scientific American,* February 1983. pp. 85–93.
13. D. A. B. Miller, S. D. Smith, and C. T. Seaton, "Optical Bistability in Semiconductors," *IEEE Journal of Quantum Electronics,* **17**(3), 1981, pp. 312–317.
14. P. W. Smith (Ed.), "Special Issue on Optical Bistability," *IEEE Journal of Quantum Electronics,* **17**(3), 1981.
15. E. Garmire (Ed.), "Special Issue on Optical Bistability," *IEEE Journal of Quantum Electronics,* **21**(9), 1985.

APPENDIX A

The Periodic Lens Waveguide

A lens waveguide is composed of a series of lenses that are arranged in a periodic array. The simplest one, and the one we will consider, is composed of a single type of lens and the lenses are equally spaced a distance, d, apart. Clearly, the period for this lens waveguide is d.

For each of the three cases considered below the incoming ray to the zeroth lens is a paraxial ray. Thus, in all instances $s_0 = \infty$, $s'_0 = f$, and the slope of the ray emerging from the zeroth lens, r'_1, is $-r_0/f$. In the first case where $d = 2f$, $r_1 = -r_0$, $s_1 = f$, and $s'_1 = \infty$. As a result, $r'_2 = 0$. The higher order values for r_n and r'_n are listed in Table A-1, and the ray path is traced in Fig. A-1*a*. As we can see, the ray path repeats itself every fourth lens.

In a similar way the ray path for the case $d = 3f$ can be traced. The data are listed in Table A-2. There it can be seen that the period for the ray path is three-lens spacings, and this is also illustrated in Fig. A-1*b*.

When $d = 4f$, there is no period as both $|r_0|$ and $|r'_0|$ continue to increase as they pass through each succeeding lens. This is shown in Table A-3. This lens system, which is illustrated in Fig. A-1*c*, is therefore unstable. Thus, it appears that for $d < d_{\text{crit}}$ the ray motion is periodic and for $d > d_{\text{crit}}$ the ray path is unstable.

This can be demonstrated mathematically by noting that r_{n+1} and r'_{n+1} can be written in terms of r_n and r'_n. Assuming the lens has no thickness, it has no effect on r_n. The lens, however, does change the slope by $-r_n/f$. The lens operator is therefore

$$\begin{pmatrix} 1 & 0 \\ \dfrac{-1}{f} & 1 \end{pmatrix} \tag{A-1}$$

The ray changes its position by an amount $r'_n d$ between the lenses, but there is no change in r'_n. Thus the space operator is

$$\begin{pmatrix} 1 & d \\ 0 & 1 \end{pmatrix} \tag{A-2}$$

TABLE A-1 RAY TRACING FOR THE CASE $d = 2f$

No.	s_n	s'_n	r_n	r'_n
0	∞	f	r_0	0
1	f	∞	$-r_0$	$-r_0/f$
2	∞	f	$-r_0$	0
3	f	∞	r_0	r_0/f
4	∞	f	r_0	0

Since for this lens waveguide the period is composed of a lens and a space between the lens, the period operator is

$$\begin{pmatrix} 1 & d \\ 0 & 1 \end{pmatrix} \begin{pmatrix} 1 & 0 \\ \dfrac{-1}{f} & 1 \end{pmatrix} = \begin{pmatrix} 1 - \dfrac{d}{f} & d \\ \dfrac{-1}{f} & 1 \end{pmatrix} \tag{A-3a}$$

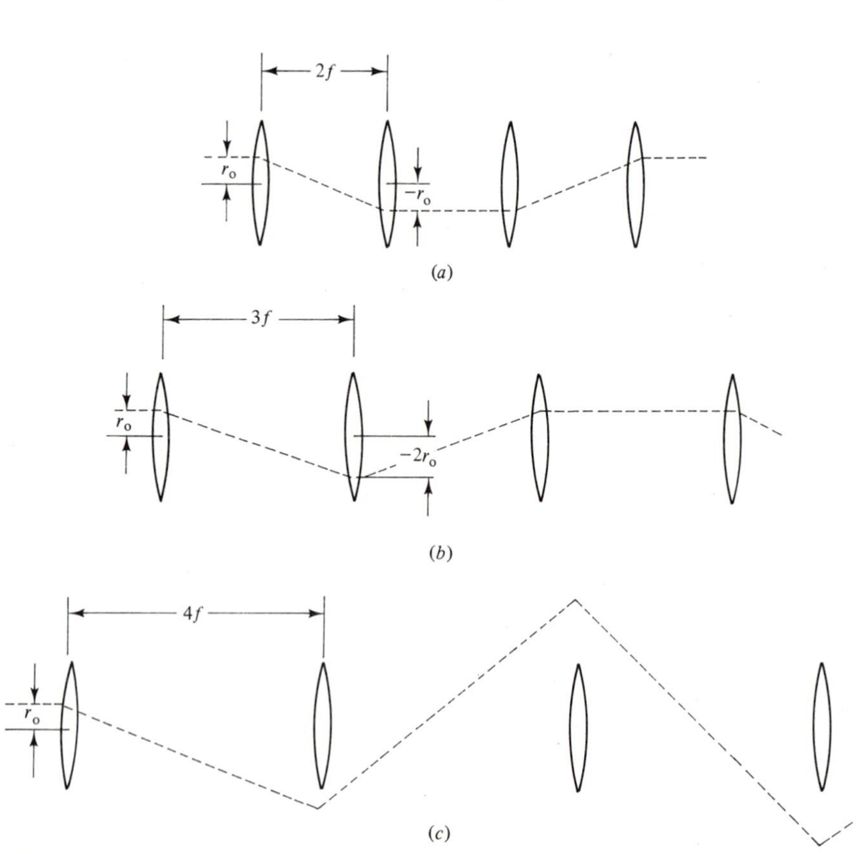

Figure A-1 Simple periodic lens waveguides with periods (a) 2f, (b) 3f, and (c) 4f.

TABLE A-2 RAY TRACING FOR THE CASE $d = 3f$

No.	s_n	s'_n	r_n	r'_n
0	∞	f	r_0	0
1	$2f$	$2f$	$-2r_0$	$-r_0/f$
2	f	∞	r_0	r_0/f
3	∞	f	r_0	0
4	$2f$	$2f$	$-2r_0$	$-r_0/f$

For any lens waveguide the general equation is

$$\begin{matrix} r_{n+1} \\ r'_{n+1} \end{matrix} = \begin{pmatrix} A & B \\ C & D \end{pmatrix} \begin{matrix} r_n \\ r'_n \end{matrix} \tag{A-3b}$$

From Eq. A-3b

$$r_{n+1} = Ar_n + Br'_n \tag{A-4}$$

Or rearranging

$$r'_n = \frac{1}{B}(r_{n+1} - Ar_n) \tag{A-5a}$$

which can also be written

$$r'_{n+1} = \frac{1}{B}(r_{n+2} - Ar_{n+1}) \tag{A-5b}$$

Also from Eq. A-3b

$$r'_{n+1} = \mathrm{C}r_n + \mathrm{D}r'_n = \mathrm{C}r_n + \frac{D}{B}r_{n+1} - \frac{DA}{B}r_n \tag{A-6}$$

Equating Eqs. A-5b and A-6 and reducing

$$r_{n+2} - (A + D)r_{n+1} + (AD - BC)r_n = 0 \tag{A-7a}$$

Since $AD - BC$ always equals 1 (This can be verified for Eq. A-3a)

$$r_{n+2} - 2br_{n+1} + r_n = 0 \tag{A-7b}$$

From the difference equation

$$\nabla^2 r_{n+1} = (r_{n+2} - r_{n+1}) - (r_{n+1} - r_n) = r_{n+2} - 2r_{n+1} + r_n \tag{A-8}$$

we note that Eq. A-7b can be written

$$\nabla^2 r_{n+1} + 2(1 - b)r_{n+1} = 0 \tag{A-7c}$$

TABLE A-3 RAY TRACING FOR THE CASE $d = 4f$

No.	s_n	s'_n	r_n	r'_n
0	∞	f	r_0	0
1	$3f$	$3f/2$	$-3r_0$	$-r_0/f$
2	$5f/2$	$5f/3$	$5r_0$	$2r_0/f$
3	$7f/3$	$7f/4$	$-7r_0$	$-3r_0/f$

As is true for a differential equation, the solution to the second order difference equation is sinusoidal when $b > 1$ and exponential when $b < 1$. When the former condition prevails, the waveguide is stable with a periodic ray path, and when $b < 1$, the system is unstable. For the simple lens waveguide discussed above $b = 1 - (d/2f)$, $d = d_{crit}$ when $b = 1$, so that $d_{crit} = 4f$. This is what was suggested above.

A solution for a stable waveguide is

$$r_n = A \sin (n\theta + \phi) \tag{A-9}$$

Thus

$$A = \frac{r_0}{\sin \phi} \tag{A-10}$$

where ϕ is found from

$$\tan \phi = \frac{\sqrt{(4f/d) - 1}}{1 + f(r'_0/r_0)} \tag{A-11}$$

A is also given by the equation

$$A^2 = \frac{4f}{4f - d}(r_0^2 + 2fr_0r'_0 + dfr'^2_0) \tag{A-12}$$

Substituting Eq. A-9 into Eq. A-7b, we obtain

$$A \sin (2\theta + \phi) - 2bA \sin (\theta + \phi) + A \sin \phi = 0 \tag{A-13}$$

Expanding and rearranging yields

$$(\sin \theta \cos \theta - b \sin \theta) \cos \phi + (\cos^2 \theta - 2b \cos \theta) \sin \phi = 0$$

Since ϕ can be any value, the coefficients in θ must be zero. This yields

$$\cos \theta = b \tag{A-14}$$

The interested student is invited to apply these equations to the three cases studied in the first paragraphs.

APPENDIX B

Modes in a Waveguide

By using ray theory and the fact that the x component of the ray in a planar dielectric must form a standing wave, we showed in Section 2-2.2 that only certain transverse modes could propagate. Moreover, there is a maximum number that can propagate, and this number decreases as either the dimensions of the waveguide decrease or the wavelength of the ray increases. In this appendix these ideas are put on a firmer mathematical basis by considering wave propagation in a perfectly conducting rectangular waveguide of a wave that has the familiar $\exp[j(\omega t - \beta z)]$ dependence. Having a perfectly conducting waveguide does not conceptually affect the results; it only makes the boundary conditions, and therefore the algebra, simpler.

When there is no loss in the propagating medium, Eq. 1-6c for the general case becomes

$$\nabla^2 \bar{\xi} = \mu\epsilon \frac{\partial^2 \bar{\xi}}{\partial t^2} \tag{B-1a}$$

which for the z component is

$$\frac{\partial^2 \xi_z}{\partial x^2} + \frac{\partial^2 \xi_z}{\partial y^2} + \frac{\partial^2 \xi_z}{\partial z^2} = \mu\epsilon \frac{\partial^2 \xi_z}{\partial t^2} \tag{B-1b}$$

and it reduces to

$$\frac{\partial^2 \xi_z(x, y)}{\partial x^2} + \frac{\partial^2 \xi_z(x, y)}{\partial y^2} + (\mu\epsilon\omega^2 - \beta^2)\xi_z(x, y) = 0 \tag{B-1c}$$

when the $\exp[j(\omega t - \beta z)]$ dependence is inserted.

We now use Eq. B-1c to solve for the transverse magnetic, TM, modes for the rectangular waveguide with the cross-sectional dimensions of a and b that are shown in Fig. B-1. A TM mode is defined by the fact $H_z = 0$. Eq. B-1c can be solved by the separation of variables

$$\frac{\xi''(x)}{\xi(x)} = -(\mu\epsilon\omega^2 - \beta^2) - \frac{\xi''(y)}{\xi(y)} = -k_c^2 - \frac{\xi''(y)}{\xi(y)} = -k_x^2 \tag{B-2}$$

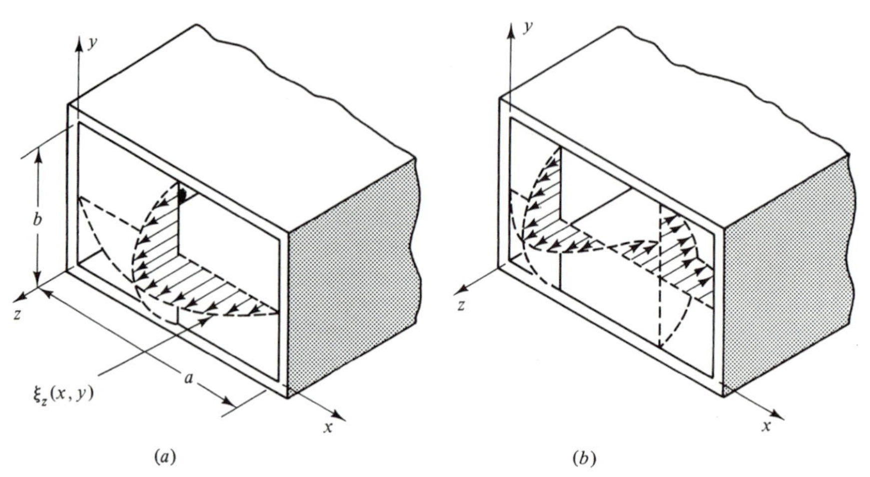

Figure B-1 The (*a*) TM_{11} and (*b*) TM_{21} modes.

The solution to this equation is

$$\xi(x) = C_1 \sin k_x x + C_2 \cos k_x x \tag{B-3}$$

For a perfect conductor

$$\xi(0) = \xi(a) = 0 \tag{B-4}$$

Thus

$$\xi(0) = C_2 = 0 \tag{B-5a}$$

$$\xi(a) = C_1 \sin k_x a = 0 \tag{B-5b}$$

or

$$k_x = \frac{m\pi}{a} \tag{B-5c}$$

In the same way we find that

$$\xi(y) = C_3 \sin k_y y \tag{B-6}$$

where

$$k_y^2 = k_c^2 - k_x^2 = \left(\frac{n\pi}{b}\right)^2 \tag{B-7}$$

Thus

$$\xi_z = \xi_{z0}(m, n) e^{j(\omega t - \beta z)} \sin \left(\frac{m\pi}{a}\right) x \sin \left(\frac{n\pi}{b}\right) y \tag{B-8}$$

and $\xi_z(x, y)$ is illustrated in Fig. B-1 for $m = n = 1$ and $m = 2$ and $n = 1$. These modes are the TM_{11} and TM_{21} modes.

From the definition of k_c in Eq. B-2,

$$\beta^2 = \mu\epsilon\omega^2 - \left(\frac{m\pi}{a}\right)^2 - \left(\frac{n\pi}{b}\right)^2 \tag{B-9}$$

For $\mu\epsilon\omega^2 < (m\pi/a)^2 + (n\pi/b)^2$, β is imaginary so that the wave attenuates; it does not propagate. Thus, for a given set of dimensions there is a frequency below which a wave will not propagate. This critical angular frequency, ω_c, is

$$\omega_c = \frac{\left[\left(\frac{\pi}{a}\right)^2 + \left(\frac{\pi}{b}\right)^2\right]^{1/2}}{c} \tag{B-10a}$$

$$= \frac{\sqrt{2}\pi}{ac} \tag{B-10b}$$

when $a = b$.

When $\mu\epsilon\omega^2 > (m\pi/a)^2 + (n\pi/b)^2$, β is real so that the wave will propagate, but it propagates in a fashion quite different from that of a plane wave. The wavelength, group velocity—the velocity at which electromagnetic energy propagates, and the phase velocity are functions of the frequency, and ξ_x, ξ_y, H_x, and H_y have different x and y dependencies.

As before

$$\lambda = \frac{2\pi}{k} \tag{B-11}$$

From Eq. B-9, k can be written

$$k = \beta = \sqrt{\mu\epsilon}\,\omega\left\{1 - \frac{1}{\mu\epsilon\omega^2}\left[\left(\frac{m\pi}{a}\right)^2 + \left(\frac{n\pi}{b}\right)^2\right]\right\}^{1/2}$$

$$= k^0[1 - (k_c/k^0)^2]^{1/2} \tag{B-12}$$

Thus, k increases from 0 at $\omega = \omega_c$ and approaches k^0, the plane wave value of k, for $\omega \gg \omega_c$. This is illustrated in Fig. B-2. The wavelength accordingly decreases from ∞ at $\omega = \omega_c$ to the plane wave value for $\omega \gg \omega_c$.

The phase velocity, v_p, is defined the same way it is for the velocity of a plane wave, and it is

$$v_p = \frac{\omega}{k} = \frac{\omega}{k_0[1 - (k_c/k^0)^2]^{1/2}} = \frac{c}{[1 - (k_c/k^0)^2]^{1/2}} \tag{B-13}$$

On the other hand, the group velocity, v_g, is

$$v_g = \frac{d\omega}{dk} = c[1 - (k_c/k^0)^2]^{1/2} \tag{B-14}$$

Thus, v_p decreases and v_g increases toward c for $\omega \gg \omega_c$, and this is illustrated in Fig. B-2.

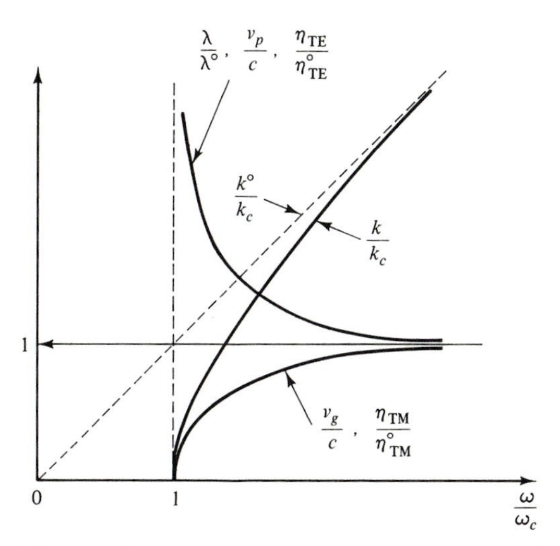

Figure B-2 The ratios of a number of parameters for the waveguide and their values for a plane wave (°) plotted as a function of w/w_c.

It is not necessary to solve other differential equations to find ξ_x, ξ_y, H_x, and H_y because they can be expressed in terms of ξ_z and H_z. From

$$\bar{\nabla} \times \bar{\xi} = -\mu \frac{\partial \bar{H}}{\partial t} \tag{1-2a}$$

$$\frac{\partial \xi_z}{\partial y} - \frac{\partial \xi_y}{\partial z} = \frac{\partial \xi_z}{\partial y} + j\beta \xi_y \quad = -j\omega\xi H_x \tag{B-15}$$

$$\frac{\partial \xi_x}{\partial z} - \frac{\partial \xi_z}{\partial x} = -j\beta \xi_x - \frac{\partial \xi_z}{\partial x} = -j\omega\mu H_y \tag{B-16}$$

$$\frac{\partial \xi_y}{\partial x} - \frac{\partial \xi_x}{\partial y} = \qquad -j\omega\mu H_z \tag{B-17}$$

Also from

$$\bar{\nabla} \times \bar{H} = \epsilon \frac{\partial \xi}{\partial t} \tag{1-4a}$$

when there is no current,

$$\frac{\partial H_z}{\partial y} + j\beta H_y = j\omega\epsilon \xi_x \tag{B-18}$$

$$-j\beta H_x - \frac{\partial H_z}{\partial x} = j\omega\epsilon \xi_y \tag{B-19}$$

$$\frac{\partial H_y}{\partial x} - \frac{\partial H_x}{\partial y} = j\omega\epsilon \xi_z \tag{B-20}$$

Rearranging Eq. B-19 yields

$$H_x = -\frac{1}{j\beta}\left(\frac{\partial H_z}{\partial x} + j\omega\epsilon \xi_y\right) \tag{B-21}$$

By plugging this expression into Eq. B-15 and rearranging, one obtains

$$\xi_y = \frac{1}{k_c^2}\left(-j\beta \frac{\partial \xi_z}{\partial y} + j\omega\mu \frac{\partial H_z}{\partial x}\right) \tag{B-22}$$

Using similar procedures one finds that

$$\xi_x = -\frac{1}{k_c^2}\left(j\beta \frac{\partial \xi_z}{\partial x} + j\omega\mu \frac{\partial H_z}{\partial y}\right) \tag{B-23}$$

$$H_y = -\frac{1}{k_c^2}\left(j\omega\epsilon \frac{\partial \xi_z}{\partial x} + j\beta \frac{\partial H_z}{\partial y}\right) \tag{B-24}$$

$$H_x = \frac{1}{k_c^2}\left(j\omega\epsilon \frac{\partial \xi_z}{\partial y} - j\beta \frac{\partial H_z}{\partial x}\right) \tag{B-25}$$

Using Eq. B-23 for the TM modes

$$\xi_x = -\frac{j\beta}{k_c^2}\frac{\partial \xi_z}{\partial x} = -\frac{j\beta}{k_c^2}\frac{m\pi}{a}\,\xi_z(m, n) \cos\left(\frac{m\pi}{a}\right)x \sin\left(\frac{n\pi}{b}\right)y \tag{B-26}$$

Also

$$\xi_y = -\frac{j\beta}{k_c^2}\frac{\partial \xi_z}{\partial y} = -\frac{j\beta}{k_c^2}\frac{n\pi}{b}\,\xi_z(m, n) \sin\left(\frac{m\pi}{a}\right)x \cos\left(\frac{n\pi}{b}\right)y \tag{B-27}$$

$$H_x = \frac{j\omega\epsilon}{k_c^2}\frac{\partial \xi_z}{\partial y} = \frac{j\omega\epsilon}{k_c^2}\frac{n\pi}{b}\,\xi_z(m, n)\sin\left(\frac{m\pi}{a}\right)x\cos\left(\frac{n\pi}{b}\right)y \tag{B-28}$$

$$H_y = -\frac{j\omega\epsilon}{k_c^2}\frac{\partial \xi_z}{\partial x} = -\frac{j\omega\epsilon}{k_c^2}\frac{m\pi}{a}\,\xi_z(m, n)\cos\left(\frac{m\pi}{a}\right)x\sin\left(\frac{n\pi}{b}\right)y \tag{B-29}$$

We end the discussion of the TM mode by noting that the intrinsic wave impedance, η, is also a function of the frequency.

$$\eta_{\text{TM}} = \frac{\xi_x}{H_y} = \frac{k}{\omega\epsilon} = \eta^0\left[1 - \left(\frac{k_c}{k^0}\right)^2\right]^{1/2} \tag{B-30}$$

which has the same frequency dependence as v_g.

For the transverse electric, TE, modes $\xi_z = 0$. The differential equation that must be solved is

$$\frac{\partial^2 H_z}{\partial x^2} + \frac{\partial^2 H_z}{\partial y^2} + \frac{\partial^2 H_z}{\partial z^2} = \mu\epsilon\frac{\partial^2 H_z}{\partial t^2} \tag{B-31}$$

which reduces to

$$\frac{\partial^2 H_z(x, y)}{\partial x^2} + \frac{\partial^2 H_z(x, y)}{\partial y^2} + (\mu\epsilon\omega^2 - \beta^2)H_z(x, y) = 0 \tag{B-32}$$

Again the differential equation can be solved by the separation of variables. Thus

$$H(x) = C_5 \sin k_x x + C_6 \cos k_x x \tag{B-33}$$

The boundary conditions, however, are different since $\xi_T = 0$, not $H_T = 0$. From Eq. B-22

$$\xi_y(0) = \xi_y(a) = 0 = j\omega\mu\frac{\partial H_z(0)}{\partial x} = j\omega\mu\frac{\partial H_z(a)}{\partial x} \tag{B-34}$$

Therefore

$$\frac{\partial H(0)}{\partial x} = k_x C_5 = 0 \tag{B-35}$$

$$\frac{\partial H(a)}{\partial x} = k_x C_6 \sin k_x a = 0 \tag{B-36}$$

Thus

$$k_x = \frac{m\pi}{a} \tag{B-37}$$

In much the same way

$$H(y) = C_7 \sin k_y y + C_8 \cos k_y y \tag{B-38}$$

$$\frac{\partial H(b)}{\partial y} = k_y C_8 \sin k_y b = 0$$

and

$$k_y = \frac{n\pi}{b} \tag{B-39}$$

Thus

$$H_z = H_{z0}(m, n)e^{j(\omega t - \beta z)}\cos\left(\frac{m\pi}{a}\right)x\cos\left(\frac{n\pi}{b}\right)y \tag{B-40}$$

$$\xi_x = -\frac{j\omega\mu}{k_c^2}\frac{\partial H_z}{\partial y} = \frac{j\omega\mu}{k_c^2}\frac{n\pi}{b}\,H_z(m, n)\cos\left(\frac{m\pi}{a}\right)x\sin\left(\frac{n\pi}{b}\right)y \tag{B-41}$$

$$\xi_y = \frac{j\omega\mu}{k_c^2}\frac{\partial H_z}{\partial x} = -\frac{j\omega\mu}{k_c^2}\frac{m\pi}{a}\,H_z(m, n)\sin\left(\frac{m\pi}{a}\right)x\cos\left(\frac{n\pi}{b}\right)y \tag{B-42}$$

$$H_x = -\frac{j\beta}{k_c^2}\frac{\partial H_z}{\partial x} = \frac{j\beta}{k_c^2}\frac{m\pi}{a} H_z(m, n) \sin\left(\frac{m\pi}{a}\right)x \cos\left(\frac{n\pi}{b}\right)y \tag{B-43}$$

$$H_y = -\frac{j\beta}{k_c^2}\frac{\partial H_z}{\partial y} = \frac{j\beta}{k_c^2}\frac{n\pi}{b} H_z(m, n) \cos\left(\frac{m\pi}{a}\right)x \sin\left(\frac{n\pi}{b}\right)y \tag{B-44}$$

On further analysis one finds the same relationships for k, λ, v_p, and v_g for a propagating wave when $\omega > \omega_c$, and the same relationship for β when the beam attenuates for $\omega < \omega_c$. η differs, however, since

$$\eta_{TE} = \frac{\xi x}{H_y} = \frac{\eta^0}{[1 - (k_c/k^0)^2]^{1/2}} \tag{B-45}$$

It has the same functional dependence as the phase velocity. The TE modes also differ from the TM modes in that there is a wave for $n = m = 0$—the TE_{∞} mode.

APPENDIX C

Derivation of Electron and Hole Concentrations

In Section 4-3.3 we stated that

$$n = \frac{2}{h^3}(2\pi m_n^* kT)^{3/2} e^{(E_F - E_g)/kT} \tag{4-16a}$$

and

$$p = \frac{2}{h^3}(2\pi m_p^* kT)^{3/2} e^{-E_F/kT} \tag{4-16b}$$

In this appendix these expressions will be derived.

The number of electrons per unit volume occupying a given number of states per unit volume is the sum of the products of the number of states per unit volume with an energy between E and $E + dE$ and the probability a state with that energy is occupied. Thus, the number of electrons per unit volume in the conduction band is the sum of the products of the density of states per unit volume times the Fermi–Dirac distribution function. Stated mathematically

$$n = \int_{E_g}^{E_{TC}} \rho(E) f(E)\, dE \tag{C-1}$$

where E_{TC} represents the energy at the top of the conduction band. Substituting Eqs. 4-54 and 7-19 into Eq. C-1 yields

$$n = \int_{E_g}^{E_{TC}} \frac{4\pi}{h^3} \frac{(2m_n^*)^{3/2}(E - E_g)^{1/2}\, dE}{e^{(E-E_F)/kT} + 1} \tag{C-2}$$

If we make the assumptions that $E_{TC} \gg E_g$ so it can be replaced by ∞ and $E \gg E_F$, Eq. C-2 becomes

$$n = \int_{E_g}^{\infty} \frac{4\pi}{h^3} (2m_n^*)^{3/2}(E - E_g)^{1/2} e^{(E_F - E)/kT}\, dE \tag{C-3}$$

The first assumption is good and, in fact, introduces less error than assuming that the density of states equation is valid for all energies. The second assumption is valid for all but heavily ($\gtrsim 10^{18}$ cm^{-3}) n-type doped semiconductors. For heavily doped material E_F is too close to the bottom of the conduction band for the assumption $\exp[(E - E_F)/kT] \gg 1$ to be valid. These semiconductors are said to be degenerate. Physically, the problem is that there are so many electrons in the conduction band that the probability a state is occupied is relatively large. Thus, the probability that a second electron will want to occupy a state that is already occupied cannot be ignored. This means that the Fermi–Dirac distribution function, which takes Pauli's principle into account, cannot be replaced by the Maxwell–Boltzmann distribution function, which does not. We assumed that this substitution could be made. When $\exp[(E - E_F)/kT] \gg 1$, the probability that a state has an electron in it is so small that ignoring the probability that a second electron would attempt to occupy the same state introduces essentially no error.

Making the substitution $x = E - E_g$ and making the appropriate change in the limits of integration, Eq. C-3 becomes

$$n = \frac{4\pi}{h^3}(2m_n^*)^{3/2}e^{(E_F-E_g)/kT}\int_0^\infty x^{1/2}e^{-x/kT}\,dx \tag{C-4}$$

Using the definite integral

$$\int_0^\infty x^{1/2}e^{-ax}\,dx = \frac{\sqrt{\pi}}{2a^{3/2}} \tag{C-5}$$

Eq. C-5 becomes

$$n = \frac{2}{h^3}(2\pi m_n^* kT)^{3/2}e^{(E_g-E_F)/kT} \tag{4-16a}$$

The probability a state in the valence band is occupied by a hole is $1 - f(E)$, and the density of states equation for holes in the valence band is

$$\rho(E) = \frac{2}{h^3}(2m_p^*)^{3/2}(-E)^{1/2} \tag{C-6}$$

since the top of the valence band is the zero point energy for homojunctions, and the energy of the holes is positive downward. Thus

$$p = \int_{E_{BV}}^0 \frac{2}{h^3}\frac{(2m_p^*)^{3/2}(-E)^{1/2}\,dE}{1 + e^{(E_F-E)/kT}} \tag{C-7}$$

where E_{BV} is the energy of the bottom of the valence band. Assuming that $E_{BV} \to -\infty$ and that $\exp[(E_F - E)/kT] \gg 1$, Eq. C-7 becomes

$$p = \int_{-\infty}^0 \frac{2}{h^3}(2m_p^*)^{3/2}(-E)^{1/2}e^{(E-E_F)/kT}\,dE \tag{C-8}$$

Again, the second assumption is valid unless the material is degenerate p-type. Making the substitution $x = -E$, making the appropriate changes in the limits of the integration, and reversing the limits of integration while changing $-dx$ to $+dx$,

$$p = \frac{2}{h^3}(2m_p^*)^{3/2}e^{-E_F/kT}\int_0^\infty x^{1/2}e^{-x/kT}\,dx \tag{C-9}$$

which with Eq. C-5 becomes

$$p = \frac{2}{h^3}(2\pi m_p^* kT)^{3/2}e^{-E_F/kT} \tag{4-16b}$$

APPENDIX D

List of Symbols

a acceleration
A area
A_{ij} Einstein coefficient for spontaneous emission
B magnetic induction
recombination rate constant
base transport factor
B_{ij} Einstein coefficient for stimulated emission and absorption
c speed of light in a material
c_0 speed of light in a vacuum (2.998×10^8 m/sec)
C capacitance
proportionality constant for the energy levels in the hydrogen atom
D electric displacement
diffusion coefficient
D_n electron diffusion coefficient
D_p hole diffusion coefficient
E energy
E_c energy of the bottom of the conduction band
E_F Fermi energy
E_{Fm} Fermi energy of a metal
E_{Fn} Fermi energy of an *n*-type semiconductor
E_{Fp} Fermi energy of a *p*-type semiconductor
E_i Fermi energy of an intrinsic semiconductor
E_n energy of *n*th energy level
E_v energy of the top of the valence band
f focal length
F finesse
g thermal generation rate
$g(\nu)$ line shape function
g_{op} optical generation rate

G gain
$\mathcal{G}$ thermal conductance
h Planck's constant (6.625×10^{-34} J · sec)
h_{FE} common emitter gain coefficient
H magnetic field
i current
angle of incidence
i_B Brewster angle
I intensity
moment of inertia
j square root of -1
j_f free current per unit length
j_s surface current per unit length
J current density
molecular rotation quantum number
total angular momentum quantum number
J_{Dp} hole diffusion current density
J_{Dn} electron diffusion current density
J_{Ep} hole electric field current density
J_{En} electron electric field current density
J_{op} optically induced current density
J_s reverse saturation current density
k wave vector, $2\pi/\lambda$
electrostatic force constant = 8.99×10^9 N · m²/C² = $\frac{1}{4}\pi\epsilon_0$
Boltzmann's constant (1.380×10^{-23} J/K)
k_f force constant
l_i direction cosines
l angular momentum quantum number for electron
L inductance
angular momentum
orbital angular momentum quantum number for atom
length
L_z z component of the angular momentum
m magnification
z component of the angular momentum quantum number
mass of a free electron (9.107×10^{-31} kg)
maximum number of modes in a waveguide
m_n^* effective mass of electron
m_p^* effective mass of hole
m_s spin quantum number of electron
M magnetization
M_s spin quantum number of atom
n index of refraction
principal quantum number
electron concentration
n' real part of the index of refraction
n'' imaginary part of the index of refraction
n_a acceptor concentration
n_d donor concentration
n_i intrinsic carrier concentration

n_0 equilibrium electron concentration
n_r concentration of electron recombination centers
n_t threshold population inversion concentration
N_0 Avogadro's number (6.025×10^{23} atoms/g · molecular wt)
p dipole moment
hole concentration
p_0 equilibrium hole concentration
P polarization
power
probability
q electronic charge (1.602×10^{-19} C)
time varying charge on a capacitor plate
q_0 amplitude of time varying charge
q_{st} amplitude of time varying charge when $\omega \longrightarrow 0$
Q_l quality factor
Q_p total injected hole charge in the *n*-type material
r angle of reflection
ratio of the magnitudes of the electric field for the reflected and incident rays
recombination rate
r^* complex conjugate of r
r_n *n*th Bohr radius
R reflectance
resistance
radius of curvature
$\mathcal{R}$ responsivity
s object distance
s' image distance
S Poynting vector
surface vector
S_{ij} spontaneous transition rate
t angle of transmission
time
ratio of the magnitudes of the electric field for the transmitted and incident ray
t^* complex conjugate of t
t_{st} transit time
T transmissivity
temperature
elapsed time during one period of a periodic function
u energy per unit volume
$u(\nu)$ energy per unit volume per unit frequency
v velocity
vibration quantum number
v_d drift velocity
v_T thermal velocity
V voltage
volume
normalized film thickness
V_{br} breakdown voltage
V_0 contact potential
W depletion layer width

W_b base width
W_i stimulated emission or absorption rate
x_i width of the intrinsic layer in a *pin* structure
x_n depletion layer width in the *n*-type material
x_0 amplitude of an oscillating dipole
x_p depletion layer width in the *p*-type material
x_{st} static separation distance of a dipole
Z ionic charge
α absorption coefficient
polarizability
thermal coefficient of resistivity
current transfer ratio
α_g gain coefficient
β propagation constant
β_l frictional force constant
γ injection efficiency
Γ fraction of electromagnetic energy in active region of a heterojunction laser
δ phase angle
Δ_i screening factor
$\bar{\nabla}$ gradient
ϵ electric permittivity
ϵ' real part of the electric permittivity
ϵ'' imaginary part of the electric permittivity
ϵ_0 electric permittivity of free space (8.854×10^{-12} C·m/V)
ϵ_r relative permittivity or dielectric constant
η intrinsic wave impedance
efficiency
η_i internal quantum efficiency
η_{ex} external quantum efficiency
θ_i critical insertion angle
λ wavelength
λ_0 wavelength in free space
Λ molecular angular momentum quantum number
energy loss per pass in an optical cavity
μ magnetic permeability
μ_0 magnetic permeability of free space $4\pi \times 10^{-7}$ W/A·m
ν frequency
ν_0 natural frequency
$\delta\nu$ frequency separation between adjacent longitudinal laser modes
$\delta\nu^*$ bandwidth of a longitudinal laser mode
$\Delta\nu$ emission peak width
ξ electric field
ϕ lag angle
ϕ_b potential barrier height at a metal semiconductor junction
ϕ_c critical angle
ϕ_m angle of incidence of the m_{th} mode
work function of a metal
ρ charge density
ρ_i density of states of the *i*th energy level
$\rho(\nu)$ density of states with frequency between ν and $\nu + d\nu$

σ conductivity
cross section
σ_F free charge per unit area
σ_i intrinsic conductivity
σ_p polarization charge per unit area
τ time constant
τ_n lifetime of an electron
τ_p lifetime of a hole
τ_{sp} time constant for spontaneous decay
χ electric susceptibility
electron affinity
χ_m magnetic susceptibility
ω angular frequency
ω_0 natural angular frequency

INDEX